Hyperspectral Remote Sensing

Hyperspectral Remote Sensing

Michael T. Eismann

SPIE PRESS

Bellingham, Washington USA

Library of Congress Cataloging-in-Publication Data

Eismann, Michael Theodore, 1964-
  Hyperspectral remote sensing / Michael Eismann.
      p. cm.
  Includes bibliographical references and index.
  ISBN 978-0-8194-8787-2
  1. Remote sensing. 2. Multispectral photography. 3. Image processing. I. Title.
  G70.4.E39 2012
  621.36'78--dc23

                                                              2011046489

Published by
SPIE
P.O. Box 10
Bellingham, Washington 98227-0010 USA
Phone: +1 360.676.3290
Fax: +1 360.647.1445
Email: Books@spie.org
Web: http://spie.org

Publication Year: 2012

The content of this book reflects the work and thought of the author. Every effort has
been made to publish reliable and accurate information herein, but the publisher is not
responsible for the validity of the information or for any outcomes resulting from reliance
thereon.

Printed in the United States of America.
Second printing 2017
For updates to this book, visit http://spie.org and type "PM210" in the search field.

Dedicated with love to Michelle, Maria, and Katie

Contents

Preface .... **xv**

List of Acronyms .... **xvii**

Chapter 1   Introduction **1**

  1.1   Hyperspectral Remote Sensing 2
  1.2   Elements of Hyperspectral Sensing 7
        1.2.1   Material spectroscopy 8
        1.2.2   Radiative transfer 9
        1.2.3   Imaging spectrometry 14
        1.2.4   Hyperspectral data processing 20
  1.3   Examples of Remote Sensing Applications 21
  1.4   Summary 33
  References 33

Chapter 2   Optical Radiation and Matter **37**

  2.1   Propagation of Electromagnetic Radiation 37
        2.1.1   Propagation in free space 38
        2.1.2   Propagation in dense media 39
        2.1.3   Plane waves in dense media 41
  2.2   Complex Index of Refraction 44
        2.2.1   Relationship with the complex dielectric constant 44
        2.2.2   Lorentz oscillator model 45
        2.2.3   Drude theory of strong conductors 52
  2.3   Propagation through Homogenous Media 53
  2.4   Reflection and Transmission at Dielectric Interfaces 56
  2.5   Reflection and Transmission at Conductor Interfaces 61
  2.6   Radiometry 69
        2.6.1   Point sources 71
        2.6.2   Lambertian sources 72
        2.6.3   Spherical scatterers 73
  2.7   Propagation through Scattering Media 74
        2.7.1   Mie scattering theory 76
        2.7.2   Rayleigh scattering theory 78

2.8   Summary 80
2.9   Further Reading 81
References 81

Chapter 3   Atomic and Molecular Spectroscopy **83**

3.1   Quantum Mechanics 83
       3.1.1   Stationary states of a quantum mechanical system 85
       3.1.2   Interaction with electromagnetic radiation 86
       3.1.3   Born–Oppenheimer approximation 90
3.2   Electromagnetic Absorption and Emission 91
       3.2.1   Einstein coefficients 92
       3.2.2   Line broadening 93
3.3   Electronic Spectroscopy of Atoms 95
       3.3.1   Single-electron atoms 96
       3.3.2   Polyelectronic atoms 101
3.4   Rotational Spectroscopy of Molecules 107
3.5   Vibrational Spectroscopy of Molecules 111
       3.5.1   Diatomic molecules 111
       3.5.2   Polyatomic molecules 120
3.6   Electronic Spectroscopy of Molecules 123
3.7   Summary 130
3.8   Further Reading 130
References 131

Chapter 4   Spectral Properties of Materials **133**

4.1   Apparent Spectral Properties 133
       4.1.1   Homogenous absorbing layer 134
       4.1.2   Opaque scattering layer 138
       4.1.3   Transparent scattering layer 141
       4.1.4   Multiple absorbing layers 142
       4.1.5   Multilayer dielectric thin films 144
       4.1.6   Rough-surface reflectance 146
       4.1.7   Emissivity and Kirchhoff's law 150
4.2   Common Remote Sensing Materials 154
       4.2.1   Atmospheric gases 154
       4.2.2   Liquid water 158
       4.2.3   Vegetation 163
       4.2.4   Minerals 172
       4.2.5   Soils 179
       4.2.6   Road materials 186
       4.2.7   Metals 189
       4.2.8   Paints and coatings 191

4.3 Summary 193
4.4 Further Reading 195
References 196

Chapter 5   Remotely Sensed Spectral Radiance **199**

5.1 Radiative Transfer Modeling.. 199
     5.1.1 Atmospheric modeling.. 201
     5.1.2 Moderate-resolution atmospheric transmission and radiation code 206
     5.1.3 Atmospheric path transmission............................ 209
     5.1.4 Atmospheric path radiance................................. 217
     5.1.5 Downwelling radiance 221
5.2 Remote Sensing Models 227
     5.2.1 Facet model for a solid surface 228
     5.2.2 Gaseous effluent model...................................... 233
     5.2.3 Shallow-water model.. 235
5.3 Summary 241
References 241

Chapter 6   Imaging System Design and Analysis........................ **243**

6.1 Remote Imaging Systems... 243
6.2 Optical System Design... 247
     6.2.1 Point spread function.. 247
     6.2.2 Optical aberrations 250
     6.2.3 Modulation transfer function 256
     6.2.4 Lens design 257
6.3 FPA Materials and Devices.. 266
     6.3.1 Quantum detectors.. 268
     6.3.2 Photoconductors... 270
     6.3.3 Photodiodes 272
     6.3.4 Detector materials... 275
     6.3.5 Detector noise 281
     6.3.6 Detector performance... 285
6.4 Radiometric Sensitivity... 286
     6.4.1 Signal and background radiance........................... 288
     6.4.2 Focal plane irradiance 289
     6.4.3 Photoelectronic conversion................................. 291
     6.4.4 Total system noise 292
     6.4.5 Total system performance 294
6.5 Spatial Sampling... 296
6.6 Spatial Resolution 300
     6.6.1 Ground-resolved distance 301

6.6.2   System modulation transfer function 302
6.7   Image Quality 307
6.8   Summary 310
6.9   Further Reading 310
References 311

Chapter 7   Dispersive Spectrometer Design and Analysis 313

7.1   Prism Spectrometers 314
7.1.1   Prism dispersion 315
7.1.2   Prism spectrometer design 319
7.2   Grating Spectrometers 324
7.2.1   Grating diffraction 325
7.2.2   Grating spectrometer design 331
7.3   Imaging Spectrometer Performance 338
7.3.1   Spatial and spectral mapping 338
7.3.2   Spatial and spectral response functions................. 340
7.3.3   Radiometric sensitivity 346
7.4   System Examples 348
7.4.1   Airborne Visible/Infrared Imaging Spectrometer. 349
7.4.2   Hyperspectral Digital Imagery Collection
Experiment.. 351
7.4.3   Hyperion 353
7.4.4   Compact Airborne Spectral Sensor 354
7.4.5   Spatially Enhanced Broadband Array Spectrograph System.. 357
7.4.6   Airborne Hyperspectral Imager 357
7.5   Summary 360
References 361

**Chapter 8   Fourier Transform Spectrometer Design and
Analysis 363**

8.1   Fourier Transform Spectrometers............................... 364
8.1.1   Interferograms.. 366
8.1.2   Spectrum reconstruction.................................... 368
8.1.3   Spectral resolution.. 369
8.1.4   Spectral range.. 370
8.1.5   Apodization 372
8.1.6   Uncompensated interferograms 373
8.2   Imaging Temporal Fourier Transform Spectrometers 373
8.2.1   Off-axis effects 375
8.2.2   Additional design considerations 376

8.3 Spatial Fourier Transform Spectrometers 377
8.4 Radiometric Sensitivity 380
 8.4.1 Signal-to-noise ratio 380
 8.4.2 Noise-equivalent spectral radiance 382
 8.4.3 Imaging spectrometer sensitivity comparison 384
8.5 System Examples 387
 8.5.1 Field-Portable Imaging Radiometric Spectrom-
    eter Technology 387
 8.5.2 Geosynchronous Imaging Fourier Transform
    Spectrometer 388
 8.5.3 Spatially Modulated Imaging Fourier
    Transform Spectrometer 390
 8.5.4 Fourier Transform Hyperspectral Imager 391
8.6 Summary 393
References 393

Chapter 9 Additional Imaging Spectrometer Designs 395

9.1 Fabry–Pérot Imaging Spectrometer 395
9.2 Acousto-optic Tunable Filter 400
9.3 Wedge Imaging Spectrometer 403
9.4 Chromotomographic Imaging Spectrometer 407
 9.4.1 Rotating direct-view prism spectrometer 408
 9.4.2 Multi-order diffraction instrument 413
9.5 Summary 415
References 415

Chapter 10 Imaging Spectrometer Calibration 417

10.1 Spectral Calibration 417
 10.1.1 Spectral mapping estimation 418
 10.1.2 Spectral calibration sources 419
 10.1.3 Spectral-response-function estimation 420
 10.1.4 Spectral calibration example 421
10.2 Radiometric Calibration 423
 10.2.1 Nonuniformity correction of panchromatic
    imaging systems 424
 10.2.2 Radiometric calibration sources 431
 10.2.3 Dispersive imaging spectrometer calibration 435
 10.2.4 Imaging Fourier transform spectrometer
    calibration 444
10.3 Scene-Based Calibration 445
 10.3.1 Vicarious calibration 446
 10.3.2 Statistical averaging 446

10.4   Summary 449
References 449

Chapter 11 Atmospheric Compensation 451

11.1   In-Scene Methods 452
            11.1.1   Empirical line method 454
            11.1.2   Vegetation normalization 458
            11.1.3   Blackbody normalization 467
            11.1.4   Temperature–emissivity separation 476
11.2   Model-Based Methods 485
            11.2.1   Atmospheric Removal Program 485
            11.2.2   Fast line-of-sight atmospheric analysis of
                         spectral hypercubes 490
            11.2.3   Coupled-subspace model 492
            11.2.4   Oblique projection retrieval of the atmosphere 495
11.3   Summary 498
References 500

Chapter 12 Spectral Data Models 503

12.1   Hyperspectral Data Representation 503
            12.1.1   Geometrical representation 504
            12.1.2   Statistical representation 507
12.2   Dimensionality Reduction 510
            12.2.1   Principal-component analysis 510
            12.2.2   Centering and whitening 513
            12.2.3   Noise-adjusted principal-components analysis 516
            12.2.4   Independent component analysis 520
            12.2.5   Subspace model 523
            12.2.6   Dimensionality estimation 526
12.3   Linear Mixing Model 529
            12.3.1   Endmember determination 531
            12.3.2   Abundance estimation 535
            12.3.3   Limitations of the linear mixing model 537
12.4   Extensions of the Multivariate Normal Model 539
            12.4.1   Local normal model 540
            12.4.2   Normal mixture model 544
            12.4.3   Generalized elliptically contoured distributions 549
12.5   Stochastic Mixture Model 551
            12.5.1   Discrete stochastic mixture model 553
            12.5.2   Estimation algorithm 555
            12.5.3   Examples of results 558

12.6   Summary 558
12.7   Further Reading 560
References 560

Chapter 13 Hyperspectral Image Classification **563**

13.1   Classification Theory 563
13.2   Feature Extraction 569
         13.2.1   Statistical separability 570
         13.2.2   Spectral derivatives 572
13.3   Linear Classification Algorithms 573
         13.3.1   *k*-means algorithm 580
         13.3.2   Iterative self-organizing data analysis technique .. 584
         13.3.3   Improved split-and-merge clustering 586
         13.3.4   Linear support vector machine 589
13.4   Quadratic Classification Algorithms 593
         13.4.1   Simple quadratic clustering 594
         13.4.2   Maximum-likelihood clustering 595
         13.4.3   Stochastic expectation maximization 598
13.5   Nonlinear Classification Algorithms 605
         13.5.1   Nonparametric classification 606
         13.5.2   Kernel support vector machine 612
13.6   Summary 614
13.7   Further Reading 615
References 615

Chapter 14 Hyperspectral Target Detection **617**

14.1   Target Detection Theory 617
         14.1.1   Likelihood ratio test 618
         14.1.2   Multivariate normal model 620
         14.1.3   Generalized likelihood ratio test 622
14.2   Anomaly Detection 624
         14.2.1   Mahalanobis distance detector 624
         14.2.2   Reed–Xiaoli detector 630
         14.2.3   Subspace Reed–Xiaoli detector 633
         14.2.4   Complementary subspace detector 635
         14.2.5   Normal mixture model detectors 638
14.3   Signature-Matched Detection 646
         14.3.1   Spectral angle mapper 646
         14.3.2   Spectral matched filter 649
         14.3.3   Constrained energy minimization 657
         14.3.4   Adaptive coherence/cosine estimator 658
         14.3.5   Subpixel spectral matched filter 662

14.3.6 Spectral matched filter with normal mixture model 666

14.3.7 Orthogonal subspace projection 669

14.4 False-Alarm Mitigation 674

14.4.1 Quadratic matched filter 674

14.4.2 Subpixel replacement model 676

14.4.3 Mixture-tuned matched filter 678

14.4.4 Finite target matched filter 680

14.4.5 Least-angle regression 682

14.5 Matched Subspace Detection 684

14.5.1 Target subspace models 685

14.5.2 Subspace adaptive coherence/cosine estimator 689

14.5.3 Joint subspace detector 692

14.5.4 Adaptive subspace detector 694

14.6 Change Detection 695

14.6.1 Affine change model 696

14.6.2 Change detection using global prediction 698

14.6.3 Change detection using spectrally segmented prediction 703

14.6.4 Model-based change detection 707

14.7 Summary 712

14.8 Further Reading 714

References 714

Index **717**

Preface

Hyperspectral imaging is an emerging field of electro-optical and infrared remote sensing. Advancements in sensing and processing technology have reached a level that allows hyperspectral imaging to be more widely applied to remote sensing problems. Because of this, I was asked roughly six years ago to serve as an adjunct faculty member at the Air Force Institute of Technology in Ohio to construct and teach a graduate course on this subject as part of their optical engineering program. As I searched for a suitable textbook from which to teach this course, it became apparent to me that there were none that provided the comprehensive treatment I felt the subject required. Hyperspectral remote sensing is a highly multidisciplinary field, and I believe that a student of this subject matter should appreciate and understand all of its major facets, including material spectroscopy, radiative transfer, imaging spectrometry, and hyperspectral data processing. There are many resources that suitably cover these areas individually, but none that are all inclusive. This book is my attempt to provide an end-to-end treatment of hyperspectral remote sensing technology.

I have been using this textbook in manuscript form to teach a one-quarter class at the graduate level, with Masters and Ph.D. students taking the course as an elective and subsequently performing their research in the hyperspectral remote sensing field. The amount of material is arguably too much to fit within a single quarter and would ideally be spread over a semester or two quarters if possible. The content of the book is oriented toward the physical principles of hyperspectral remote sensing as opposed to applications of hyperspectral technology, with the expectation that students finish the class armed with the required knowledge to become practitioners in the field; be able to understand the immense literature available in this technology area; and apply their knowledge to the understanding of material spectral properties, the design of hyperspectral systems, the analysis of hyperspectral imagery, and the application of the technology to specific problems.

There are many people I would like to thank for helping me complete this book. First, I would like to thank the Air Force Research Laboratory for their support of this endeavor, and my many colleagues in the hyperspectral remote sensing field from whom I have drawn knowledge and inspiration during the 15 years I have performed research in this area. I would like to thank all of my OENG 647 Hyperspectral Remote Sensing students at the Air Force Institute of Technology who suffered through early versions of this manuscript and provided invaluable feedback to help improve it. In particular, I owe great thanks to Joseph Meola of the Air Force Research Laboratory, who performed a very thorough review of the manuscript, made numerous corrections and suggestions, and contributed material to Chapters 10 and 14, including participating in useful technical discussions concerning nuances of signal processing theory. I am very grateful for thorough, insightful, and constructive reviews of my original manuscript performed by Dr. John Schott of the Rochester Institute of Technology and Dr. Joseph Shaw of Montana State University on behalf of SPIE Press, as well as Tim Lamkins, Dara Burrows, and their staff at SPIE Press for turning my manuscript into an actual book. Additionally, I would like to acknowledge the support of Philip Maciejewski of the Air Force Research Laboratory for performing vegetation spectral measurements, the National Aeronautics and Space Agency (NASA) for the Hyperion data, the Defense Intelligence Agency for the HYDICE data, John Hackwell and the Aerospace Corporation for the SEBASS data, Patrick Brezonik and the University of Minnesota for the lake reflectance spectra, Joseph Shaw of Montana State University for the downwelling FTIR measurements, Bill Smith of NASA Langley Research Center for the GIFTS schematic and example data, and others acknowledged throughout this book for the courtesy of using results published in other books and journals.

Finally, this book would not have been possible were it not for the help and support of my wife Michelle and daughters Maria and Katie, who provided great patience and encouragement during the many hours that their husband and father was preparing, typing, and editing this book instead of giving time to them and attending to other things around our home. Now that this immense undertaking is completed, I hope to make up for some of what was lost.

Michael T. Eismann
Beavercreek, Ohio
March 2012

List of Acronyms

ACE	adaptive coherence/cosine estimator
ADC	analog-to-digital conversion
AHI	Airborne Hyperspectral Imager
AIRIS	Adaptive Infrared Imaging Spectroradiometer
AIS	Airborne Imaging Spectrometer
amu	atomic mass unit
AOTF	acousto-optic tunable filter
AR	antireflection (coating)
ARCHER	Airborne Real-Time Cueing Hyperspectral Enhanced Reconnaissance
ARM	Atmospheric Radiation Measurement (site)
ASD	adaptive subspace detector
ATREM	atmospheric removal program
AUC	area under (the ROC) curve
AVIRIS	Airborne Visible/Infrared Imaging Spectrometer
BI	bare-soil index
BLIP	background-limited performance
BRDF	bidirectional reflectance distribution function
CBAD	cluster-based anomaly detector
CCD	charge-coupled device
CCSMF	class-conditional spectral matched filter
CDF	cumulative distribution function
CEM	constrained energy minimization (detector)
CFAR	constant false-alarm rate
CMOS	complementary metal-oxide semiconductor
COMPASS	Compact Airborne Spectral Sensor
CSD	complementary subspace detector
CTIS	chromotomographic imaging spectrometer
DDR-SDRAM	double-data-rate synchronous dynamic random access memory
DFT	discrete Fourier transform
DHR	directional hemispherical reflectance
DIRSIG	Digital Image and Remote Sensing Image Generation
DISORT	multiple-scattering discrete-ordinate radiative transfer program for a multilayered plane-parallel medium
DN	data number
DOP	degree of polarization

ED	Euclidian distance (detector)
ELM	empirical line method
EM	expectation maximization (algorithm)
EO/IR	electro-optical and infrared
FAM	false-alarm mitigation
FAR	false-alarm rate
FCBAD	fuzzy cluster-based anomaly detector
FFT	fast Fourier transform
FIRST	Field Portable Imaging Radiometric Spectrometer Technology (spectrometer)
FLAASH	fast line-of-sight atmospheric analysis of spectral hypercubes
FOV	field of view
FPA	focal plane array
fps	frames per second
FTHSI	Fourier Transform Hyperspectral Imager
FTIR	Fourier transform infrared (spectrometer)
FTMF	finite target matched filter
FTS	Fourier transform spectrometer
FWHM	full-width at half-maximum
GIFTS	Geosynchronous Imaging Fourier Transform Spectrometer
GIQE	general image-quality equation
GLRT	generalized likelihood ratio test
GMM	Gaussian mixture model
GMRX	Gaussian mixture Reed–Xiaoli (detector)
GRD	ground-resolved distance
GSD	ground-sample distance
HDR	hemispherical directional reflectance
HICO^{TM}	Hyperspectral Imager for the Coastal Ocean
HITRAN	high-resolution transmission molecular absorption
HYDICE	Hyperspectral Digital Imagery Collection Experiment
ICA	independent component analysis
IFOV	instantaneous field of view
IS	integrating sphere
ISAC	in-scene atmospheric compensation
ISMC	improved split-and-merge clustering
ISODATA	iterative self-organizing data analysis technique
JPL	Let Propulsion Lab
JSD	joint subspace detector
kNN	k nearest neighbor
KS	Kolmogorov–Smirnov (test)

LARS	least-angle regression
LBG	Linde–Buzo–Gray (clustering)
LMM	linear mixing model
LOS	line of sight
LRT	likelihood ratio test
LVF	linear variable filter
LWIR	longwave infrared
MBCD	model-based change detector
MD	Mahalanobis distance
ML	maximum likelihood (algorithm)
MLE	maximum-likelihood estimate
MNF	maximum (or minimum) noise fraction
MODTRAN	moderate-resolution atmospheric transmission and radiance code
MSE	mean-squared error
MTF	modulation transfer function
MTMF	mixture-tuned matched filter
MWIR	midwave infrared
NA	numerical aperture
NAPC	noise-adjusted principal component
NCM	normal compositional model
NDVI	Normalized Differential Vegetation Index
NEI	noise-equivalent irradiance
NEL	noise-equivalent radiance
NEP	noise-equivalent power
NESR	noise-equivalent spectral radiance
NIIRS	Normalized Image Interpretability Rating Scale
NIR	near infrared
NIST	National Institute of Standards and Technology
NVIS	night vision imaging spectrometer
OPD	optical path difference
OPRA	oblique projection retrieval of the atmosphere
OSP	orthogonal subspace projection
PALM	pair-wise adaptive linear matched (filter)
PC	principal component
PCA	principal-component analysis
PPI^{TM}	Pixel Purity Index^{TM}
ppm	parts per million
PSF	point spread function
QSF	quadratic spectral filter
QTH	quartz tungsten

QUAC	quick atmospheric compensation
RMS	root mean square
ROC	receiver operating characteristic
ROIC	readout integrated circuit
RX	Reed–Xiaoli (detector)
SAM	spectral angle mapper
SCR	signal-to-clutter ratio
SEBASS	Spectrally Enhanced Broadband Array Spectrograph System
SEM	stochastic expectation maximization
SMF	spectral matched filter
SMIFTS	Spatially Modulated Imaging Fourier Transform Spectrometer
SMM	stochastic mixing model
SNR	signal-to-noise ratio
SRF	spectral response function
SS	subspace (detector)
SS-ACE	subspace adaptive coherence/cosine estimator
SSD	subpixel subspace detector
SSRX	subspace Reed–Xiaoli (detector)
SVD	singular-value decomposition
SVM	support vector machine
SWIR	shortwave infrared
SWIR1	short-wavelength end of the SWIR spectral region
SWIR2	long-wavelength end of the SWIR spectral region
TMA	three-mirror anastigmatic (design)
USGS	United States Geological Survey
VIS	visible
VNIR	visible and near-infrared
ZPD	zero path difference

Chapter 1
Introduction

Remote sensing has been defined as "the field of study associated with extracting information about an object without coming into physical contact with it" (Schott, 2007). In that regard, we use remote sensors to capture specific information from which we make decisions. For example, a weather satellite can take sophisticated measurements of global atmospheric parameters that ultimately drive someone's decision whether to carry an umbrella to work or not. Often, the sensors do not directly measure the particular information of interest, but simply provide data from which the desired information can be extracted based on some observables of the remotely viewed objects. When viewed from this perspective, it is important to consider the four fundamental remote sensing questions posed in Fig. 1.1. That is, we must understand the observables that carry the information of interest, what the sensor actually measures relative to this inherent information content, how well the sensor preserves this information in the data it provides, and ultimately how the information can be extracted by subsequent data processing. This book addresses one particular field of remote sensing, that of hyperspectral imaging, and attempts to provide a thorough treatment of the scientific principles underlying this remote sensing discipline, so that the reader will have a good appreciation of each of these four fundamental aspects of the technology.

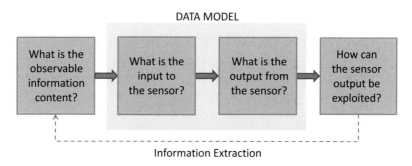

Figure 1.1 Four fundamental remote sensing questions.

1.1 Hyperspectral Remote Sensing

Hyperspectral remote sensing combines two sensing modalities familiar to most scientists and engineers: imaging and spectrometry. An imaging system captures a picture of a remote scene related to the spatial distribution of the power of reflected and/or emitted electromagnetic radiation integrated over some spectral band. This is most commonly (but not always) the visible band. On the other hand, spectrometry measures the variation in power with the wavelength or frequency of light, capturing information related to the chemical composition of the materials measured. The instrumentation used to capture such spectral information is called a spectrometer. Before delving further into specific issues associated with how these two attributes are combined in a hyperspectral remote sensor, it is beneficial to first discuss the broader field of electro-optical and infrared (EO/IR) remote sensing, clarify some terminology, and set an appropriate context for the more-detailed treatment that this book provides. For further background reading on the more-general field of EO/IR remote sensing, Schott (2007) provides an excellent reference.

The EO/IR spectral region is a portion of the electromagnetic spectrum that nominally ranges from 0.4 to 14 µm wavelength (20- to 750-THz frequency) and is widely used for remote imaging. The field of EO/IR remote sensing is extremely diverse in terms of the types of sensors employed, and encompasses a wide variety of applications. This diversity is constrained in this book by restricting it to remote sensing systems that are passive in nature, provide image data over relatively broad areas, view objects on or near the earth from airborne or space-based platforms, and are typically far away (typically many kilometers) from the objects being imaged. The term *passive*, in this context, refers to sensing systems that rely completely on natural sources of electromagnetic radiation, such as the sun and moon. That is, they do not incorporate a source such as a laser for illuminating the scene, as is the case for *active* remote sensors such as laser radars.

There are two primary sources of radiation to consider with respect to passive EO/IR remote sensing systems: reflected sunlight and thermal emission from objects in the scene. Figure 1.2 illustrates a nominal strength of these two sources that might be captured by a remote sensor for typical terrestrial objects during the day as a function of optical wavelength. According to the figure, the EO/IR spectrum can be segregated into five basic spectral regions defined as follows: visible (VIS) from 0.4 to 0.7 µm, near infrared (NIR) from 0.7 to 1.1 µm, shortwave infrared (SWIR) from 1.1 to 3.0 µm, midwave infrared (MWIR) from 3 to 5 µm, and longwave infrared (LWIR) from 5 to 14 µm. Reflected sunlight is the dominant radiation source in the VIS, NIR, and SWIR regions, and thermal emission dominates in the LWIR region. In terms of radiance (a

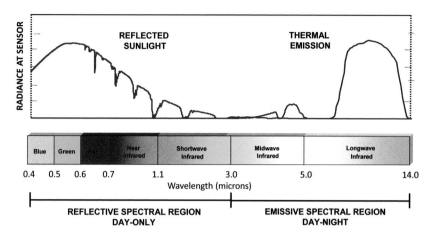

Figure 1.2 Primary natural sources of EO/IR radiation and definition of specific spectral regions.

measure of the optical power per unit area scattered in the direction of the sensor), the sources are roughly comparable in the MWIR, especially on the short-wavelength end, so that both must be considered for sensors operating in this spectral region. This assumes that the sources are at nominal terrestrial temperatures in the 290 to 300 K range; the comparison would differ significantly for unusually cold or hot objects. At night, direct solar illumination disappears, and VIS, NIR, and SWIR radiation is due to a combination of natural sources such as the moon, stars, and airglow, and manmade sources such as street lights. The magnitude of the radiation from such sources is generally too small to support hyperspectral remote sensing at night and is negligible relative to the solar component during the day, so such sources are ignored in this book. There are, however, potential applications of hyperspectral imaging in this spectral region at night, when interest is solely in the characterization of bright sources such as lamps and fires.

   Most conventional EO/IR sensors are imaging systems and operate by forming images of the remote scene by integrating collected radiation over a sufficiently broad spectral range to achieve adequate sensitivity (i.e., the image signal is large compared to the aggregate sensor noise level). Such *panchromatic* imaging is conventionally performed over the visible and near-infrared (VNIR), MWIR, or LWIR spectral regions, with examples depicted in Fig. 1.3. The VNIR spectral region is preferred for daytime imaging, as it coincides with the peak solar spectral radiance, is a region where sensor technology is very mature, and most closely coincides with human vision. The latter regions are preferred for night imaging, with the MWIR emerging as the spectral region of choice for long-range imaging systems due to its better performance with comparable-sized optics and

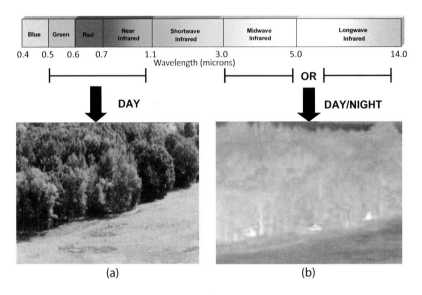

Figure 1.3 Illustrative examples of panchromatic imagery, including both (a) a daytime VIS and (b) nighttime LWIR image of a tree line with parked vehicles (three bright objects in the LWIR image).

a more mature detector technology. The physical basis of image content in VNIR imagery is a combination of broadband reflectance differences between objects (shape) and illumination variation due to the 3D nature of the scene (shading). In the MWIR and LWIR, variation in object temperature and emissivity is the source of image structure. In either case, information extracted from broadband image analysis is primarily spatial in nature.

The broadband imaging concept can be extended to *multispectral* imaging, as depicted in Fig. 1.4, by simultaneously capturing images in multiple spectral bands where the pixels are spatially registered, so that they view identical parts of the scene. Each band of a multispectral image roughly contains the same or similar information content as a broadband image. Differences between bands, however, provide additional information that depends on how the reflectance or emissivity of the materials within an image pixel coarsely varies as a function of wavelength. A multiband display will accentuate these differences in terms of color, with the term *true color* referring to a multiband display coincident with the human visual response (that is, bands centered in the 450-, 550-, and 650-nm range corresponding to blue, green, and red colors), and *false color* referring to a general mapping of multispectral bands into a visible color display. Multispectral imagery supports not only enhanced display of scene content, but also quantitative analysis based on the intrinsic spectral characteristics of imaged objects (Schott, 2007).

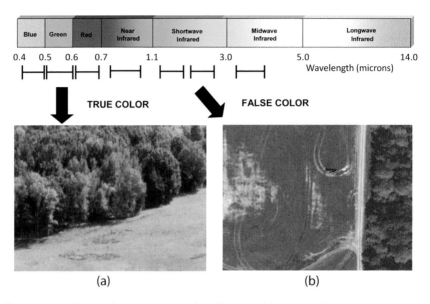

Figure 1.4 Illustrative examples of multispectral imagery, including (a) a true-color VIS image of the tree line from Fig. 1.2 and (b) a false-color VNIR image of a similar scene from a more overhead vantage point. Note that the vegetation is red in the false-color image because of the higher reflectance of vegetation in the NIR band mapped to the red display.

For example, multispectral imagery can support image classification or thematic mapping into regions based on spectral (and hence material) content, or object detection based on its underlying spectral characteristics. Such analysis is limited, however, to the spectral information captured by the small number of spectral bands. Therefore, the optimum selection of bands for a particular application can become very important.

The Landsat satellite is a good example of a multispectral sensor. Since the early 1970s, Landsat satellites have mapped the surface of the earth using a multispectral EO/IR sensor system that provides imagery in the following bands: 0.45–0.52, 0.52–0.60, 0.63–0.69, 0.76–0.90, 1.55–1.75, 2.08–2.35, and 10.4–12.5 μm. The first three bands roughly represent the visible blue, green, and red bands, the fourth is an NIR band, the fifth and sixth bands fall in the SWIR region, and the last band covers much of the LWIR. Landsat was developed primarily as an earth-observing satellite, with large 30-m ground pixels but global area coverage capabilities. As such, information pertaining to terrestrial surface characteristics is extracted largely through the relative intensity of imagery in the various bands; this intensity is related to the underlying spectral reflectance characteristics composing the surface. A variety of both human visualization and automated processing methods have been developed for multispectral sensors such as Landsat to accentuate particular surface features, classify surface materials, and extract quantitative information to improve our understanding of the earth.

Enabled by recent advances in the manufacturing of large, 2D optical detector arrays, *hyperspectral* imaging sensors are able to simultaneously capture both the spatial and spectral content of remote scenes with excellent spatial and spectral resolution and coverage. These performance characteristics are defined in Chapter 6. The resulting data product is sometimes called a hypercube which, as depicted in Fig. 1.5, is a 3D dataset composed of layers of grayscale images, each pixel of which contains a finely sampled spectrum whose features are related to the materials contained within it. A *spectrum* is simply defined here as the distribution of some measure of light power or material property with optical wavelength, such as the radiance spectrum shown in the figure. The spectrum contains information related to the chemical composition of the material composing the spatial pixel of the imaging spectrometer, information that arises through physical relationships between the spectrum of reflected or emitted light, the vibrational and electronic resonances of molecules composing the material, and microscopic surface and volumetric properties. This is the same physical underpinning as the exploitation of multispectral imagery to extract scene information; in this case, however, the finer spectral sampling results in a greater wealth of information from which to ascertain the physical characteristics of

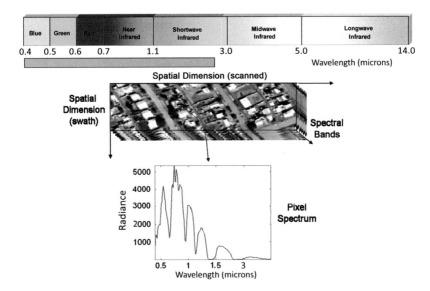

Figure 1.5 Illustrative examples of hyperspectral imagery across the VNIR/SWIR spectral region. Finely sampled spectral bands across the continuous spectral region can be assembled into a 3D hypercube, where a trace in the spectral direction for each pixel contains a spectrum related to the pixel material content.

the imaged objects. It is its ability to capture a continual spectrum—as opposed to a more discrete set of often-discontinuous spectral bands—that distinguishes a hyperspectral sensor from a multispectral sensor.

From an information-processing perspective, one can simply extend the visualization and processing approaches from multispectral imaging to the point where the number of bands becomes very large and the width of the bands becomes very small. This viewpoint often leads to methods that simply extract subsets of the hyperspectral bands to support various applications. This approach, however, arguably overemphasizes the imaging aspect of hyperspectral sensing and fails to properly exploit its nature as a spectrometer. An alternate perspective is to emphasize the spectral sampling, resolution, and range required to capture intrinsic spectral characteristics of scene materials to support an information-processing application, such as material detection, classification, or identification, in the presence of uncertainties due to unknown illumination and environmental conditions, the background environment in which the objects of interest are located, and sensor noise. Such information-processing problems become analogous to communications signal processing, where one typically wants to extract some sort of signal from interference and noise. Here, the material spectral signature is the signal, and the illumination/environmental variations and background spectral signatures are the interference. The key is to provide an overdetermined set of measurements to support separation of the signal from the interference and noise. The term *hyper*, a mathematical prefix to denote four or more dimensions, can then be associated with capturing enough spectral measurement dimensions to support such signal separation for a given application.

1.2 Elements of Hyperspectral Sensing

When hyperspectral sensing is viewed in the manner suggested, the importance of understanding all elements of the sensing chain—from material spectral phenomenology through image data processing and analysis—is clear. This book attempts to provide the basis for such an end-to-end understanding based on the four primary technical disciplines that compose the field of hyperspectral remote sensing: material spectroscopy, radiative transfer, imaging spectrometry, and hyperspectral data processing.

The general organization of the book is illustrated in Fig. 1.6 and can be seen to correlate with the four fundamental remote sensing questions. A brief introduction to these four disciplines is given in this introduction, and the remainder of the book expands on the relevant technical principles of each discipline and specific applications for the design, development,

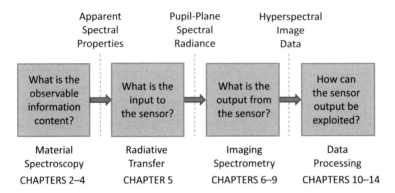

Figure 1.6 Organization of this book with respect to the fundamental elements of hyperspectral remote sensing.

and characterization of hyperspectral remote sensing and information processing.

1.2.1 Material spectroscopy

Material spectroscopy refers to the science of the interaction of electromagnetic radiation with matter, particularly with respect to the wavelength dependence of observable material features, and provides the physical basis of hyperspectral sensing in terms of the direct relationship between observable (or apparent) material spectral features and the inherent compositional characteristics. In its purest form, spectroscopy relates to electromagnetic interaction at the atomic and molecular level and requires a strictly quantum mechanical treatment. Such a treatment provides great insight into the origins of observable spectral features that are the foundation for the information extracted from a hyperspectral sensor. However, since most practical sensing applications involve the observation of materials composed of very complex mixtures of molecules, a more empirical understanding of spectroscopy that is specifically focused on apparent spectral properties, such as reflectance and emissivity spectra, is sometimes sufficient and often more practical. This book covers both to some extent. Chapters 2 and 3 examine spectroscopic principles from a more theoretical perspective, with Chapter 2 covering classical electromagnetic theory and Chapter 3 covering the quantum mechanical theory. Chapter 4 addresses spectroscopy from the standpoint of measured spectral properties.

A simple example of quartz LWIR spectral properties is perhaps helpful in conveying the physical relationships and differences between intrinsic and apparent spectral properties of materials. Quartz is a crystalline or amorphous form of SiO_2 molecules that are extremely common in silicate-based soils prevalent all over the world. A physical chemist would recognize SiO_2 as a triatomic molecule, with three vibrational degrees of freedom and three corresponding resonant absorptions of electromagnetic radiation: symmetric and asymmetric stretch resonances that roughly coincide, and a bending resonance that occurs at a lower electromagnetic frequency. As is typical for such materials, these resonances correspond to frequencies in the LWIR spectral region and are responsible for the *intrinsic spectral properties* of crystalline or amorphous quartz. Intrinsic spectral properties relate to the fundamental material characteristics that describe how the material interacts with electromagnetic radiation of different wavelengths. These intrinsic spectral properties are captured by the spectral nature of the real and imaginary parts of the complex index of refraction $n(\lambda)$ and $\kappa(\lambda)$, shown in Fig. 1.7 for both crystalline and amorphous forms.

The intrinsic spectral properties are unique to a material and therefore can be considered to provide a signature from which the material type can be determined. However, the complex index of refraction is not a directly measurable quantity and must be inferred from other measurements, such as spectral *reflectance* or *transmittance*. If we make measurements as a function of wavelength, they are called *apparent spectral properties* and represent the fundamental observable signatures of the material. Through a physical model of the material, it is possible to relate the apparent spectral properties to the intrinsic spectral properties, so that they carry the same information about material type.

For example, Fig. 1.8 compares the apparent reflectance for a quartz surface in air based on such a model to an empirical spectral reflectance measurement with a laboratory instrument. While the overall reflectance predicted by the model is much higher than the empirical measurement, the underlying spectral shape is very similar and exhibits spectral features at the same resonant wavelengths. The spectral features around 9 and 10.5 μm are directly related to the fundamental resonances of the SiO_2 molecule and are characteristic of all silicate-based minerals and soils. Differences in the expression of these features in real materials relate to differences in surface roughness, volumetric properties, and mixtures with other substances—aspects that are difficult to model properly.

1.2.2 Radiative transfer

A fundamental difference between hyperspectral sensing and traditional spectroscopy is the desire in the former case to observe these fundamental features remotely and without a controlled source. Radiative transfer is the science of the propagation of electromagnetic radiation from various sources to a sensor, including interactions with objects and their environment, as well as the atmospheric. Radiative transfer is another critical factor that must be carefully understood because it provides the

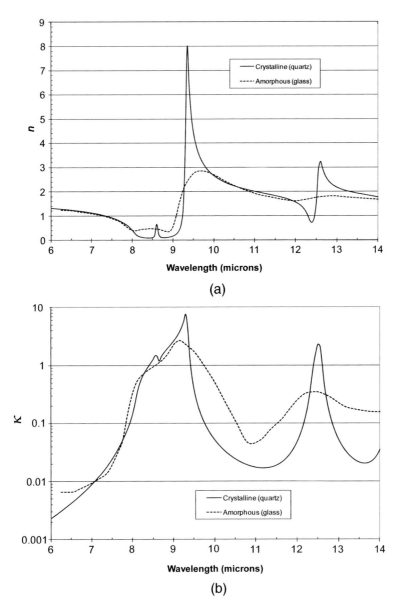

Figure 1.7 Intrinsic spectral properties of quartz in amorphous and crystalline forms. (a) The real part of the complex index of refraction $n(\lambda)$ impacts the material refractive properties, while (b) the imaginary part $\kappa(\lambda)$ impacts its internal absorptive properties; however, both influence its apparent reflectance distribution.

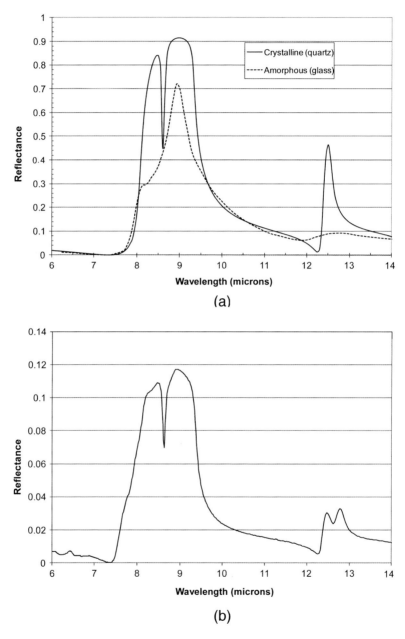

Figure 1.8 Comparison of the apparent reflectance distribution of quartz based on (a) simple modeling from the intrinsic spectral properties in Fig. 1.5 to (b) an empirical reflectance measurement.

physical relationship between the measured spectrum, typically in the form of a spectral radiance measurement at the sensor location, to the apparent material characteristic, such as its reflectance distribution. To illustrate this importance, we consider an example in the reflective spectral region where solar illumination is the dominant source.

The underlying spectral properties of materials are often characterized in terms of their apparent reflectance distribution as a function of wavelength. Strictly, the reflectance is dependent on both the illumination and observation orientation angles relative to the material surface, but it is common in hyperspectral remote sensing to use a diffuse scattering assumption that ignores these dependencies. A typical imaging scenario is shown in Fig. 1.9, where material is illuminated by the sun. Because of the intervening atmosphere, the illumination actually includes two components: the direct solar illumination reduced by the atmospheric attenuation, and an indirect component due to energy scattered by atmospheric constituents such as aerosols, clouds, and surrounding objects. Compared to the direct component that impinges on the object from one distinct direction, the indirect component is diffuse in incident angular dependence. Under some approximations, it is sufficient to describe the diffuse downwelling illumination in terms of its surface

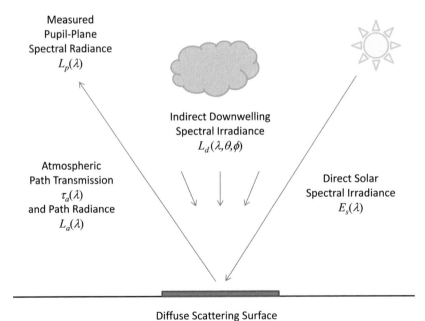

Figure 1.9 Basic scenario for radiative transfer under direct and indirect solar downwelling radiance.

irradiance integrated over the hemisphere above the surface:

$$E_d(\lambda) = \int_0^{2\pi} \int_0^{\pi/2} L_d(\lambda, \theta, \varphi) \sin\theta \cos\theta \, d\theta \, d\varphi, \qquad (1.1)$$

where we define $L_d(\lambda, \theta, \phi)$ as the indirect spectral radiance, a radiometric quantity that maintains the angular dependence of propagating radiation, and $E_d(\lambda)$ as the integrated spectral irradiance. (These radiometric quantities are more precisely defined in Chapter 2.) This diffuse illumination can be influenced by light scattered either directly or indirectly from adjacent objects in the scene.

If we characterize the atmospheric path from the surface to the remote sensor by its path transmission $\tau_a(\lambda)$ and path radiance $L_a(\lambda)$, then the pupil-plane spectral radiance $L_p(\lambda)$ at the sensor aperture can be modeled as

$$L_p(\lambda) = \frac{\tau_a(\lambda)\rho(\lambda)}{\pi} [E_s(\lambda) + E_d(\lambda)] + L_a(\lambda). \qquad (1.2)$$

Like the downwelling radiance, the path radiance $L_a(\lambda)$ can also be influenced by scattering from the local region around the surface of interest, commonly known as the adjacency effect. The pupil-plane spectral radiance represents the radiant flux per unit area, solid angle, and wavelength that can be sensed at the remote location for a given location in the scene. An ideal hypercube would represent the spatial variation in spectral radiance across the scene, corresponding to the spectral reflectance of each scene pixel.

The effects of the atmosphere and environment on the remotely measured spectral radiance relative to the material reflectance are demonstrated by considering the VNIR hyperspectral imaging scenario depicted in Fig. 1.10. The scene consists of four painted panels in a grass field in front of a tree line, with the measured laboratory reflectance spectra shown in Fig. 1.11. Each spectrum exhibits a distinctive shape that makes it easily distinguished from the other. One can compare these spectra to the output of an imaging spectrometer for each panel location in Fig. 1.12. These are now displayed in pupil-plane spectral radiance by the sensor response function, which is computed from raw data numbers through a sensor calibration process described in Chapter 10. Even after such calibration, these measurements still include the unknown illumination, atmospheric transmission, and atmospheric path radiance terms in Eq. (1.2) that not only dramatically alter the shape of the measured spectra relative to the underlying reflectance spectra, but also impart numerous fine features due to atmospheric absorption lines. Some of these absorption lines are identified in a modeled atmospheric transmission plot

Figure 1.10 Scene composed of four painted panels in a field. From left to right, the panels are black, green, tan, and bare aluminum.

shown in Fig. 1.13. While the measured spectra are just as distinct from each other as the laboratory spectra, matching them to the laboratory spectra requires some knowledge of these illumination and atmospheric influences. This is the domain of radiative transfer that is further detailed in Chapter 5.

1.2.3 Imaging spectrometry

Imaging spectrometry refers to the art and science of designing, fabricating, evaluating, and applying instrumentation capable of simultaneously capturing spatial and spectral attributes of a scene with enough fidelity to preserve the fundamental spectral features that provide for object detection, classification, identification, and/or characterization. A variety of optical techniques can be employed to develop a sensor capable of capturing hyperspectral imagery, including dispersive prism or grating spectrometers, Michelson Fourier transform spectrometers, spatial Fourier transform spectrometers, scanning Fabry–Perot etalons, acousto-optic tunable filters, and dielectric filters. However, the most common method from the perspective of remote sensing implementation from airborne or space-based platforms is the dispersive imaging spectrometer, which can map spectral information along one direction of a 2D detector array. At any point in time, the spectrometer collects a frame composing a single spatial slice of the scene at multiple wavelengths of light. Then the other spatial dimension is expanded by scanning and collecting multiple frames. Prism spectrometers (one class of dispersive spectrometers) have been demonstrated for airborne operation over the reflective and emissive spectral regions. Grating spectrometers can offer some advantage in terms of spectral resolution and compactness and have been successfully demonstrated over the VIS through the SWIR region for both airborne and space-based platforms.

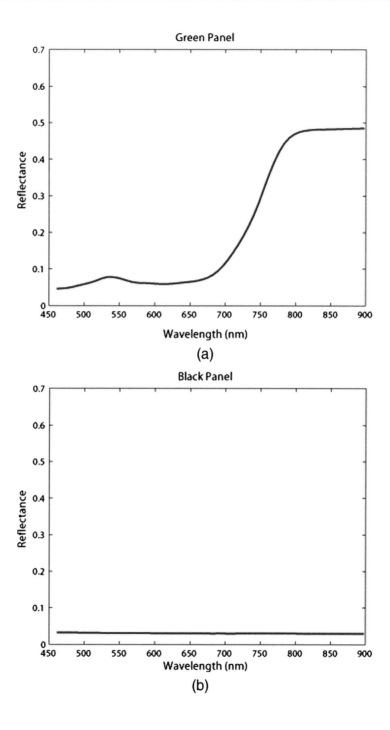

Figure 1.11 Laboratory reflectance spectra for the four panels depicted in Fig. 1.10: (a) green, (b) black, (c) tan, and (d) aluminum.

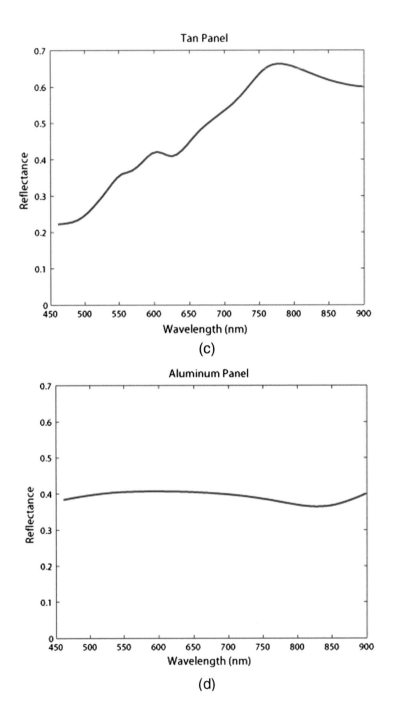

Figure 1.11 (*continued*)

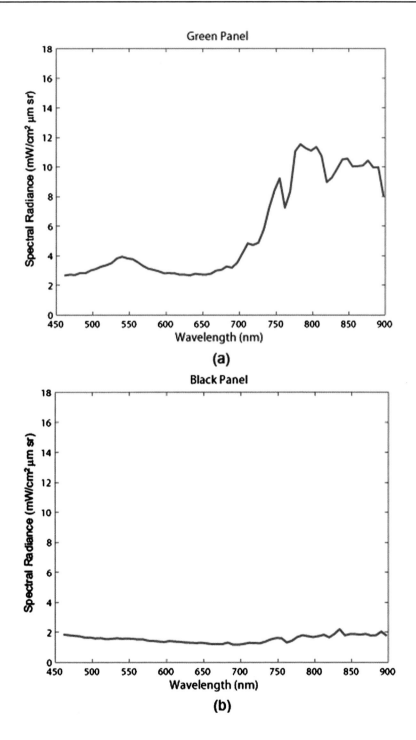

Figure 1.12 Measured spectra for the four panels depicted in Fig. 1.10: (a) green, (b) black, (c) tan, and (d) aluminum.

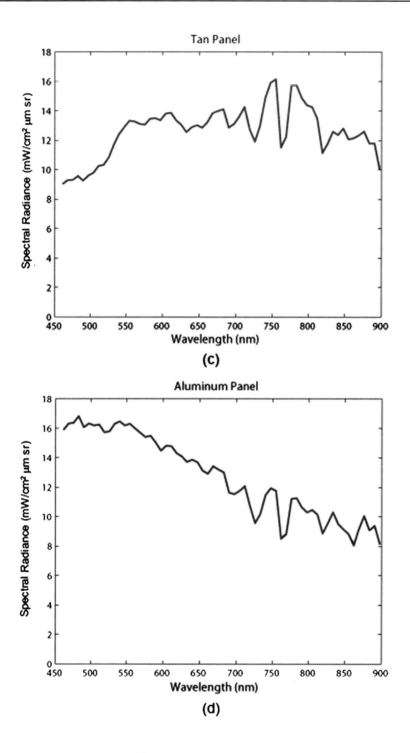

(c)

(d)

Figure 1.12   (*continued*)

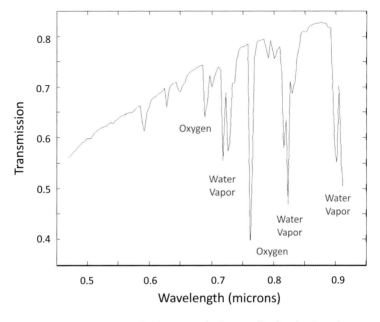

Figure 1.13 Modeled atmospheric transmission profile for the imaging scenario associated with the measured spectra depicted in Fig. 1.7 indicating specific atmospheric absorption lines.

A typical grating spectrometer arrangement is depicted in Fig. 1.14. Light entering the grating spectrometer from a remote location (from the left in the figure) is first focused by an imaging lens assembly to form an image on the slit plane. The slit is an opaque surface that blocks all light except for a rectangular area one pixel high (in the dimension shown) and many pixels wide (in the orthogonal direction). This surface is reimaged by the three spectrometer mirrors onto the 2D detector array, so that the slit height nominally matches the detector element size and the slit width matches the detector array width. If there are no other elements, the 2D detector array simply produces a line image of the portion of the scene that passes through the slit. However, the spectrometer also includes a periodic blazed grating on the surface of the middle mirror of the reimaging optics that disperses light across the detector array dimension shown. Thus, the arrangement forms an optical spectrum for each spatial location in the entrance slit across this dimension of the array. To understand the design and performance of such spectrometers, it is necessary to have a sound understanding of the design and performance of EO/IR imaging systems in general. A review of EO/IR imaging system design and performance is provided in Chapter 6, which can serve as a refresher for readers already familiar with this subject or as a tutorial for readers new to the subject. The principles, design, and performance of prism and grating spectrometers are then further detailed in Chapter 7.

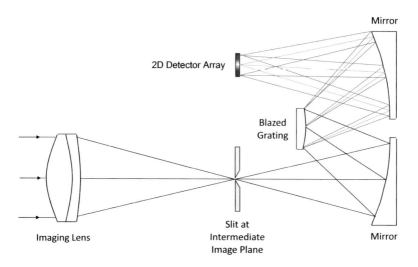

Figure 1.14   Basic optical diagram of a grating imaging spectrometer.

Fourier transform spectrometers use interferometry to capture spectral information and have particular advantages in terms of their ability to obtain very fine spectral resolution. An *imaging* Fourier transform spectrometer basically incorporates an imaging detector array into the interferometer to simultaneously capture interferograms for a number of detectors viewing different locations in the remote scene. The principles, design, and performance of Fourier transform spectrometers are detailed in Chapter 8, followed by a variety of other imaging spectrometer approaches in Chapter 9. This is followed in Chapter 10 by a description of the important issues concerning spectrometer calibration, the method where raw sensor data is converted into physically meaningful units.

1.2.4 Hyperspectral data processing

After hyperspectral data have been calibrated, the question becomes how to extract and utilize the information content within the data. Unlike conventional panchromatic imagery or even multispectral imagery, the information content in hyperspectral imagery does not readily lend itself to human visualization and requires computer processing to extract it. Thus, significant attention in the field of hyperspectral remote sensing research is devoted to developing algorithmic techniques to detect, classify, identify, quantify, and characterize objects and features of interest in the captured data. Roughly speaking, two general approaches are taken toward such processing: physical modeling and statistical signal processing. Model-based approaches using the radiative transfer methods described in Chapter 5 are applied to construct an underlying physical model of the scene composition, illuminating sources, and propagation of radiation through the atmosphere to the remote sensor. Typical methods employ a linear mixing model for the radiance spectra and a sophisticated radiation transport code such as MODTRAN to capture the atmospheric transmission, radiation, and scattering processes. These methods are often focused beyond the detection and classification of materials to a more quantitative estimation of constituent material abundances and physical properties. For example, coastal applications might involve the estimation of properties such as chlorophyll concentration, gelbstoff content, turbidity, depth, and bottom type.

When applying hyperspectral remote sensing to object detection and classification problems, it is sometimes advantageous to use statistical processing methods or some combination of physical modeling and statistical signal processing, as these methods are better able to capture the randomness of natural and manmade variability in spectral characteristics that generally exists in real imagery and is difficult to physically model. A variety of such methods have been developed for hyperspectral imaging applications, many of which evolved from techniques used in fields of communication and radar signal processing for the detection of signals in the presence of structured noise. Many of these methods become very specific to the application of interest and are better addressed within the context of the application. However, most methods have three basic functions in common: (1) means of compensating for the unknown influences of the atmosphere and environment to allow spectral radiance data, captured by the imaging spectrometer, to be compared in some quantitative manner to apparent spectral properties of materials of interest, such as their reflectance or emissivity distributions; (2) an underlying model for the statistical variance within the imagery, along with mathematical methods to fit the data to pertinent model parameters; and (3) some approach for detecting, classifying, and identifying specific materials of interest in the scene based on their unique spectral characteristics. Detection, classification, and identification are typically based on hypothesis testing using appropriate statistical models for the spectral data. Fundamental methods for performing these three data processing and analysis functions are successively reviewed in Chapters 11 through 14.

1.3 Examples of Remote Sensing Applications

The purpose of this book is to provide a fairly comprehensive treatment of the fundamental principles of hyperspectral remote sensing as opposed to an in-depth review of the applications of the technology. However, there is clearly some benefit to at least anecdotally discussing a few application examples to establish some context for the ways in which the tools of hyperspectral sensing can be applied to real-world problems. This section merely attempts to provide such context, and the reader is encouraged to consult other sources, some of which are referenced at the end of this chapter and elsewhere throughout this book, to broaden the context achieved here.

The early development of hyperspectral imaging systems was primarily oriented toward earth remote sensing applications and was initially based on the Airborne Imaging Spectrometer (AIS) and the Airborne Visible/Infrared Imaging Spectrometer (AVIRIS) built by the NASA Jet Propulsion Laboratory (Goetz *et al.*, 1985; Green *et al.*, 1998). By simultaneously capturing both spatial and spectral features of a remote scene at fine resolution, the data were useful for distinguishing subtle material differences and generating thematic maps that provide the spatial distribution of surface material content. This unique ability of hyperspectral sensors to remotely extract features related to material content on a pixel-by-pixel basis has led to its widespread application to a variety of other remote sensing fields, including geology (Kruse and Lefkoff, 1993; Kruse *et al.*, 2003), vegetation science (Goodenough *et al.*, 2003; Pu *et al.*, 2003; Townsend *et al.*, 2003; Kumar *et al.*, 2001), agriculture (Haboudane, 2004), land use monitoring (Bachman *et al.*, 2002; Landgrebe, 2005), urban mapping (Benediktsson *et al.*, 2005; Roessner *et al.*, 2001), coastal analysis (Mosher and Mitchell, 2004; Davis *et al.*, 2002; Brando and Dekker, 2003; Dekker *et al.*, 2001), bathymetry (Adler-Golden *et al.*, 2005), mine detection (Zare *et al.*, 2008), law enforcement (Elerding *et al.*, 1991; Daughtry and Walthall, 1998), and chemical agent detection and characterization (Kwan *et al.*, 2006).

Geological applications arise directly out of the unique spectral characteristics of minerals in geological formations. From remotely sensed spectral information, the hyperspectral analyst attempts to estimate the mineral content of surface materials from the diagnostic spectral features of the silicate, carbonate, sulfate, hydroxide, and other mineral types, as previously discussed in the quartz example. An example of this process is depicted in Fig. 1.15, which illustrates mineral mapping results from an airborne LWIR imaging spectrometer called the Spectrally Enhanced Broadband Array Spectrograph System (SEBASS) that collected hyperspectral imagery over the Comstock Mining District near Reno, Nevada (Vaughan *et al.*, 2003). This particular region was a gold, silver, and mercury mining area in the mid-nineteenth century and is covered with mine tailings composed of quartz, pyrite, kaolinite clay, and iron oxides of pyrite such as hematite, goethite, and jarosite. To determine the presence of these minerals, image spectra are first calibrated and atmospherically corrected to estimate surface spectral *emissivity*, a quantity related to the spectral reflectance and approximated as one minus the reflectance for opaque objects that have zero net

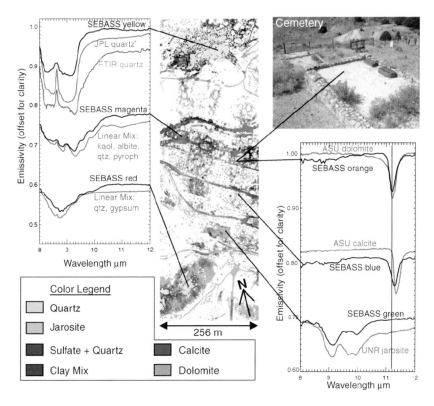

Figure 1.15 Use of SEBASS LWIR hyperspectral imagery to perform mineral mapping of the Comstock Mining District near Reno, Nevada [reprinted with permission from Vaughan *et al.* (2003) © 2003 Elsevier].

transmission. Figure 1.15 indicates matching between the image-derived spectral emissivity estimates and laboratory data from a mineral library. Note the reasonable matching in regions of the characteristic mineral features such as the quartz reststrahlen feature discussed earlier. A color-coded mineral classification map such as that shown can be produced based on some quantitative metrics for the quality of the spectral match to spatially indicate the presence of surface minerals. In this example, the expected quartz, kaolinite clay, and jarosite tailings are identified along with dolomite and calcite from marble gravestones in a cemetery present in the scene.

Such mineralogy analysis based on LWIR hyperspectral imagery can be extended to map larger regions for the study of geological formations to infer the potential presence of subsurface materials of interest such as precious minerals or fossil fuels. Another example is depicted in Fig. 1.16, where SEBASS imagery was again employed in September 1999 to spectrally identify a larger range of minerals over the Yerington District in Nevada by similar spectral matching methods (Cudahy *et al.*, 2000). The

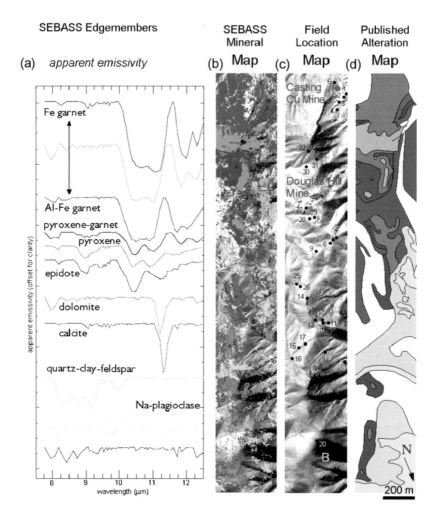

Figure 1.16 A geological map of the Yerington District in Nevada formed from SEBASS LWIR hyperspectral imagery (Cudahy *et al.*, 2000) (reprinted courtesy of Commonwealth Scientific and Industrial Research Organization, Australia).

mineral matching results in this case were compared to a ground survey map of surface minerals to characterize the efficacy of the method.

Given inertial and pointing data from an airborne sensor, along with digital terrain elevation data of the imaged site, such image-derived mineral map data can be overlaid onto a 3D orthorectified surface map, as depicted in Fig. 1.17 for LWIR SEBASS hyperspectral imagery of Cuprite, Nevada (Huneycutt, 2009). This particular image consists of surface material classified into six surface material types, each corresponding to a different display color: red quartzite (yellow), glauconitic sandstone (violet), gray quartzite (dark blue), gray slate (green), phyllite (light blue),

Figure 1.17 Merging of a SEBASS LWIR hyperspectral image with digital terrain elevation data to produce a 3D orthorectified geological map. The various colors relate to surface material classes with different mineral content, and the resulting classification map is draped on a digital elevation map for context (Huneycutt, 2009) (reprinted courtesy of The Aerospace Corporation).

and oolitic limestone (red). The MWIR and LWIR spectral regions are particularly useful for such geological analysis because most minerals have characteristic spectral features in this part of the infrared spectrum.

This basic concept can also be applied to detecting and distinguishing vegetation types in hyperspectral imagery, largely due to the unique spectral characteristics of chlorophyll in the VNIR spectral region. Not only does the presence of chlorophyll make healthy vegetation readily detectable in an image, but more-subtle differences in the particular types and quantity of chlorophyll pigments, as well as differences in water content, cellulose, lignin, and other constituent concentrations in various vegetation types, can also support plant classification. When collected over wide areas of farmland, such information can be extremely useful in predicting crop yields. Other vegetation properties such as leaf area, chlorophyll content, moisture content, and biomass can be extracted from remotely sensed spectra and can provide indicators of plant disease and vegetation stress, possibly enabling more-effective remediation to maximize crop production. Such analysis can find similar environmental applications in nonagricultural areas, such as monitoring the health and evolution of rainforests, wetlands, and forests.

As an example of vegetation mapping methodology and results, consider the crop classification maps depicted in Fig. 1.18 from the Summitville abandoned mine in San Luis Valley, Colorado (Clark *et al.*, 1995). VNIR/SWIR hyperspectral images were collected over this area in September 1993 by the AVIRIS instrument and were spectrally analyzed to produce a vegetation map. The resulting product is shown in Fig. 1.18(b) and compared to the truth map in Fig. 1.18(a). The colors correspond

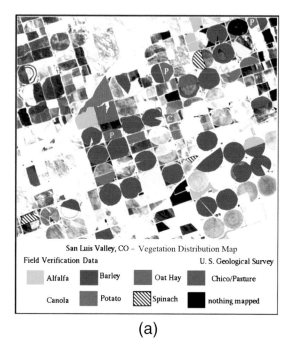

(a)

(b)

Figure 1.18 Crop classification map of San Luis, Colorado based on spectral remote sensing data: (a) truth map and (b) classification map based on AVIRIS VNIR/SWIR hyperspectral imagery (Clark *et al.*, 1995).

to the crop species as indicated in the figures. Crop classification was performed by a spectral classification algorithm called Tricorder (Clark and Roush, 1984) that performs a least-squares-based mapping of chlorophyll, cellulose, and water absorption bands to library reflectance spectra corresponding to potential vegetation species that make up the scene. This particular example resulted in 96% classification accuracy in the regions of the image for which there was truth data.

The geological and agricultural applications described involve some sort of mapping function, ranging from classification of land cover materials over presumably large areas, or further quantitative estimation of specific parameters such as mineral or chlorophyll content over such an area. There are other applications that involve searching for specific objects in a scene based on their unique spectral characteristics. We refer to such applications as target detection, where the target could range from a combat vehicle (military application) to a trace gas leaking from a facility (environmental safety application). One particular target detection example where hyperspectral remote sensing has been applied concerns civilian search and rescue (Eismann *et al.*, 2009). The Civil Air Patrol, an Air Force auxiliary whose primary function is to locate and rescue downed aircraft in the continental United States, developed a system called Airborne Real-Time Cueing Hyperspectral Enhanced Reconnaissance (ARCHER) that uses a VNIR hyperspectral remote sensor to automatically find indications of downed aircraft and cue an onboard operator. ARCHER functionally operates in three ways: (1) it searches for areas that are spectrally anomalous to a surrounding background that can be indicative of a crash site; (2) it searches for specific target material signatures (i.e., laboratory reflectance distribution) that are known to be present in a particular crash site; or (3) it searches for specific changes caused by a crash in a region that has been previously imaged. As depicted in Fig. 1.19, ARCHER exploits the spectral differences between the aircraft materials and the natural background to locate (or detect) specific locations in the scene where a crash might be present. The onboard operator then needs only to examine these cued locations as opposed to exhaustively search, at high spatial resolution, the entire search volume. This method was effectively employed to locate materials associated with an F-15 aircraft crash in Dent County, Missouri. Some of the results are displayed in Figs. 1.20 and 1.21.

Hyperspectral target detection is not limited to the detection of solid objects by virtue of their reflectance spectra. It has also been widely used for detecting gaseous effluents based on their unique spectral transmission properties. Gaseous materials tend to have distinct and narrow absorption bands in the MWIR and LWIR spectral regions that often make their detection, even at small concentrations, very tractable. A simple example

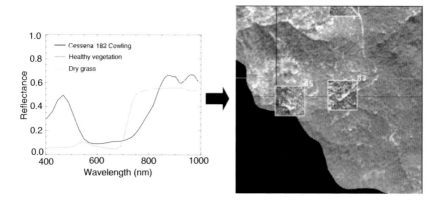

Figure 1.19 Illustration of the use of hyperspectral imagery to support civilian search and rescue. The spectral differences between aircraft materials and natural background are used for finding specific locations where a crash site is most likely to be present. This cues an onboard operator for closer examination (Eismann *et al.*, 2009).

Figure 1.20 False-color georectified map produced from the ARCHER hyperspectral remote sensor indicating specific locations where the spectral detection processor found matches to F-15 paint. The primary wreckage site is in the center of the image, while some of the other cues correspond to additional materials such as the ejection seat and vertical stabilizer (Eismann *et al.*, 2009).

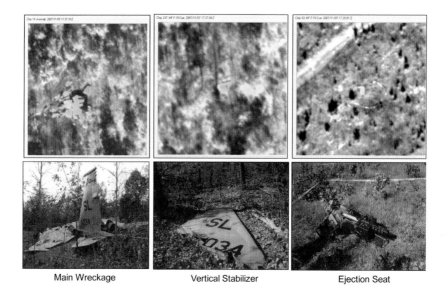

Main Wreckage   Vertical Stabilizer   Ejection Seat

Figure 1.21 High-resolution image chips produced by fusing the ARCHER hyperspectral and panchromatic imagery for spectrally cued regions. The bottom row provides corresponding ground photos from the Air Force Accident Investigation Board Report (Eismann *et al.*, 2009).

is shown in Fig. 1.22, where a release of small amounts of SF_6 is identified from LWIR SEBASS imagery using a spectral matching technique similar to that used for solid minerals, but adapted to the gaseous nature of the object of interest. Specifically, the quantity of interest in the case is the spectral *absorbance*, which is expressed through spectral transmission of the gas. In this case, SF_6 has a very simple absorption spectrum, making detection straightforward. More-complex gaseous molecules can have more-complicated absorption spectra, but the uniqueness of their spectral properties is the key element in spectral detection. Such techniques can be extended to the detection of specific gaseous molecules in a complex mixture of different molecules, and even to the quantification of the concentrations of the gaseous components. Applications range from the detection of gaseous leaks, such as Freon® or propane for environmental and physical safety reasons, to the quantification of pollutant gases in industrial emissions to assess compliance to environmental protection laws.

Hyperspectral remote sensing also provides capabilities to extract information from bodies of water such as lakes, rivers, and oceans. For these applications, the visible spectral region is particularly important, because water is transparent in this region. Some applications involve the detection and characterization of submerged objects and suspended material, such as the detection and quantification of algal plumes for

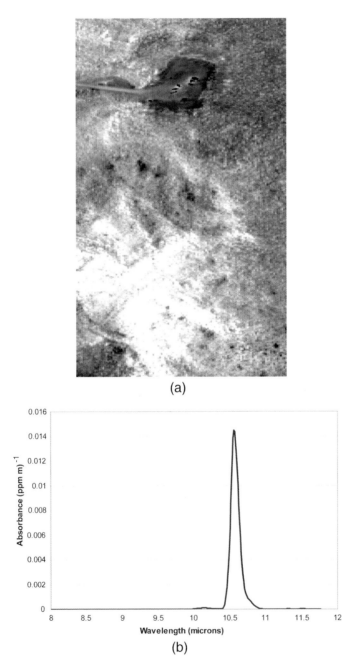

(a)

(b)

Figure 1.22 Product of SEBASS LWIR hyperspectral image processing to detect the presence of SF_6 with the given absorption spectrum. (a) Single-band image with a blue overlay indicating the pixels that were spectrally detected as SF_6. (b) SF_6 absorbance spectrum in units of $(ppm\text{-}m)^{-1}$, where ppm is the concentration in parts per million.

environmental health monitoring. Other applications aim to quantify specific characteristics of water bodies such as the concentration of chlorophyll and other suspended or dissolved matter, water turbidity, bottom sediment type, and water depth. In 2009, the Hyperspectral Imager for the Coastal Ocean (HICO™) was launched onto the International Space Station to perform environmental characterization of coastal ocean regions based on the measurement and estimation of such parameters (Corson *et al.*, 2010). Examples of the type of remote sensing products that such a remote sensing system can provide are illustrated in Fig. 1.23. Specifically, this example is based on a collection over the Florida Keys, a color image of which is provided in Fig. 1.23(a) based on three HICO™ imaging spectrometer bands. Figure 1.23(b) provides a false-color depiction of the estimated chlorophyll concentration, indicating benthic algae and sea grasses in the shallow regions of the coast. The corresponding bathymetry, or water depth, profile estimated from hyperspectral imagery is depicted in Fig. 1.23(c).

The final example relates to the application of a hyperspectral remote sensor to a disaster management scenario involving quantitative physical analysis of temperature. In this example, a VNIR/SWIR spectrometer called the night vision imaging spectrometer (NVIS) was flown over the World Trade Center in New York eight days after the September 11, 2001 attack by Al-Qaeda hijackers. Fire rescue was still trying to locate hot spots flaring up at the disaster site and extinguish the remaining fires hampering rescue and recovery efforts and causing safety

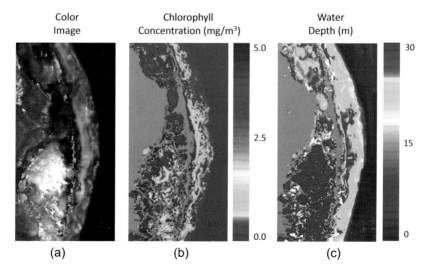

Figure 1.23 Image products derived from the Hyperspectral Imager for the Coastal Ocean: (a) color image, (b) chlorophyll concentration, and (c) bathymetry map [reprinted with permission from Corson *et al.* (2010) © 2010 IEEE].

risks. Conventional overhead imaging systems were employed to aid in hot-spot identification, but were thwarted by the heavy smoke and dust that still permeated the air. Figure 1.24(a) indicates the smoke at the disaster site that limited visibility to the ground. Because smoke

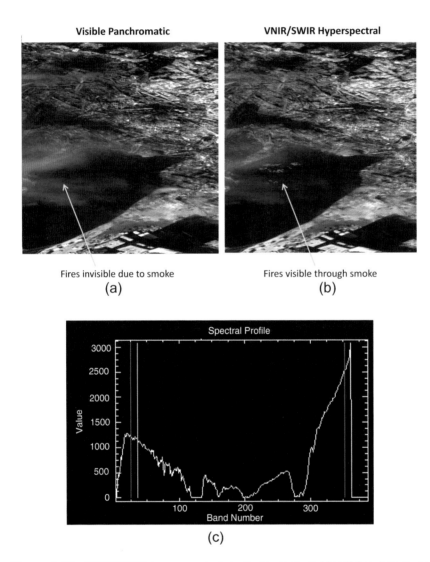

Figure 1.24 VNIR/SWIR hyperspectral analysis product identifying hot stops to aid in disaster response after the World Trade Center attack on September 11, 2001: (a) visible panchromatic image, (b) color composite from VNIR/SWIR hyperspectral image, and (c) spectral profile of a fire pixel indicating the color composite bands. The SWIR spectral capability allowed visibility through the smoke and dust, identification of high temperature regions, and estimation of fire temperatures. (Images courtesy of the U. S. Army Research, Development, and Engineering Command Night Vision and Electronic Sensors Directorate.)

and dust scatters much more heavily at shorter wavelengths, the SWIR end of the hyperspectral imager provided much higher air-to-ground transmission. Further, the fires were so hot that their thermal emission could be detected as an increase in intensity at the long-wavelength end of the VNIR/SWIR spectrum. Thus, a false-color map derived from the hyperspectral image, shown in Fig. 1.24(b), was used to accentuate the hot spots. Furthermore, temperature was estimated by matching the SWIR intensity profile to blackbody radiation functions of varying temperature after compensating for atmospheric transmission and path radiance. This is seen in the example fire spectrum in Fig. 1.24(c) that was captured from the hyperspectral image. The blue, green, and red lines indicate the three spectral bands used to form the false-color composite image.

1.4 Summary

Hyperspectral remote sensing combines spectrometry with imaging to provide the capability to extract material information from remotely sensed objects in a scene. Such capabilities have extensive applications to fields ranging from geology and earth science to defense and disaster relief. To fully appreciate hyperspectral remote sensing systems, it is necessary to understand not only spectrometer instrumentation and data processing methods, but also the physical underpinnings of material spectroscopy and radiative transfer. The remainder of this book delves into these topics in succession to provide such an end-to-end understanding of this emerging remote sensing field.

References

Adler-Golden, S. M., Acharya, P. K., Berk, A., Matthew, M. W., and Gorodetzky, D., "Remote bathymetry of the littoral zone from AVIRIS, LASH, and QuickBird imagery," *IEEE T. Geosci. Remote* **43**(2) 337–347 (2005).

Bachman, C. M., Donato, T., Lamela, G. M., Rhea, W. J., Bettenhausen, M. H., Fosina, R. A., DuBois, K. R., Porter, J. H., and Truitt, B. R., "Automatic classification of land cover on Smith Island, VA using HyMAP imagery," *IEEE T. Geosci. Remote* **40**(10), 2313–2330, (2002).

Benediktsson, J. A., Palmason, J. A., and, Svieinsson, J. R., "Classification of hyperspectral data from urban areas based on extended morphological profiles," *IEEE T. Geosci. Remote* **43**, 480–491 (2005).

Brando, V. E. and Dekker, A. G., "Satellite hyperspectral remote sensing for estimating estuarine and coastal water quality," *IEEE T. Geosci. Remote* **41**, 1378–1387 (2003).

Clark, R. N. and Roush, T. L., "Reflectance spectroscopy: quantitative analysis techniques for remote sensing applications," *J. Geophys. Res.* **89**, 6329–6340 (1984).

Clark, R. N, King, T. V. V., Ager, C., and Swayze, G. A., "Initial vegetation species and senescence/stress mapping in the San Luis Valley, Colorado, using imaging spectrometer data," *Proc. Summitville Forum '95*, Colorado Geological Survey Special Publication **38**, 64–69 (1995) (http://speclab.cr. usgs.gov, last accessed Sept. 2011).

Corson, M. R., Lucke, R. L., Davis, C. O., Bowles, J. H., Chen, D. J., Gai, B.-C., Korwan, D. R., Miller, W. D., and Snyder, W. A., "The Hyperspectral Imager for the Coastal Ocean (HICO^{TM}) environmental littoral imaging from the International Space Station," *Proc. IEEE Geosci. Remote Sens. Symp.*, 3752–3755 (2010).

Cudahy, T. J., Okada, K., Yamato, Y., Maekawa, M, Hackwell, J. A., and Huntington, T. F., *Mapping Skarn and Porphyry Alteration Mineralogy at Yerington, Nevada Using Airborne Hyperspectral TIR SEBASS Data*, Commonwealth Scientific and Industrial Research Organization, Division of Exploration and Mining, Australia, Open File Report No. 734R (2000).

Daughtry, C. S. T., and Walthall, C. L., "Spectral discrimination of *Cannibas sativa L.* leaves and canopies," *Remote Sens. Environ.* **64**, 192–201 (1998).

Davis, C. O., Cardner, K. L., and Lee, Z., "Using hyperspectral imaging to characterize the coastal environment," *Proc. IEEE Aerospace Conf.* **3**, 1515–1521 (2002).

Dekker, A. G., Brando, V. E., Anstee, J. M., Pinnel, N., Kutser, T., Hoogenboom, E. J., Peters, S., Pasterkamp, R., Vos, R., Olbert, C., and Malthus, T. J. M., "Imaging Spectrometry of Water," Ch. 5, pp. 306–359 in *Imaging Spectrometry: Basic Principles and Prospective Applications*, van der Meer, F. D., and DeJong, S. M. (Eds.), Kluwer Academic, Dordrecht, Netherlands (2001).

Eismann, M. T., Stocker, A. D., and Nasrabadi, N. M., "Automated hyperspectral cueing for civilian search and rescue," *Proc. IEEE* **97**, pp. 1031–1055 (2009).

Elerding, G. T., Thunen, J. G., and Woody, J. M., "Wedge imaging spectrometer: application to drug and pollution law enforcement," *Proc. SPIE* **1479**, 380–392 (1991) [doi:10.1117/12.44546].

Goetz, A. F. H., Vane, G., Solomon, J., and Rock, B., "Imaging spectrometry for earth remote sensing," *Science* **228**, 1147–1153 (1985).

Goodenough, D. G., Dyk, A., Niemann, K. O., Pearlman, J. S., Hao, C., Han, T., Murdoch, M., and West, C., "Processing Hyperion and ALI for forest classification," *IEEE T. Geosci. Remote* **41**(6), 1321–1331 (2003).

Green, R. O., Eastwood, M. L., Sarture, C. M., Chrien, T. G., Aronsson, M., Chippendale, B. J., Faust, J. A., Pauri, B. E., Chovit, C. J., Solis, M., Olah, M. R., and Williams, O., "Imaging spectroscopy and the Airborne Visible/Infrared Imaging Spectrometer (AVIRIS)," *Remote Sens. Environ.* **65**, 227–248 (1998).

Haboudane, D. D., "Hyperspectral vegetation indices and novel algorithms for predicting green LAI of crop canopies: modeling and validation in the context of precision agriculture," *Remote Sens. Environ.* **90**, 337–352 (2004).

Huneycutt, A. J., *Gallium-Doped Silicon Detectors for Space-based Imaging Sensors*, Aerospace Report ATR-2009(8278)-1, Los Angeles, CA, The Aerospace Corporation (2009).

Kruse, F. A. and Lefkoff, A. B., "Knowledge-based geologic mapping with imaging spectrometers," *Remote Sens. Environ.*, **8**, 3–28 (1993).

Kruse, F. A., Boardman, J. W., and Huntington, J. F., "Comparison of airborne hyperspectral data and the EO-1 Hyperion for mineral mapping," *IEEE T. Geosci. Remote* **41**, 1388–1400 (2003).

Kumar, L., Schmidt, K., Dury, S., and Skidmore, A., "Imaging Spectrometry and Vegetation Science," Ch. 5, pp. 306–359 in *Imaging Spectrometry: Basic Principles and Prospective Applications*, van der Meer, F. D., and DeJong, S. M. (Eds.), Kluwer Academic, Dordrecht, Netherlands (2001).

Kwan, C., Ayhan, B., Chen, G., Wang, J., Ji, B., and Chang, C.-I., "A novel approach for spectral unmixing, classification, and concentration estimation of chemicals and biological agents," *IEEE T. Geosci. Remote* **44**, 409–419 (2006).

Landgrebe, D. A., "Multispectral land sensing: where from, where to?" *IEEE T. Geosci. Remote* **43**, 414–421 (2005).

Mosher, T. and Mitchell, M., "Hyperspectral imager for the coastal ocean (HICO)," *Proc. IEEE Aerospace Conf.*, pp. 6–13, March (2004).

Pu, R., Gong, P., Biging, G. S., and Larrieu, M., "Extraction of red-edge optical parameters from Hyperion data for estimation of forest leaf area index," *IEEE T. Geosci. Remote* **41**(6), 916–921 (2003).

Roessner, S., Segl, K., Heiden, V., and Kaufmann, H., "Automated differentiation of urban surfaces based on airborne hyperspectral imagery," *IEEE T. Geosci. Remote* **39**(7), 1525–1532 (2001).

Schott, J. R., *Remote Sensing: The Image Chain Approach*, Second Edition, Oxford University Press, Oxford, New York (2007).

Townsend, P. A., Foster, J. R., Chastain, R. A., and Currie, W. S., "Application of imaging spectroscopy to mapping canopy nitrogen in forests of the central Appalachian mountains using Hyperion and AVIRIS," *IEEE T. Geosci. Remote* **41**(1) 1347–1354 (2003).

Vaughan, R. G., Calvin, W. M., and Taranik, J. V., "SEBASS hyperspectral thermal infrared data: surface emissivity measurement and mineral mapping," *Remote Sens. Environ.* **85**, 48–63 (2003).

Zare, A., Bolton, J., Gader, P., and Schatten, M., "Vegetation mapping for landmine detection using long-wave hyperspectral imagery," *IEEE T. Geosci. Remote* **46**(1), 172–178 (2008).

Chapter 2
Optical Radiation and Matter

The physical basis of hyperspectral remote sensing is the presence of features in the observed spectral properties of materials that are characteristic to their chemical and physical make-up. To establish this physical basis, one must explore the interaction of optical radiation with matter to understand how material properties are expressed in radiometric observations. Since materials are composed of atoms and molecules whose physical understanding requires a quantum mechanical treatment, a full understanding of the spectroscopy of materials is derived from quantum-mechanical principles. However, classical electrodynamics offers several useful insights into the underlying physics and provides foundational principles of significant utility with regard to the remote sensing field.

This chapter reviews the fundamental principles of the interaction of optical radiation with matter that arise from the classical theory of electromagnetic radiation. The focus of the treatment is interaction and propagation through dense media, and the characterization of such media by a complex index of refraction. It is shown that the complex index of refraction is an intrinsic spectral property of the material that governs the transmission and reflectance at material interfaces, as well as transmission through uniform media. A classical model for the complex index of refraction that can be further extended by quantum mechanics relates the spectral locations and shapes of resonant spectral features to the electrical and mechanical characteristics of the atoms and molecules from which the material is composed.

2.1 Propagation of Electromagnetic Radiation

The physics of optical radiation is governed by Maxwell's equations (Maxwell, 1954). In free space, the differential form of these fundamental laws of electromagnetic radiation is

$$\nabla \cdot \mathbf{E} = 0, \tag{2.1}$$

$$\nabla \cdot \mathbf{H} = 0, \tag{2.2}$$

$$\nabla \times \mathbf{E} = -\mu_0 \frac{\partial \mathbf{H}}{\partial t}, \tag{2.3}$$

and

$$\nabla \times \mathbf{H} = \varepsilon_0 \frac{\partial \mathbf{E}}{\partial t}, \tag{2.4}$$

where $\mathbf{E}$ is the electric field amplitude, $\mathbf{H}$ is the magnetic field amplitude, $\varepsilon_0 = 8.854 \times 10^{-12}$ F/m is the electric permittivity of free space, and $\mu_0 = 4\pi \times 10^{-7}$ H/m is the magnetic permeability of free space. Boldface characters are used to denote vector quantities. The divergence is defined in rectangular coordinates by the dot product

$$\nabla \cdot \mathbf{A} = \frac{\partial \mathbf{A}}{\partial x}\hat{x} + \frac{\partial \mathbf{A}}{\partial y}\hat{y} + \frac{\partial \mathbf{A}}{\partial z}\hat{z}, \tag{2.5}$$

and the curl is defined by the cross product

$$\nabla \times \mathbf{A} = \begin{vmatrix} \hat{x} & \hat{y} & \hat{z} \\ \dfrac{\partial}{\partial x} & \dfrac{\partial}{\partial y} & \dfrac{\partial}{\partial z} \\ A_x & A_y & A_z \end{vmatrix}. \tag{2.6}$$

The propagation of electromagnetic radiation arises out of the coupling of the electric and magnetic fields dictated by Eqs. (2.3) and (2.4). That is, changes in the electric field alter the magnetic field and vice versa, such that the electromagnetic wave propagates through space and time.

2.1.1 Propagation in free space

It is customary to characterize electromagnetic waves in terms of the time- and space-varying electric field, partly due to the fact that the electric field dominates interactions with charged particles in a material. To that end, Maxwell's equations are reformulated into a differential wave equation for only the electric field. Taking the curl of Eq. (2.3),

$$\nabla \times \nabla \times \mathbf{E} = \nabla \times \left(-\mu_0 \frac{\partial \mathbf{H}}{\partial t} \right). \tag{2.7}$$

By applying the vector identity

$$\nabla \times \nabla \times \mathbf{A} = \nabla(\nabla \cdot \mathbf{A}) - \nabla^2 \mathbf{A}, \tag{2.8}$$

Eq. (2.7) can be reduced to

$$\nabla(\nabla \cdot \mathbf{E}) - \nabla^2 \mathbf{E} = \mu_0 \frac{\partial}{\partial t}(\nabla \times \mathbf{H}). \tag{2.9}$$

By substituting Eqs. (2.1) and (2.4) into Eq. (2.9), we arrive at the wave equation in free space,

$$\nabla^2 \mathbf{E} = \frac{1}{c^2}\frac{\partial^2 \mathbf{E}}{\partial t^2}, \tag{2.10}$$

where the speed of light in a vacuum c is related to the physical constants ε_0 and μ_0 according to

$$c = \frac{1}{\sqrt{\varepsilon_0 \mu_0}} = 2.998 \times 10^8 \text{ m/sec.} \tag{2.11}$$

The simplest solutions to Eq. (2.8) are monochromatic, or single-frequency, plane waves of the form

$$\mathbf{E}(\mathbf{r}, t) = \mathbf{E}_0 e^{-i(\mathbf{k}\cdot\mathbf{r}-\omega t)}, \tag{2.12}$$

where $\mathbf{E}_0$ represents the wave amplitude and polarization direction, $\mathbf{r}$ is a spatial position vector, t is time, ω is the angular frequency of oscillation ($\omega = 2\pi v$, where v is the temporal frequency in hertz), and $\mathbf{k}$ is a wave vector that points in the direction of phase propagation and whose amplitude dictates the spatial wavelength λ of the propagating phase fronts according to

$$k = |\mathbf{k}| = \frac{\omega}{c} = \frac{2\pi}{\lambda}. \tag{2.13}$$

The temporal frequency of oscillation v and the spatial wavelength λ are then related by

$$c = \lambda v. \tag{2.14}$$

These relationships are depicted in Fig. 2.1.

2.1.2 Propagation in dense media

When an electromagnetic field propagates in anything other than free space, the solutions must conform to the boundary conditions of the environment. In many situations, it is common to do this by expressing

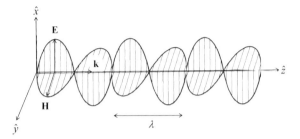

Figure 2.1   Electromagnetic wave propagation.

the general solution as a superposition of elemental solutions, such as plane waves or other wave solutions propagating in various directions, and solving for the coefficients of the superposition to match the boundary conditions. In this chapter, we are specifically interested in the interaction of electromagnetic radiation with dense media that include free or bound charged particles, electrons, and protons, in some atomic and molecular arrangement. In such a system, the electromagnetic radiation produces Lorentz forces on the charged particles according to

$$\mathbf{F} = q\mathbf{E} + q\,\mathbf{v} \times \mathbf{H}, \tag{2.15}$$

where q is the electronic charge and $\mathbf{v}$ is the charge velocity. This alters the state of the charges (i.e., makes them move), which consequently produce electrical and magnetic fields of their own that perturb the propagating electromagnetic radiation. Instead of solving all boundary conditions microscopically within the dense medium, this interaction is considered from a macroscopic perspective, where electrodynamics is treated by spatially averaging the local effects. Based on this approach, the macroscopic form of Maxwell's equations is given by

$$\nabla \cdot \mathbf{D} = \rho, \tag{2.16}$$

$$\nabla \cdot \mathbf{B} = 0, \tag{2.17}$$

$$\nabla \times \mathbf{E} = -\frac{\partial \mathbf{B}}{\partial t}, \tag{2.18}$$

and

$$\nabla \times \mathbf{H} = \mathbf{J} + \frac{\partial \mathbf{D}}{\partial t}, \tag{2.19}$$

where $\mathbf{D}$ is the electric displacement, $\mathbf{B}$ is the magnetic induction, ρ is the free-charge density, and $\mathbf{J}$ is the free-charge current density. The relationship between these newly introduced vector quantities is captured by the constitutive relations

$$\mathbf{D} = \varepsilon \, \mathbf{E}, \tag{2.20}$$

$$\mathbf{B} = \mu \, \mathbf{H}, \tag{2.21}$$

and

$$\mathbf{J} = \sigma \, \mathbf{E}. \tag{2.22}$$

The material-dependent characteristics are captured by the electric permittivity of medium ε, magnetic permeability of medium μ, and conductivity of medium σ. All of these material parameters can vary both spatially and spectrally. The spectral dependence describes the property change with free-space wavelength λ.

In a general anisotropic medium, the electric permittivity, magnetic permeability, and conductivity are dependent on the direction of propagation. This is handled mathematically by treating ε, μ, and σ as tensors, causing the vector directions of $\mathbf{D}, \mathbf{B}$, and $\mathbf{J}$ to not necessarily coincide with $\mathbf{E}$ and $\mathbf{H}$. In isotropic materials, however, all three of these properties are treated as scalar quantities. Furthermore, the magnetic permeability is equal to μ_0 for nonmagnetic materials. For dielectric materials, conductivity is further assumed to be zero; however, we find that conductivity is actually a key characteristic in terms of understanding the spectral properties of materials. The remainder of this chapter further explores how these material properties, specifically electric permittivity and conductivity, behave as a function of electromagnetic frequency or wavelength, and what impact they have on observable properties such as reflectance and transmission spectra. Isotropic materials are assumed throughout and, while general magnetic permeability is carried through the analysis, magnetic materials are not explicitly addressed.

2.1.3 Plane waves in dense media

The constitutive relations can be substituted into the third and fourth Maxwell's equations to provide

$$\nabla \times \mathbf{E} = -\mu \frac{\partial \mathbf{H}}{\partial t} \tag{2.23}$$

and

$$\nabla \times \mathbf{H} = \sigma \mathbf{E} + \varepsilon \frac{\partial \mathbf{E}}{\partial t}. \tag{2.24}$$

Again using Eqs. (2.7) and (2.8), it can be shown that

$$\nabla^2 \mathbf{E} - \nabla(\nabla \cdot \mathbf{E}) = \mu\varepsilon\frac{\partial^2 \mathbf{E}}{\partial t^2} + \mu\sigma\frac{\partial \mathbf{E}}{\partial t}. \tag{2.25}$$

For a charge-free medium, the second term on the left-hand side of Eq. (2.25) is obviously zero by Eq. (2.1). In a dense, conductive medium, however, this is not so obvious, as Eq. (2.16) suggests that it can be nonzero. Thus, it requires further analysis (Jackson, 1962). By taking the divergence of Eq. (2.19) and substituting Eqs. (2.16) and (2.17), the differential equation for the charge density follows:

$$\frac{\partial \rho}{\partial t} = -\frac{\sigma}{\varepsilon}\rho, \tag{2.26}$$

which has the solution

$$\rho = \rho_0 e^{-t/\tau}, \tag{2.27}$$

where $\tau = \varepsilon/\sigma$ is the relaxation time of the conductor. Typical relaxation times are on the order of 10^{-18} s, so any induced electric charge density will decay so rapidly that ρ can practically be assumed to be zero. This reduces the wave equation in dense media to

$$\nabla^2 \mathbf{E} = \mu\varepsilon\frac{\partial^2 \mathbf{E}}{\partial t^2} + \mu\sigma\frac{\partial \mathbf{E}}{\partial t}. \tag{2.28}$$

Comparing the free-space wave calculation in Eq. (2.10) to this dense medium form, the existence of a damping term is recognized in the latter. This means that the electric field decays in time or space (or both) due to the conductivity of the material. This manifests itself as material absorption, which is a physical characteristic fundamental to hyperspectral remote sensing that is considered further.

Again, we consider elemental, monochromatic plane wave solutions to the wave equation in Eq. (2.28) of the form

$$\mathbf{E} = \mathbf{E}_0 e^{-i\omega t}, \tag{2.29}$$

where the entire temporal dependence is assumed to be a complex exponential with an angular frequency ω. Substituting Eq. (2.29) into Eq. (2.28) results in the spatial differential equation

$$\nabla^2 \mathbf{E} = -\omega^2 \mu\varepsilon \mathbf{E} - i\omega\mu\sigma \mathbf{E}. \tag{2.30}$$

This can be reduced to the simple form

$$\nabla^2 \mathbf{E} + \tilde{k}^2 \mathbf{E} = 0, \tag{2.31}$$

by defining a complex propagation vector

$$\tilde{k}^2 = \tilde{\mathbf{k}} \cdot \tilde{\mathbf{k}} = \frac{\omega^2}{c^2} \left(\frac{\varepsilon}{\varepsilon_0} + i \frac{\sigma}{\omega \varepsilon_0} \right). \tag{2.32}$$

A tilde is used to explicitly denote complex quantities. The complex propagation vector takes the place of the real-valued propagation vector introduced in Eq. (2.10) for free-space solutions. The real part relates to wave propagation direction and wavelength (as before), while the imaginary part relates to the direction and strength of the wave decay.

It is common to define a complex dielectric constant

$$\tilde{\varepsilon} = \varepsilon_1 + i\varepsilon_2, \tag{2.33}$$

where

$$\varepsilon_1 = \frac{\varepsilon}{\varepsilon_0} \tag{2.34}$$

and

$$\varepsilon_2 = \frac{\sigma}{\omega \varepsilon_0}, \tag{2.35}$$

such that

$$\tilde{k}^2 = \frac{\omega^2}{c^2} \tilde{\varepsilon}. \tag{2.36}$$

Thus, the complex dielectric constant, which is a function of electromagnetic radiation frequency or wavelength, carries all of the (nonmagnetic) material-dependent properties about the propagation of electromagnetic radiation within a material.

The solutions to Eq. (2.31) depend on the spatial boundary conditions in the dense medium. In an unbounded medium, plane wave solutions exist in the form

$$\mathbf{E}(\vec{r}, t) = \mathbf{E}_0 e^{i(\tilde{\mathbf{k}} \cdot \mathbf{r} - \omega t)}, \tag{2.37}$$

where **r** represents the spatial coordinate, and the complex propagation constant is written in vector form. These are of the same form as the plane wave solutions of the free-space wave equation, except that now the wave vector is complex valued. The directions of the real and imaginary components of the complex propagation vector correspond to the phase front normal and spatial direction of decay, respectively. When these directions are aligned, the plane wave is called homogenous; otherwise, it is termed inhomogeneous. Even though they are ideal mathematical constructions, plane waves provide significant insight into the properties of optical radiation. They also form a complete set of elemental solutions to the wave equation. Therefore, any solution can be described as a spatial and temporal superposition of these elementary solutions. In the field of guided-wave optics, for example, such an approach is used to develop solutions to fairly complicated spatial structures. In the field of spectrometry, however, the interest lies more in the temporal or spectral domain than the spatial domain.

2.2 Complex Index of Refraction

In the field of optics, materials tend to be characterized in terms of their index of refraction as opposed to their electric permittivity (Fowles, 1989). The complex index of refraction $\tilde{N}$ is defined such that

$$\tilde{k} = \frac{\omega}{c}\tilde{N} = \frac{2\pi}{\lambda}\tilde{N}. \tag{2.38}$$

We explicitly define the real and imaginary parts of the complex index of refraction as

$$\tilde{N} = n + i\kappa, \tag{2.39}$$

where it should be clear from Eq. (2.37) that the real index n affects the phase propagation, and the imaginary index κ affects the decay properties. These are typically called the optical constants of the material and, like the complex dielectric constant, they fully capture the inherent propagation characteristics of the material. More specifically relevant to hyperspectral remote sensing, their spectral dependence defines the intrinsic spectral properties of the material.

2.2.1 Relationship with the complex dielectric constant

As both the complex index of refraction and the complex dielectric constant fully represent the intrinsic electromagnetic properties of materials, they must be directly related to each other. Combining Eqs. (2.32), (2.33) and (2.38),

$$\tilde{N}^2 = \frac{c^2}{\omega^2}\tilde{k}^2 = \varepsilon_1 + i\varepsilon_2. \tag{2.40}$$

Now inserting Eq. (2.39) into Eq. (2.40), we find that

$$(n + i\kappa)^2 = \varepsilon_1 + i\varepsilon_2. \tag{2.41}$$

Equating the real and imaginary parts of Eq. (2.41) leads to two equations:

$$n^2 - \kappa^2 = \varepsilon_1 \tag{2.42}$$

and

$$2n\kappa = \varepsilon_2, \tag{2.43}$$

which can be solved for the optical constants in terms of the real and imaginary parts of the complex dielectric constant:

$$n = \left[\frac{1}{2} \left(\sqrt{\varepsilon_1^2 + \varepsilon_2^2} + \varepsilon_1 \right) \right]^{1/2} \tag{2.44}$$

and

$$\kappa = \left[\frac{1}{2} \left(\sqrt{\varepsilon_1^2 + \varepsilon_2^2} - \varepsilon_1 \right) \right]^{1/2}. \tag{2.45}$$

2.2.2 Lorentz oscillator model

At this point, we recognize that the frequency (or wavelength) dependence of the complex index of refraction (or alternatively, the complex dielectric constant) dictates the intrinsic spectral properties of materials, but we have as yet no insight into what those properties might be. That is, they are thus far merely scalar functions of frequency or wavelength. It turns out, however, that the functional form of the optical constants is directly related to the chemical makeup of the material. Conceptually, the picture to consider is as follows: A material is composed of charged particles that are either free to move around (conductor) or bound within the atomic and molecular structure. If we consider the atomic and molecular structure as a mechanical system, then the compliance of the bound charges to oscillate with electromagnetic radiation at a certain temporal frequency depends on the underlying mechanics of the system. That is, the charged particles are bound by atomic and molecular forces, but perturbed by the external Lorentz forces from electromagnetic radiation. In this classical mechanical view, the material strongly absorbs radiation at electromagnetic frequencies matched to the natural resonances of the atomic and molecular bonds, coupling electromagnetic energy into mechanical oscillations. At frequencies not matched to such resonances, absorption is weak. The Lorentz oscillator model presented here adopts this classical mechanical approach and forms a model of the optical constants on which it is based. Strictly, a quantum mechanical treatment is needed to accurately model atomic and molecular characteristics, but this classical approach provides significant insight into the spectral nature of the optical constants and, in fact, is a very useful semi-empirical basis for modeling the optical constants of solids and liquids.

The Lorentz model (Lorentz, 1952) is based on treating a molecular bond as a forced, damped harmonic oscillator. The oscillating electric field associated with the incident electromagnetic radiation forces the bound electrons to behave as a system of coherent oscillating dipoles that absorb and reradiate incident energy. The equation of motion for a forced, damped harmonic oscillator consisting of a single, bound electron (as depicted in Fig. 2.2) is

$$m_e \frac{d^2x}{dt^2} = -e\,\mathcal{E}(t) - m_e\gamma\frac{dx}{dt} - Kx, \qquad (2.46)$$

where the electric field polarization is assumed to be in the x direction and defined by the scalar $\mathcal{E}(t)$, $e = 1.6 \times 10^{-19}$ C is the electron charge, $m_e = 0.91 \times 10^{30}$ kg is the electron rest mass, K is the restoring force associated with the atomic or molecular bond, and γ is the associated damping coefficient. The charged particles in the figure can be considered to be either electrons in an atom or atoms in a molecule if Eq. (2.46) is adjusted by replacing e with the appropriate charge differential and m_e with the appropriate reduced mass,

$$m_r = \frac{m_1 m_2}{m_1 + m_2}, \qquad (2.47)$$

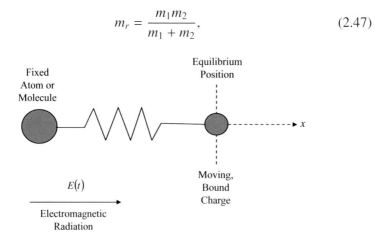

Figure 2.2 Lorentz oscillator model for a single, bound charge.

where m_1 and m_2 are the rest masses of the bound particles. The natural, resonant frequency of the single-electron harmonic oscillator is

$$\omega_0 = \sqrt{\frac{K}{m_e}}. \tag{2.48}$$

Making this substitution, the equation of motion for the time-harmonic forcing field in Eq. (2.37) becomes

$$\frac{d^2x}{dt^2} + \gamma\frac{dx}{dt} + \omega_0^2 x = -\frac{e}{m_e}\mathcal{E}_0 e^{-i\omega t}. \tag{2.49}$$

A scalar electric field amplitude $\mathcal{E}_0$ is used in this case due to the 1D analysis. A scripted variable is used to distinguish it from the nonscripted E used throughout this book to denote irradiance. The solution to Eq. (2.49) is of the form

$$x(t) = \left[\frac{-\left(\frac{e\mathcal{E}_0}{m_e}\right)}{\omega_0^2 - \omega^2 + i\omega\gamma}\right] e^{-i\omega t}. \tag{2.50}$$

If M is the number of oscillators (in terms of number of electrons) per unit volume, then the macroscopic polarization due to the forcing field is

$$P = -Mex(t) = \left[\frac{\left(\frac{Me^2}{m_e}\right)}{\omega_0^2 - \omega^2 + i\omega\gamma}\right]\mathcal{E}_0 e^{-i\omega t}. \tag{2.51}$$

However, the electric displacement in the macroscopic form of Maxwell's equations is related to local polarization due to the oscillating charges according to the definition

$$\mathbf{D} = \varepsilon\,\mathbf{E} = \varepsilon_0\,\mathbf{E} + \mathbf{P}. \tag{2.52}$$

Therefore, it follows that scalar polarization is given by

$$P = (\varepsilon - \varepsilon_0)\,\mathcal{E}_0 e^{-i\omega t}. \tag{2.53}$$

Comparing Eqs. (2.51) and (2.53), the functional form of the complex electric permittivity is predicted by the Lorentz oscillator model to be

$$\varepsilon = \varepsilon_0 + \frac{\left(\frac{Me^2}{m_e}\right)}{\omega_0^2 - \omega^2 + i\omega\gamma}. \tag{2.54}$$

By separating the real and imaginary parts and dividing by ε_0, the complex dielectric constant becomes

$$\tilde{\varepsilon} = 1 + \left(\frac{Me^2}{m_e\varepsilon_0}\right)\frac{\omega_0^2 - \omega^2}{(\omega_0^2 - \omega^2)^2 + \omega^2\gamma^2} + i\left(\frac{Me^2}{m_e\varepsilon_0}\right)\frac{\omega\gamma}{(\omega_0^2 - \omega^2)^2 + \omega^2\gamma^2}.$$

(2.55)

Defining the plasma frequency (whose significance is discussed later) as

$$\omega_p^2 = \frac{Me^2}{m_e\varepsilon_0},$$

(2.56)

it is apparent from Eq. (2.55) that

$$\varepsilon_1 = 1 + \frac{\omega_p^2(\omega_0^2 - \omega^2)}{(\omega_0^2 - \omega^2)^2 + \omega^2\gamma^2}$$

(2.57)

and

$$\varepsilon_2 = \frac{\omega_p^2\omega\gamma}{(\omega_0^2 - \omega^2)^2 + \omega^2\gamma^2}.$$

(2.58)

The frequency dependence of the optical constants for this simple harmonic oscillator can then be computed from Eqs. (2.57) and (2.58) by Eqs. (2.44) and (2.45).

At radiation frequencies near the material resonance

$$|\omega - \omega_0| \ll \omega$$

(2.59)

it is easy to show that Eqs. (2.57) and (2.58) can be approximated by

$$\varepsilon_1 \approx 1 + \frac{\omega_p^2(\omega_0 - \omega)/2\omega_0}{(\omega_0 - \omega)^2 + (\gamma/2)^2}$$

(2.60)

and

$$\varepsilon_2 \approx \frac{\omega_p^2\gamma/4\omega_0}{(\omega_0 - \omega)^2 + (\gamma/2)^2}.$$

(2.61)

These expressions can be used along with Eqs. (2.42) and (2.43) to determine the frequency dependence of the optical constants. In general, these expressions are too complicated to reveal any underlying behavior.

For a weak absorber ($\varepsilon_2 \ll 1$ everywhere), however, the optical constants can be approximated by

$$n \approx 1 + \frac{1}{2}(\varepsilon_1 - 1) \tag{2.62}$$

and

$$\kappa \approx \frac{1}{2}\varepsilon_2. \tag{2.63}$$

In this case, the optical constants exhibit the same dependence with frequency, known as dispersion, as the components of the complex dielectric constant described by Eqs. (2.60) and (2.61). In a more general case, the optical constants are relatively similar to the complex dielectric constant in spectral behavior, with greater differences for highly absorbing materials.

Figure 2.3 depicts the wavelength dependence of optical constants near resonance for both weak and strong absorbers by using Eq. (2.13) to convert frequency to wavelength. The two most important features to notice are the large Lorentzian peak in the absorptive term κ at the resonant frequency, and the large rate of change in the real part n near resonance at 9.5 µm. The latter feature is often called anomalous dispersion, since the real index increases rapidly with wavelength instead of exhibiting the slight decrease with wavelength that always occurs in regions of transparency between resonant wavelengths. Note the distorted shape of both the real and imaginary indices in the case of the strong absorber.

Real materials exhibit multiple resonances, one for each mechanical degree of freedom for which there is a dipole moment (or differential charge), and each resonance can have multiple overtones due to nonlinearities not captured by the linear Lorentz model. A discussion of the physical basis for the number, frequencies, and strengths of these resonances is deferred to the quantum mechanical treatment in Chapter 3. But the Lorentz model can be extended to accommodate J resonances by extending the expression for the complex dielectric constant to

$$\tilde{\varepsilon} = 1 + \omega_p^2 \sum_{j=1}^{J} \left(\frac{f_j}{\omega_j^2 - \omega^2 - i\omega\gamma_j} \right), \tag{2.64}$$

where f_j, ω_j, and γ_j respectively represent the oscillator strength, resonance frequency, and damping constant for the j'th resonance. The

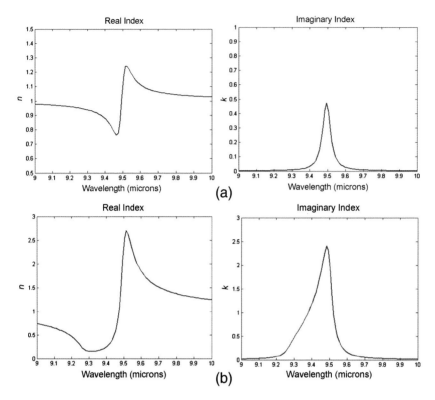

Figure 2.3 Spectral behavior of optical constants near resonance for (a) weak and (b) strong absorbers.

oscillator strengths are normalized according to

$$\sum_{j=1}^{J} f_j = 1. \tag{2.65}$$

The parameters in Eq. (2.64) are generally derived by fitting experimental measurements (Palik, 1985). Again, the optical constants can be computed based on this semi-empirical model using Eqs. (2.44) and (2.45).

The index of refraction behavior as a function of wavelength is qualitatively depicted in Fig. 2.4. At very short wavelengths (high frequency), $n = 1$ and $\kappa = 0$. At very long wavelengths (low frequency),

$$n^2 = 1 + \omega_p^2 \sum_{j=1}^{J} \frac{f_j}{\omega_j^2}, \tag{2.66}$$

and $\kappa = 0$. In between, n decreases slightly with wavelength between resonances and exhibits anomalous dispersion at resonant wavelengths.

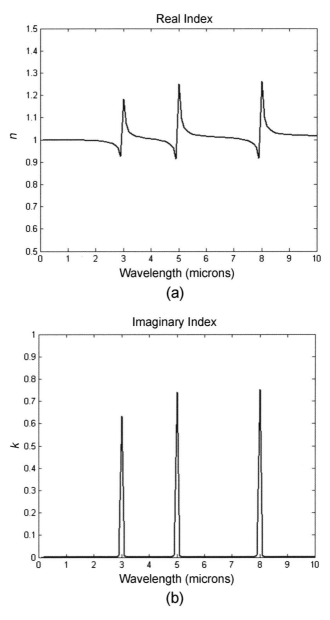

Figure 2.4 Qualitative behavior of the complex index of refraction at different optical wavelengths: (a) real index n and (b) imaginary index k.

The imaginary part κ is near zero between resonances (nonabsorbing) and exhibits Lorentzian peaks at resonant wavelengths. The resonant wavelengths correspond to bond strengths through Eq. (2.48), such that the strong electronic bonds in an atom correspond to the visible and ultraviolet part of the optical spectrum, while weaker molecular bonds correspond to the infrared. The particular set of resonant wavelengths, oscillator strengths, and damping coefficients can be unique to a particular material type and serve as a basis for uniquely identifying a material. This is the fundamental basis of the application of spectral sensing to material detection, classification, and identification.

2.2.3 Drude theory of strong conductors

The Lorentz model is based on an assumption of bound charges and therefore does not account for the free charges in strong conductors such as metals. It can be extended, however, based on the Drude theory (Drude, 1900). According to this theory, free charges are modeled in a similar manner to the Lorentz model, but with a natural resonant frequency of zero and a damping constant of $1/\tau$, where τ represents the relaxation time of the metal. The equation of motion is then given by

$$\frac{d^2x}{dt^2} + \frac{1}{\tau}\frac{dx}{dt} = -\frac{e}{m_0}\mathcal{E}_0 e^{-i\omega t}. \tag{2.67}$$

Using the same analysis as in the previous section, it is seen that the real and imaginary parts of a complex dielectric constant are given by

$$\varepsilon_1 = 1 - \frac{\omega_p^2 \tau^2}{1 + \omega^2 \tau^2} \tag{2.68}$$

and

$$\varepsilon_2 = \frac{\omega_p^2 \tau/\omega}{1 + \omega^2 \tau^2}. \tag{2.69}$$

These terms are added to the composite model in Eq. (2.64) for weak conductors, resulting in an aggregate form of the complex dielectric constant, given by

$$\tilde{\varepsilon} = 1 - \frac{\omega_p^2}{\omega^2 - i\omega/\tau} + \omega_p^2 \sum_{j=1}^{J}\left(\frac{f_j}{\omega_j^2 - \omega^2 - i\omega\gamma_j}\right). \tag{2.70}$$

For strong conductors, Eqs. (2.68) and (2.69) adequately model the material, since free charges dominate interactions with the electromagnetic radiation. Typical relaxation times of metals are on the order of 10^{13} s, while their plasma frequencies are on the order of 10^{15} Hz in the visible to near-ultraviolet region. Figure 2.5 illustrates the typical behavior of n and κ as a function of frequency for a metal near the plasma frequency. At radiation frequencies significantly above the plasma frequency ($\omega \gg \omega_p$), the imaginary index κ approaches zero, and the metal becomes transparent, because the frequency is too high for free charges to respond. At radiation frequencies significantly below the plasma frequency ($\omega \ll \omega_p$), the imaginary index κ dominates the real index n, and the metal becomes extremely attenuating. The real index is less than unity for a broad region near the plasma frequency.

2.3 Propagation through Homogenous Media

In an optical regime, the frequency of radiation is too high for the electric field to be directly measured. Instead, optical detectors respond only to the power of optical radiation incident on them. The specific quantity discussed in this section is irradiance, which is defined as the power incident on or passing through a unit surface area. As optical radiation flows through a material, we expect irradiance to decrease due to the absorption of energy by molecules of which the material is composed. Further, we also expect the amount of decrease to be highly wavelength dependent, especially near resonant wavelengths where the molecules are

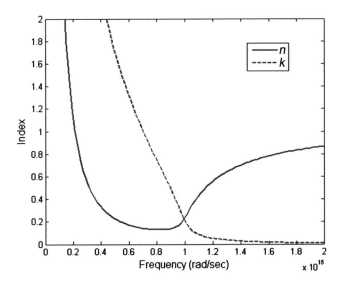

Figure 2.5 Index of refraction behavior near the plasma frequency. The real and imaginary indices are modeled in this example for $\omega_p = 10^{15}$ rad/sec and $\tau = 10^{14}$ sec.

more absorptive. Therefore, it is useful to consider irradiance as a function of wavelength, which is a quantity referred to as an irradiance spectrum.

Consider a plane wave traveling through a homogenous material in the z direction. In this case, the plane wave solution reduces to the form

$$\mathbf{E}(\mathbf{r}, t) = \mathbf{E}_0 e^{i(\tilde{k}z - \omega t)}, \tag{2.71}$$

where a scalar complex propagation constant can be used instead of a vector quantity. From Eq. (2.40), the complex propagation constant can be expressed as

$$\tilde{k} = \frac{\omega}{c}\tilde{N} = \frac{2\pi}{\lambda}\tilde{N}. \tag{2.72}$$

The plane wave solutions then take the form

$$\mathbf{E}(\mathbf{r}, t) = \mathbf{E}_0 e^{i\frac{2\pi}{\lambda}(\tilde{N}z - ct)}. \tag{2.73}$$

Because of the linearity of wave equations, superpositions of elemental solutions are also solutions. Therefore, a general solution for a homogenous wave propagating in the z direction is given by

$$\mathbf{E}(\mathbf{r}, t) = \int \mathbf{E}_0(\lambda)\, e^{i\frac{2\pi}{\lambda}(\tilde{N}z - ct)} d\lambda. \tag{2.74}$$

Equation (2.74) makes the connection between the temporal evolution of the field $\mathbf{E}(\mathbf{r}, t)$ and the spectral distribution of the field amplitude $\mathbf{E}_0(\lambda)$.

The flow of energy in an electromagnetic wave is characterized in terms of the Poynting vector,

$$\mathbf{S} = \frac{1}{2}(\mathbf{E} \times \mathbf{H}). \tag{2.75}$$

At optical frequencies, only time-averaged quantities are measurable. In particular, irradiance (power per unit area) on a surface is defined as the time average of the magnitude of the Poynting vector, which we denote here as

$$E = S_{avg} = \langle |\mathbf{S}| \rangle. \tag{2.76}$$

Be careful not to confuse irradiance E with electric field amplitude $\mathcal{E}$, which unfortunately share the same letter in standard notation in the literature. The brackets in Eq. (2.76) denote the time average over a period of wave oscillation. For the generalized plane wave case in Eq. (2.71), this reduces to

$$E(\lambda) = \frac{1}{2} \varepsilon c \, \mathcal{E}^2(\lambda) \, e^{-\frac{4\pi\kappa}{\lambda} z}. \tag{2.77}$$

Defining the irradiance spectrum at $z = 0$ as

$$E(\lambda, 0) = \frac{1}{2} \varepsilon c \, \mathcal{E}^2(\lambda), \tag{2.78}$$

the irradiance after a propagation distance z is given by

$$E(\lambda, z) = E(\lambda, 0) \, e^{-\beta_a z}, \tag{2.79}$$

where an absorption coefficient β_a is defined as

$$\beta_a = \frac{4\pi\kappa}{\lambda}. \tag{2.80}$$

Equation (2.79) is recognized as Beer's law (Beer, 1865) and makes clear the fact that the imaginary part of the optical constant κ is the absorptive part of the complex index of refraction, and that irradiance decreases more rapidly as a function of propagation distance near the resonant wavelengths, where κ exhibits peaks according to the Lorentz model. From Eq. (2.79), we can define the irradiance transmission spectrum for a material thickness z as

$$T(\lambda) = e^{-\beta_a z}. \tag{2.81}$$

A typical transmission spectrum near resonance is shown in Fig. 2.6. As the product of the absorption coefficient and thickness increases, the depth of the absorption feature increases (i.e., the minimum transmission decreases) until the minimum transmission becomes essentially zero. Beyond that point, the absorption band widens with greater thickness. A quantity called optical depth, denoted by δ, is defined as

$$\delta = \beta_a z. \tag{2.82}$$

By approximating Eq. (2.81) as

$$T(\lambda) \approx 1 - \beta_a z, \tag{2.83}$$

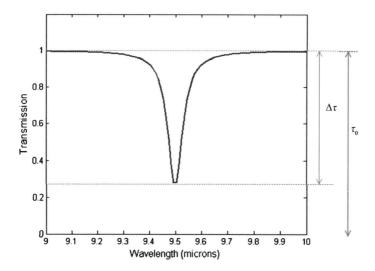

Figure 2.6   Expression of a single resonant absorption feature on a transmission spectrum.

the optical depth for weak absorbers can be approximated as

$$\delta = \frac{\Delta\tau}{\tau_0},\qquad(2.84)$$

where $\Delta\tau$ is the absorption feature depth and τ_0 is the mean transmission away from the resonance feature.

2.4 Reflection and Transmission at Dielectric Interfaces

When optical radiation confronts boundaries between homogenous media, it is partially reflected back into the medium of incidence and partially transmitted and refracted into the adjacent medium. The case of dielectric media ($\kappa = 0$) is presented in this section and then extended in the following section to conductive media. Consider the situation in Fig. 2.7, where a plane wave in homogenous material with an index of refraction n_1 is incident at an angle θ_i to the normal vector of a smooth, planar interface with a material having an index of refraction n_2. At the interface, part of the optical radiation is reflected at angle θ_r, and the remainder is transmitted at an angle θ_t. The challenge is to determine these directions, as well as the relative reflected and transmitted energy, as a function of incident angle and wavelength.

The direction of the reflected and transmitted energy is determined by the requirement of phase continuity at the interface. This is described mathematically by

$$\mathbf{k}_t \cdot \hat{\mathbf{s}} = \mathbf{k}_r \cdot \hat{\mathbf{s}} = \mathbf{k}_i \cdot \hat{\mathbf{s}},\qquad(2.85)$$

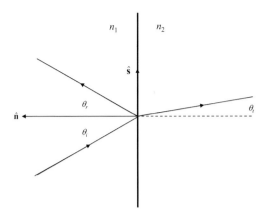

Figure 2.7 Reflection and transmission at a smooth, planar boundary between two dielectrics.

where the propagation vectors $\mathbf{k}_t$, $\mathbf{k}_r$, and $\mathbf{k}_i$ correspond to the transmitted, reflected, and incident waves, respectively, and $\hat{\mathbf{s}}$ is a unit vector along the surface in the plane of incidence. Equation (2.85) leads to the familiar law of reflection,

$$\theta_r = \theta_i, \tag{2.86}$$

and Snell's law for refraction,

$$n_2 \sin \theta_t = n_1 \sin \theta_i. \tag{2.87}$$

The law of reflection is independent of material properties, while Snell's law depends on the index of refraction. Since the index is wavelength dependent, the refracted angle is also wavelength dependent, a property known as dispersion. As an example, consider an interface between air ($n = 1$) and a material with an index of refraction shown in Fig. 2.8, with light incident at angles of $0, 30$, and 60 deg from the surface normal. The dispersion properties for the light transmitted into the material are shown in Fig. 2.9. There is no dispersion at normal incidence. The dispersion increases at higher angles of incidence and maintains a shape that is opposite that of the index of refraction. That is, light is dispersed to larger angles (away from normal) when the index of refraction decreases, and it is dispersed to smaller angles (toward the normal) when the index of refraction increases. Normal dispersion refers to the spectral region where the index decreases monotonically with wavelength. In this region, shorter wavelengths are dispersed to larger angles than are longer wavelengths. Anomalous dispersion refers to the opposite case.

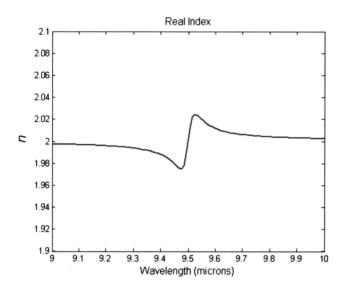

Figure 2.8 Index of refraction assumed for dielectric interface reflectivity and transmissivity examples.

The relative amount of reflected and transmitted energy is determined by the requirement that the tangential electric and magnetic fields are continuous at the interface, boundary conditions that vary depending on the polarization of the incident radiation (Lorrain and Corson, 1962). By decomposing the incident radiation into components for which the electric fields are polarized orthogonal to the plane of incidence (s-polarized or transverse electric) and in the plane of incidence (p-polarized or transverse magnetic), it becomes sufficient to understand the reflectance and transmittance properties in these two cases. *Amplitude reflectivity* r_s and r_p for s- and p-polarized cases, respectively, are defined as the ratio of the reflected-to-incident electric field amplitude. These can be shown to be

$$r_s = \frac{n_1 \cos\theta_i - n_2 \cos\theta_t}{n_1 \cos\theta_i + n_2 \cos\theta_t} \tag{2.88}$$

and

$$r_p = \frac{n_1 \cos\theta_t - n_2 \cos\theta_i}{n_1 \cos\theta_t + n_2 \cos\theta_i}. \tag{2.89}$$

Power reflectivity is defined as the ratio of the reflected-to-incident irradiance and can be determined from Eqs. (2.88) and (2.89) for both polarization cases as

$$R_s(\lambda) = |r_s|^2 \tag{2.90}$$

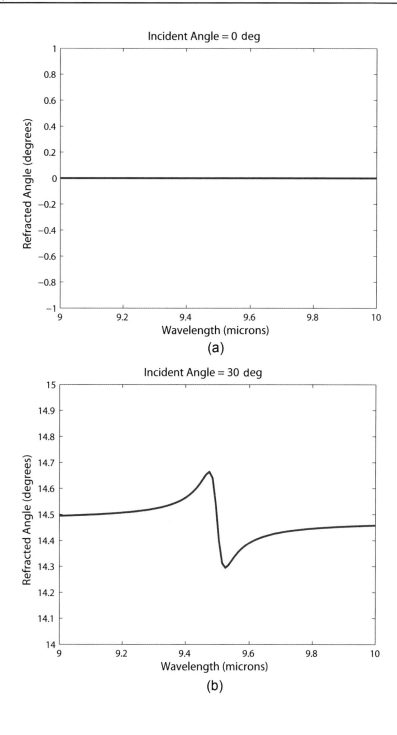

Figure 2.9 Dispersion of refracted light from air into a dielectric material with the index of refraction shown in Fig. 2.8 for three incident angles: (a) 0, (b) 30, and (c) 60 deg.

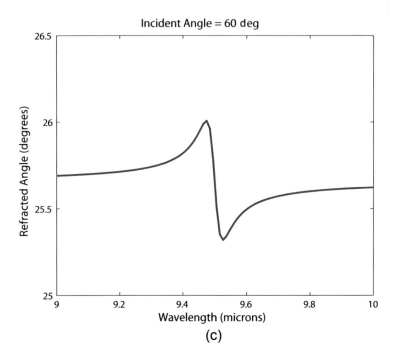

Figure 2.9   (continued)

and

$$R_p(\lambda) = |r_p|^2. \tag{2.91}$$

By conservation of energy, the corresponding *power transmissivity* across the interface for the two polarizations is

$$T_s(\lambda) = 1 - R_s(\lambda) \tag{2.92}$$

and

$$T_p(\lambda) = 1 - R_p(\lambda). \tag{2.93}$$

These properties are spectrally dependent to the extent that the indices of refraction of the materials are spectrally dependent. In the case of normal incidence, power reflectivity is independent of polarization and given by

$$R(\lambda) = \left|\frac{n_2 - n_1}{n_2 + n_1}\right|^2. \tag{2.94}$$

For unpolarized incident radiation not at normal incidence, total surface reflectivity is

$$R(\lambda) = \frac{R_s + R_p}{2},$$ (2.95)

and total surface transmissivity is

$$T(\lambda) = 1 - \frac{R_s + R_p}{2}.$$ (2.96)

Even when the incident radiation is unpolarized, the reflected and transmitted radiation can be polarized.

Consider again the air–dielectric interface example corresponding to the index of refraction depicted in Fig. 2.8. Figure 2.10 illustrates the interface reflectivity and transmissivity spectra for both *s*- and *p*-polarizations, corresponding to the same three angles of incidence modeled in Fig. 2.9. Spectral behavior is minimal, but reflectivity and transmissivity are highly dependent on the angle of incidence and polarization. With regard to spectral sensing, the implication is that, while the wavelength location of spectral features in the reflected spectrum will not change with illumination and viewing angle, the magnitude of the spectral features will vary.

2.5 Reflection and Transmission at Conductor Interfaces

The analysis in the preceding section can be extended to the case where one or more of the materials are conductive by replacing the real indices of refraction with complex indices of refraction, and by treating the angles as complex quantities (Born and Wolf, 1980). Thus, the relevant equations become

$$\tilde{N}_2 \sin \tilde{\theta}_t = \tilde{N}_1 \sin \tilde{\theta}_i,$$ (2.97)

$$r_s = \frac{\tilde{N}_1 \cos \tilde{\theta}_i - \tilde{N}_2 \cos \tilde{\theta}_t}{\tilde{N}_1 \cos \tilde{\theta}_i + \tilde{N}_2 \cos \tilde{\theta}_t},$$ (2.98)

and

$$r_p = \frac{\tilde{N}_1 \cos \tilde{\theta}_t - \tilde{N}_2 \cos \tilde{\theta}_i}{\tilde{N}_1 \cos \tilde{\theta}_t + \tilde{N}_2 \cos \tilde{\theta}_i}.$$ (2.99)

While the use of complex angles is mathematically correct, it requires some interpretation. To determine the real angle of phase propagation and wave decay, one must compute the complex propagation vector based on

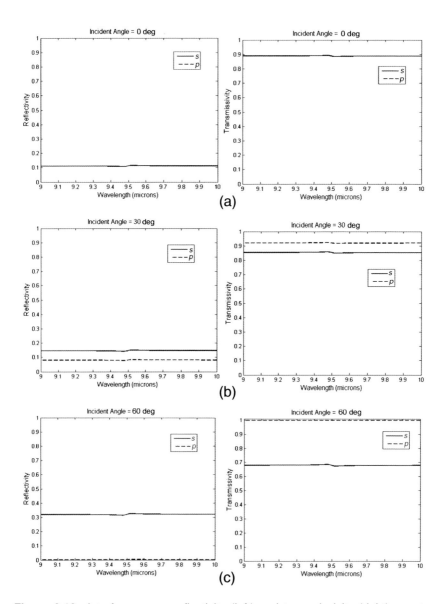

Figure 2.10 Interface power reflectivity (left) and transmissivity (right) spectra from air into a dielectric material with the index of refraction shown in Fig. 2.8 for various incidence angles: (a) 0, (b) 30, and (c) 60 deg (s = s-polarization and p = p-polarization).

Eq. (2.97) and then calculate the angles from the normal and tangential components of the real and imaginary parts.

Consider the case where the incident material is a dielectric such that its index of refraction and incident angle are both real valued. This is typical in remote sensing situations where the incident medium is often air. In this case, Eq. (2.97) becomes

$$(n_2 + i\kappa_2) \sin \tilde{\theta}_t = n_1 \sin \theta_i. \tag{2.100}$$

From Eq. (2.100), it follows that

$$\sin \tilde{\theta}_t = \frac{n_1(n_2 - i\kappa_2) \sin \theta_i}{(n_2^2 + \kappa_2^2)} \tag{2.101}$$

and

$$\cos \tilde{\theta}_t = \sqrt{1 - \frac{n_1^2(n_2^2 - \kappa_2^2) \sin^2 \theta_i}{(n_2^2 + \kappa_2^2)^2} + i\frac{2n_1^2 n_2 \kappa_2 \sin^2 \theta_i}{(n_2^2 + \kappa_2^2)^2}}. \tag{2.102}$$

For convenience, define a and b such that

$$\cos \tilde{\theta}_t = ae^{ib}. \tag{2.103}$$

Then it can be shown that

$$a = \left[\left(1 - \frac{n_1^2(n_2^2 - \kappa_2^2) \sin^2 \theta_i}{(n_2^2 + \kappa_2^2)^2}\right)^2 + \left(\frac{2n_1^2 n_2 \kappa_2 \sin^2 \theta_i}{(n_2^2 + \kappa_2^2)^2}\right)^2\right]^{1/4} \tag{2.104}$$

and

$$b = \frac{1}{2} \tan^{-1}\left[\frac{2n_1^2 n_2 \kappa_2 \sin^2 \theta_i}{(n_2^2 + \kappa_2^2)^2 - n_1^2(n_2^2 - \kappa_2^2) \sin^2 \theta_i}\right]^{1/4}. \tag{2.105}$$

It follows that the complex propagation vector is given by

$$\tilde{\mathbf{k}} = \frac{\omega}{c}[(n_1 \sin \theta_i)\,\hat{\mathbf{s}} - a(n_2 \cos b - \kappa_2 \sin b)\,\hat{\mathbf{n}}]$$
$$- i\frac{\omega}{c}[a(\kappa_2 \cos b + n_2 \sin b)\,\hat{\mathbf{n}}]. \tag{2.106}$$

The imaginary part of the complex propagation vector contains only a component in the surface normal direction. Therefore, the wave decays in a direction perpendicular to the interface. The real part of the propagation vector has components in both tangential and normal directions, so the transmitted wave is inhomogeneous. The real angle of refraction in the conductive medium can be determined from the tangential and normal components in Eq. (2.106) that provide the relation

$$\tan\theta_t = \frac{n_1 \sin\theta_i}{a(n_2 \cos b - \kappa_2 \sin b)}. \tag{2.107}$$

As an example, we consider the same situation analyzed for the air–dielectric interface in the previous section, but now include an imaginary index for the second material shown in Fig. 2.11 to compute the dispersion characteristics shown in Fig. 2.12. In cases of off-axis incidence, dispersion is affected by the presence of an absorptive term in the complex index of refraction. Therefore, both real and imaginary components factor into the refractive properties of the material.

To compute the interface reflectivity and transmissivity spectra, it is convenient to define

$$\tilde{N}_2 \cos\tilde{\theta}_t = u + iw, \tag{2.108}$$

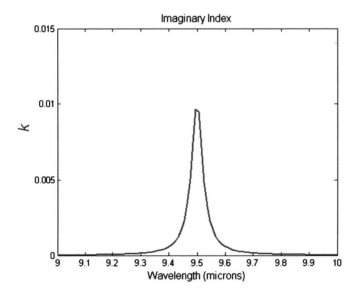

Figure 2.11 Imaginary index k assumed for conductive interface reflectivity and transmissivity examples. The corresponding real index n is shown in Fig. 2.8.

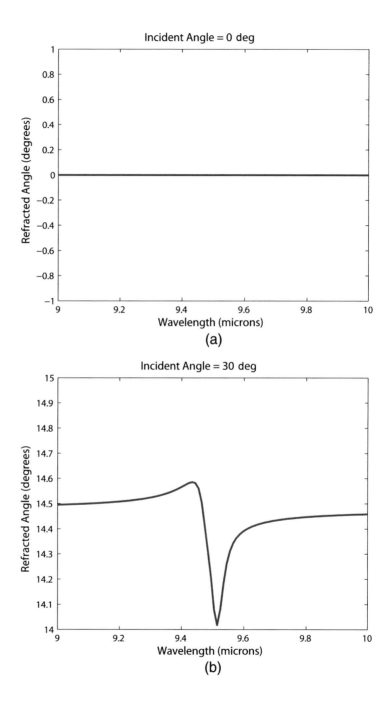

Figure 2.12 Dispersion of refracted light from air into a conductive material, with optical constants shown in Figs. 2.8 and 2.11 for various incident angles: (a) 0, (b) 30, and (c) 60 deg.

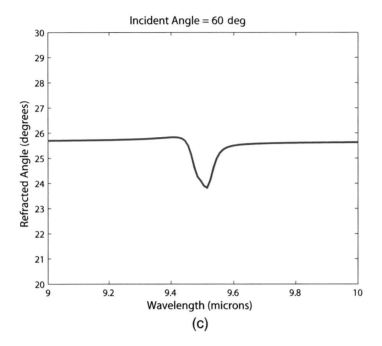

Figure 2.12 (*continued*)

where u and w are real valued. From Eq. (2.95),

$$(u + iw)^2 = \tilde{N}_2^2 - n_1^2 \sin^2 \theta_i. \qquad (2.109)$$

From Eq. (2.109) and the definition of the complex index of refraction, the real and imaginary parts of Eq. (2.108) can be determined as

$$u = \left[\frac{\sqrt{(n_2^2 - \kappa_2^2 - n_1^2 \sin^2 \theta_i)^2 + 4n_2^2 \kappa_2^2} + (n_2^2 - \kappa_2^2 - n_1^2 \sin^2 \theta_i)}{2} \right]^{1/2} \qquad (2.110)$$

and

$$w = \left[\frac{\sqrt{(n_2^2 - \kappa_2^2 - n_1^2 \sin^2 \theta_i)^2 + 4n_2^2 \kappa_2^2} - (n_2^2 - \kappa_2^2 - n_1^2 \sin^2 \theta_i)}{2} \right]^{1/2}. \qquad (2.111)$$

Equation (2.108) can be inserted into Eqs. (2.98) and (2.99) to compute the amplitude reflectivity for the two polarization cases as a function of u and ω. Power reflectivity can then be computed using Eqs. (2.90) and (2.91).

The results are

$$R_s = \frac{(n_1 \cos \theta_i - u)^2 + w^2}{(n_1 \cos \theta_i + u)^2 + w^2} \qquad (2.112)$$

and

$$R_p = \frac{[(n_2^2 - \kappa_2^2) \cos \theta_i - n_1 u]^2 + [2n_2\kappa_2 \cos \theta_i - n_1 w]^2}{[(n_2^2 - \kappa_2^2) \cos \theta_i + n_1 u]^2 + [2n_2\kappa_2 \cos \theta_i + n_1 w]^2}. \qquad (2.113)$$

The total power reflectivity and transmittivity spectra in the case of unpolarized incident radiation can be determined using Eqs. (2.95) and (2.96). As in the dielectric case, reflected and transmitted radiation is polarized, even when incident radiation is unpolarized. To fully understand their polarization properties, it is important to consider not only the power reflectivity but also the phase change that occurs for both components at the interface.

At normal incidence, the power reflectance is polarization independent and becomes

$$R = \frac{(n_2 - n_1)^2 + \kappa_2^2}{(n_2 + n_1)^2 + \kappa_2^2}. \qquad (2.114)$$

When $\kappa_2 = 0$, this simplifies further to Eq. (2.94). As the optical constants are spectrally dependent, power reflectance is also spectrally dependent. Of particular interest is the behavior near material resonances, where reflectance can change dramatically. Near such a resonance, κ_2 can dominate Eq. (2.84) such that the reflectance becomes much higher than at optical wavelengths away from resonance. For very strong resonances, reflectance can approach unity over a band of optical wavelengths. This is denoted as a *reststrahlen* band.

Consider again the air–conductor interface example corresponding to the optical constants depicted in Figs. 2.8 and 2.11. Figure 2.13 illustrates the interface reflectivity and transmissivity spectra for both s- and p-polarizations corresponding to the same three angles of incidence modeled in Fig. 2.12. In this case, the imaginary part of the index of refraction dominates the spectra characteristics of transmissivity and reflectivity. The strong reflectivity peak coincides with resonance for all angles of incidence. The optical depth of the transmissivity feature is roughly independent of incident angle and polarization. With respect to real materials, the particular κ modeled in this example is very high but corresponds to very strong absorption bands that might be characteristics

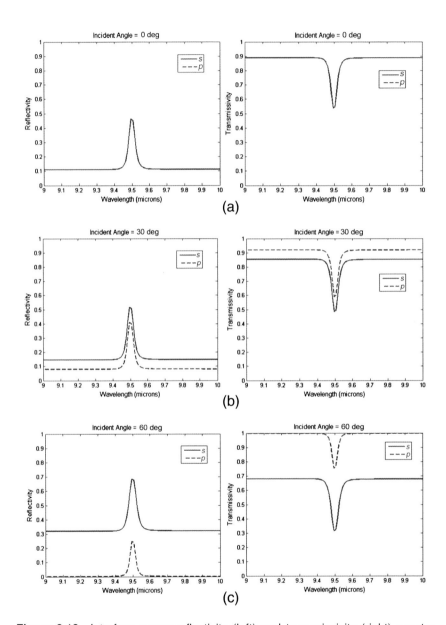

Figure 2.13 Interface power reflectivity (left) and transmissivity (right) spectra from air into a dielectric material with the index of refraction shown in Fig. 2.8 for various incidence angles: (a) 0, (b) 30, and (c) 60 deg (s = s-polarization and p = p-polarization).

of pure minerals such as quartz. Features such as this are seen in real reflectance spectra described in Chapter 4.

2.6 Radiometry

The mathematical models outlined in the prior sections are sufficient to understand the spectral reflectance and transmission properties of stratified, homogenous materials with smooth interfaces. These properties are based on the optical constants of the media on which they are composed, and the plane wave propagation of electromagnetic radiation. When materials exhibit rough surfaces or interfaces, or when they contain inhomogeneities such as distributed volume scatterers, then the directionality of the propagating plane waves is significantly altered and a scattering analysis is required. To model scattering, it is necessary first to formally model the directional power flow of radiation, a topic called radiometry.

We begin the discussion of radiometry by formally introducing the radiometric quantities defined in Table 2.1. *Irradiance*, designated here by the letter E, was already introduced previously by the time average of the Poynting vector. Irradiance represents the electromagnetic power per unit area incident on, reflected from, or flowing through a surface, and is commonly given in units of $\mu W/cm^2$. When referring to the power per unit area emitted from a source of radiation, irradiance is commonly referred to as *exitance* and is designated by the letter M. Irradiance is a band-integrated quantity, meaning that there must be a finite spectral bandwidth for there to be a power flow. When considering the spectral nature of irradiance (that is, the distribution of irradiance with frequency or wavelength), it is useful to quantify the irradiance per unit spectral bandwidth. This is known as the *spectral irradiance*, commonly designated as E_λ or M_λ, where the subscript denotes a spectral quantity and has units of $\mu W/cm^2$ μm. The irradiance (or exitance) is computed from the spectral irradiance (or exitance) by integrating it over the spectral band.

Table 2.1 Definitions of radiometric quantities.

Quantity	Designation	Relationship	Units
Irradiance	E	$E = \frac{\partial P}{\partial A} = \int E_\lambda \, d\lambda$	$\mu W/cm^2$
Spectral irradiance	E_λ	$E_\lambda = \frac{\partial P}{\partial A \partial \lambda} = \int L_\lambda \, d\Omega$	$\mu W/cm^2 \, \mu m$
Radiance	L	$L = \frac{\partial P}{\partial A \partial \Omega} = \int L_\lambda \, d\lambda$	$\mu W/cm^2 \, sr$
Spectral radiance	L_λ	$L_\lambda = \frac{\partial P}{\partial A \partial \Omega \partial \lambda}$	$\mu W/cm^2 \, \mu m \, sr$
Intensity	I	$I = \frac{\partial P}{\partial \Omega} = \int L \, dA$	$\mu W/sr$
Power	P	$P = \int E \, dA = \int I \, d\Omega$	μW

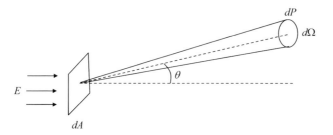

Figure 2.14 Relationship between irradiance and radiance. The radiance is the distribution of power per unit area scattered in a differential solid angle at a given angular orientation.

To model the directional distribution of scattered radiation, one must consider the relative power scattered as a function of angle in three dimensions. To do so, a quantity called *radiance* is defined. With reference to Fig. 2.14, radiance L is defined as the differential power dP from an incremental area dA that flows in the direction θ into an incremental solid angle $d\Omega$, or

$$dP = L \, dA \, d\Omega \, \cos\theta. \qquad (2.115)$$

Radiance is, in general, angularly dependent and commonly expressed in units of $\mu W/cm^2$ sr. The total power flowing through the differential area is given by the hemispherical integral

$$P = dA \int_{2\pi} L \cos\theta \, d\Omega = E \, dA. \qquad (2.116)$$

Equation (2.116) provides the relationship between irradiance and radiance. Note the cosine of the angle from normal in the integral; this cosine accounts for the smaller apparent differential area when power flows at an angle from the normal. As illustrated in Fig. 2.15, the irradiance

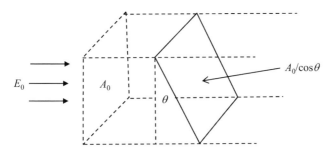

Figure 2.15 Illustration of the reduction in irradiance on a slanted surface by the cosine of the angle between the surface normal and the direction of energy flow.

on the slanted surface is reduced to

$$E = E_0 \cos \theta, \tag{2.117}$$

where E_0 is the irradiance on a surface whose normal is parallel to the direction of electromagnetic energy flow because of the increase in illuminated surface area at non-normal incidence. The quantity *spectral radiance L_λ* is defined in an analogous manner as spectral irradiance and, as the radiance per unit spectral bandwidth, it has units $\mu W/(cm^2\ \mu m\ sr)$. Note that the subscript is often dropped for notational simplicity. If an irradiance or radiance is given as a function of wavelength, then it is assuredly a spectral irradiance or radiance quantity, even if not explicitly specified as such.

2.6.1 Point sources

Another radiometric quantity defined in Table 2.1 is the *intensity I*, which is the spatially averaged radiance over a surface. This quantity is often used to quantify the radiometric output of small or very distant point sources, such as stars. The geometry of point source irradiance is illustrated in Fig. 2.16. A point source radiates uniformly in all directions; therefore, the intensity is

$$I = \frac{P}{4\pi}. \tag{2.118}$$

The total power that flows through the spherical, differential areas dA_1 and dA_2 in the figure must be equal due to conservation of energy. Therefore, the irradiance on the spherical surface of radius r from the surface is

$$E = \frac{dP}{dA} = \frac{dP}{r^2 d\theta\, d\phi} = \frac{dP}{r^2 d\Omega} = \frac{I}{r^2}. \tag{2.119}$$

Irradiance decreases in inverse proportion to the square of the point source distance. The integration of a solid angle in computing irradiance from

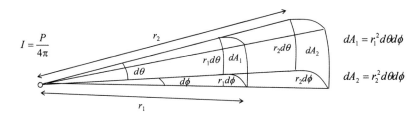

Figure 2.16 Geometry of a point source to compute irradiance on a spherical surface.

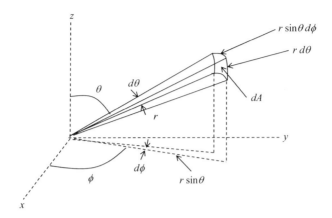

Figure 2.17   Solid angle geometry for integrating radiance to compute irradiance, or integrating intensity to compute power.

radiance or intensity is typically performed in spherical coordinates using the geometry in Fig. 2.17, where the differential solid angle is

$$d\Omega = \frac{dA}{r^2} = \frac{r\,d\theta\,r\sin\theta\,d\theta}{r^2} = \sin\theta\,d\theta\,d\phi. \qquad (2.120)$$

From this derivation, it is clear that the solid angle of a sphere is

$$\Omega_{sphere} = \int_0^{2\pi}\int_0^{\pi} \sin\theta\,d\theta\,d\varphi = 4\pi, \qquad (2.121)$$

as expected.

2.6.2 Lambertian sources

While a point source represents one extreme of radiometric sources, the opposite extreme is the Lambertian source or scatterer (depicted in Fig. 2.18), defined as a source whose radiance distribution is independent of angle. For a given solid angle $d\Omega$ oriented at an angle θ from the surface normal, the differential irradiance is given by

$$dE = L\cos\theta\,d\Omega. \qquad (2.122)$$

The cosine factor again appears due to the fundamental definition of radiance in Eq. (2.115). The total irradiance emitted or scattered from the surface into the hemisphere is then computed as

$$E = \int_0^{2\pi}\int_0^{\pi/2} L\cos\theta\sin\theta\,d\theta\,d\varphi = \pi L, \qquad (2.123)$$

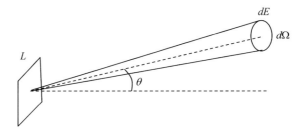

Figure 2.18   Differential radiance for a Lambertian source or scatterer.

which might be unexpected, because the total solid angle of a hemisphere is 2π and not merely π. This apparent discrepancy is explained by the cosine factor fundamental to the relationship between radiance and irradiance.

2.6.3 Spherical scatterers

With these radiometric principles in hand, we can consider the situation in Fig. 2.19, where directional optical radiation with irradiance E_i interacts with a single scatterer, the starting point for understanding the optical properties of distributed scattering media. Suppose that the total scattered power is P_s. A quantity called a scattering cross-section σ_s is defined such that it equals the effective cross-sectional area in a plane normal to the incident radiation, through which this amount of power P_s would flow. Mathematically,

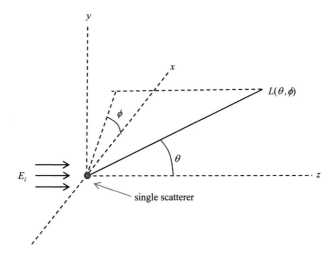

Figure 2.19   Geometry for determining the radiance distribution for a single scatterer.

$$\sigma_s = \frac{P_s}{E_i}. \tag{2.124}$$

If this power were scattered uniformly across a sphere of radius R centered on the scatterer like a point source, that scattered irradiance would be

$$E_s = \frac{\sigma_s}{4\pi R^2} E_i. \tag{2.125}$$

If there are M identical scatterers per unit volume within a total scattering volume V, the total scattered power is given by

$$P_s = MV\sigma_s E_i. \tag{2.126}$$

A normalized metric for the scattering cross-section is the scattering efficiency, which is the ratio of the scattering cross-section to the cross-sectional area it presents to the incident radiation. For a spherical scatterer of radius a, the scattering efficiency Q_s is given by

$$Q_s = \frac{\sigma_s}{\pi a^2}. \tag{2.127}$$

Since real scatterers are not point sources, the angular dependence of the scattered radiation must also be considered. To do so, a scattering phase function $p(\theta, \phi)$ is defined that models the relative radiance distribution as a function of the spherical coordinates (θ, ϕ), normalized such that

$$\iint_{4\pi} p(\theta, \varphi) \, d\Omega = 1. \tag{2.128}$$

The radiance distribution is then related to the scattering cross-section and the scattering phase function through the intensity distribution

$$I(\theta, \phi) = \int L(\theta, \phi) \, dA = MV\sigma_s \, E_i p(\theta, \phi). \tag{2.129}$$

In general, the scattering cross-section and phase function are spectrally dependent. In that case, the expressions in Eqs. (2.125), (2.126) and (2.129) relate the scattered spectral irradiance and radiance to the incident spectral irradiance.

2.7 Propagation through Scattering Media

When light propagates through a scattering medium, it can be absorbed by the combination of scatterers and the medium in which they are embedded (if not free space), as well as scattered away from the direction of propagation. Both are loss mechanisms. Absorption is treated in the same way as scattering through homogenous media, but now the absorption coefficient β_a represents the combined absorption loss. This can be represented by an absorption cross-section σ_a with the same interpretation as the scattering cross-section σ_s. If there are M scatterers per unit volume, each with an absorption coefficient σ_a, then the differential change in power over a propagation distance dz for a scattering volume V with cross-sectional area dA (as depicted in Fig. 2.20) is

$$dP = -MV\sigma_a E = M\sigma_a E dA dz. \tag{2.130}$$

From Eq. (2.130), we arrive at a differential equation for the irradiance,

$$\frac{dE}{dz} = -M\sigma_a E, \tag{2.131}$$

from which the Beer's law results as a solution:

$$E = E_0 e^{-M\sigma_a z}. \tag{2.132}$$

Comparing to the Beer's law expression in Eq. (2.81), the absorption coefficient and the absorption cross-section are related by

$$\beta_a = M\sigma_a. \tag{2.133}$$

When the spectral dependence of the absorption cross-section and absorption coefficient is included, Eq. (2.132) characterizes the decrease in spectral irradiance with propagation through the medium.

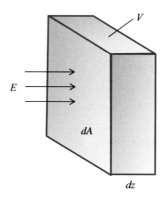

Figure 2.20 Irradiance propagation through a differential scattering volume.

A scattering coefficient β_s can be similarly defined to represent the effective loss in power from the propagating wave scattering in directions away from the path of propagation. This scattering coefficient is related to the scattering cross-section by

$$\beta_s = M\sigma_s. \tag{2.134}$$

The combined effect of scattering and absorption is then characterized by an extinction coefficient β_e, given by

$$\beta_e = \beta_a + \beta_s = M(\sigma_a + \sigma_s). \tag{2.135}$$

From these quantities, the aggregate transmission through a thickness z of scattering medium is given by

$$T(\lambda) = e^{-\beta_e z}. \tag{2.136}$$

For gaseous media, it is common to characterize absorption, scattering, and extinction coefficients relative to the gaseous concentration, typically measured in units of parts per million (ppm). This is done because concentration can vary significantly with conditions (pressure, volume, and temperature) under which the gas is observed. Thus, an extinction coefficient β_c is used that represents the extinction per unit concentration; that is, it has units of $(\text{ppm m})^{-1}$ as opposed to the typical m^{-1} units of β_e. In such cases, Beer's law must be modified to explicitly include the concentration C of the gas, leading to the expression

$$T(\lambda) = e^{-\beta_c C z}. \tag{2.137}$$

As in Eq. (2.136), transmission is displayed as a spectral quantity, and as such, corresponds to the decrease in spectral irradiance with propagation. The extinction coefficient in each case is assumed to be a spectrally varying quantity.

2.7.1 Mie scattering theory

To determine the scattering cross-section and phase function for a particular type of scatterer, one must find solutions to Maxwell's equations that satisfy the spatial boundary conditions resulting from the presence of the scatterer. The Mie scattering theory (Mie, 1908) is the application of Maxwell's equations to the specific situation depicted in Fig. 2.19, where a plane wave is incident on an isotropic, dielectric sphere of radius a and index of refraction n. A solution is found in the far field by using separation of variables in a spherical coordinate system centered on the sphere, decomposing the plane wave into Legendre polynomials, and applying the boundary conditions that require continuity of phase and tangential fields on the surface of the sphere (Bohren and Huffman, 1998). The solution is given in terms of scattering coefficients that relate the scattered electric field components as a function of angle to the incident field components through the matrix equation

$$
\begin{bmatrix} (\mathcal{E}_\|)_s \\ (\mathcal{E}_\perp)_s \end{bmatrix} = \frac{e^{-ik(R-z)}}{-ikR} \begin{bmatrix} S_2(\theta) & 0 \\ 0 & S_1(\theta) \end{bmatrix} \begin{bmatrix} (\mathcal{E}_\|)_i \\ (\mathcal{E}_\perp)_i \end{bmatrix},
\tag{2.138}
$$

where k is the free-space propagation constant. Defining a normalized size parameter as $x = ka$, the solution for the scattering coefficients $S_1(\theta)$ and $S_2(\theta)$ is given by the following set of equations:

$$
S_1(\theta) = \sum_{m=1}^{\infty} \frac{2m+1}{m(m+1)} [a_m \pi_m + b_m \tau_m],
\tag{2.139}
$$

$$
S_2(\theta) = \sum_{m=1}^{\infty} \frac{2m+1}{m(m+1)} [b_m \pi_m + a_m \tau_m],
\tag{2.140}
$$

$$
\pi_m = \frac{1}{\sin\theta} P_m^1(\cos\theta),
\tag{2.141}
$$

$$
\tau_m = \frac{d}{d\theta} P_m^1(\cos\theta),
\tag{2.142}
$$

$$
a_m = \frac{\psi_m(x)\frac{d}{dr}[\psi_m(nx)] - n\psi_m(nx)\frac{d}{dr}[\psi_m(x)]}{\xi_m(x)\frac{d}{dr}[\psi_m(nx)] - n\psi_m(nx)\frac{d}{dr}[\xi_m(x)]},
\tag{2.143}
$$

$$
b_m = \frac{n\psi_m(x)\frac{d}{dr}[\psi_m(nx)] - \psi_m(nx)\frac{d}{dr}[\psi_m(x)]}{n\xi_m(x)\frac{d}{dr}[\psi_m(nx)] - \psi_m(nx)\frac{d}{dr}[\xi_m(x)]},
\tag{2.144}
$$

where P_m^1 corresponds to the first derivative of associated Legendre polynomials of the first kind, and ψ_m and ξ_m are Ricatti–Bessel functions, which are composed of spherical Bessel functions of the first, second, and third kind (Lebedev, 1972). More information on the derivation of these mathematical relationships is provided by Bohren and Huffman (1998).

In the case where incident radiation is unpolarized, the scattering cross-section is related to the scattering coefficients in Eq. (2.138) by

$$
\sigma_s = \frac{1}{k^2} \iint_{4\pi} \frac{1}{2} [|S_1(\theta)|^2 + |S_2(\theta)|^2] \, d\Omega
$$

$$
= \frac{2\pi}{k^2} \sum_{m=1}^{\infty} (2m+1) [|a_m|^2 + |b_m|^2],
\tag{2.145}
$$

where the 2D integral is performed in spherical coordinates over the entire sphere. A backscatter cross-section is also defined as

$$\sigma_b = \frac{2\pi}{k^2}[|S_1(180\,\text{deg})|^2 + |S_2(180\,\text{deg})|^2].$$ (2.146)

The backscatter cross-section represents radiation scattered in only a single direction opposite that of the incident irradiance and can be used to compute the backscattered irradiance as

$$E_b = \frac{MV\sigma_b}{4\pi R^2}E_i,$$ (2.147)

where R is the distance from the scatterers.

2.7.2 Rayleigh scattering theory

The Mie scattering expressions can be greatly simplified under the assumption that the size of the sphere is significantly smaller than the wavelength of the incident radiation. This is known as the Rayleigh scattering regime (Rayleigh, 1871). Under this approximation, the scattering coefficients become

$$S_1(\theta) = \frac{3}{2}a_1$$ (2.148)

and

$$S_2(\theta) = \frac{3}{2}a_1 \cos\theta,$$ (2.149)

where

$$a_1 = -i\frac{2x^3}{3}\frac{n^2 - 1}{n^2 + 2}.$$ (2.150)

Substituting these expressions into Eqs. (2.145) and (2.146), the scattering and backscatter cross-sections in the Rayleigh regime become

$$\sigma_s = \frac{64\pi^4 a^6}{3\lambda^4}\left|\frac{n^2 - 1}{n^2 + 2}\right|^2$$ (2.151)

and

$$\sigma_b = \frac{64\pi^4 a^6}{\lambda^4}\left|\frac{n^2 - 1}{n^2 + 2}\right|^2,$$ (2.152)

where a is again the radius of the spherical scatterer. It can also be shown that the normalized phase function is

$$p(\theta, \varphi) = \frac{3}{8\pi}(1 + \cos^2 \theta). \tag{2.153}$$

From the prior expressions, models can be derived for the scattering coefficient,

$$\beta_s = \frac{64M\pi^5 a^6}{3\lambda^4} \left| \frac{n^2 - 1}{n^2 + 2} \right|^2, \tag{2.154}$$

for scattered spectral intensity,

$$I_s(\theta, \varphi, \lambda) = \frac{64MV\pi^5 a^6}{3\lambda^4} \left| \frac{n^2 - 1}{n^2 + 2} \right|^2 (1 + \cos^2 \theta) E_i(\lambda), \tag{2.155}$$

and for backscattered spectral irradiance,

$$E_b(\lambda) = \frac{8MV\pi^3 a^6}{R^2 \lambda^4} \left| \frac{n^2 - 1}{n^2 + 2} \right|^2 E_i(\lambda). \tag{2.156}$$

In all cases, the scattering strength scales with λ^{-4} such that the effect at shorter wavelengths is considerably more pronounced than at longer wavelengths. This λ^{-4} dependence of Rayleigh scattering has substantial impacts on atmospheric transmission spectra in the visible spectral region.

In addition to its strong spectral dependence, the strength of Rayleigh scattering is also strongly dependent on the polarization orientation of the incident radiation. The degree of polarization (DOP) can be expressed in terms of the scattering coefficients in Eqs. (2.148) and (2.149) as

$$\text{DOP} = \frac{|S_1(\theta)|^2 - |S_2(\theta)|^2}{|S_1(\theta)|^2 + |S_2(\theta)|^2}. \tag{2.157}$$

Making the proper substitutions and solving leads to

$$\text{DOP} = \frac{1 - \cos^2 \theta}{1 + \cos^2 \theta}. \tag{2.158}$$

It is clear from Eq. (2.158) that the scattered radiation is unpolarized (DOP = 0) in the forward and backward scattered directions, but is fully polarized (DOP = 1) in the scattering direction orthogonal to the incident radiation. This prediction of a pure single Rayleigh scattering theory is not supported in nature, because molecules are not infinitesimally small and are asymmetrically shaped. The theoretical maximum DOP is about 93% based on these considerations and is actually in the 50 to 60% range due to multiple scattering effects for clear conditions and becomes lower if clouds are present (Pust and Shaw, 2008).

As a final note, some scattering problems involve only a nominal flow of radiation as opposed to a specific propagation direction. For example, when considering light transmission through a translucent medium, one often only cares about the amount of energy transmitted through the medium, irrespective of the specific forward angle of scattering. In that case, scattering in the forward hemisphere relative to the incident radiation does not reduce the transmission; only scattering in the back hemisphere does. In these cases, only the integrated energy in the backscattered hemisphere, characterized by the modified scattering coefficient,

$$\beta'_s = \frac{32M\pi^5 a^6}{3\lambda^4} \left| \frac{n^2 - 1}{n^2 + 2} \right|^2 , \tag{2.159}$$

is considered a loss with respect to the transmitted energy in the forward hemisphere.

2.8 Summary

According to classical electromagnetic theory, the complex index of refraction is the intrinsic material property that characterizes the interaction of optical radiation with matter. The spectral dependence of this material property is the fundamental signature with respect to hyperspectral remote sensing measurements. Through the Lorentz oscillator model, mechanical resonances in atomic or molecular systems are directly related to anomalous dispersion features in the real part of the index of refraction, and absorption peaks in the imaginary part. The precise wavelengths at which these features occur are related to the bond strengths within the atoms or molecules; thus, they can uniquely identify the material type. Most materials are too complex to numerically model from first principles, but the Lorentz model provides a mathematical form of the complex index of refraction that can be used to fit empirical measurement data. Such data can be found in material handbooks and databases.

The complex index of refraction is not a directly measurable quantity. Instead, it affects observable properties such as spectral transmission, reflectance, and absorbance. Beer's law governs transmission through a homogenous medium, and the Fresnel equations govern the transmissivity and reflectivity at smooth material boundaries. The interactions at rough surfaces and through scattering media are a bit more complicated to model.

Even the case of Mie scattering, involving simple spherical scatterers, is quite complex. However, such interactions can again be characterized by parameters such as cross-sections and phase functions fit to empirical data. These relationships are particularly important for understanding the observed spectral properties of vegetation and paint coatings, as examples, for which reflected radiation is often due to volume scattering, and for characterizing radiative transfer through the atmosphere (Smith, 1993).

While Maxwell's equations are fundamental laws governing electromagnetic radiation, the Lorentz oscillator model is an approximation for the interaction of such radiation with atoms and molecules for which a quantum mechanical analysis is required. As is seen in the next chapter, such an analysis predicts spectral behavior not supported by classical mechanics. On the other hand, the semi-empirical Lorentz model remains useful in practice, especially for modeling solid materials.

2.9 Further Reading

For a more thorough treatment of electromagnetic field and wave theory, the reader is referred to standard textbooks such as those written by Lorrain and Corson (1962) or Jackson (1962). Perhaps the most thorough reference concerning optical radiation is from Born and Wolf (1980), but Fowles (1989) provides a concise treatment that is more readable. A more extensive treatment of radiometry can be found in Schott (2007), while Bohren and Huffman (1998) thoroughly details scattering theory.

References

Beer, A., *Einleitung in der Elektrostatik, die Lehre von Magnetismus und die Elektrodynamik*, Friedrich Vieweg und Sohn, Braunschweig, Germany (1865).

Bohren, C. F. and Huffman, D. R., *Absorption and Scattering of Light by Small Particles*, Wiley & Sons, New York (1998).

Born, M. and Wolf, E., *Principles of Optics*, Sixth Edition, Pergamon Press Oxford, UK (1980).

Drude, P., "Zur Elektronentheorie der Metalle," *Ann. Phys.* **306**, p. 566 (1900).

Fowles, G. R., *Introduction to Modern Optics*, Dover, Mineola, NY (1989).

Jackson, J. D., *Classical Electrodynamics*, Wiley & Sons, New York (1962).

Lebedev, N. N., *Special Functions and Their Applications*, Dover, New York (1972).

Lorentz, H. A., *Theory of Electrons*, Second Edition, Dover, New York (1952).

Lorrain, P. and Corson, D., *Electromagnetic Fields and Waves*, Second Edition, W. H. Freeman, New York (1962).

Maxwell, J. C., *Treatise on Electricity and Magnetism*, Third Edition, Dover, New York (1954).

Mie, G., "Beitrage zur Optik trüber Medien speziell kolloidaler Metallösungen," *Ann. Phys.* **25**, 377–445 (1908).

Palik, E. D., *Handbook of Optical Constants of Solids*, Academic Press, Orlando, FL (1985).

Pust, N. J. and Shaw, J. A., "Digital all-sky polarization imaging of partly cloudy skies," *Appl. Optics* **47**, 190–198 (2008).

Rayleigh, Lord, "On the light from the sky, its polarization and colour," *Philos. Mag. J. Sci.* **41**, 107–112 (1871).

Schott, J. R., *Remote Sensing: The Image Chain Approach*, Second Edition, Oxford University Press, Oxford, New York (2007).

Smith, F. G. (Ed.), *The Infrared and Electro-Optical Systems Handbook: Volume 2, Atmospheric Propagation of Radiation*, SPIE Press, Bellingham, WA (1993).

Chapter 3
Atomic and Molecular Spectroscopy

A quantum mechanical model is necessary to accurately understand the spectroscopy of atoms and molecules. While the classical mechanical model provides some insight into the spectral characteristics of materials and a useful basis for semi-empirical modeling of amorphous solids and liquids, it fails to capture a variety of observed spectral features such as rotational lines in gaseous spectra, band structures of crystalline solids, and quantized emission spectra of atoms.

The quantum mechanical treatment accurately models these effects by characterizing particles such as electrons in an atom or molecule through *wave functions*. Such wave functions exist only at discrete energy levels, and interaction with optical radiation involves the allowed transitions between such energy levels. In this way, the spectral characteristics of emission and absorption depend on the specific energy levels of the atom or molecule, the dipole moment associated with the energy level between which the system transitions, and the state populations.

According to the Born–Oppenheimer approximation, a complex molecule can be analyzed by decomposing the wave function and energy levels into atomic, vibrational, and rotational components. Therefore, the mathematical treatment in this chapter separately addresses these individual components to elucidate characteristic features, and then discusses the more complicated case of aggregate spectral properties. While this chapter only scratches the surface of the field of atomic and molecular spectroscopy, the insights gained are of great importance in understanding the observed spectral properties of real materials, as well as the effects of the atmosphere on hyperspectral measurements in a remote setting.

3.1 Quantum Mechanics

From a quantum mechanical perspective, the state of a molecular system is described in terms of wave function $\Psi(x, y, z, t)$. This wave function represents either a particle or system of particles (atoms and electrons), while the effect of the local environment (bonding forces within atoms and molecules) and external fields is characterized by a potential function where the particles reside. The squared magnitude of the wave function,

$$\Psi^*(x, y, z, t)\Psi(x, y, z, t), \tag{3.1}$$

is proportional to the probability that the particle resides at specified time t and location (x, y, z). As this quantity is a probability density function, it can be normalized such that

$$\iiint \Psi^*(x, y, z, t)\Psi(x, y, z, t)dxdydz = 1. \tag{3.2}$$

All of the characteristics of the system state are defined by the wave function. Because the system state is defined probabilistically as opposed to deterministically, system characteristics can only be defined as expected values, as opposed to deterministic quantities. Three such expected values of particular interest relate to the expected location in space,

$$
\begin{aligned}
\hat{x}(t) &= \iiint \Psi^*(x, y, z, t)x\, \Psi(x, y, z, t)dxdydz \\
\hat{y}(t) &= \iiint \Psi^*(x, y, z, t)y\, \Psi(x, y, z, t)dxdydz \\
\hat{z}(t) &= \iiint \Psi^*(x, y, z, t)z\, \Psi(x, y, z, t)dxdydz,
\end{aligned}
\tag{3.3}
$$

the expected momentum,

$$
\begin{aligned}
\hat{p}_x(t) &= \iiint \Psi^*(x, y, z, t) \left[-i\hbar\frac{\partial}{\partial x}\, \Psi(x, y, z, t) \right] dxdydz \\
\hat{p}_y(t) &= \iiint \Psi^*(x, y, z, t) \left[-i\hbar\frac{\partial}{\partial y}\, \Psi(x, y, z, t) \right] dxdydz \\
\hat{p}_z(t) &= \iiint \Psi^*(x, y, z, t) \left[-i\hbar\frac{\partial}{\partial z}\, \Psi(x, y, z, t) \right] dxdydz,
\end{aligned}
\tag{3.4}
$$

and the expected energy level,

$$\hat{E}(t) = \iiint \Psi^*(x, y, z, t) \left[i\hbar\frac{\partial}{\partial t}\, \Psi(x, y, z, t) \right] dxdydz. \tag{3.5}$$

In Eq. (3.5), $\hbar = h/2\pi$, where h is Planck's constant ($h = 6.626 \times 10^{-34}$ J s). If the expected energy level is constant in time, the wave function represents a stationary state of the system.

3.1.1 Stationary states of a quantum mechanical system

The wave function must be a solution to the Schrödinger wave equation (Schrödinger, 1926),

$$\left[-\frac{\hbar^2}{2m}\nabla^2 + V(x,y,z,t)\right]\Psi(x,y,z,t) = i\hbar\frac{\partial}{\partial t}\Psi(x,y,z,t), \qquad (3.6)$$

where m is the particle mass and $V(x,y,z,t)$ is the potential function. The left-hand side of Eq. (3.6) is known as the Hamiltonian of the system, while the right-hand side is recognized as the energy operator within Eq. (3.5). Stationary states of a system can be represented by the time-harmonic solution,

$$\Psi(x,y,z,t) = \psi(x,y,z)\,e^{-iEt/\hbar}, \qquad (3.7)$$

or superpositions thereof, where the constant E represents the energy level of the stationary state. The time-independent wave function $\psi(x,y,z)$ is a solution to the time-independent Schrödinger wave equation

$$-\frac{\hbar^2}{2m}\nabla^2\psi(x,y,z) + V(x,y,z)\ \psi(x,y,z) = E\psi(x,y,z). \qquad (3.8)$$

When Eq. (3.8) is solved for the stationary states of a particular quantum mechanical system defined by $V(x,y,z)$, there are normally multiple solutions $\{\psi_i(x,y,z); i = 1,2,\ldots N\}$ that correspond to different energy levels $\{E_i\}$. Mathematically, the solutions $\{\psi_i(x,y,z)\}$ are the eigenfunctions of the system, and the corresponding energy levels E_i are the eigenvalues. While any linear combination of solutions is also a solution of the Schrödinger wave equation, one typically considers the orthonormal set of eigenfunctions that exhibit the property

$$\iiint \psi_i^*(x,y,z,t)\psi_j(x,y,z,t)dxdydz = \delta_{ij}, \qquad (3.9)$$

where δ_{ij} is the Kronecker delta function: $\delta_{ij} = 1$ if and only if $i = j$. As an example, consider the triple quantum well in Fig. 3.1, defined by the

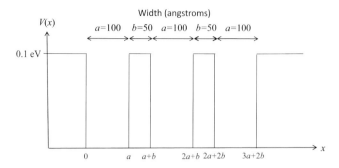

Figure 3.1   Potential function for a triple quantum well.

potential function produced by epitaxial crystal growth in the x direction:

$$V(x, y, z) = \begin{cases} V_0 & x < 0 \\ 0 & 0 \le x \le a \\ V_0 & a < x < a + b \\ 0 & a + b \le x \le 2a + b \\ V_0 & 2a + b < x < 2a + 2b \\ 0 & 2a + 2b \le x \le 3a + 2b \\ V_0 & 3a + 2b < x \end{cases} \qquad (3.10)$$

The first four wave-function solutions in terms of increasing energy level for $V_0 = 0.1$ eV, $a = 100$ Å, and $b = 50$ Å are illustrated in Fig. 3.2. A few typical characteristics of wave functions are evident from these examples. First, wave functions exhibit higher-spatial-frequency content as the energy level increases. Second, they tend to alternate between even and odd symmetry. Finally, wave functions are nonzero even in the barrier regions, meaning that particles can exist within the barriers and move through the barriers between the wells. This characteristic, known as tunneling, can only be explained through a quantum mechanical model.

3.1.2 Interaction with electromagnetic radiation

In the presence of electromagnetic radiation of angular frequency ω,

$$V(x, y, z, t) = V_0(x, y, z) + e(\mathbf{E}_f \cdot \mathbf{r}) \cos(\omega t), \qquad (3.11)$$

where $V_0(x, y, z)$ is the stationary field due to the local environment, $\mathbf{E}_f$ is the vector electric field amplitude, and $\mathbf{r}$ is a spatial vector. Equation (3.11) simply introduces the Lorentz force from the electromagnetic radiation into the potential function. Solutions to Eq. (3.6) with the potential

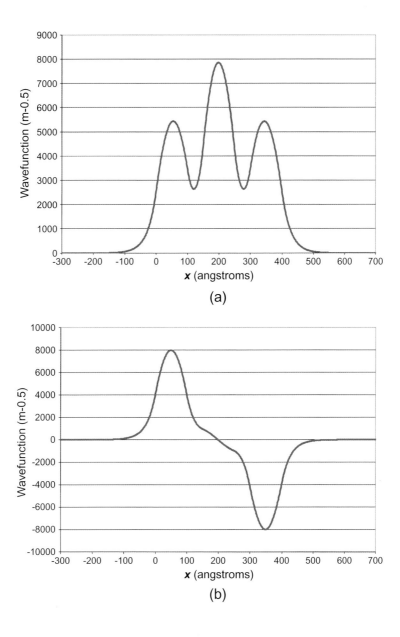

Figure 3.2 Wave functions corresponding to the lowest four energy levels of the triple quantum well: (a) first level, (b) second level, (c) third level, and (d) fourth level.

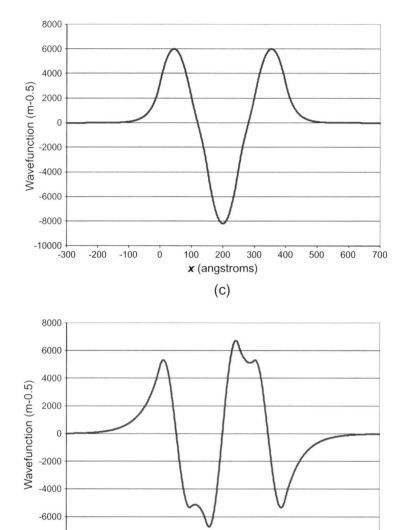

(c)

(d)

Figure 3.2   (*continued*)

function in Eq. (3.11) are of the form

$$\Psi(x, y, z, t) = c_i(t)\,\psi_i(x, y, z)\,e^{-iE_i t/\hbar} + c_j(t)\,\psi_j(x, y, z)\,e^{-iE_j t/\hbar}, \quad (3.12)$$

which can be interpreted as follows. The energy absorbed from the electromagnetic radiation causes a transition from a wave function ψ_i with energy E_i to a wave function ψ_j with energy E_j, according to coefficients $c_i(t)$ and $c_j(t)$. These coefficients are bound by the initial conditions $c_i(0) = 1$ and $c_j(0) = 0$, meaning that the particle starts in the i'th state. The probability that a state transition to ψ_j occurs at time t is given by the quantity $c_j^*(t)c_j(t)$. By inserting Eq. (3.12) into Eq. (3.6), this is shown to be

$$c_j^*(t)\,c_j(t) = \frac{|\mathbf{E}_f \cdot \mathbf{M}_{ij}|^2}{\hbar} \frac{\sin^2[(E_j - E_i - \hbar\omega)\,t/2\hbar]}{[(E_j - E_i - \hbar\omega)/2\hbar]^2}, \quad (3.13)$$

where $\mathbf{M}_{ij}$ is the transition-moment integral defined by

$$\mathbf{M}_{ij} = \iiint \psi_i^*(x, y, z)\boldsymbol{\mu}\,\psi_j(x, y, z)\,dxdydz, \quad (3.14)$$

and $\boldsymbol{\mu}$ is the electric dipole moment of the atomic or molecular system. The dipole moment is typically defined as

$$\boldsymbol{\mu} = \sum_n q_n \mathbf{r}_n, \quad (3.15)$$

where q_n and $\mathbf{r}_n$ represent the charges and positions of all charged particles in the system. In the quantum mechanical model, the dipole moment can be represented by

$$\boldsymbol{\mu} = \sum_n \iiint \psi_n^*(x, y, z)\,q_n\mathbf{r}\,\psi_n(x, y, z)\,dxdydz, \quad (3.16)$$

where ψ_n represents the wave functions for all the charged particles.

A few important observations should be drawn from Eqs. (3.13) through (3.16). First, transitions can only occur between states for which the transition-moment integral is nonzero for the polarization direction of incident electromagnetic radiation. Note that this integral is identically zero for transitions between states where both wave functions are either symmetric or antisymmetric in the polarization direction because the integrand in Eq. (3.14) is an odd function in these cases, and, therefore, the integral is identically zero. Thus, transitions can occur between the first and second levels of the triple quantum well depicted in Figs. 3.1 and 3.2, but not between the first and third levels. Such transitions are called *forbidden transitions*, while those for which the integral is nonzero are called *allowed transitions*. For allowed transitions, energy is only substantially absorbed (causing a substantial state transition probability) if the radiation frequency is such that the second term in Eq. (3.13) is large. This occurs near the transition frequency or wavelength, such that

$$\hbar\omega_{ij} = \frac{hc}{\lambda_{ij}} = E_j - E_i, \qquad (3.17)$$

which is equivalent to the resonance condition of the classical model. For forbidden transitions, energy is not absorbed irrespective of the radiation frequency.

3.1.3 Born–Oppenheimer approximation

The Hamiltonian of an atomic or molecular system represents the sum of kinetic and potential energy of the system, that is, the first and second terms of the Schrödinger wave equation. In a molecule, kinetic energy includes the motions of both the negatively charged electrons and the positively charged nuclei. Rigorously, one needs to find solutions to Eq. (3.8) for this complete system; these solutions will be very complex, even for simple atoms and molecules. Born and Oppenheimer (1927) proposed a useful approximation based on the recognition that vibrating nuclei move much slower than electrons, such that electronic states can be computed based on fixed nuclei. Mathematically, the Born–Oppenheimer approximation holds that

$$\psi(x, y, z) = \psi_e(x', y', z') \psi_n(x, y, z), \qquad (3.18)$$

where (x', y', z') are electron coordinates relative to fixed nuclei, and (x, y, z) are nuclear coordinates. It follows that

$$E = E_e + E_n, \qquad (3.19)$$

where E_e and E_n are the energy levels corresponding to the electronic and nuclear wave functions ψ_e and ψ_n, respectively.

The Born–Oppenheimer approximation can similarly be extended by factoring the nuclear wave functions into components related to the rigid rotation of the molecule as a whole, designated by ψ_r, and the vibrational motion of atoms within a molecule, designated by ψ_v, such that

$$\psi(x, y, z) = \psi_e(x'', y'', z'') \psi_v(x', y', z') \psi_r(x, y, z) \qquad (3.20)$$

and

$$E = E_e + E_v + E_r. \tag{3.21}$$

In Eq. (3.20), the double-primed coordinates correspond to fixed nuclei, and the primed coordinates correspond to nonrotating nuclei. Through this approximation, the electronic, vibrational, and rotational spectral characteristics can be addressed separately and then aggregated. The treatment in this chapter begins by considering the simpler cases of electronic spectroscopy of single atoms and rotational spectroscopy of rigid molecules, and eventually aggregates the simpler models to address the complex case where electronic, vibrational, and rotational transitions can simultaneously occur.

3.2 Electromagnetic Absorption and Emission

In the quantum mechanical model, absorption and emission of radiation occurs during transitions between electronic states, such that the energy of the radiation equals the change in kinetic energy corresponding to the electron transition. This means that the photon energy

$$E_p = \hbar\omega = \frac{hc}{\lambda} = hc\sigma \tag{3.22}$$

equals the difference in energy levels $E_p = E_j - E_i$, which is just a restatement of Eq. (3.17). It is common in spectroscopy to define a wavenumber σ that is the reciprocal of the wavelength. Typically, σ is measured in cm^{-1} units, such that:

$$\sigma\,(cm^{-1}) = \frac{10,000}{\lambda\,(\mu m)}. \tag{3.23}$$

When a charged particle transitions between states, one of the three processes depicted in Fig. 3.3 can occur: stimulated absorption, spontaneous emission, or stimulated emission. In *stimulated absorption*, an atom or molecule absorbs a quantum of radiation, or photon, and is excited from a lower state with energy E_i to a higher state with energy E_j. In *spontaneous emission*, a photon is emitted as the atom or molecule relaxes from a higher state E_j to a lower state E_i. In *stimulated emission*, the relaxation from a higher to lower state is induced by a photon, and another photon is emitted that is coherently in phase with the incident photon. This latter process is the basis of laser radiation.

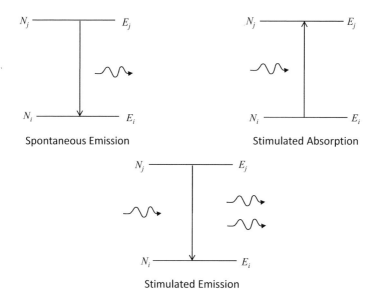

Figure 3.3   Absorption and emission processes in an atom or molecule.

3.2.1 Einstein coefficients

According to the theory developed by Einstein (1917), the rate of population change of the upper state in the two-state system depicted in Fig. 3.3 is given by

$$\frac{dN_j}{dt} = (N_i - N_j)B_{ij}\rho_\sigma(\sigma_{ij}) - N_jA_{ij}, \tag{3.24}$$

where N_i is the number density of the i'th state, N_j is the number density of the j'th state, and $\rho_\sigma(\sigma_{ij})$ is the spectral energy density of the incident radiation at the transition wavenumber corresponding to Eq. (3.17). The coefficients A_{ij} and B_{ij} are known as Einstein coefficients and are related by

$$A_{ij} = 8hc\sigma_{ij}B_{ij}. \tag{3.25}$$

The Einstein coefficients are related to the dipole moment by

$$A_{ij} = \frac{16\pi^3 n\sigma_{ij}^3}{3h\varepsilon_0}|\mathbf{M}_{ij}|^2 \tag{3.26}$$

and

$$B_{ij} = \frac{2\pi^2}{3h\varepsilon_0 n^2}|\mathbf{M}_{ij}|^2, \tag{3.27}$$

where the dipole moment can be decomposed into components along the three coordinate-axis directions as

$$|\mathbf{M}_{ij}|^2 = |(\mathbf{M}_{ij})_x|^2 + |(\mathbf{M}_{ij})_y|^2 + |(\mathbf{M}_{ij})_z|^2 . \tag{3.28}$$

The Einstein coefficients determine the transition rates for the three radiative processes shown in Fig. 3.3 and therefore determine the emitted power and absorption characteristics. The radiated spectral power of spontaneous emission is determined by the Einstein A coefficient as

$$P = hc\sigma A_{ij}\delta(\sigma - \sigma_{ij}) = \frac{16\pi^3 nc\sigma^4}{3h\varepsilon_0}|\mathbf{M}_{ij}|^2 \,\delta(\sigma - \sigma_{ij}), \tag{3.29}$$

where the Dirac delta function $\delta(\sigma - \sigma_{ij})$ indicates that power is only emitted at the transition wavenumber. The absorption coefficient is determined by the Einstein B coefficient as

$$\beta_a(\sigma) = h\sigma B_{ij}(N_i - N_j)\,\delta(\sigma - \sigma_{ij}) = \frac{2\pi^2\sigma}{3nh\varepsilon_0}(N_i - N_j)\,|\mathbf{M}_{ij}|^2 \,\delta(\sigma - \sigma_{ij}). \tag{3.30}$$

If the number density of the upper state N_j is higher than that of a lower state N_i, Equation (3.30) becomes negative, which represents a gain medium. This condition, known as population inversion, is central to the lasing process. Under stimulated absorption, $N_i > N_j$ and Eq. (3.30) can be related to the classical oscillator model, where the absorption coefficient is related to an oscillator strength f_{ij} corresponding to the material resonance through the Lorentzian profile of κ. This oscillator strength is related to the Einstein B coefficient according to

$$f_{ij} = \frac{4\varepsilon_0 m_e hc^3 \sigma_{ij}}{e^2}B_{ij}, \tag{3.31}$$

where m_e and e are again the mass and charge of an electron.

3.2.2 Line broadening

Thus far, the quantum mechanical model predicts absorption and emission features that have no spectral width, as denoted by the Dirac delta functions in Eqs. (3.29) and (3.30). That is, either the frequency of radiation perfectly matches the difference in energy levels according to Eq. (3.17) or the process does not occur. In reality, there is some inherent width to the energy levels, and the degree of match does not need to be perfect. The result is that the absorption and emission spectral bandwidth is broadened.

This broadening can be characterized by replacing the Dirac delta function by a *line shape* $g(\sigma)$, normalized such that

$$\int_0^\infty g(\sigma)\,d\sigma = 1.$$

(3.32)

Many physical mechanisms can broaden the spectral features observed in materials. In the Lorentz model, the damping factor is the driving factor in determining the width of the spectral features. In the quantum mechanical model, the ultimate limit to the spectral line width is driven by the lifetime of the upper state to which the system is excited according to the Heisenberg uncertainty principle:

$$\Delta E\,\Delta t \geq \hbar.$$

(3.33)

The interpretation is that if the particle is known to exist in the upper state to within a finite lifetime, then there is a natural uncertainty or broadening in the energy level of the upper state. This, in turn, results in spectral broadening through Eq. (3.17). Denoting the upper-state lifetime as τ, the spectral line width [full-width at half-maximum (FWHM)] can be shown to be

$$\Delta\sigma = \frac{2.78}{\pi c\tau} \approx \frac{1}{c\tau}$$

(3.34)

or

$$\Delta\lambda = \lambda^2 \Delta\sigma = \frac{2.78\lambda^2}{\pi c\tau} \approx \frac{\lambda^2}{c\tau}.$$

(3.35)

This is referred to as the natural line width and represents the fundamental limit to the width of $g(\sigma)$. The line shape of an absorption feature centered at σ_0 is typically modeled using the Lorentzian line-shape function

$$g(\sigma) = \frac{\Delta\sigma}{(\sigma - \sigma_0)^2 + (\Delta\sigma/2)^2},$$

(3.36)

which is the Fourier transform of an exponential decay function. If spontaneous emission to the i'th state is the only process by which the upper state decays, then $\tau = 1/A_{ij}$ and

$$\tau = \frac{1}{A_{ij}} = \frac{3hc\varepsilon_0}{16\pi^3 n\sigma_{ij}^3 |M_{ij}|^2}.$$

(3.37)

Typically, nonradiative decay processes such as molecular collisions and acoustic interactions further reduce the upper-state lifetime.

When multiple atoms or molecules interact with an electromagnetic field, collisions between molecules also tend to broaden the absorption line width. Equation (3.36) can be used to model collision broadening by letting τ represent the mean time between collisions. For molecules in gaseous form, the line width is a function of temperature and pressure with a dependence, given by

$$\Delta\sigma = \Delta\sigma_0 \left(\frac{P}{P_0}\right)\left(\frac{T_0}{T}\right)^n, \tag{3.38}$$

where $\Delta\sigma_0$ is in the range 0.01 to 0.1 cm^{-1} for $T_0 = 273$ K and $P_0 = 1000$ mbar, and n is empirically determined (nominally 0.5). In a liquid or solid, this collisional broadening is even greater due to the interactions between molecules in close proximity.

When molecules move at high speeds, there is a Doppler shift in the effective transition frequency. Therefore, if a gas exhibits a large random variation in the speed and direction of the gas molecules, as is the case for a gas with high kinetic energy, then the aggregate effect of the Doppler shifts is spectral broadening. This Doppler broadening is characterized by

$$g(\sigma) = \frac{1}{\alpha_D \sqrt{\pi}} e^{-(\sigma-\sigma_0)^2/\alpha_D^2}, \tag{3.39}$$

where

$$\alpha_D = \sigma_0 \sqrt{\frac{2kT}{Mc^2}} \tag{3.40}$$

and M is the molecular mass. The FWHM is

$$\Delta\sigma = 2\alpha_D \sqrt{\ln 2}. \tag{3.41}$$

The total line shape of a transition is the convolution of the natural line shape with those due to collision and Doppler broadening.

3.3 Electronic Spectroscopy of Atoms

Electronic spectroscopy deals with the transitions of electrons between orbital states within an atom. Such state transitions tend to be of high energy difference, and corresponding electronic absorption features usually occur at wavelengths in the ultraviolet and visible spectral region. Electronic states are governed by the Schrödinger wave equation but are very difficult to analyze. Only the simplest atoms, such as hydrogen, have analytical solutions or straightforward numerical solutions. The spectroscopy of the hydrogen atom is discussed in some detail and is then extrapolated qualitatively to more-complicated atoms.

3.3.1 Single-electron atoms

The hydrogen atom, consisting of one proton and one electron, provides the simplest example of the application of quantum theory to the electronic spectroscopy of atoms. One-electron solutions are also applicable to hydrogen-like ions such as He^+, Li^{2+}, Be^{3+}, and H_2^+ that vary only in the number of protons in the nucleus. Figure 3.4 depicts such a single-electron atom consisting of a nucleus with Z protons. If the electron has a mass m_e, and the nucleus a mass M, an equivalent system can be solved that has a fixed nucleus and an electron with a reduced mass:

$$m_r = \frac{M\,m_e}{M + m_e}. \tag{3.42}$$

The electrostatic potential function corresponding to such an atom is given in spherical coordinates by

$$V(r, \theta, \phi) = \frac{-Ze^2}{4\pi\varepsilon_0 r}. \tag{3.43}$$

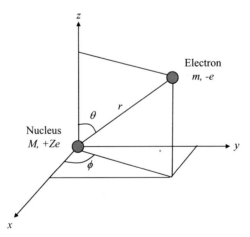

Figure 3.4   Coordinate geometry of a single-electron atomic model.

The Schrödinger wave equation is solved for this potential function in spherical coordinates where the Laplacian is defined as

$$\nabla^2 = \frac{1}{r^2 \sin\theta}\left[\sin\theta\frac{\partial}{\partial r}\left(r^2\frac{\partial}{\partial r}\right) + \frac{\partial}{\partial\theta}\left(\sin\theta\frac{\partial}{\partial\theta}\right) + \frac{1}{\sin\phi}\frac{\partial^2}{\partial\phi^2}\right], \quad (3.44)$$

with spherical coordinates (r, θ, ϕ) defined in Fig. 3.4. The wave functions corresponding to the stationary states are of the form

$$\psi(r, \theta, \phi) = R_{n,l}(r)\,\Theta_{l,m}(\theta)\,\Phi_m(\phi), \quad (3.45)$$

where $R_{n,l}(r)$ are radial wave functions, $\Theta_{l,m}(\theta)$ are associated Legendre polynomials

$$\Phi_m(\phi) = e^{im\phi}, \quad (3.46)$$

n is the principal quantum number ($n = 1, 2, \ldots$), l is the azimuthal quantum number ($l = 0, 1, \ldots, n - 1$), and m is the magnetic quantum number ($m = -l, \ldots, -2, -1, 0, 1, 2, \ldots, l$). The functional form of the low-order radial wave functions and associated Legendre polynomials are given in Tables 3.1 and 3.2, with the radial wave functions plotted in Fig. 3.5 as a function of the Bohr radius,

$$a_0 = \frac{(4\pi\varepsilon_0)\,\hbar^2}{m_r e^2}. \quad (3.47)$$

Additional details are provided by Anderson (1971) and Bransden and Joachain (2003).

The energy levels of the hydrogen atom are given by

$$E_n = \frac{-Z^2 e^4 m_r}{8\varepsilon_0^2 h^2 n^2}, \quad (3.48)$$

Table 3.1 Functional form of low-order radial wave functions.

n	l	$R_{n,l}(r)$
1	0	$\left(\frac{z}{a_0}\right)^{3/2} 2e^{-Zr/a_0}$
2	0	$\left(\frac{z}{2a_0}\right)^{3/2}\left(1 - \frac{Zr}{2a_0}\right)e^{-Zr/2a_0}$
3	1	$\frac{1}{\sqrt{3}}\left(\frac{z}{2a_0}\right)^{3/2}\left(\frac{Zr}{a_0}\right)e^{-Zr/2a_0}$
3	0	$2\left(\frac{z}{3a_0}\right)^{3/2}\left(1 - \frac{2Zr}{3a_0} + \frac{2Z^2r^2}{27a_0^2}\right)e^{-Zr/3a_0}$

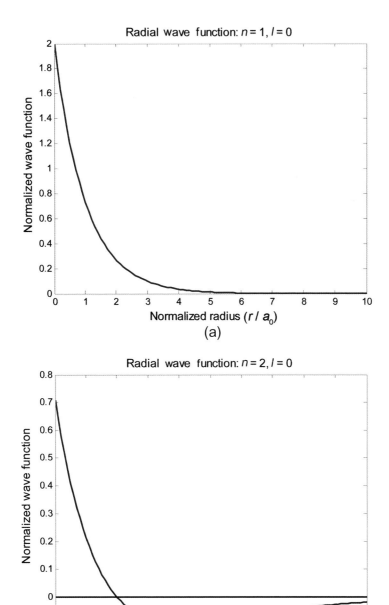

Figure 3.5   Low-order radial wave functions of the hydrogen atom: (a) $n = 1, l = 0$; (b) $n = 2, l = 0$; (c) $n = 2, l = 1$; and (d) $n = 3, l = 0$.

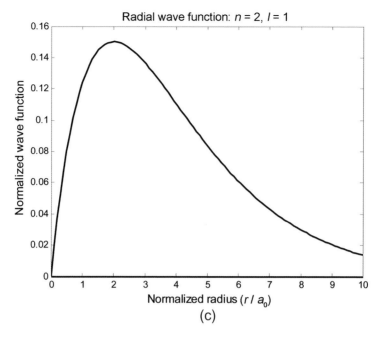

(c)

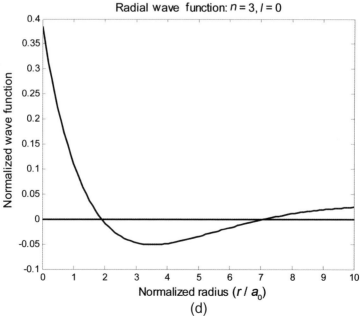

(d)

Figure 3.5 *(continued)*

Table 3.2 Functional form of low-order associated Legendre polynomials.

l	m	$\Theta_{l,m}(\theta)$
0	0	$1/\sqrt{2}$
1	0	$\left(\sqrt{6}/2\right)\cos\theta$
1	± 1	$\left(\sqrt{3}/2\right)\sin\theta$
2	0	$\left(\sqrt{10}/4\right)\left(3\cos^2\theta - 1\right)$
2	± 1	$\left(\sqrt{15}/2\right)\sin\theta\cos\theta$
2	± 2	$\left(\sqrt{15}/2\right)\sin^2\theta$

where it is apparent from Eq. (3.48) that the energy levels are independent of the azimuthal and magnetic quantum numbers and are only dependent on the principal quantum number n. Absorption and emission of radiation is possible at wavelengths corresponding to the energy differences between these states, as shown in Fig. 3.6. Of these, the Balmer series is in the visible spectral region, and the Paschen and Pfund series are in the infrared spectral region. These features are seen in the emission spectrum of the hydrogen discharge lamp illustrated in Fig. 3.7 but are typically not seen in absorption spectra because only the ground state ($n = 1$) is populated under ambient conditions. Only the Lyman series in the ultraviolet spectral region is evident in the absorption spectra.

The azimuthal quantum number l defines the orbital angular momentum of the electron state, which is given by

$$p_l = [l(l+1)]^{1/2}\hbar. \tag{3.49}$$

When placed in a magnetic field, the component of the angular momentum vector along the field direction is quantized to $m\hbar$. Electrons can also have spin states designated by the spin quantum number s, which has a value of one half and is quantized to two orientations by another quantum number $m_s = -\frac{1}{2}$ or $\frac{1}{2}$. According to the Pauli exclusion principle, no two electrons in a system can have the same set of quantum numbers. Since the energy levels are independent of l, m, and m_s, they are *degenerate*, meaning they are different states with the same energy level.

Electron states in an atom, known as electron orbitals, are labeled according to the values of n and l, where symbols s, p, d, f, etc. correspond to $l = 0, 1, 2, 3$, etc. These letters correspond to the terms *sharp*, *principal*, *diffuse*, and *fundamental* used by experimental spectroscopists. Table 3.3 provides the wave functions for the first three energy levels. The degeneracy is given by $2l(l+1)$. The angular wave functions are commonly described as real-valued, orthogonal linear combinations of $\psi_{n,l,m}(r, \theta, \phi)$.

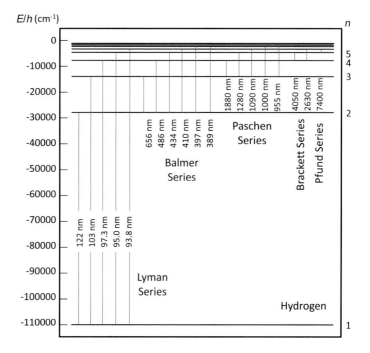

Figure 3.6   Energy band diagram of the hydrogen atom.

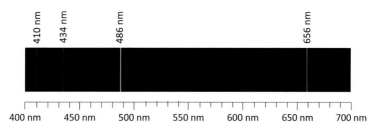

Figure 3.7   Emission spectrum of the hydrogen atom (Jircitano, 2011).

Figure 3.8 provides polar diagrams to illustrate the nature of the angular wave functions for the 1s, 2d, and 3p orbitals.

3.3.2 Polyelectronic atoms

For a polyelectronic atom, the Hamiltonian becomes

$$-\frac{\hbar^2}{2m_r}\sum_i \nabla_i^2 - \sum_i \frac{Ze^2}{4\pi\varepsilon_0 r_i} + \sum_{i<j} \frac{e^2}{4\pi\varepsilon_0 r_{ij}}, \tag{3.50}$$

where the sum covers all of the electrons in the atom. The first two terms in Eq. (3.50) are identical to those of the hydrogen atom, while the third term takes into account the Coulombic repulsion between electrons.

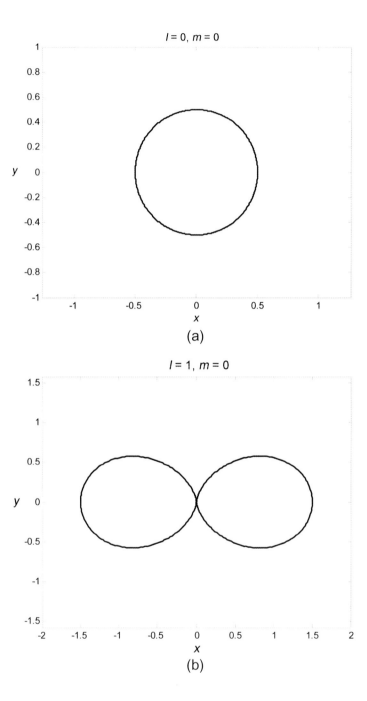

Figure 3.8 Polar diagrams for the low-order angular wave functions: (a) $l = 0, m = 0$; (b) $l = 1, m = 0$; (c) $l = 1, m = 1$; (d) $l = 2, m = 0$; (e) $l = 2; m = 1$; and (f) $l = 2, m = 2$.

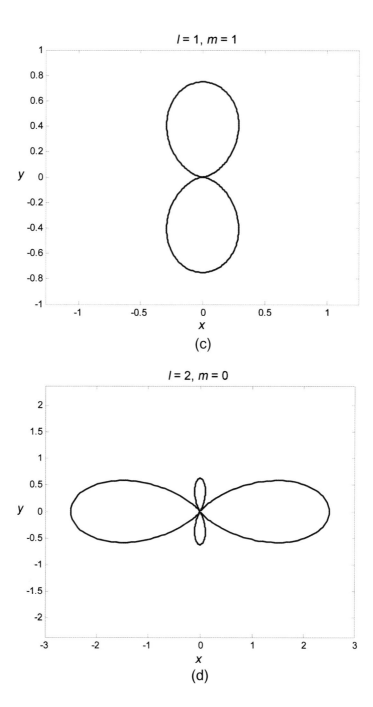

Figure 3.8   (*continued*)

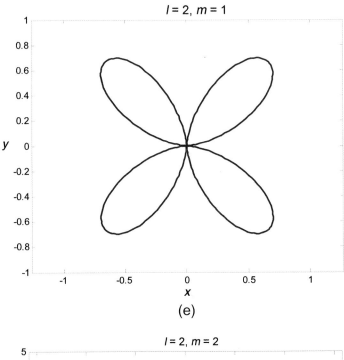

(e)

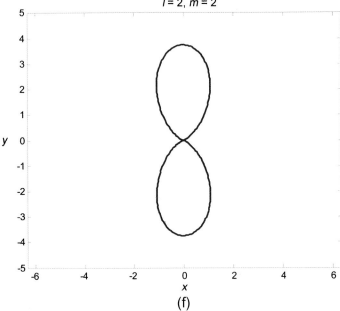

(f)

Figure 3.8   (*continued*)

Table 3.3 Wave functions for the first two hydrogen energy levels.

Orbital	n	l	m	$\psi_{n,l,m}(r, \theta, \phi)$
1s	1	0	0	$\frac{1}{\sqrt{\pi}} \left(\frac{z}{a_0}\right)^{3/2} 2e^{-Zr/a_0}$
2s	2	0	0	$\frac{1}{2\sqrt{2\pi}} \left(\frac{z}{2a_0}\right)^{3/2} \left(1 - \frac{Zr}{2a_0}\right) e^{-Zr/2a_0}$
2p$_0$	2	1	0	$\frac{1}{4\sqrt{2\pi}} \left(\frac{z}{2a_0}\right)^{3/2} \left(\frac{Zr}{a_0}\right) e^{-Zr/2a_0} \cos\theta$
2p$_{\pm1}$	2	1	±1	$\mp\frac{1}{8\sqrt{\pi}} \left(\frac{z}{3a_0}\right)^{3/2} \left(\frac{Zr}{a_0}\right) e^{-Zr/2a_0} \sin\theta\, e^{\pm i\phi}$

Because of this third term, the Schrödinger wave equation can no longer be solved analytically. An important outcome of this charge screening effect is that the degeneracy is removed, meaning that the energy levels become slightly dependent on l and m. An example is the energy level diagram of lithium depicted in Fig. 3.9. In this case, selection rules are associated with angular-wave-function symmetry, as predicted by the transition moment integral in Eq. (3.14). For the case of lithium, these quantum selection rules are $\Delta l = \pm1$. When this is not the case, the transition moment integral is identically zero. Figure 3.10 shows an energy level diagram for helium, which has an additional selection rule $\Delta s = 0$. Emission spectra for both are shown in Figs. 3.11 and 3.12. In both cases, only the principal series originating from the ground state are evident in the absorption spectrum. For helium, these all fall in the ultraviolet spectral region. For lithium, the ground state is 2s because the 1s orbital is filled. One of the lines in the principal series is in the visible region of the spectrum.

The states of polyelectronic atoms are designated based on the total orbital angular momentum L, spin S, and total momentum (angular + spin) J obtained by a summation of l, s, and j over all of the electrons. The designations S, P, D, F, G, etc. correspond to total orbital angular momentum numbers $L = 0, 1, 2, 3, 4$, etc., and orbital designators are conventionally post-superscripted by the total spin multiplicity $2S + 1$ and subscripted by the total momentum quantum number J. L can range in integers from zero to the summation of l over all electrons, S can range from zero to the summation of s over all electrons, and J can range from $|L - S|$ to $L + S$. Thus, a $^3D_1$ orbital denotes $L = 2, S = 1$, and $J = 1$. For complex atoms, the energy-band diagrams can obviously become very complicated, but the following four quantum selection rules normally apply:

1. $\Delta L = -1, 0, +1$, except that $L = 0$ cannot transition to $L = 0$.
2. The summation of l over all electrons must change between even and odd values. Even-to-even and odd-to-odd transitions are forbidden.
3. $\Delta J = -1, 0, +1$, except that $J = 0$ cannot transition to $J = 0$.
4. $\Delta S = 0$.

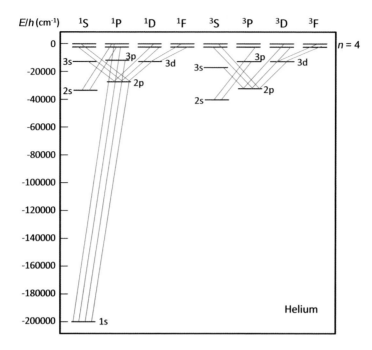

Figure 3.9 Energy-band diagram of the helium atom.

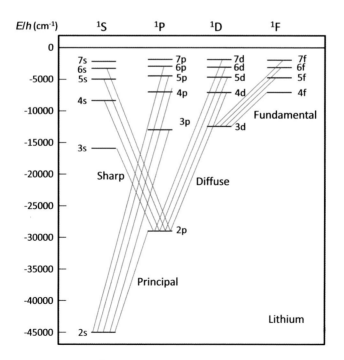

Figure 3.10 Energy-band diagram of the lithium atom.

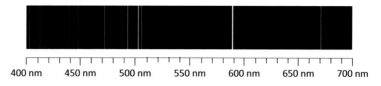

Figure 3.11 Emission spectra for the helium atom (Jircitano, 2011).

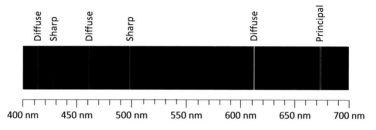

Figure 3.12 Emission spectra for the lithium atom (Jircitano, 2011).

Figures 3.13 and 3.14 illustrate emission spectra for nitrogen and oxygen atoms, each of which contain numerous visible emission lines; however, the spectra of nitrogen and oxygen molecules, N_2 and O_2, respectively, are more important from a remote sensing perspective and are discussed later.

3.4 Rotational Spectroscopy of Molecules

Attention is now turned to the situation of a rigid molecule that rotates in the presence of electromagnetic radiation. From the classical picture, such motion is expected to be considerably lower in energy and correspond to very low-frequency or long-wavelength spectral characteristics. Indeed, purely rotational molecular features are generally observed only at radio

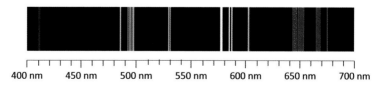

Figure 3.13 Emission spectra for the nitrogen atom (Jircitano, 2011).

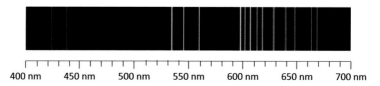

Figure 3.14 Emission spectra for the oxygen atom (Jircitano, 2011).

frequency wavelengths; however, they are observed in conjunction with vibrational characteristics in the infrared spectral region. The treatment in this section, therefore, is a stepping stone to the subject of vibrational molecular spectroscopy to follow.

Consider a rigid molecule for which there is a nonzero dipole moment μ. The axes of the molecule are defined such that the origin lies at the center of gravity of the molecule. In general, there are three rotational degrees of freedom, and for each of them the molecule exhibits a moment of inertia that can be computed as

$$I = \sum_i m_i r_i^2, \tag{3.51}$$

where m_i and r_i are the mass and distance, respectively, of each atom from each respective rotational axis. The moments of inertia about the three principal rotational axes are designated as $I_a, I_b,$ and I_c, where

$$I_c \geq I_b \geq I_a. \tag{3.52}$$

For a diatomic molecule with atoms of mass m_1 and m_2, such as the carbon monoxide (CO) molecule depicted in Fig. 3.15,

$$I_c = I_b = m_r r^2, \tag{3.53}$$

$I_a = 0, r$ is the internuclear distance, and m_r is the reduced mass:

$$m_r = \frac{m_1 m_2}{m_1 + m_2}. \tag{3.54}$$

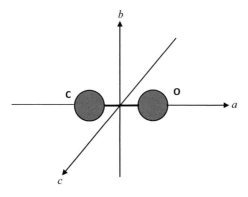

Figure 3.15 Principal rotational axes for the carbon monoxide molecule.

Solution of the Schrödinger wave equation for the rigid diatomic rotor produces energy levels given by

$$E = \frac{h^2}{8\pi^2 I} J(J+1),\qquad(3.55)$$

where J is a rotational quantum number with values of $0, 1, 2$, etc. Defining a rotational constant,

$$B = \frac{h^2}{8\pi^2 I}.\qquad(3.56)$$

Equation (3.55) simplifies to $E = BJ(J+1)$.

To absorb or emit radiation, the transition moment of the form shown in Eq. (3.14), composed of the wave functions of the rigid body rotor, must be nonzero. This occurs when the following two conditions are satisfied:

1. The molecule has a nonzero dipole moment μ in the orthogonal plane to a principal rotational axis with a nonzero moment of inertia.

2. $\Delta J = -1$ or $+1$, where ΔJ corresponds to the difference between upper and lower states of transition.

For purely rotational transitions, $\Delta J = -1$ has no physical meaning since it is always defined in terms of the upper state minus the lower state (even for emission); however, $\Delta J = -1$ becomes important for vibrational–rotational transitions.

For purely rotational transitions, the second quantum selection rule means that

$$\Delta E = B(J+1)(J+2) - BJ(J+1) = 2B(J+1),\qquad(3.57)$$

where J represents the rotational quantum number of the lower state. The rotational lines at $\sigma = \Delta E/hc$ are linearly spaced in wavenumbers, as shown in Fig. 3.16 for a CO molecule. This analysis, however, assumes that the rotating molecule is perfectly rigid. As the bond in real molecules exhibits some elasticity, the molecule tends to stretch as it rotates, an effect known as centrifugal distortion. The resultant effect is to perturb the energy levels so that they are no longer linearly spaced. These distorted energy levels can be modeled by the equation

$$E = BJ(J+1) - DJ^2(J+1)^2,\qquad(3.58)$$

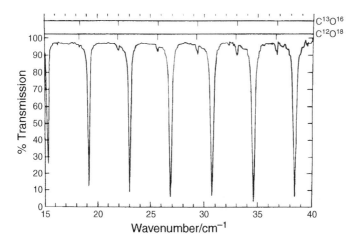

Figure 3.16 Rotational spectrum for the carbon monoxide molecule [reprinted with permission from Fleming and Chamberlain (1974) © 1974 Elsevier].

where the centrifugal distortion constant D depends on the stiffness of the bond according to

$$D = \frac{2B^2}{\omega_e^2 I},\qquad(3.59)$$

ω_e is the natural resonant frequency of the bond

$$\omega_e = \sqrt{\frac{K}{m_r}},\qquad(3.60)$$

and K is the bond strength as defined for the Lorentz oscillator model. As K increases, D becomes smaller and the rotational lines become more linearly spaced. As K decreases, D increases. The spacing between energy levels becomes larger as J increases.

For polyatomic molecules, rotation is possible in up to three directions, and the moments of inertia can be different in each direction. In such cases, three rotational constants A, B, and C are defined for the a, b, and c principal axes of rotation, each of which has the same form as Eq. (3.56) with the corresponding moments of inertia I_a, I_b, and I_c in the denominator. Rotational quantum numbers are defined for each axis under the constraint that the quantum numbers for the a and b axes must not exceed those of the b and c axes, respectively. Symmetry often requires that transitions are only allowed for $\Delta K = 0$, where K is the quantum number for the a or b axis, and the transition frequencies are unchanged from those based on the moment of inertia about the c axis alone.

3.5 Vibrational Spectroscopy of Molecules

To investigate the vibrational spectroscopy of molecules, we assume that each atom in the molecule is a rigid body and model the mechanics of the system of atoms. We first consider the simple case of a diatomic molecule, and then extend the treatment to more-complex polyatomic molecules. The possibility of combined vibrational–rotational transitions is incorporated through application of the Born–Oppenheimer approximation.

3.5.1 Diatomic molecules

Consider the case of a diatomic molecule composed of two atoms of mass m_1 and m_2 at an equilibrium distance r_e. The wave functions of the stationary states for this system can be determined by solving the time-independent Schrödinger wave equation in spherical coordinates based on the geometry shown in Fig. 3.17 and the harmonic oscillator approximation of the molecular bond. Under the harmonic oscillator approximation, the potential function is given by

$$V(r, \theta, \phi) = \frac{1}{2} K (r - r_e)^2, \qquad (3.61)$$

where K measures the bond strength and r_e is the internuclear distance at equilibrium. The coordinate system used is fixed relative to the first molecule such that the wave equation is again solved using the relative mass m_r defined in Eq. (3.54).

The harmonic oscillator wave function solutions take the form

$$\psi(r, \theta, \phi) = R(r)\,\Theta(\theta)\,\Phi(\phi), \qquad (3.62)$$

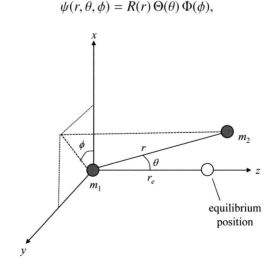

Figure 3.17 Diatomic-molecule coordinate geometry.

where

$$R(\alpha) = \sqrt{\frac{1}{2^v v! \pi^{1/2}}} \, H_v(\alpha) \, e^{-\alpha^2/2}, \tag{3.63}$$

$$\Theta(\theta) = L_J(\cos\theta), \tag{3.64}$$

$$\Phi(\phi) = \frac{1}{\sqrt{2\pi}} e^{im\phi}, \tag{3.65}$$

and

$$\alpha = \sqrt{\frac{2\pi m_r \omega_e}{h}} (r - r_e). \tag{3.66}$$

In Eqs. (3.63) to (3.66), H_v are Hermite polynomial functions, L_J are Legendre polynomial functions, and ω_e is the resonance frequency defined in Eq. (3.60). The solutions are indexed by quantum numbers v, J, and m that are constrained to

$$v = 0, 1, 2, \ldots, \tag{3.67}$$

$$J = 0, 1, 2, \ldots, \tag{3.68}$$

and

$$m = 0, \pm 1, \ldots, \pm J. \tag{3.69}$$

The quantum number v relates to the vibrational state, while the quantum numbers J and m relate to the rotational state. Table 3.4 provides equations for the low-order Hermite polynomials, and Fig. 3.18 illustrates the low-order wave functions in the radial direction. The energy levels of the diatomic molecule under the harmonic oscillator approximation are given by

$$E_{v,J} = \left(v + \frac{1}{2}\right)\hbar\omega_e + BJ(J+1) - DJ^2(J+1)^2, \tag{3.70}$$

where B and D are defined in Eqs. (3.56) and (3.59). The first term in Eq. (3.70) corresponds to vibrational states, and the latter two correspond to the rotational states discussed in the previous section.

The harmonic oscillator approximation in Eq. (3.61) is inadequate to accurately model the energy levels in a diatomic molecule. Instead, the potential function corresponding to such a bond is better modeled by the Morse potential depicted in Fig. 3.19 and described mathematically by

$$V(r, \theta, \phi) = D_e[1 - e^{-\beta(r-r_e)}]^2, \tag{3.71}$$

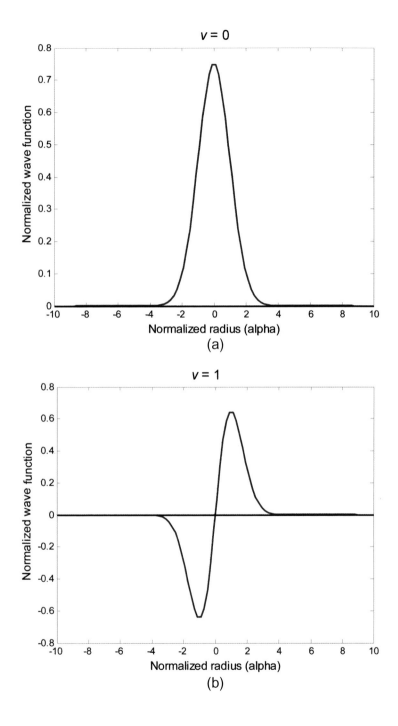

Figure 3.18 Low-order radial wave functions of the harmonic oscillator: (a) $v = 0$, (b) $v = 1$, (c) $v = 2$, (d) $v = 3$, (e) $v = 4$, and (f) $v = 5$.

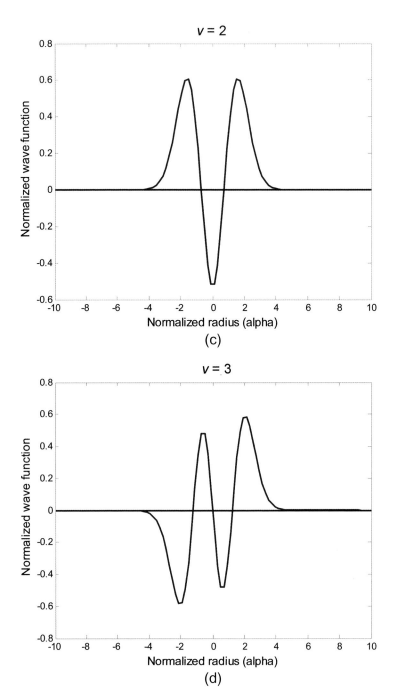

Figure 3.18   (*continued*)

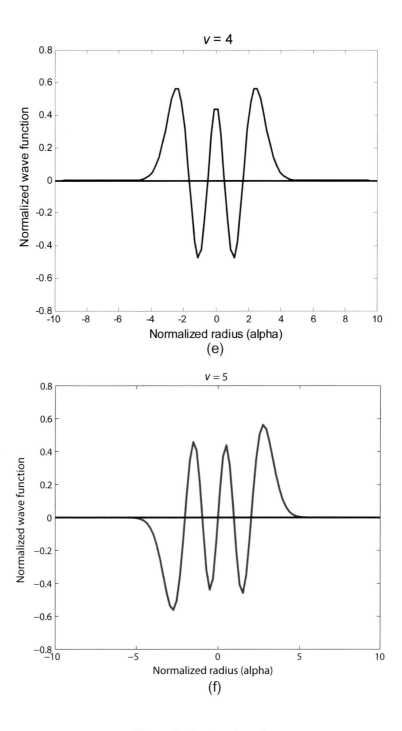

Figure 3.18 (*continued*)

Table 3.4 Functional form of low-order Hermite polynomials.

v	$H_v(\alpha)$
0	1
1	2α
2	$4\alpha^2 - 1$
3	$8\alpha^3 - 12\alpha$
4	$16\alpha^4 - 48\alpha^2 + 12$
5	$32\alpha^5 - 160\alpha^3 + 120\alpha$

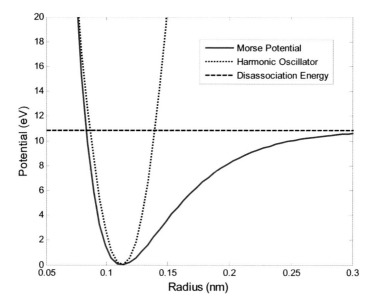

Figure 3.19 Morse potential well for CO compared to the harmonic-oscillator approximation.

where β represents the bond strength and D_e represents the disassociation energy required to break the bond and separate the two molecules. The wave functions for the stationary states of this potential function are considerably more complicated than those for a harmonic oscillator, but the energy levels can be given approximately by

$$E_{v,J} = \left(v + \frac{1}{2}\right)\hbar\omega_e - \left(v + \frac{1}{2}\right)^2 \hbar\omega_e\chi_e + BJ(J + 1)$$

$$-DJ^2(J + 1)^2 - \alpha_e\left(v + \frac{1}{2}\right)J(J + 1), \qquad (3.72)$$

where

$$\omega_e = \beta \sqrt{2D_e/m_r}, \tag{3.73}$$

$$\chi_e = \hbar\omega_e/4D_e, \tag{3.74}$$

$$\alpha_e = \frac{3\hbar^3 \omega_e}{16\pi^2 m_r r_e^2 D_e} \left(\frac{1}{\beta r_e} - \frac{1}{\beta^2 r_e^2} \right), \tag{3.75}$$

and B and D are again given by Eqs. (3.55) and (3.58).

Considering the energy levels in Eq. (3.72), electromagnetic radiation can be absorbed when the radiation frequency is such that

$$\hbar\omega = \frac{hc}{\lambda} = E_{v+\Delta v, J+\Delta J} - E_{v,J} \tag{3.76}$$

for allowed transitions between specified states. For a diatomic molecule, the transition moment is given by

$$M_{v,v+\Delta v} = \iiint \psi_v^*(r, \theta, \phi) \, \boldsymbol{\mu} \, \psi_{v+\Delta v}(r, \theta, \phi) \, r^2 \sin\theta \, dr \, d\theta \, d\phi. \tag{3.77}$$

The dipole moment can be approximated by a Taylor series expansion about the equilibrium point as

$$\boldsymbol{\mu} = \boldsymbol{\mu}_e + \left[\frac{\partial\boldsymbol{\mu}}{\partial(r - r_e)} \right]_e (r - r_e) + \left[\frac{\partial^2\boldsymbol{\mu}}{\partial(r - r_e)^2} \right]_e (r - r_e)^2 + \cdots, \tag{3.78}$$

where the subscript e designates evaluation of the term at the equilibrium position. Substituting Eq. (3.78) into Eq. (3.77), the first term becomes zero due to the orthogonality of the wave functions. If quadratic and higher terms are ignored, the transition moment is approximated as

$$\mathbf{M}_{v,v+\Delta v} = \left[\frac{\partial\boldsymbol{\mu}}{\partial(r - r_e)} \right]_e \iiint \psi_v^*(r, \theta, \phi) \, (r - r_e)$$
$$\times \psi_{v+\Delta v}(r, \theta, \phi) \, r^2 \sin\theta \, dr \, d\theta \, d\phi. \tag{3.79}$$

This is nonzero if the molecule has a nonzero dipole moment and

$$\Delta v = \pm 1. \tag{3.80}$$

An additional selection rule,

$$\Delta J = \pm 1, \tag{3.81}$$

comes from the rotational analysis of the prior section. The situation where $\Delta v = -1$ has no physical meaning when considering only molecular transitions but comes into play in conjunction with electronic transitions. In Eq. (3.81), $\Delta J = +1$ is referred to as the R branch, and $\Delta J = -1$ is referred to as the P branch. Each continues a number of transitions at approximately linearly spaced transition wavenumbers, increasing in wavenumber from the vibrational transition wavenumber in the case of the R branch and decreasing in the case of the P branch. This is illustrated in Fig. 3.20. Depending on symmetry, polyatomic molecules can also exhibit a Q branch, corresponding to allowed transitions with $\Delta J = 0$. Note that while vibrational transitions can occur in homonuclear diatomic molecules such as O_2, they do not result in electromagnetic absorption or radiation because the dipole moment is zero.

An important additional factor that must be considered in determining the absorption or emission of electromagnetic energy is the distribution of energy of molecules composing a material. That is, the probability of transition between two states and the strength-corresponding absorption for the frequency specified in Eq. (3.34) is proportional to the number of molecules that occupy the initial state of the transition. The state distribution is governed by the Maxwell–Boltzmann distribution:

$$\frac{N_j}{N_i} = e^{-(E_j - E_i)/kT}, \tag{3.82}$$

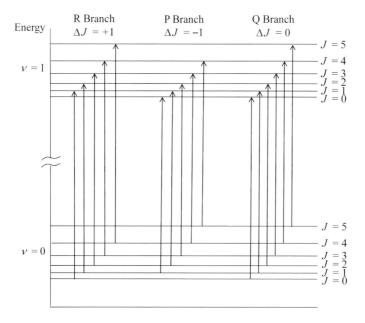

Figure 3.20 Vibrational–rotational energy-band diagram illustrating P, R, and Q branches.

Table 3.5 Characteristics of
the carbon monoxide molecule.

Characteristic	Value
m_1	12 amu
m_2	16 amu
m_r	6.86 amu
K	1902 N/m
$\hbar\omega_e$	0.2692 eV
χ_e	6.203×10^{-3}
B	0.2397 meV
α_e	2.169 μeV
D_e	10.86 eV
β	23.37 nm^{-1}
r_e	0.1128 nm

where N_i and N_j represent the molecular number densities of two respective energy states, k is Boltzmann's constant ($k = 1.38 \times 10^{-23}$ J/K), and T is the absolute temperature. For the vibrational states of a diatomic molecule,

$$\frac{N_v}{N_0} = e^{-v\hbar\omega_e/kT}. \tag{3.83}$$

At typical ambient temperatures, molecules are predominately in the ground state $v = 0$. For the rotational states of a diatomic molecule,

$$\frac{N_J}{N_0} = (2J + 1)\,e^{-BJ(J+1)/kT}. \tag{3.84}$$

The factor preceding the exponential accounts for the degeneracy of the rotational states according to Eq. (3.69). Because of this, distribution does not decrease monotonically as in Eq. (3.83) but peaks for some nonzero value of J.

This analysis can be directly applied to the case of a CO molecule to assess the nature of its absorption spectrum. The characteristics of the CO molecule are given in Table 3.5, where 1 eV equals 1.602×10^{-19} J, and one atomic mass unit (1 amu) equals 1.66×10^{-27} kg. The resulting absorption spectrum for the vibrational transition between the ground state and the first excited state is depicted in Figs. 3.21 and 3.22 for two different pressure–temperature conditions. As the temperature increases, the peaks of the R and P branches shift away from the wavelength of the Q branch (that does not exist), as predicted by Eq. (3.84). The change in pressure causes the individual spectral lines to broaden to the point where they begin to merge.

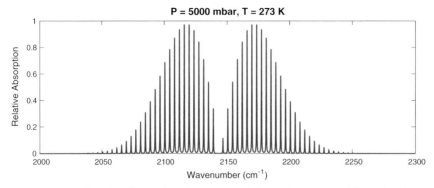

Figure 3.21 Vibrational–rotational structure of the carbon monoxide molecule at $P = 5000$ mbar and $T = 273$ K.

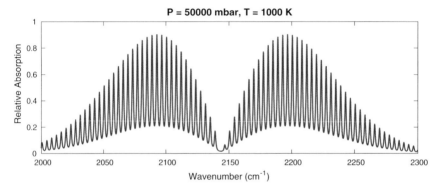

Figure 3.22 Vibrational–rotational structure of the carbon monoxide molecule at $P = 50000$ mbar and $T = 1000$ K.

3.5.2 Polyatomic molecules

Spectroscopy of polyatomic molecules is significantly more complex than spectroscopy of diatomic molecules, and a full theoretical or even quantitative analysis is generally impractical. However, their spectroscopic nature can be understood through a combination of empirical measurements and analysis of the symmetry properties of the molecule. In general, a molecule composed of n atoms will exhibit $3n - 6$ modes of vibrational motion, each of which can result in a resonance similar to that detailed for the diatomic molecule, including rotational modes. For a linear molecule, $3n - 5$ modes exist.

Molecular vibration is characterized in terms of the normal modes of motion as opposed to the motion of each individual molecule or pair of molecules. Figures 3.23 and 3.24 illustrate the normal modes for water (H_2O) and carbon dioxide (CO_2). In the H_2O case, three normal modes exist: a symmetric stretching mode ($\sigma_1 = 3657$ cm^{-1}), a symmetric bending mode ($\sigma_2 = 1595$ cm^{-1}), and an asymmetric stretching mode

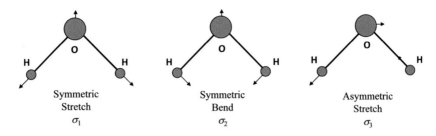

Figure 3.23   Normal vibrational modes of the water molecule.

(σ_3 = 3756 cm^{-1}). A similar molecule such as sulfur dioxide (SO_2) exhibits similar spectroscopic characteristics but at different resonance frequencies due to different bond strengths (σ_1 = 1151 cm^{-1}, σ_2 = 519 cm^{-1}, and σ_3 = 1361 cm^{-1}). Note the higher frequency and energy of the stretching modes relative to the bending modes. Due to the linear nature of the CO_2 molecule, four modes exist: a symmetric stretching mode (σ_1 = 1337 cm^{-1}), two degenerate bending modes with orthogonal polarization (σ_2 = 667 cm^{-1}), and an asymmetric stretching mode (σ_3 = 2349 cm^{-1}).

The given frequencies correspond to the transition frequency from the ground state (v = 0) to the first excited state (v = 1) for each particular vibrational mode, referred to as the *fundamental frequency*. Transitions between rotational modes can also occur simultaneously, so each fundamental absorption feature can exhibit the P and R branch characteristics depicted in Fig. 3.20. Also, transitions can occur between the ground state and higher-level excited states (e.g., v = 2). These transitions occur nominally at multiples of the fundamental frequency and are called *overtones*. Finally, multiple vibrations can be simultaneously excited; for example, simultaneous excitation from the ground state into the first excited states of modes 1 and 2 would result in a resonance at σ = $\sigma_1 + \sigma_2$. Such features only occur if they are allowed for a particular field polarization. The transition probability is governed by the distribution of states and the transition moment integral for the particular wave functions of the states involved. For complex molecules, the latter is determined empirically rather than quantitatively.

Figure 3.25 illustrates a measured absorption spectrum over the 2- to 14-μm spectral range for sulfur dioxide. Both asymmetric and symmetric stretch features are seen in the 7- to 10-μm range. The asymmetric stretch feature contains a Q-branch feature, while the symmetric stretch feature does not. The rotational features in the R and P branches are not resolved by the measurement instrument, so only the overall distribution function is apparent. Overtones for these vibrational resonances occur in the 3- to 5-μm range, of which only the stretch–stretch interaction at 4 μm is of adequate strength to see in this plot. Bending features for this molecule occur in the 20-μm range.

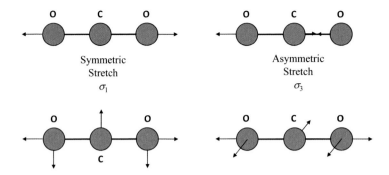

Figure 3.24   Normal vibrational modes of the carbon dioxide molecule.

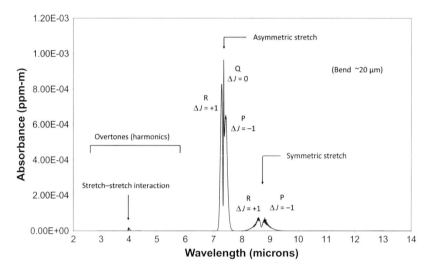

Figure 3.25   Measured absorption spectrum of sulfur dioxide.

The analysis of more-complex molecules can be approached from the perspectives of molecular symmetry and molecular decomposition. Molecular symmetry is useful in determining which transitions between the normal vibrational modes of a molecule are allowed and which are forbidden, as well as which modes are degenerate. In general, the spectroscopic characteristics of molecules become simpler as they possess greater symmetry. Also, molecules exhibiting similar symmetry exhibit similar spectral characteristics. Detailed spectroscopic analysis is often performed using the tools of chemical group theory, a subject that is not detailed here.

Molecular decomposition is another technique often employed for spectroscopic analysis. Although the number of normal modes of large molecules can be quite extensive, the vibrational spectra often possess features associated with particular bonds in the molecule whose vibration

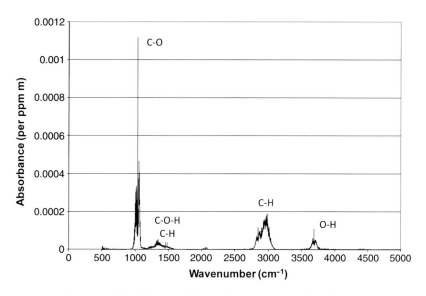

Figure 3.26 Measured absorption spectrum of methanol.

characteristics are localized. This often occurs in terminal groups of large molecules. For example, consider the methanol molecule CH_3–O–H, whose absorption spectrum is provided in Fig. 3.26. Starting at the high-frequency end, the primary absorption features correspond to a strong O–H stretching mode at 3650 cm^{-1}, two methyl C–H stretching modes at 2900 cm^{-1}, a broad C–O–H bending mode underlying sharper C–H stretching modes in the 1250- to 1500-cm^{-1} region, and a C–O stretching mode at 1030 cm^{-1}. Table 3.6 provides some typical stretching and bending group vibration modes (Hollas, 2004).

3.6 Electronic Spectroscopy of Molecules

To understand the electronic spectral features of molecules, it is necessary to model how the electronic wave functions are altered by the complex arrangement of nuclei and other electrons. One analytical approach that provides some qualitative insight into such behavior is the molecular orbital approach. This model constructs molecular orbitals that are constrained to be linear combinations of the atomic orbitals of the atoms from which the molecule is formed. Therefore, the molecular orbitals are similar to the atomic orbitals in the region near each atom, but they smoothly transition in the region between the atoms. Not all atomic orbitals, however, are reasonable candidates for molecular orbitals. Instead, they must possess certain properties; specifically, they must be similar in energy level, exhibit significant overlap, and have the same symmetry properties with respect to the directions of molecular symmetry.

Table 3.6 Typical stretching and bending mode for molecular groups.

Stretch	ω (cm^{-1})	Stretch	ω (cm^{-1})	Bend	ω (cm^{-1})
$\equiv$C—H	3300	—C$\equiv$N	2100	$\equiv$C—H	700
$=$C$<^{\text{H}}$	3020	$=$C—F	1100	$=$C$<^{\text{H}}_{\text{H}}$	1100
O$=$C$<^{\text{H}}$	2800	$=$C—Cl	650	—C$=^{\text{H}}_{\text{H}}$	1000
$=$C—H	2960	$=$C—Br	560	$>$C$<^{\text{H}}_{\text{H}}$	1450
—C$\equiv$C—	2050	$=$Cl—I	500	C$\equiv$C—C	300
$>$C$=$C$<$	1650	—O—H	3600		
$=$C—C$=$	900	$>$N—H	3350		
$=$Si—Si$=$	430	$=$P$=$O	1295		
$>$C$=$O	1700	$>$S$=$O	1310		

For example, consider the simple case of a symmetric diatomic molecule such as O_2 or N_2. Choosing the same quantum index n for the atomic orbitals of both atoms certainly satisfies the similar energy condition. The 1s atomic orbitals are a poor choice because they are confined near the nucleus and exhibit little overlap. The 2s orbitals are better candidates in this respect and also provide the right symmetry properties. If χ_1 and χ_2 represent atomic orbitals (i.e., wave functions) and E_A represents the atomic energy level, it can be shown that the molecular energy levels are given approximately by

$$E = E_A \pm \iiint \chi_1^*(x, y, z)\, H\, \chi_2(x, y, z)\, dx\, dy\, dz, \qquad (3.85)$$

where H is the Hamiltonian for the entire molecule. The molecular orbitals in this case are simply

$$\psi(x, y, z) = \frac{1}{\sqrt{2}}[\chi_1(x, y, z) \pm \chi_2(x, y, z)]. \qquad (3.86)$$

Molecular orbitals for symmetric diatomic molecules are depicted in Fig. 3.27 for various combinations of identical atomic orbitals (Hollas, 2004). The designations σ, π, u, and g correspond to symmetry properties of the molecule.

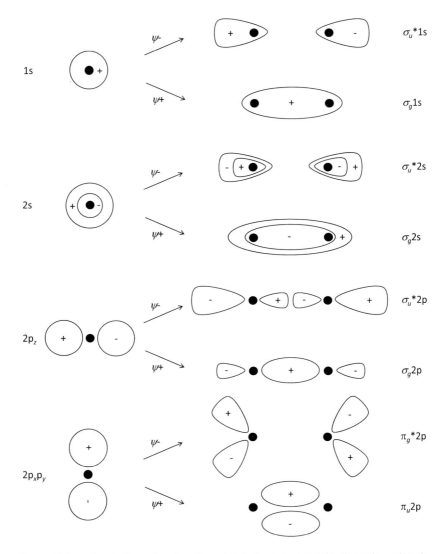

Figure 3.27 Illustration of molecular orbitals for a symmetric diatomic molecule.

Electronic states in a diatomic molecule are designated in a similar fashion as those in an atom, but with a new taxonomy for quantum numbers. Specifically, the quantum number Λ corresponds to the total angular momentum along the internuclear axis and can take on values $\Lambda = 0, 1, 2, 3$, etc. The quantum number Σ relates to the component of S along the internuclear axis and can take on integer values between $-S$ and S. The quantum number Ω represents the total (orbital plus spin) angular momentum along the internuclear axis and equals $|\Lambda + \Sigma|$. Electronic states are designated as $\Sigma, \Pi, \Delta, \Phi, \Gamma$, etc., corresponding to $\Lambda = 0, 1, 2, 3, 4$, etc., with the designator pre-superscripted by $2S + 1$ and subscripted by Ω. For

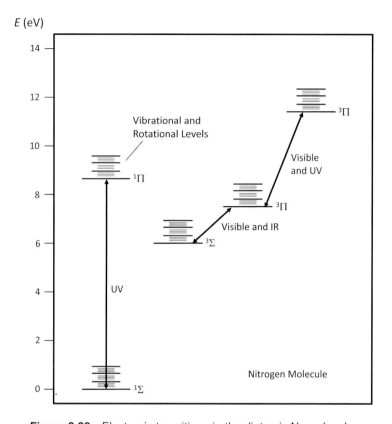

Figure 3.28 Electronic transitions in the diatomic N_2 molecule.

example, a Π molecular orbital with $\Lambda = 1$ and $\Sigma = 1, 0, -1$ has three components: $^3\Pi_2, {}^3\Pi_1$, and $^3\Pi_0$. The electronic transitions in diatomic molecules usually correspond to the following selection rules (with some exceptions): $\Delta\Lambda = -1, 0, +1$; $\Delta S = 0$; and $\Delta\Sigma = 0$. Examples of allowed transitions include $^1\Sigma$ to $^1\Pi$, $^2\Pi$ to $^2\Delta$, and $^3\Pi$ to $^3\Sigma$. Figure 3.28 illustrates a band diagram showing allowed transitions for the nitrogen molecule N_2.

When an electron transitions into an excited molecular orbital, the transition can be accompanied by a change in the vibrational–rotational state of the molecule as well. This is indicated by the vibrational and rotational levels in the band diagram for N_2 in Fig. 3.24. Another physical picture is given in Fig. 3.29, where the electron transitions from one vibrational–rotational potential well to another. In electronic transitions in molecules, known as *vibronic transitions*, there are no restrictions on the values that Δv can take. Thus, all transitions between vibrational levels are allowed. However, the strength of the absorption or emission is still dependent on the transition moment integral, which favors certain vibrational-state changes. Under the Born–Oppenheimer approximation,

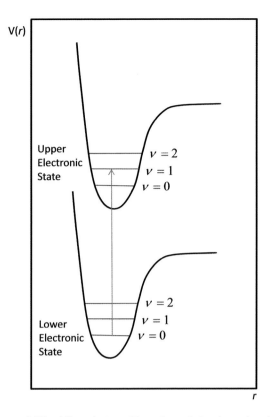

Figure 3.29 Vibronic transitions in a diatomic molecule.

the transition-moment integral can be factored into

$$\mathbf{M} = \mathbf{M}_e \iiint \psi_v^*(x, y, z)\, \psi_{v+\Delta v}(x, y, z)\, dx\, dy\, dz, \qquad (3.87)$$

where $\mathbf{M}_e$ is the electronic-transition-moment integral, ψ_v is the vibrational wave function for the lower state, and $\psi_{v+\Delta v}$ is the vibrational wave function for the upper state. The integral in Eq. (3.87) is known as the Franck–Condon factor and is dependent on the overlap of the vibrational wave functions. Since the potential well for the upper electronic state can be shifted relative to that of the lower well (as illustrated in Fig. 3.30), the transition-moment integral can peak for particular nonzero values of Δv. These correspond to the observed vibronic absorption and emission peaks of the molecule.

The basic principles of electronic spectroscopy of diatomic molecules extend to polyatomic molecules, but the molecular orbitals can become quite complex and result in numerous allowed transitions. At high vibrational energy, these transitions can become crowded within a small

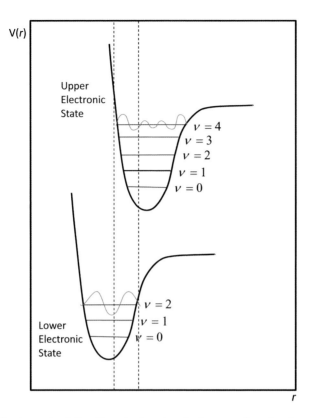

Figure 3.30 Impact of the Franck–Condon factor on vibronic absorption and emission peaks.

energy range to the point where the energy differences become smaller than the line widths. At this point, energy levels can form energy bands, and the absorption and emission spectra can become diffuse; that is, they exhibit broad regions of absorption and emission. This situation is nominally depicted in Fig. 3.31.

In crystalline semiconductors, where the atoms are arranged in a periodic lattice, electrons in the outermost orbital are influenced by the neighboring atoms when atomic separation is small. As in the case of molecular orbitals, this gives rise to overall wave functions of the lattice with energy levels that split relative to those of the individual atoms. To understand this effect, consider the triple quantum well in Fig. 3.1, where barrier width is varied. The energy levels of the first six states are shown in Fig. 3.32 as a function of barrier width. When the barrier width is very large, the wave functions for each well are non-overlapping, and the energy levels are degenerate. That is, there are three identical lower-energy states, one for each well, and three higher-energy states. As the barrier decreases in width, the wave functions begin to overlap (as depicted in Fig. 3.2) and

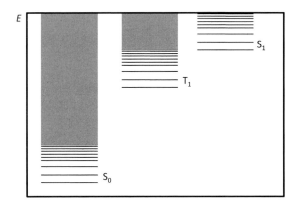

Figure 3.31 Diffuse spectra in polyatomic molecules.

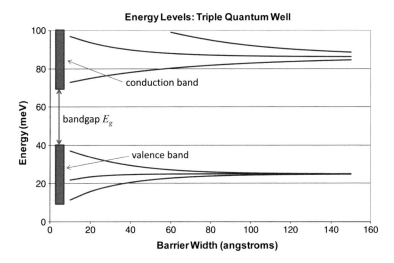

Figure 3.32 Energy level splitting into bands for a triple quantum well. For a large number of wells, the levels merge into valence and conduction bands as shown.

the energy levels split into bands. If the number of wells were increased to N, there would be N levels packed into the same energy range within each band. In a crystalline semiconductor, the shape of the potential wells is not square, but the periodicity of the lattice produces the same effect. The electrons are delocalized and described by Bloch wave functions of the form

$$\psi(\mathbf{r}) = u_k(\mathbf{r})e^{i\mathbf{k}\cdot\mathbf{r}}, \qquad (3.88)$$

where $u_k(\mathbf{r})$ has the periodicity of the lattice and $\mathbf{k}$ is the momentum vector of the electron. The valence band is defined as the highest filled band, and the conduction band is the next-highest band. Transitions between valence and conduction bands dominate the electro-optical behavior of semiconductors. One result of this is that their absorption spectra exhibit a cutoff wavelength λ_c beyond which absorption is zero because the incident photon has insufficient energy to cause a transition across the bandgap E_g from the top of the valence band to the bottom of the conduction band. This cutoff wavelength is defined by

$$\lambda_c = \frac{hc}{E_g} \tag{3.89}$$

and is a very important characteristic of photon detectors. An example is silicon, for which the cutoff wavelength is 1.1 μm. The implications of this characteristic of semiconductors are further examined when considering optical materials and optoelectronic devices such as detectors in Chapter 6.

3.7 Summary

While the quantum mechanical treatment of atomic and molecular spectroscopy might appear to bear little resemblance to the classical Lorentz oscillator model, it does predict some similar spectral features such as fine absorption peaks at specific resonant wavelengths. In the quantum mechanical picture, such wavelengths are associated with allowed energy-state transitions, with absorption strengths dictated by transition moments and state occupancy levels. The quantum mechanical treatment also predicts features such as the rotational structure of molecular absorption and various atomic features that defy classical mechanics. The former, in particular, are extremely important for understanding and characterizing the spectral behavior of gaseous media, such as atmospheric transmission and radiation spectra.

The Born–Oppenheimer approximation is a very useful construct for separating the complex radiative interactions into simpler atomic, molecular vibrational, and molecular rotational components. Nevertheless, analytical and even numerical modeling of anything beyond relatively simple diatomic and triatomic molecules is very difficult (if even possible) from first principles. The application of symmetry, group theory, and molecular decomposition helps in understanding the spectroscopy of more-complex molecules, and much work in this field is driven by precise laboratory measurements.

3.8 Further Reading

Much of the discussion of the spectroscopic principles covered in this chapter follows the more extensive treatment by Hollas (2004). For a more in-depth coverage of quantum mechanics, the reader can consult an undergraduate text such as Anderson (1971). To explore spectroscopic principles with more rigor from a chemical analysis standpoint, Harris and Bertolucci (1989) provides an interesting reference. For an extremely thorough treatment of the physics of spectroscopy, the reader is encouraged to consult Bransden and Joachain (2003). Further information about the various special mathematical functions used in this chapter can be found in Lebedev (1972).

References

Anderson, E. E., *Modern Physics and Quantum Mechanics*, Saunders, Philadelphia (1971).

Born, M. and Oppenheimer, R., "Zur Quantentheorie der Moleküle," *Ann. Phys. Chem.* **22**, 291–294 (1927).

Bransden, B. H. and Joachain, C. J., *Physics of Atoms and Molecules*, Second Edition, Prentice Hall, Harlow, UK (2003).

Einstein, A., "On the quantum theory of radiation," *Z. Phys.* **18**, p. 121 (1917).

Fleming, J. W. and Chamberlain, J., "High resolution far infrared Fourier transform spectrometry using Michelson interferometers with and without collimation," *Infrared Phys.* **14**, 277–292 (1974).

Harris, D. C. and Bertolucci, M. D., *Symmetry and Spectroscopy: An Introduction to Vibrational and Electronic Spectroscopy*, Dover, New York (1989).

Hollas, J. M., *Modern Spectroscopy*, Fourth Edition, Wiley, Hoboken, NJ (2004).

Jircitano, A. J., "Click on the symbol to see the atomic emission spectrum," http://chemistry.bd.psu.edu/jircitano/periodic4.html (last accessed Sep. 2011).

Lebedev, N. N., *Special Functions and Their Applications*, Dover, New York (1972).

Schrödinger, E., "An undulatory theory of the mechanics of atoms and molecules," *Phys. Rev.* **28**, 1049–1070 (1926).

Chapter 4
Spectral Properties of Materials

The previous two chapters dealt with the relationship between the *intrinsic* spectral properties of materials, such as the absorption coefficient and complex index of refraction as they relate to the chemical composition of the medium. Such properties are not directly measurable. Instead, what one is able to measure, either remotely or in a laboratory setting, are *apparent* spectral properties such as reflectance and transmission. Such properties are related not only to the chemical composition, but also the physical makeup of the material, including features such as physical dimensions, scatterer concentrations and size distributions, and surface roughness characteristics.

This chapter proceeds in two major sections. The first section is largely theoretical in nature, establishing the relationships between the intrinsic and apparent spectral properties of materials. This begins with the simple case of a single homogenous layer with smooth surfaces but continues to deal with scattering layers, multiple layers, rough surfaces, and surface emission. The second section provides a more empirical investigation of the measured properties of common remote sensing materials, such as atmospheric gases, liquid water, vegetation, minerals, soils, and manmade materials.

4.1 Apparent Spectral Properties

Apparent spectral properties provide the observable signature for hyperspectral remote sensors and are measurable quantities. These include spectral reflectance, transmittance, and emissivity. Spectral reflectance is defined here as the ratio of spectral irradiance reflected off a material to spectral irradiance incident on the material. Similarly, spectral transmittance is the ratio of spectral irradiance transmitted through the material to the incident irradiance. An inferred quantity that is related to these two properties through conservation of energy is the spectral absorbance, which is the ratio of irradiance absorbed in the material to the incident irradiance upon it. Spectral emissivity, which is the ratio of emitted spectral radiance to that of an ideal blackbody, is also related to absorbance. Note that the simple definitions given here do not account for the directionality of reflected and transmitted irradiance, an important issue when dealing with rough surfaces and scattering media. More-precise definitions are provided later in this chapter to account for this increased complexity.

4.1.1 Homogenous absorbing layer

Consider the simple case depicted in Fig. 4.1 of a single homogenous layer of an absorbing material with an index of refraction $n_2 + i\kappa_2$ with smooth, flat, parallel surfaces bounded by different absorbing media $n_1 + i\kappa_1$ and $n_3 + i\kappa_3$. It is common to have light incident from air where $n_1 = 1$ and $\kappa_1 = 0$, but we maintain generality in this respect. For the time being, however, light is assumed to propagate normal to the surfaces of the layer. The apparent reflectance $\rho(\lambda)$, absorbance $\alpha(\lambda)$, and transmittance $\tau(\lambda)$ are defined as the relative irradiance reflected, absorbed, and transmitted by the layer, as depicted in the figure. By conservation of energy, these apparent material properties are related by

$$\rho(\lambda) + \alpha(\lambda) + \tau(\lambda) = 1. \tag{4.1}$$

Apparent material properties are dependent on the intrinsic spectral properties and geometry of the material. Absorbance through a single pass

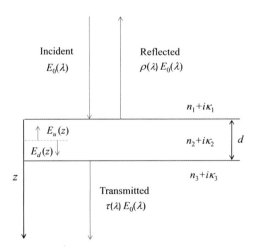

Figure 4.1 Geometry for determining apparent spectral properties of a homogenous absorbing layer.

is related to the absorption coefficient $\beta(\lambda)$ through Beer's law,

$$\alpha_{pass}(\lambda) = 1 - e^{-\beta_a d}, \tag{4.2}$$

where $\beta_a(\lambda)$ is the absorption coefficient and d is the material thickness. From Chapter 2, the absorption coefficient is related to the complex index of refraction according to

$$\beta_a = \frac{4\pi\kappa_2}{\lambda}. \tag{4.3}$$

For optically thin materials,

$$\alpha(\lambda) \approx \beta_a \, d. \tag{4.4}$$

Power reflectivity at the two surfaces is also dependent on the complex index of refraction according to the Fresnel equations at normal incidence,

$$R_{12} = R_{21} = \frac{(n_1 - n_2)^2 + (\kappa_1 - \kappa_2)^2}{(n_1 + n_2)^2 + (\kappa_1 + \kappa_2)^2} \tag{4.5}$$

and

$$R_{23} = R_{32} = \frac{(n_2 - n_3)^2 + (\kappa_2 - \kappa_3)^2}{(n_2 + n_3)^2 + (\kappa_2 + \kappa_3)^2}. \tag{4.6}$$

These are both extended from the dielectric–conductor equations given in Chapter 2 to accommodate conductor–conductor interfaces at normal incidence. Note that the extension to nonnormal propagation is not quite as straightforward. For each surface, power transmissivity is given by $1 - R$. Note that the terms reflectivity and transmissivity are associated with the reflection and transmission of a single plane wave of a specific direction according to the law of reflection and Snell's law, while the terms reflectance and transmission relate to the transfer of irradiance irrespective of direction. We begin the analysis, however, considering the simple case of on-axis incidence for a material with smooth surfaces, and then generalize from there to arbitrary-incidence angles and ultimately rough surfaces, for which radiation is scattered into a broad range of angles.

One might be tempted to assume that reflectance equals either R_{12} or a sum of R_{12} and R_{23}, and that absorbance equals the single-pass absorbance given by Eq. (4.2), with the transmission then computed from Eq. (4.1). But this is erroneous because it accounts for only first- (or first- and second-) surface reflectance and not for multiply-reflected energy inside the material that ultimately transmits out of the material in either the reflected or transmitted directions. A proper method of analysis is to propagate a ray into the material and sum all of the irradiance components that leak out of the material at both interfaces, as indicated in Fig. 4.2. These irradiance components account for front- and back-surface reflectance/transmittance, as well as Beer's-law absorption as the irradiance propagates back and forth in the material. This summation results in an infinite geometric series for both apparent reflectance and apparent transmittance. From the conservation-of-energy relationship in Eq. (4.1), effective absorbance can also be computed.

An alternate analytical approach employed throughout this chapter is the two-stream method based on coupled differential equations for the downwelling-irradiance $E_d(z)$ components and upwelling-irradiance $E_u(z)$ components inside the material, as depicted in Fig. 4.1. Solving these differential equations in concert with the surface boundary conditions provides the upwelling and downwelling irradiance as a function of z inside the material. The external irradiance is related to the internal irradiance through the surface reflectance. In this way the apparent spectral properties are derived: reflectance based on the external upwelling irradiance, and transmittance based on the external downwelling irradiance.

For the single homogenous layer depicted in Fig. 4.1, the two-stream differential equations are simply

$$\frac{dE_d(z)}{dz} = -\beta_a E_d(z) \qquad (4.7)$$

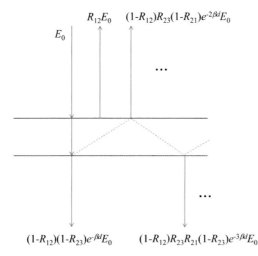

Figure 4.2   Light propagation in a homogenous material layer.

and

$$\frac{dE_u(z)}{dz} = +\beta_a E_u(z). \tag{4.8}$$

Note that the right-hand side of Eq. (4.8) is positive due to radiation propagating and decaying in the $-z$ direction. In this simple case, the equations are not actually coupled, and the solutions are simply

$$E_d(z) = Ae^{-\beta_a z} \tag{4.9}$$

and

$$E_u(z) = Be^{+\beta_a z}, \tag{4.10}$$

where A and B are constants of integration. The boundary conditions are

$$E_d(0) = (1 - R_{12})E_0 + R_{21}E_u(0) \tag{4.11}$$

and

$$E_u(d) = R_{23}E_d(d). \tag{4.12}$$

Substituting Eqs. (4.9) and (4.10) into Eq. (4.12), it is easily shown that

$$B = AR_{23}e^{-2\beta_a d}. \tag{4.13}$$

Substituting this result along with Eq. (4.10) into Eq. (4.11) yields the constants

$$A = \frac{(1 - R_{12})}{1 - R_{21}R_{23}e^{-2\beta_a d}}E_0 \tag{4.14}$$

and

$$B = \frac{(1 - R_{12})R_{23}e^{-2\beta_a d}}{1 - R_{21}R_{23}e^{-2\beta_a d}}E_0. \tag{4.15}$$

The apparent spectral properties of the material can now be solved based on the internal irradiance solutions. First, reflectance is given by

$$\rho(\lambda) = R_{12} + (1 - R_{21})\frac{E_u(0)}{E_0}. \tag{4.16}$$

From Eq. (4.10),

$$E_u(0) = B, \tag{4.17}$$

such that the apparent reflectance becomes

$$\rho(\lambda) = \frac{R_{12} + (1 - R_{12} - R_{21})R_{23}e^{-2\beta_a d}}{1 - R_{21}R_{23}e^{-2\beta_a d}}. \tag{4.18}$$

Similarly, transmittance is given by

$$\tau(\lambda) = (1 - R_{23})\frac{E_d(d)}{E_0}. \tag{4.19}$$

Since

$$E_d(d) = Ae^{\beta_a d}, \tag{4.20}$$

according to Eq. (4.9), the transmittance becomes

$$\tau(\lambda) = \frac{(1 - R_{12})(1 - R_{23})\,e^{-\beta_a d}}{1 - R_{21}R_{23}e^{-2\beta_a d}}. \tag{4.21}$$

From Eq. (4.1), the effective absorbance can be derived from Eqs. (4.18) and (4.21) as

$$\alpha(\lambda) = \frac{1 - R_{12} - (1 - R_{12})(1 - R_{23})\,e^{-\beta_a d} - (1 - R_{12})R_{23}\,e^{-2\beta_a d}}{1 - R_{21}R_{23}e^{-2\beta_a d}}. \tag{4.22}$$

4.1.2 Opaque scattering layer

Now consider the same situation depicted in Fig. 4.1, but with a scattering medium incorporated in the middle layer. In this case, the layer is characterized by scattering coefficient β_s that accounts for energy scattered from the downwelling irradiance into the upwelling direction (i.e., backscattered hemisphere), as defined in Chapter 2. It is assumed that this same scattering coefficient applies for upwelling-to-downwelling scattering. We again consider normal incidence but now include the total upwelling-and-downwelling irradiance, irrespective of exact direction, when computing reflectance and transmittance. Under these assumptions, the two-stream model results in the coupled differential equations,

$$\frac{dE_d(z)}{dz} = -(\beta_a + \beta_s)E_d(z) + \beta_s E_u(z) \tag{4.23}$$

and

$$\frac{dE_u(z)}{dz} = (\beta_a + \beta_s)E_u(z) - \beta_s E_d(z). \tag{4.24}$$

These are sometimes known as the Kubelka–Munk equations (Kubelka and Munk, 1931). By solving Eq. (4.23) for $E_u(z)$, taking the derivative with respect to z, and substituting both into Eq. (4.24), the coupled equations can be reduced to a second-order differential equation for the downwelling irradiance $E_d(z)$ of the form

$$\frac{d^2 E_d(z)}{dz^2} - \beta_a(\beta_a + 2\beta_s)E_d(z) = 0, \tag{4.25}$$

which has the general solution

$$E_d(z) = Ae^{-\beta_e z} + Be^{+\beta_e z}, \tag{4.26}$$

where an effective attenuation coefficient β_e is defined as

$$\beta_e = \sqrt{\beta_a(\beta_a + 2\beta_s)}. \tag{4.27}$$

The upwelling irradiance can then be solved from Eq. (4.23) as

$$E_u(z) = \left(\frac{\beta_a + \beta_s - \beta_e}{\beta_s}\right)Ae^{-\beta_e z} + \left(\frac{\beta_a + \beta_s + \beta_e}{\beta_s}\right)Be^{+\beta_e z}. \tag{4.28}$$

The constants of integration A and B are derived based on the boundary conditions, and the apparent spectral properties are then determined by Eqs. (4.16) and (4.19). A limiting case is when the combination of material absorption and thickness is large enough such that the downwelling irradiance fully decays before reaching the lower boundary; that is, the material layer is opaque. Practical examples include thick paint layers and deep bodies of water. Under this opacity approximation, the boundary conditions are

$$E_d(0) = (1 - R_{12})E_0 + R_{21}E_u(0) \tag{4.29}$$

and

$$E_d(d) = E_u(d) = 0. \tag{4.30}$$

It is clear by the opacity assumption that B and hence $\tau(\lambda)$ are both zero. A *volume reflectance* R_{vol} is defined to characterize the fractional backscattering of the downwelling irradiance from the volume of material below the first interface. When the material thickness is effectively infinite, as is assumed in this case, the volume reflectance is often called the *albedo*. This volume reflectance, given by

$$R_{vol} = \frac{E_u(0)}{E_d(0)} = \frac{\beta_a + \beta_s - \beta_e}{\beta_s} \tag{4.31}$$

and depicted in Fig. 4.3, is only dependent on the ratio of the absorption and scattering coefficients β_a/β_s. From Eq. (4.29), the constant A can be solved as

$$A = E_d(0) = \frac{1 - R_{12}}{1 - R_{21}R_{vol}}E_0 \tag{4.32}$$

and then substituted into Eq. (4.16) to solve for the apparent reflectance as

$$\rho(\lambda) = \frac{R_{12} + (1 - R_{12} - R_{21})R_{vol}}{1 - R_{21}R_{vol}}. \tag{4.33}$$

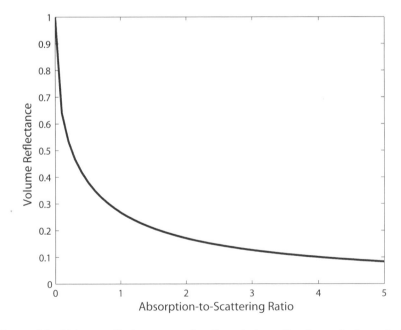

Figure 4.3 Volume reflectance as a function of absorption-to-scattering ratio for an opaque scattering layer.

It follows that the effective absorbance is

$$\alpha(\lambda) = \frac{1 - R_{12} - (1 - R_{12})R_{vol}}{1 - R_{21}R_{vol}}.$$ (4.34)

4.1.3 Transparent scattering layer

If the scattering layer is not sufficiently thick or absorptive enough to employ the opacity approximation, then the boundary conditions change to

$$E_d(0) = (1 - R_{12})E_0 + R_{12}E_u(0)$$ (4.35)

and

$$E_u(d) = R_{23}E_d(d).$$ (4.36)

Substituting Eqs. (4.26) and (4.22) into Eq. (4.36), an equation relating the constants A and B is derived:

$$B = -\left(\frac{\beta_s R_{23} - \beta_a - \beta_s + \beta_e}{\beta_s R_{23} - \beta_a - \beta_s - \beta_e}\right) e^{-2\beta_e d} A.$$ (4.37)

Defining R_{vol} again as the ratio of upwelling irradiance to downwelling irradiance at $z = 0$ as in Eq. (4.31), it follows that

$$R_{vol} = \left(\frac{\beta_a + \beta_s - \beta_e}{\beta_s}\right) \left[\frac{1 - \left(\frac{\beta_a + \beta_s + \beta_e}{\beta_a + \beta_s - \beta_e}\right)\left(\frac{\beta_s R_{23} - \beta_a - \beta_s + \beta_e}{\beta_s R_{23} - \beta_a - \beta_s - \beta_e}\right) e^{-2\beta_e d}}{1 - \left(\frac{\beta_s R_{23} - \beta_a - \beta_s + \beta_e}{\beta_s R_{23} - \beta_a - \beta_s - \beta_e}\right) e^{-2\beta_e d}}\right].$$ (4.38)

When d becomes infinite, Eq. (4.38) simplifies to Eq. (4.31) as expected.

Spectral reflectance is again given by Eq. (4.33), but this time using the more complicated volume reflectance in Eq. (4.38). The transmittance can be computed by substituting Eqs. (4.32) and (4.37) into Eq. (4.26), and then inserting this result into Eq. (4.19). The result is

$$\tau(\lambda) = \frac{(1 - R_{12})(1 - R_{23})}{1 - R_{21}R_{vol}} \left[1 - \left(\frac{\beta_s R_{23} - \beta_a - \beta_s + \beta_e}{\beta_s R_{23} - \beta_a - \beta_s - \beta_e}\right)\right] e^{-\beta_e d}.$$ (4.39)

As the thickness becomes infinite, transmittance goes to zero. The volume reflectance in Eq. (4.38) can again be expressed in terms of the absorption-to-scattering ratio β_a/β_s but is now also dependent on the normalized thickness $\beta_a d$. Figure 4.4 illustrates how the volume reflectance varies with these parameters.

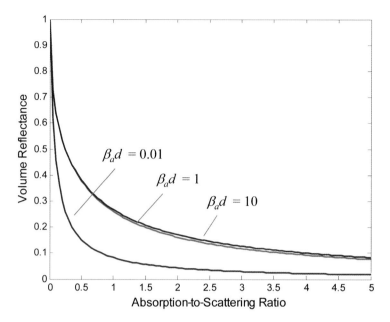

Figure 4.4   Volume reflectance as a function of absorption-to-scattering ratio and normalized thickness for a transparent scattering layer.

4.1.4 Multiple absorbing layers

The two-stream analysis approach can be extended to materials composed of multiple absorbing (but nonscattering) layers by solving the coupled differential equations within each layer and enforcing the appropriate boundary conditions between them. An example of this situation might be coated manmade or even natural materials. Consider an individual absorbing layer as depicted in Fig. 4.5, with the downwelling and upwelling irradiance defined directly above each layer and denoted with the superscript d or u. From the preceding analysis, these are related according to

$$E_k^d = (1 - R_k)\, e^{-\beta_k d}\, E_{k-1}^d + R_k\, e^{-2\beta_k d}\, E_k^u \qquad (4.40)$$

and

$$E_{k-1}^u = (1 - R_k)\, e^{-\beta_k d}\, E_k^u + R_k\, E_{k-1}^d. \qquad (4.41)$$

These can be arranged to compute the downwelling and upwelling radiance at the k'th interface based on those of the $(k-1)$'th, and written

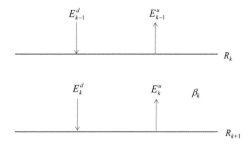

Figure 4.5 Definitions for analysis of multiple absorbing layers.

in transfer matrix form as

$$
\begin{bmatrix} E_k^d \\ E_k^u \end{bmatrix} = \begin{bmatrix} \dfrac{1 - 2R_k}{1 - R_k}e^{-\beta_k d} & \dfrac{R_k}{1 - R_k}e^{-\beta_k d} \\ \dfrac{-R_k}{1 - R_k}e^{\beta_k d} & \dfrac{1}{1 - R_k}e^{\beta_k d} \end{bmatrix} \begin{bmatrix} E_{k-1}^d \\ E_{k-1}^u \end{bmatrix}. \tag{4.42}
$$

Defining the transfer matrix as $\mathbf{M}_k$, Eq. (4.42) can be rewritten as

$$
\begin{bmatrix} E_k^d \\ E_k^u \end{bmatrix} = \mathbf{M}_k \begin{bmatrix} E_{k-1}^d \\ E_{k-1}^u \end{bmatrix}. \tag{4.43}
$$

If there are N layers, then an equation of the form of Eq. (4.43) can be written for each layer, and these equations can be combined into the aggregate relationship

$$
\begin{bmatrix} E_N^d \\ E_N^u \end{bmatrix} = \mathbf{M}_N \mathbf{M}_{N-1} \cdots \mathbf{M}_2 \mathbf{M}_1 \begin{bmatrix} E_0^d \\ E_0^u \end{bmatrix}. \tag{4.44}
$$

There is also a terminal boundary condition,

$$
E_N^u = R_{N+1} E_N^u, \tag{4.45}
$$

such that Eq. (4.44) can be solved for the total reflected irradiance E_0^u given a known incident irradiance E_0^d. The apparent reflectance of the composite material is then simply

$$
\rho(\lambda) = \frac{E_0^u}{E_0^d}. \tag{4.46}
$$

Similarly, E_N^d can be computed from Eq. (4.44) to provide the apparent transmittance

$$\tau(\lambda) = \frac{(1 - R_{N+1})E_N^d}{E_0^d}.$$

(4.47)

This transfer-matrix approach can be further extended to scattering layers by applying the analysis from the previous section to determine the transfer matrix of each layer, and then proceeding with the methodology described in this section. This is left as an exercise for the interested reader.

4.1.5 Multilayer dielectric thin films

A special (but important) case of multilayered media is the situation where the layers are dielectrics (i.e., nonabsorbing) and the interfaces are optically smooth and flat. Under such circumstances, it is insufficient to apply the two-stream irradiance model because the light coherently interferes. That is, the planar phase of the electric field is preserved and must be considered in the analysis. An example of this situation is thin-film-coated optics. This was ignored in the prior analysis because, in normal situations, the lack of optical quality of the interfaces destroys the interference, allowing an incoherent optical analysis. For thin-film coatings for which these assumptions are invalid, one must account for the phase of the electric field as it propagates between interfaces and superpose the scalar or vector electric field and not just irradiance.

A transfer-matrix approach can be applied for coherent thin-film analysis, but the matrices in this case relate to electric-field amplitude and phase as opposed to irradiance. For the single layer depicted in Fig. 4.6, amplitude reflectivity r and amplitude transmissivity t can be computed by

$$\begin{bmatrix} 1 \\ n_{k-1} \end{bmatrix} + \begin{bmatrix} 1 \\ -n_{k-1} \end{bmatrix} r = \mathbf{M}_k \begin{bmatrix} 1 \\ n_{k+1} \end{bmatrix} t,$$

(4.48)

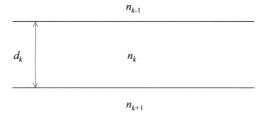

Figure 4.6 Definitions for analysis of single dielectric thin films.

where $\mathbf{M}_k$ is the layer-transfer matrix given by

$$
\mathbf{M}_k = \begin{bmatrix} \cos\left(\dfrac{2\pi d_k}{\lambda}\right) & -\dfrac{i}{n_k}\sin\left(\dfrac{2\pi d_k}{\lambda}\right) \\[2mm] -in_k\sin\left(\dfrac{2\pi d_k}{\lambda}\right) & \cos\left(\dfrac{2\pi d_k}{\lambda}\right) \end{bmatrix}.
\tag{4.49}
$$

If there are N layers, as shown in Fig. 4.7, then Eq. (4.48) still applies, but $\mathbf{M}$ is replaced by the composite transfer matrix

$$
\mathbf{M} = \mathbf{M}_1\mathbf{M}_2\cdots\mathbf{M}_{N-1}\mathbf{M}_N = \begin{bmatrix} A & B \\ C & D \end{bmatrix}.
\tag{4.50}
$$

Note that amplitude reflectivity r and transmissivity t are complex-valued quantities. The apparent reflectance and transmittance at normal incidence, in terms of irradiance, are related to the composite system matrix elements according to

$$
\rho(\lambda) = \left|\frac{A + Bn_{N+1} - C - Dn_{N+1}}{A + Bn_{N+1} + C + Dn_{N+1}}\right|^2
\tag{4.51}
$$

and

$$
\tau(\lambda) = \left|\frac{2}{A + Bn_{N+1} + C + Dn_{N+1}}\right|^2.
\tag{4.52}
$$

By properly optimizing the layer thicknesses and indices, it is possible to produce coatings with very specific spectral characteristics, including antireflection (AR) coatings, dielectric mirrors, edge and bandpass spectral filters, and narrow line filters (i.e., interference filters). Many have direct applicability to the design and fabrication of optical sensors in general.

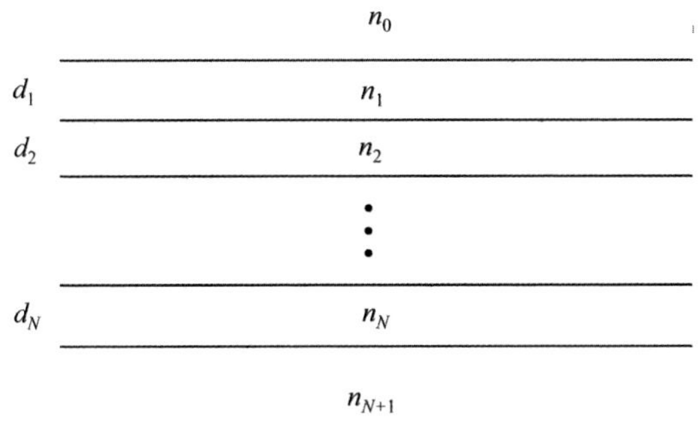

Figure 4.7 Definitions for analysis of multiple dielectric thin films.

Spectral filters based on thin-film dielectrics are particularly relevant to hyperspectral-imaging sensors. Further details are deferred to later chapters that specifically address these topics.

4.1.6 Rough-surface reflectance

The analysis to this point has addressed the apparent reflectance and transmittance of materials without regard to the directionality of the reflected and transmitted irradiance. In some cases, such as materials composed of smooth surfaces with no volume scattering, the direction of reflected and transmitted light from a directional source is also highly directional, with propagation directions conforming to the law of reflection and Snell's law. Such reflection is referred to as *specular*. When surfaces are very rough or there is significant volume scattering, reflected and transmitted light is scattered uniformly in all directions. This is referred to as *diffuse* reflectance and scattering. If scattering is perfectly uniform, it is known as Lambertian. In general cases, there can be a specular lobe of concentrated energy along with a more diffuse distribution of light over a broad angular range. Each of these situations is depicted in Fig. 4.8.

To account for directionality, it is necessary to be more particular in the definitions of reflectance and transmission when dealing with general cases involving rough surfaces and volume scattering. To do so, consider the situation depicted in Fig. 4.9, where such a material is illuminated from a direction (θ_i, ϕ_i), defined in spherical coordinates relative to the surface normal with a spectral irradiance $E(\theta_i, \phi_i, \lambda)$. The spectral radiance reflected off the surface in a direction (θ_r, ϕ_r) is given by $L(\theta_r, \phi_r, \lambda)$. The material property that relates the reflected radiance to the incident irradiance is called the *bidirectional reflectance distribution function*

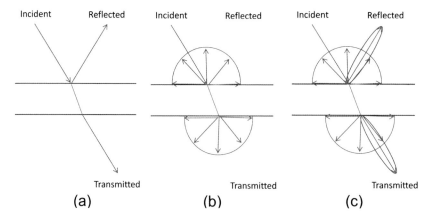

Figure 4.8 Range of possible reflectance and transmittance cases from rough and scattering materials: (a) specular case, (b) diffuse case, and (c) general case.

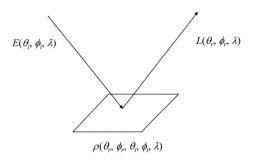

Figure 4.9 Definition of the BRDF.

(BRDF), defined as

$$\rho_{BRDF}(\theta_r, \phi_r, \theta_i, \phi_i, \lambda) = \frac{L(\theta_r, \phi_r, \lambda)}{E(\theta_i, \phi_i, \lambda)}.$$ (4.53)

Unlike the dimensionless reflectance terms defined so far, the BRDF has units of sr^{-1} since it relates radiance to irradiance. The total reflected irradiance into the hemisphere above the surface is determined by integrating the reflected radiance over the hemisphere, as described in Chapter 2. The *directional hemispherical reflectance* (DHR) is defined as the ratio of this total reflected irradiance to the incident irradiance from a particular direction, and is related to the BRDF by

$$\rho_{DHR}(\theta_i, \phi_i, \lambda) = \int_0^{\pi/2} \int_0^{2\pi} \rho_{BRDF}(\theta_r, \phi_r, \theta_i, \phi_i, \lambda) \cos \theta_r \sin \theta_r d\phi_r d\theta_r.$$ (4.54)

Similarly, we can define a *hemispherical directional reflectance* (HDR) that characterizes the reflected irradiance in a particular direction from perfectly diffuse illumination, which is related to the BRDF according to

$$\rho_{HDR}(\theta_r, \phi_r, \lambda) = \int_0^{\pi/2} \int_0^{2\pi} \rho_{BRDF}(\theta_r, \phi_r, \theta_i, \phi_i, \lambda) \cos \theta_i \sin \theta_i d\phi_i d\theta_i.$$ (4.55)

DHR and HDR are dimensionless quantities of a dual nature. Although DHR and HDR are strictly dependent on the incident or reflected angle, respectively, this dependence is often ignored. For a truly Lambertian surface, this angular dependence does not exist, and the DHR and HDR are equivalent.

The dependence of the BRDF with reflected direction (θ_r, ϕ_r) is characteristic of the roughness of the surface. This can be understood by considering the surface as a composition of microfacets, as depicted in Fig. 4.10. When a light ray strikes a facet, it is reflected according to the microfacet surface normal. As the roughness of the surface implies a distribution of facet normal directions, the law of reflection dictates that the reflected light from a particular region will be angularly distributed by twice the facet normal distribution. Part of the light also transmits through the surface and can exhibit volume scattering in the material; therefore, the distribution of the reflected energy can also be impacted by the scattering phase function of the volume scatterers.

For a perfectly smooth surface, the BRDF is a delta function in the direction specified by the law of reflection; that is, the reflectance is specular. One measure that has been used to characterize a smooth surface is

$$\Delta h < \frac{1}{8} \frac{\lambda}{\cos \theta_i}, \tag{4.56}$$

where Δh is a measure of the root-mean-square surface irregularity; however, there is not a standard definition for specularity. A diffuse or Lambertian surface is one for which the energy is reflected uniformly in all directions; i.e., the BRDF is constant with (θ_r, ϕ_r). In this case, the BRDF is related to the DHR according to

$$\rho_{BRDF}(\theta_r, \phi_r, \theta_i, \phi_i, \lambda) = \frac{\rho_{DHR}(\theta_i, \phi_i, \lambda)}{\pi}. \tag{4.57}$$

Another measure of surface roughness is the microfacet surface slope standard deviation σ_s, which is more directly pertinent to the shape of the BRDF.

A general material is neither perfectly specular nor perfectly diffuse, and the full BRDF must be used to characterize the reflected radiance. Measurement of the BRDF as a function of wavelength is extremely laborious. Often, the BRDF is estimated by measuring and fitting parameters to an underlying model such as those developed by Sandford

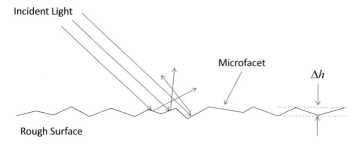

Figure 4.10   Microfacet model for rough surface reflection.

and Robertson (1985), Torrance and Sparrow (1967), and Maxwell *et al.* (1973). The particular model outlined here is a modification of the Beard–Maxwell model, which is based on microfacet surface depiction and is widely used in remote sensing and synthetic image generation applications. Consider the geometry shown in Fig. 4.11, where the surface is assumed to lay in the (x, y) plane. The BRDF is modeled as a linear combination of a specular component ρ_s, a diffuse surface component ρ_d, and a volumetric scattering component ρ_v, with the functional form

$$\rho_{BRDF}(\theta_r, \phi_r, \theta_i, \phi_i, \lambda) = \rho_s(\theta_r, \phi_r, \theta_i, \phi_i, \lambda) + \rho_d(\lambda)$$
$$+ \frac{2\rho_v(\lambda)}{\cos\theta_i + \cos\theta_r}, \tag{4.58}$$

where the angular dependence of the specular component is modeled by

$$\rho_s(\theta_r, \phi_r, \theta_i, \phi_i, \lambda) = B\frac{R\left[\theta_s, n(\lambda), \kappa(\lambda)\right]}{4\cos\theta_n\cos\theta_r\cos\theta_i}\frac{1}{\sigma_s^2 + \tan^2\theta_n}S(\theta_s, \theta_n). \tag{4.59}$$

In Eq. (4.59), B is a normalization constant, $R[\theta_s, n(\lambda), \kappa(\lambda)]$ is the Fresnel reflection for the surface boundary, σ_s is the surface slope standard deviation,

$$\theta_s = \frac{1}{2}\cos^{-1}[\cos\theta_i\cos\theta_r + \sin\theta_i\sin\theta_r\cos(\phi_r - \phi_i)] \tag{4.60}$$

is the effective specular reflectance angle, and

$$\theta_n = \cos^{-1}\left[\frac{\cos\theta_i + \cos\theta_r}{2\cos\theta_s}\right] \tag{4.61}$$

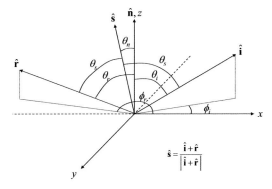

Figure 4.11 Geometry used for the Beard–Maxwell model of the BRDF.

is the angle between the surface normal and the effective specular normal. The function $S(\theta_s, \theta_n)$ takes into account the shadowing of facets by adjacent facets, and takes the form

$$S(\theta_s, \theta_n) = \frac{1 + \frac{\theta_n}{\Omega} e^{-2\theta_s/\tau}}{1 + \frac{\theta_n}{\Omega}}, \qquad (4.62)$$

where τ and Ω are fitting parameters.

To provide an example of how Beard–Maxwell parameters relate to the shape of the BRDF, polar diagrams depicting the modeled angular dependence in the (x, y) plane of the three BRDF components are shown in Fig. 4.12 for a 30-deg incident angle, one degree surface slope standard deviation, $\tau = 0.1$, and $\Omega = \pi$. A composite BDRF is also shown. In practice, the BRDF is characterized as follows. First, the DHR is directly measured for all wavelengths by placing the material in an integrating sphere, illuminating at a particular incident angle, and measuring the integrated hemispherical reflectance with a spectrometer. Second, the material is placed on a goniometer, and the BRDF is measured for a grid of incident and reflected angles in both θ and ϕ directions, covering the entire hemisphere using directional laser illumination. This limits the spectral BRDF measurements to a small number of laser lines, commonly $0.325, 0.6328, 1.06, 3.39,$ and 10.6 μm. Next, the Beard–Maxwell fit parameters $n(\lambda), \kappa(\lambda), \rho_d(\lambda), \rho_v(\lambda), \sigma_s, \tau,$ and Ω are all estimated based on the combination of DHR and BRDF measurements (Montanaro *et al.*, 2007). Finally, the complete spectral BRDF is fully defined according to Eqs. (4.58) through (4.62) by these fit parameters.

4.1.7 Emissivity and Kirchhoff's law

In the infrared, it is necessary to consider the thermal emission of electromagnetic energy in any radiometric analysis, in addition to reflected and transmitted light. The spectral radiance distribution of thermally emitted radiation is defined for a *blackbody*, an ideal radiating surface, as

$$B(\lambda, T) = \frac{2hc^2}{\lambda^5} \frac{1}{e^{hc/\lambda kT} - 1}, \qquad (4.63)$$

where k is Boltzmann's constant ($k = 1.38 \times 10^{-23}$ J/K) and T is the absolute surface temperature. The function in Eq. (4.63) is known as a blackbody radiance function, or a Planck function, and results from a quantum mechanical analysis of blackbody emission performed by Planck (1957). Figure 4.13 illustrates blackbody spectral radiance distributions for a few select temperatures. Note the change in the scale and the shift in

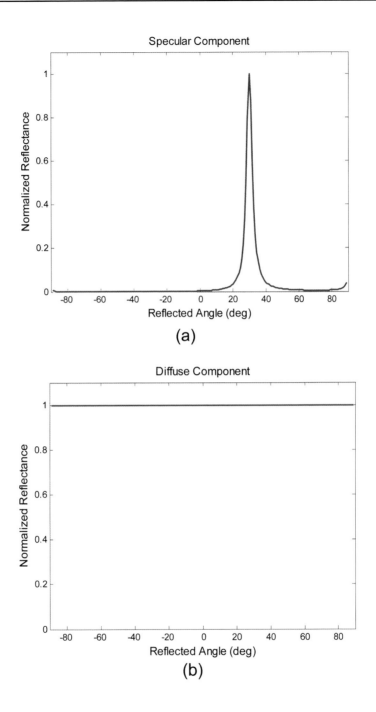

Figure 4.12 Beard–Maxwell model components and composite BRDF for $\theta_i = 45$ deg, $\sigma_s = 1$ deg, $\tau = 0.1$, and $\Omega = \pi$: (a) specular component, (b) diffuse component, (c) volumetric component, and (d) composite BRDF.

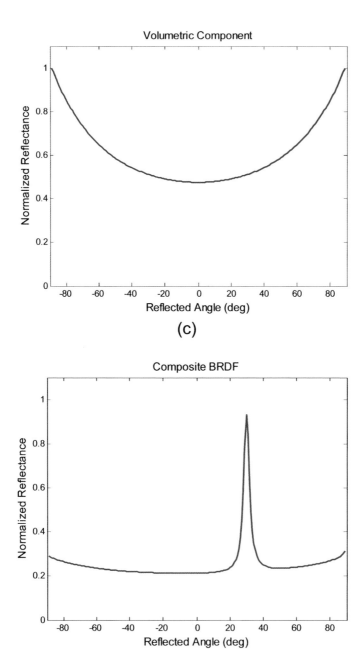

(c)

(d)

Figure 4.12   (*continued*)

peak wavelength as the temperature increases. The peak wavelength of the distribution is given by the Wien displacement law (Heald, 2003) as

$$\lambda_{peak} = \frac{2898\,\mu m\ K}{T}.$$

(4.64)

The integrated radiance over all wavelengths is given by the Stefan–Boltzmann law (Stefan, 1879; Boltzmann, 1884),

$$M = \sigma T^4.$$

(4.65)

where M is the total emitted irradiance, or exitance, and $\sigma = 5.67 \times 10^{-8}$ W/m^2 K^4 is the Stefan–Boltzmann constant. For typical ambient temperatures ($\sim$300 K), blackbody radiance is significant in the MWIR/LWIR spectral regions, but insignificant at shorter wavelengths. For high-temperature objects (e.g., fires, lighting, moon, sun, etc.), blackbody radiation must be considered at the shorter spectral regions. The 2800 and 5800 K temperatures shown correspond nominally to tungsten lamps and the sun, respectively.

The spectral radiance emitted from a material relative to that of a blackbody is a material property known as *spectral emissivity*, defined by

$$\varepsilon(\lambda) = \frac{L(\lambda, T)}{B(\lambda, T)}.$$

(4.66)

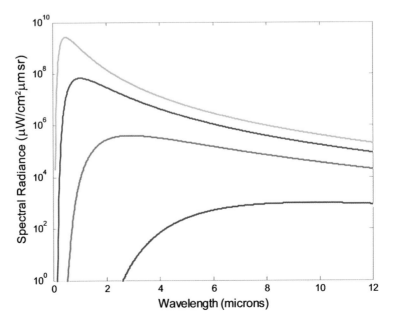

Figure 4.13 Blackbody spectral radiance distributions.

It directly follows that the emitted spectral radiance from a surface with emissivity $\varepsilon(\lambda)$ is

$$L(\lambda, T) = \frac{2hc^2}{\lambda^5} \frac{\varepsilon(\lambda)}{e^{hc/\lambda kT} - 1}. \tag{4.67}$$

The emissivity is related to the absorbance of a material. That is, for every allowed transition in a molecule that can lead to absorption of electromagnetic radiation, radiation can also be spontaneously emitted. Recalling that the transition rate for absorption is a function of the relative number density of the initial state, the same is true for emission. In this case, however, the initial and final states are switched.

When a material is in thermal equilibrium, the density of states is in steady state, and the relationship between emissivity and absorbance is given by Kirchhoff's law (Kirchhoff, 1860) as

$$\varepsilon(\lambda) = \alpha(\lambda). \tag{4.68}$$

While the thermal equilibrium condition is not strictly observed in general, it is a fair assumption for many materials and is frequently employed in remote sensing. For opaque objects [i.e., $\tau(\lambda) = 0$], Kirchhoff's law implies that

$$\varepsilon(\lambda) = 1 - \rho(\lambda). \tag{4.69}$$

Therefore, the spectral characteristics of opaque objects are uniquely defined by the reflectance characteristics under this common assumption. In fact, the emissivity exhibits directional characteristics that are similar to those of the BRDF. For optically thin materials [i.e., $\rho(\lambda) \sim 0$],

$$\varepsilon(\lambda) = 1 - \tau(\lambda). \tag{4.70}$$

This is a common assumption for clear gases and liquids that exhibit negligible backscatter.

4.2 Common Remote Sensing Materials

4.2.1 Atmospheric gases

Atmospheric gases have a substantial influence on all remote sensing applications. Therefore, an understanding of their spectral characteristics is of paramount importance to the field of hyperspectral remote sensing. The composition of the primary gases that make up the earth's atmosphere is summarized in Table 4.1. The concentrations in percent volume or ppm are nominal values since they are all dependent on altitude and various environmental impacts. The latter is particularly true with respect to water vapor, which is highly weather dependent.

Several gases (N_2, Ar, Ne, He, and H_2) have negligible effect on the spectral properties of the atmosphere due to the lack of molecular absorption, as discussed in Chapter 3, and therefore are not discussed further. As a symmetric diatomic molecule, oxygen does not have an electrical dipole moment from which to produce a vibrational response to absorb infrared radiation. It is the only simple diatomic molecule, however, that exhibits a magnetic moment due to two unpaired electrons in molecular orbitals that possess the same spin. Through this strong paramagnetism, the molecule exhibits rotational lines in the microwave region, a weak vibrational resonance near 6 μm, and electronic absorption lines in the ultraviolet, visible, and NIR parts of the spectrum (Bransden and Joachain, 2003). The visible and NIR lines occur at 688.4, 762.1, 1065, and 1269 nm. These are denoted in the modeled transmission spectrum shown in Fig. 4.14. Oxygen also absorbs ultraviolet radiation, which produces a disassociation of the O=O bond. The resulting oxygen atoms typically form ozone, again through ultraviolet absorption, the primary production mechanism of stratospheric ozone.

Water vapor has a substantial impact in defining available spectral regions for remote sensing where the atmosphere is inadequately transmissive. The water molecule exhibits three vibration modes: a symmetric stretching mode ($\sigma_1 = 3657$ cm^{-1}), a symmetric bending mode ($\sigma_2 = 1595$ cm^{-1}), and an asymmetric stretching mode ($\sigma_3 = 3756$ cm^{-1}). The infrared absorption spectrum shown in Fig. 4.15 illustrates the bending mode at 6.27 μm and the fundamental stretching modes at 2.67 and 2.74 μm. The fine structure of the P and R branches corresponding to the bending modes are marginally resolved in this measured spectrum. Due to molecular symmetry, no Q branch exists. Several overtones result

Table 4.1 Nominal concentration levels of atmospheric gases.

Atmospheric gas	Concentration
Nitrogen (N_2)	78.1%
Oxygen (O_2)	20.9%
Water vapor (H_2O)	0 to 2%
Argon (Ar)	0.9%
Carbon dioxide (CO_2)	351 ppm
Neon (Ne)	18 ppm
Helium (He)	5 ppm
Ozone (O_3)	0 to 7 ppm
Methane (CH_4)	1.6 ppm
Hydrogen (H_2)	0.5 ppm

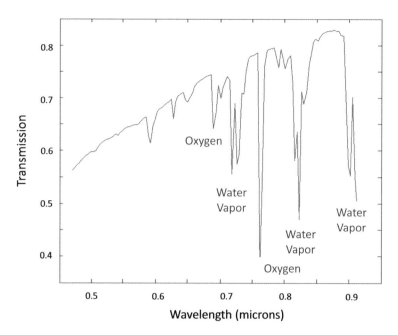

Figure 4.14 Modeled atmospheric transmission profile indicating oxygen absorption lines.

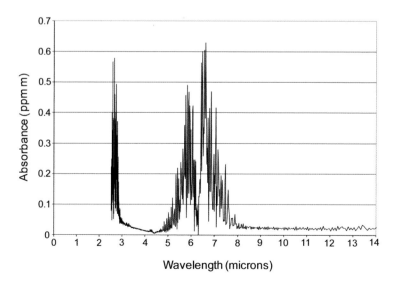

Figure 4.15   Measured absorption spectrum of water vapor (H_2O).

in absorption features in the visible through shortwave infrared spectral region. Of these, NIR and SWIR absorptions are most significant. These are centered at 1.89 μm $(a\sigma_1 + \sigma_2 + b\sigma_3; a + b = 1)$; 1.35 μm $(a\sigma_1 + b\sigma_3; a + b = 2)$; 1.11 μm $(a\sigma_1 + \sigma_2 + b\sigma_3; a + b = 2)$; 0.90 μm $(a\sigma_1 + b\sigma_3; a + b = 3)$; and 0.79 μm $(a\sigma_1 + \sigma_2 + b\sigma_3; a + b = 3)$.

The carbon dioxide molecule exhibits four vibrational modes: a symmetric stretching mode $(\sigma_1 = 1337 \text{ cm}^{-1})$, two degenerate bending modes $(\sigma_2 = 667 \text{ cm}^{-1})$, and an asymmetric stretching mode $(\sigma_3 = 2349 \text{ cm}^{-1})$. The infrared absorption spectrum in Fig. 4.16 illustrates the bending mode at 15 μm and the fundamental asymmetric stretch at 4.2 μm. There is no absorption due to the symmetric stretch because of molecular symmetry. In the log-scaled plot, overtones in the SWIR are also visible.

Ozone is formed when UV light breaks down a normal molecule of oxygen (O_2), and the single oxygen atoms combine with other disassociated O_2 molecules to form ozone. Ozone exists in significant quantities in the stratosphere from roughly 20- to 30-km altitude; however, ozone also exists in small quantities in the troposphere. Ozone is a strong absorber of UV radiation below 320 nm, a region known as the Hartley band. Absorption in this region results in the production of O_2 from ozone by a reverse reaction of that previously discussed. Ozone also absorbs visible light in the 602-nm range, a region known as the Chappuis band. The UV and visible absorption bands are responsible for an atmospheric absorption edge at roughly 320 nm, below which almost all sunlight is absorbed. The ozone molecule also exhibits a fundamental bending mode at 528 cm^{-1} (σ_2) and fundamental stretching modes at 1355 cm^{-1} (σ_1) and 1033 cm^{-1} (σ_3). These occur at 18.9, 7.39, and 9.67 μm, respectively.

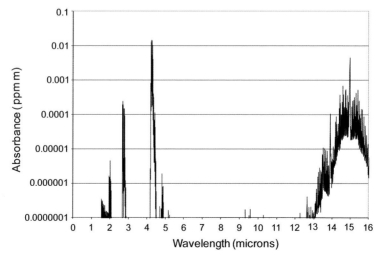

Figure 4.16 Measured absorption spectrum of carbon dioxide (CO_2).

Overtones and combination bands occur at 9.47 μm ($2\sigma_2$); 6.6 μm (σ_2 + $\sigma_3, 3\sigma_2$, and $2\sigma_3$); 4.7 μm ($2\sigma_2$ + σ_3 and $4\sigma_2$); and 3.7 μm ($2\sigma_1$). Of these, the most important absorption band in terms of remote sensing is the combination of the fundamental stretching mode and the overtone in the 9.4- to 9.8-μm region. The fundamental at 7.39 μm and overtone at 4.7 μm are also significant. These infrared absorption bands are shown in Fig. 4.17.

Methane, nitrous oxide, and carbon monoxide also exist in significant enough amounts to impact spectral properties of the atmosphere. Their infrared absorption spectra are shown in Figs. 4.18–4.20. Note that these are all displayed on a log scale to accentuate lower cross-section overtone and combination bands. The methane spectrum is similar to the methanol spectrum discussed earlier, although the locations of the vibrational bands are modified by the different molecular structure. Nitrous oxide exhibits two large vibrational stretching modes in the 7.5- to 9.0-μm region, as well as many overtone and combination bands throughout the SWIR and MWIR spectral range. The carbon monoxide molecule possesses a fundamental vibrational mode at 2170 cm^{-1}. The vibrational–rotational structure of this band centered at 4.6 μm is clearly evident, along with the overtone at 2.3 μm.

4.2.2 Liquid water

The main stretching bands of liquid water are shifted to a higher wavelength or lower wavenumber (σ_3 = 3490 cm^{-1} and σ_1 = 3280 cm^{-1}),

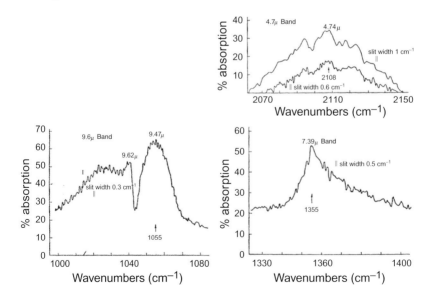

Figure 4.17 Measured absorption spectrum of ozone (O$_2$) in multiple spectral regions (Gerhard, 1932).

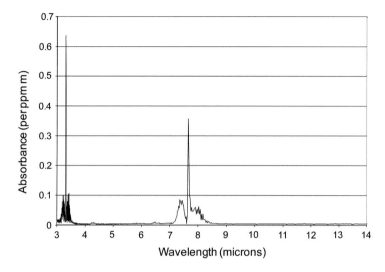

Figure 4.18   Measured absorption spectrum of methane (CH_4).

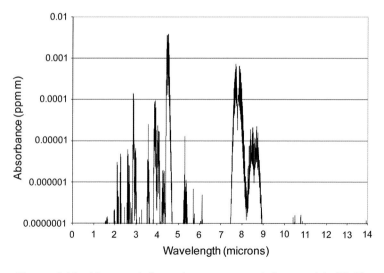

Figure 4.19   Measured absorption spectrum of nitrous oxide (N_2O).

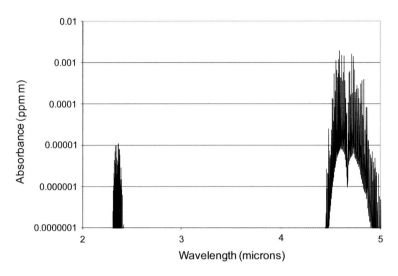

Figure 4.20 Measured absorption spectrum of carbon monoxide (CO).

and the bending band is shifted to a lower wavelength or higher wavenumber ($\sigma_2 = 1645$ cm^{-1}) due to hydrogen bonding that occurs as the molecules are more densely packed in liquid form. This frequency shifting continues to occur as water changes phase into ice and is also somewhat temperature dependent. The absorption spectra are also complicated by the addition of restricted rotations (rocking motions) due to the hydrogen bonds and the overtones and combinations that occur from these additional degrees of freedom. One important result of this is the change in spectral locations of the fundamental, overtone, and combination bands that occur in the NIR and SWIR spectral range, relative to water vapor. These spectral shifts between gas and liquid phase water are summarized in Table 4.2.

The real and imaginary parts of the complex index of refraction of liquid water are shown in Figs. 4.21 and 4.22 [figures based on data from (Palik, 1985)]. Fundamental resonances are identified by the anomalous dispersion in the real part, and the overtone and combination bands in the NIR and SWIR are evident in the imaginary part. Surface reflectance at

Table 4.2 NIR and SWIR absorption peaks of gas and liquid phase water.

Gas phase	Liquid phase
2.74 μm	2.94 μm
2.67 μm	2.87 μm
1.89 μm	1.94 μm
1.35 μm	1.47 μm
1.11 μm	1.20 μm
0.90 μm	0.98 μm

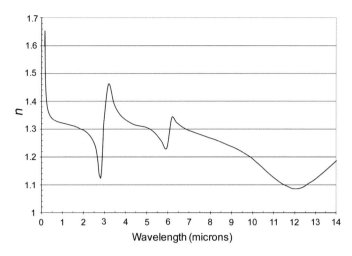

Figure 4.21   Real part of the complex index of refraction for pure liquid water.

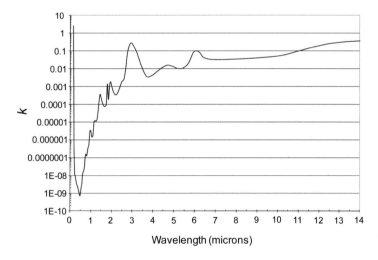

Figure 4.22   Imaginary part of the complex index of refraction for pure liquid water.

an air–water boundary (as computed from the data in Figs. 4.21 and 4.22) is shown in Fig. 4.23. Reflectance is roughly 2% from the visible through MWIR spectral range and 1% in the LWIR. The absorption coefficient (as computed from the imaginary part) is shown in Fig. 4.24. This indicates the high transmission of water in the visible spectral range and extremely high absorbance in the infrared. Even very thin layers of water are opaque in the infrared, making emissivity on the order of 98 to 99%. That is, water acts very much like a blackbody in the MWIR and LWIR spectral regions.

The characteristics depicted in the preceding figures are for pure water. Naturally occurring water, however, usually contains a variety of organic and inorganic impurities that significantly impact its spectral properties in

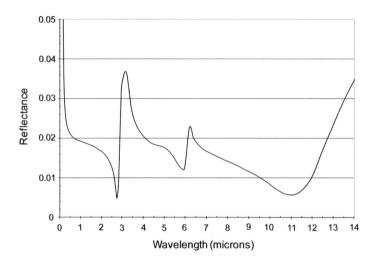

Figure 4.23   Surface reflectance for pure liquid water.

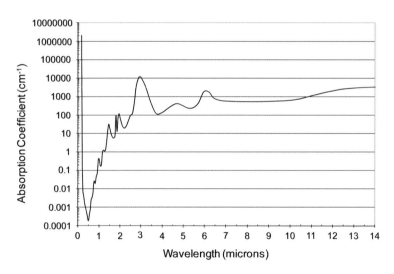

Figure 4.24   Absorption coefficient for pure liquid water.

the visible and NIR spectral range where it is otherwise transparent. The spectral properties of the impurities express themselves through absorption and scattering coefficients that characterize the volume reflectance of the natural water body. For example, reflectance in the visible and NIR spectral regions can be greatly influenced by the distribution and biomass of aquatic macrophytes, especially in shallow waters (Dekker *et al.*, 2001). Figure 4.25 shows such examples in terms of the measured apparent spectral reflectance for fresh-water lakes. The set of spectra depicted in Fig. 4.25(a) are for three lakes within the state of Minnesota [figure based on data from (Menken *et al.*, 2006)]. In this case, spectral reflectance is determined by the presence of algal activity, including suspended algae known as phytoplankton, other inorganic suspended particles such as clay, and dissolved natural organic matter. Differences in spectral reflectance are due to the varying concentrations of these constituents. Specifically, the Francis Lake spectrum with a higher reflectance and sharp rise in reflectance at 700 nm indicates larger amounts of algae. The results in Fig. 4.25(b) are for locations within the Cefni Reservoir in North Wales, United Kingdom, specifically chosen to illustrate the impact of various species of submerged, floating, and emergent macrophytes, or aquatic plants, on the apparent spectral properties (Malthus and George, 1997). In this case, high NIR reflectance is a result of emergent vegetation and other materials in the water and is not a characteristic of the water itself. The relationship between the apparent spectral properties and the constituents of bodies of water is further explored from a semi-empirical perspective in Chapter 5.

4.2.3 Vegetation

Vegetation is sensitive to optical radiation from the ultraviolet through infrared spectral range and is optimized to absorb solar energy in the visible spectrum to drive the biological process of photosynthesis necessary for plant growth. Absorbed solar energy is apportioned in roughly three ways: absorption of infrared radiation through water content for transpiration (nominally 70% total energy), absorption of visible radiation by leaf pigments to support photosynthesis (nominally 28% total energy), and absorption of ultraviolet radiation, which leads to fluorescence at longer wavelengths (nominally 2% total energy). The spectral absorbance, reflectance, and transmittance of a typical leaf are shown in Fig. 4.26. This particular example is from an evergreen shrub known as *Nerium oleander*. The physical processes underlying the characteristics shown are described starting with the visible spectral region and working toward the infrared. Further information is provided by Kumar *et al.* (2001).

The visible spectral region is characterized by low reflectance and transmittance due to the strong absorption by foliar pigments. Pigment absorption is associated with electronic state transitions and dominates the visible spectral characteristics. The bands are fairly broad because they represent diffuse spectra of complex pigment molecules. Such activated states last for only a fraction of a second, and some of the activation energy released through the decay to the ground state is consumed by photochemical events that make up the photosynthetic process. During photosynthesis, chloroplasts use the absorbed energy to convert carbon dioxide and water into carbohydrates, which is further converted by

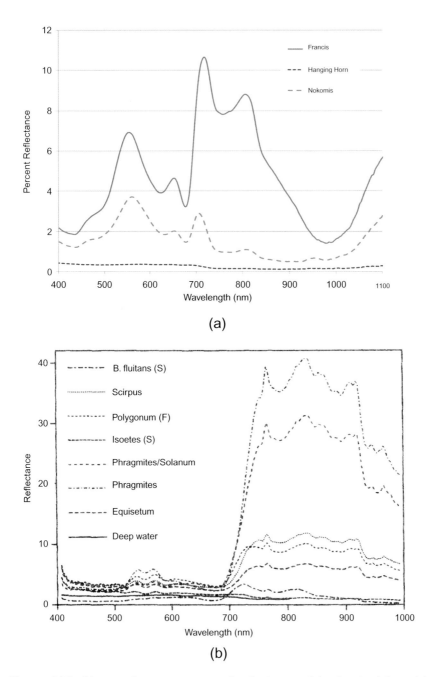

Figure 4.25 Measured apparent spectral reflectance of freshwater lakes: (a) three lakes in Minnesota and (b) various locations in Cefni Reservoir in North Wales with different submerged (S), floating (F), and emergent (others) macrophyte species [Fig. (b) reprinted with permission from (Malthus and George, 1997) © 1997 Elsevier].

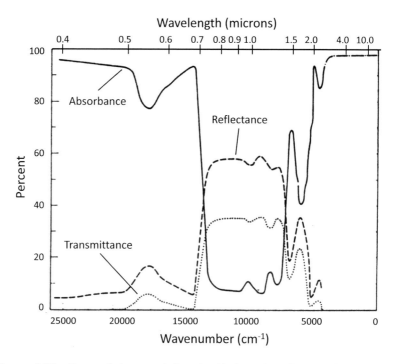

Figure 4.26 Spectral characteristics of a *Nerium oleander* leaf (Gates, 1970).

cellular respiration into adenosine triphosphate (ATP) to fuel plant maintenance and growth.

Operative pigments in chloroplasts that absorb visible energy include the primary pigment chlorophyll (65%), xanthophylls such as beta-carotene (29%), and other accessory pigments such as phycobiliproteins. The normalized absorbance spectra of a few varieties of these pigments are shown in Fig. 4.27 based on data published by Karp (2009) and Sadava *et al.* (2009), including a typical aggregate absorption spectrum compared to photochemical efficiency, a measure of the rate of photosynthesis as a function of the wavelength of incident radiation. All plants have chlorophyll *a* and most have chlorophyll *b*, both of which absorb strongly in the blue and red. This leads to the characteristic green color of vegetation indicated by the reflectance peak in Fig. 4.26 near 500 nm. It is theorized that chlorophyll absorption decreases at 500 nm because plants evolved from aquatic species that competed with aquatic bacteria for sunlight. Bacteriorhodopsin, the purple pigment responsible for light absorption in these aquatic species, peaks near 500 nm and exhibits a spectrum that is complementary to that of chlorophyll. Other types of chlorophyll also exist, and there are different forms of chlorophyll *a*. Such variations in the specific types and concentrations of foliar pigments result in diverse vegetation color. In all cases, however, pigments do not

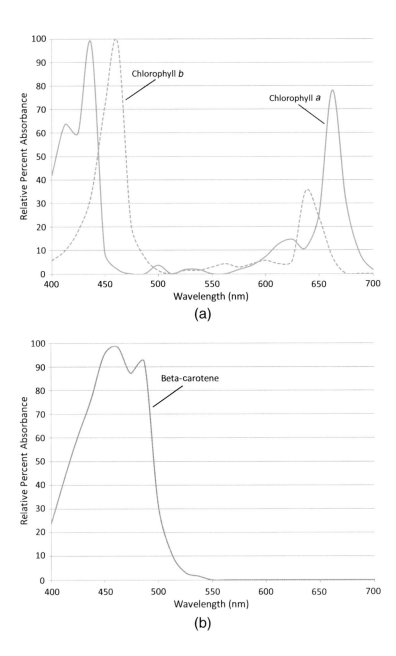

Figure 4.27 Spectral properties of foliar pigments: (a) chlorophyll absorbance, (b) xanthophyll absorbance, (c) phycobiliprotein absorbance, and (d) aggregate absorbance and photochemical efficiency.

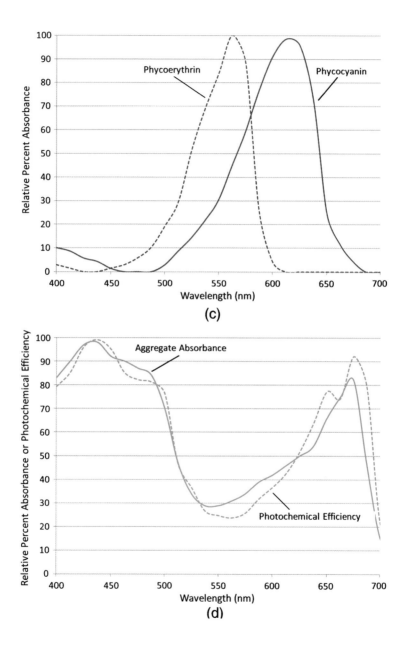

Figure 4.27 (*continued*)

absorb radiation at wavelengths above 700 nm and typically exhibit a sharp absorption edge near this wavelength.

The visible spectral characteristics of vegetation also vary by season, largely due to changes in pigment characteristics. As leaves senesce, chlorophylls degrade more quickly than carotenes, such that the spectral characteristics become dominated by carotenes and xanthophylls. This results in decreased reflectance in the 400- to 500-nm range, giving leaves a yellow appearance. As leaves die, brown pigments called tannins appear, which decrease the leaf reflectance and transmittance over the entire visible range.

A diagnostic feature of healthy vegetation is the sharp transition from low to high reflectance that occurs in the 700-nm range. This is known as the *red edge* and is due to the absorption cutoff of chlorophyll, leaving a high reflectance characterized by the water content and internal structure of the leaf. Several metrics based on the red-edge feature have been used to spectrally characterize the character, health, and state of vegetation. The Normalized Differential Vegetation Index (NDVI), given by

$$\text{NDVI} = \frac{\rho(860\,\text{nm}) - \rho(660\,\text{nm})}{\rho(860\,\text{nm}) + \rho(660\,\text{nm})}, \tag{4.71}$$

is an indicator of the vegetative character of a spectrum. Red-edge reflectance,

$$\rho_{red} = \frac{\rho(670\,\text{nm}) + \rho(780\,\text{nm})}{2}, \tag{4.72}$$

can be indicative of the health of vegetation. Finally, red-edge inflection,

$$\lambda_{red} = 700\,\text{nm} + \frac{\rho_{red} - \rho(700\,\text{nm})}{\rho(740\,\text{nm}) - \rho(700\,\text{nm})} 40\,\text{nm}, \tag{4.73}$$

can also be indicative of the state of vegetation.

The NIR spectral range is characterized by high transmittance, high reflectance, and low absorbance. Reflectance is due primarily to volume scattering within the complex internal structures of the leaves; therefore, spectral characteristics in this region are heavily influenced by the distribution of air spaces and the arrangement, size, and shape of cells within the leaf structures. For example, more-compact mesophyll layers and/or higher water content reduces the concentration of air spaces, resulting in reduced scattering and higher transmission (i.e., lower volume reflectance). Vegetation in this region of the spectrum is translucent but not transparent.

In the SWIR, spectral characteristics are dominated by strong liquid-water absorption features and influenced by foliar contents such as cellulose (major component of leaf cell walls), lignin (complex polymer that encrusts cellular fibrils), and pectin. As described in the previous section, pure liquid water exhibits absorption peaks at 0.98, 1.2, 1.47, and 1.94 μm. These bands are displaced slightly from the atmospheric water vapor bands, as depicted in Fig. 4.28, such that they are observable even through a long atmospheric path. This displacement results in regions of decreased reflectance centered on these band locations. The regions of reduced reflectance both deepen and widen as water content increases. 1.47- and 1.94-μm water absorption bands are particularly indicative of the water content of leaves; these bands disappear when vegetation is dry (Knipling, 1970). This is illustrated by the set of reflectance spectra illustrated in Fig. 4.29 that represent how spectral characteristics change after a *Philodendron hedaraceum* leaf has been removed from its stem. While this is a particularly hardy species that perhaps maintains water longer than typical vegetation, the reflectance spectra indicate a loss of water-absorption features as the leaf finally dries up after about four weeks, with more-subtle changes in the first few weeks due to vegetation stress. In addition to the change in liquid water features, there is also a dramatic change to the red edge as the chlorophyll breaks down and is ultimately replaced by tannins. After the leaf completely dries up, its spectral characteristics are due to cellulose, lignin, and pectin, which dominate the SWIR characteristics of dry plant materials and have diagnostic features at

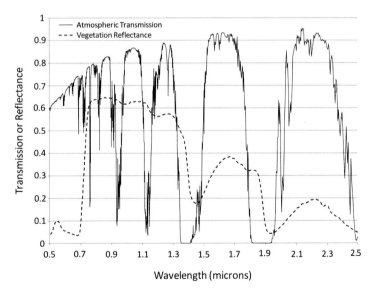

Figure 4.28 Comparison of water-absorption bands in the atmosphere and vegetation.

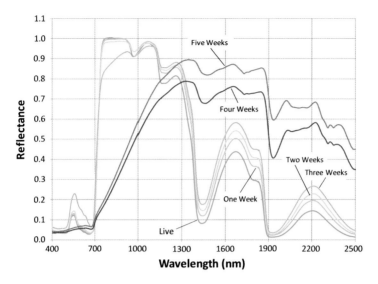

Figure 4.29 Changes in spectral reflectance of a philodendron leaf as it dries up and dies after being removed from the stem. The time periods indicate measurement time after removal.

2.09 μm and in the 2.3-μm range. Examples of cellulose, lignin, and pectin spectra are provided in Fig. 4.30.

In the thermal infrared (MWIR and LWIR), healthy vegetation exhibits high absorbance and low reflectance due primarily to liquid-water absorption. Cellulose and lignin also influence spectral characteristics,

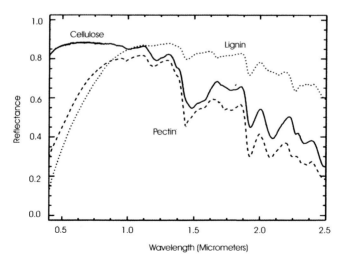

Figure 4.30 Spectral reflectance of dry plant constituents: cellulose, lignin, and pectin [reprinted with permission from Van der Meer and DeJong (2001) © 2001 Springer].

especially in dry plant materials. Figure 4.31 illustrates the spectral reflectance of dried and green grass. The dry-grass features in the 4- to 6-μm and 10- to 12-μm spectral regions correspond to cellulose. Differing types of vegetation exhibit infrared spectral behavior very similar to that shown in Fig. 4.32, since the dominant influences are not a strong function of plant type. In the VNIR/SWIR spectral region, however, variations in pigment concentrations and leaf structure cause more-significant variation, as shown by the examples given in Fig. 4.32.

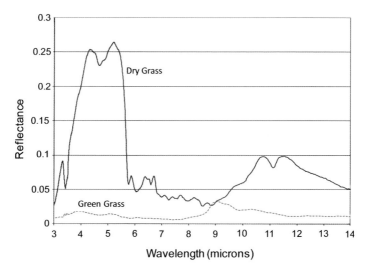

Figure 4.31 Spectral reflectance of green and dry grass in the emissive spectral region.

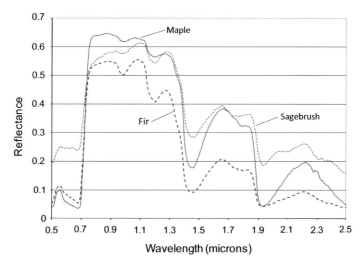

Figure 4.32 Spectral reflectance of various plants in the reflective spectral region.

Spectral reflectance of a vegetative canopy can be greatly influenced by multiple scattering that can occur between leaf layers. This generally has the effect of decreasing the apparent reflectance, as incident light becomes trapped between layers and is ultimately absorbed to a higher degree compared with the absorption level associated with a single leaf. In the NIR, reflectance can be enhanced by volume scattering due to higher leaf transmittance.

4.2.4 Minerals

The spectral characteristics of minerals are well understood and have been used for many years for remotely determining compositional information of the earth's surface. As a great diversity of minerals make up the surface of the earth, comprehensive treatment of the spectral characteristics of its minerals would be quite extensive. Fortunately, much insight into mineral spectral characteristics can be gained from looking at four common mineral constituents: silicates, carbonates, sulfates, and hydroxides. Figure 4.33 illustrates the spectral reflectance in the visible-through-SWIR spectral range for mineral types composed of these primary constituents. In the visible-to-NIR spectral range, spectral characteristics are influenced primarily by electronic transition and charge-transfer processes associated with transition-metal ions such as Fe, Cr, Co, and Ni. In a crystal field, such ions exhibit split orbital energy states and can absorb optical radiation at frequencies associated with electronic transitions between these states. Also, absorption bands can result from charge transfer (e.g., Fe^{2+} to Fe^{3+}). The goethite example in Fig. 4.33(d) exhibits characteristic spectral features in this range due to its iron content.

Spectral reflectance in the SWIR is heavily influenced by vibrational features of water, hydroxyl, carbonate, and sulfate molecules. The vibrational features are observed as reflectance minima because they are a result of volume scattering. Water characteristics have been discussed in previous sections and are evident in the calcite example. The hydroxyl ion (OH) has only one stretching mode, but its spectral location is dependent on the ion (typically a metal ion) to which it is attached. This hydroxyl feature typically occurs in the 2.7- to 2.8-μm range, but can range from 2.67 to 3.45 μm. A bending mode also exists near 10 μm, resulting in a diagnostic combination mode (bend/stretch) in the 2.2-μm (when combined with Al) to 2.3-μm (when combined with Mg) range. There is also a weak absorption band near 1.4 μm. These features are visible as reflectance dips in the goethite spectrum. The remaining SWIR features for this mineral are attributed to water absorption. The carbonate ion (CO_3) exhibits four vibrational modes in a free state: a symmetric stretching mode ($\sigma_1 = 1063$ cm^{-1}), a bending mode ($\sigma_2 = 879$ cm^{-1}), and two degenerate modes ($\sigma_3 = 1415$ cm^{-1} and $\sigma_4 = 680$ cm^{-1}). Combination

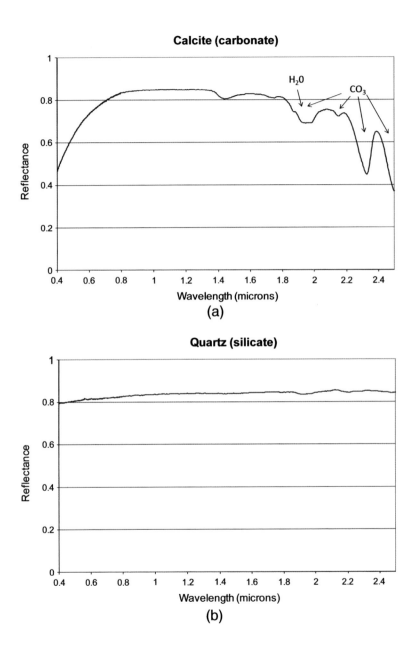

Figure 4.33 Spectral reflectance of (a) carbonate, (b) silicate, (c) sulfate, and (d) hydroxide minerals in the VNIR/SWIR spectral region.

Gypsum (sulfate)

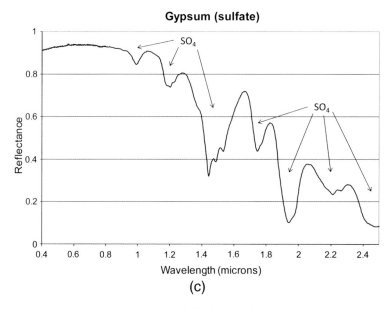

(c)

Goethite (hydroxide)

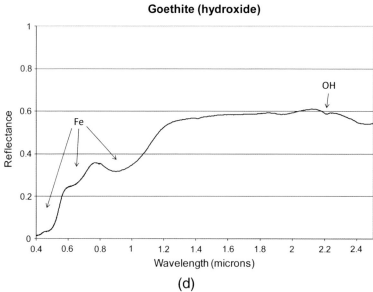

(d)

Figure 4.33   *(continued)*

and overtone bands occur at $1.9, 2.0, 2.16, 2.35$, and 2.55 μm. All of these bands are visible to some extent in the calcite example shown in Fig. 4.33(a), although the first two manifest as a perturbation to the 1.94-μm water band. The sulfate ion (SO_4) has additional degrees of freedom compared to carbonate, such that its absorption characteristics are a bit more complex. However, characteristic overtone and combination bands are clearly visible throughout the NIR and SWIR spectral range in the gypsum example shown in Fig. 4.33(c). In contrast, the silicate quartz example is relatively featureless in this spectral region.

The MWIR and LWIR behavior of these same minerals is illustrated in Fig. 4.34 and is more complicated than the shorter-wavelength characteristics. The LWIR spectral region, in particular, is often the spectral region of choice for remote geological studies because of the existence of large, identifiable reflectance features useful for mineral typing. The spectral features of minerals in the MWIR spectral region are not as unique or readily identifiable. Most minerals in the MWIR are significantly transmissive, such that absorption features manifest themselves as reflectance minima because volume reflectance dominates as in the VNIR/SWIR. Carbonates, sulfates, and nitrates exhibit characteristic absorption features in this region. Carbonate features visible in the calcite spectrum occur at $3.5, 4.0$, and 4.7 μm, with the former two associated with overtones of a strong C–O fundamental stretch. Similarly, sulfates absorb at 3.0 and 4.5 μm, as seen in the gypsum example. Silica exhibits a feature near 4.7 μm that is an overtone of the reststrahlen feature described later. Finally, nitrates are normally found in arid regions and possess absorption features near $3.6, 4.0, 4.8, 5.3$, and 5.7 μm.

In the LWIR, fundamental vibrational resonances influence the spectral characteristics of minerals and can result in very distinct regions of high reflectance. As the minerals tend to be more absorptive in this region, surface reflectance becomes the dominant interaction mechanism, such that absorption bands manifest as reflectance peaks as opposed to minima. This is described further in a detailed look at the silicate reststrahlen band. The carbonate fundamental resonances at $7.1, 9.4, 11.4$, and 14.7 μm correspond to peaks in the calcite reflectance spectrum. The exact location and shape of these peaks are influenced by the crystal field in which the carbonate ions reside, so an exact correspondence should not be expected. Not all of the spectral characteristics of the calcite example, however, can be attributed to carbonate.

Gypsum is primarily composed of calcium sulfate ($CaSO_4$), and its LWIR spectral reflectance is dominated by a reststrahlen band near the 8- to 9-μm range, associated with a fundamental stretching mode near 8.9 μm. Similar behavior in the LWIR occurs for both quartz (silicate) and goethite (hydroxide) examples. In the quartz case, a strong reststrahlen band in the

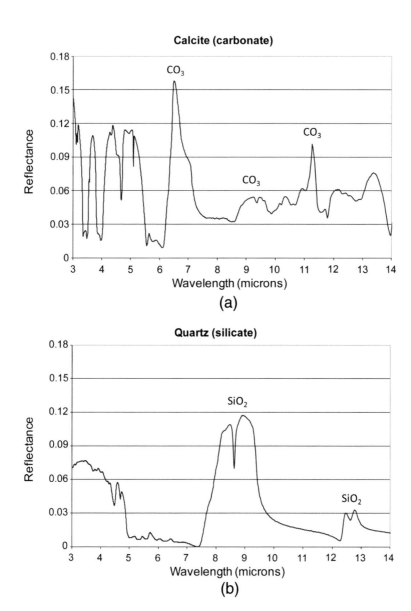

Figure 4.34 Spectral reflectance of (a) carbonate, (b) silicate, (c) sulfate, and (d) hydroxide minerals in the MWIR/LWIR spectral region.

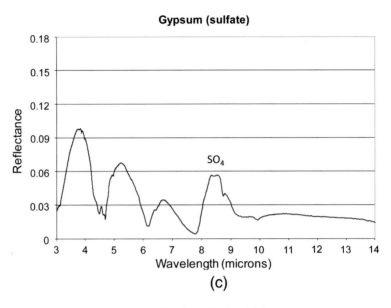

(c)

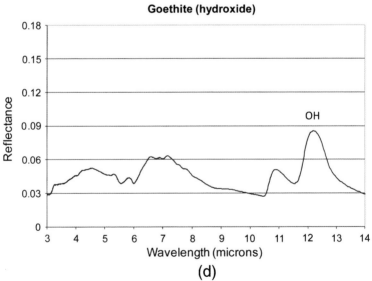

(d)

Figure 4.34 *(continued)*

8- to 9.5-μm range occurs, along with a weaker band in the 12.5- to 13-μm range. The goethite band appears as a doublet in the 11- to 13-μm range. Note that in all of these cases, there is a reflectance minimum at the short-wavelength side of each reststrahlen band. This feature is called a Christiansen peak, since it is observed as a peak in the transmission spectrum.

The characteristics of a reststrahlen band can be understood by examining a silicate mineral in more detail. Silicates are the most abundant constituent of rocks, and the Si=O bonds in silicates dominate their spectral characteristics. Silicate minerals are classified according to the way in which the oxygen molecules are shared in their solid-state form. Quartz (SiO_2) is tektosilicate, the silicate class in which all four oxygen molecules between adjacent SiO_4 tetrahedra are shared. Its optical constants, illustrated over the thermal infrared spectral range in Figs. 4.35 and 4.36 [figures based on data from (Palik, 1985)], exhibit two major vibrational resonances: a broad resonance centered at about 9.2 μm and a narrower resonance centered at 12.5 μm. The former is an overlap of asymmetric stretching modes of O–Si–O and Si–O–Si, while the latter corresponds to a symmetric Si–O–Si stretching mode. The exact location of these resonances shifts slightly to longer wavelengths as the mineral silicate content decreases. Also, the strength of the resonances is clearly enhanced in crystalline form relative to amorphous form.

Figure 4.37 provides the surface reflectance computed from the optical constants given in Figs. 4.35 and 4.36. The overlapping resonances near 9.2 μm result in a band of very high reflectance with a sharp minimum feature near 8.6 μm, a diagnostic silicon feature that is exploited in many remote sensing applications. A secondary reststrahlen feature occurs

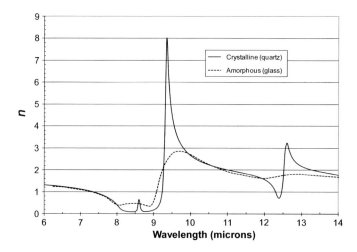

Figure 4.35   Real part of the index of refraction for SiO_2.

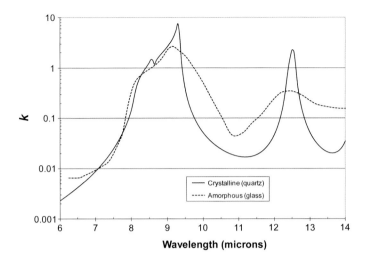

Figure 4.36 Imaginary part of the index of refraction for SiO_2.

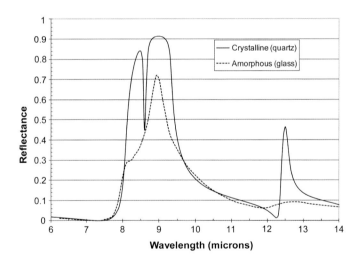

Figure 4.37 Surface reflectance SiO_2 predicted by indices of refraction.

for the 12.5-μm resonance. The aforementioned Christiansen peaks are evident as reflectance minima at 7.4 and 12.25 μm. These are called peaks because front-surface transmittance is almost perfect at these wavelengths since the index of refraction is nominally equal to one, resulting in reflectance minima due to almost all of the incident radiation being passed into the material and ultimately being absorbed.

4.2.5 Soils

The spectral characteristics of soils are dependent on both their composition (the minerals of which they are composed) and their structure, especially grain size. Many different soils share common characteristics because they are composed of common constituents such as water, silicates, carbonates, sulfates, hydroxides, and nitrates. Spectral features of minerals show up in aggregate soil in strengths dependent on their concentration.

Soil types are very diverse, and it is thus helpful to first discuss basic soil nomenclature and taxonomy (US Department of Agriculture, 1999). Loam is a soil consisting of various proportions of clay, silt, and sand. Clay is composed of fine particles of hydrous aluminum silicates and other minerals. Silt is sediment composed of very fine-grain particles and is rich in organic material. Finally, sand is primarily quartz. Soil types are organized in terms of the 12 major orders outlined in Table 4.3. While each soil order can be very diverse in terms of mineral composition, the table provides some construct with which to discuss soil spectral characteristics.

Figures 4.38 and 4.39 illustrate spectral reflectance characteristics of a few examples of soils in both the VNIR/SWIR and MWIR/LWIR spectral regions. VNIR reflectance tends to be monotonically increasing over the 0.4- to 1.2-μm range, similar to the calcite and goethite mineral spectra. One would expect oxisols to exhibit strong iron-absorption features similar to goethite. The SWIR spectral characteristics are primarily influenced by the hydroxyl, carbonate, silicate, and water features that have all been previously discussed. Organic material and water content tend to produce a reduction in reflectance across this entire spectral range, as well as a reduction in the depth of mineral features.

In the MWIR, soil spectra can be somewhat distinguished based on the relative composition of carbonate and silicate minerals. Water content also heavily influences the gross shape of MWIR spectral reflectance. In the examples shown, LWIR characteristics are dominated by the strong silicate reststrahlen features. Particularly in this spectral region, grain size, level of organic matter, and soil moisture can have a considerable effect on spectral behavior. As an illustration of grain-size impact, consider Fig. 4.40, which illustrates measured apparent reflectance spectra for wet- and dry-sieved soils from Camp Lejeune, North Carolina, for different sieve sizes (Johnson, 1998). First note the larger expression of the silicate reststrahlen feature for the larger particles of the soil, which in this case is composed primarily of fine quartz sands and kaolinite clay. The difference between the wet- and dry-sieved results for the larger size ranges indicates the presence of finer particles in the dry-sieved samples. All samples were oven dried to remove moisture before making measurements. The finer grain sizes result in a depression of the silicate reststrahlen feature due to the surface cavity effect that results; that is, a greater fraction of incident photons are multiply reflected off of soil grains before being scattered off the surface. This effect can be viewed as an indicator of disturbed soil,

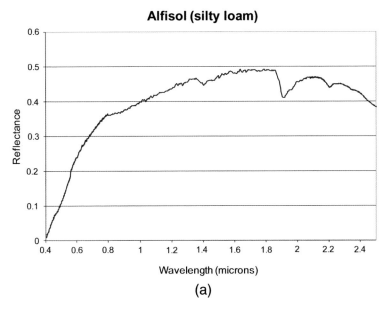

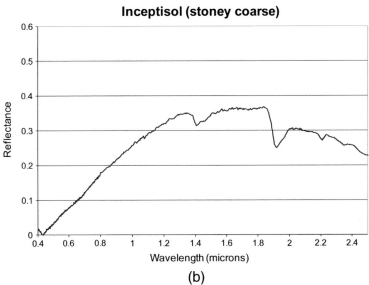

Figure 4.38 Spectral reflectance of four major soil types in the VNIR/SWIR spectral region: (a) alfisol, (b) inceptisol, (c) entisol, and (d) mollisol.

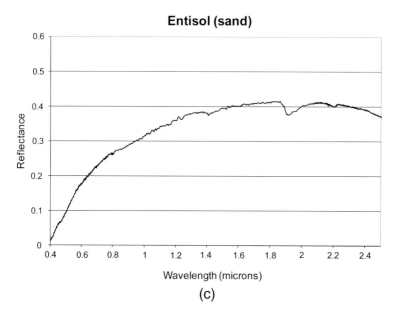

(c)

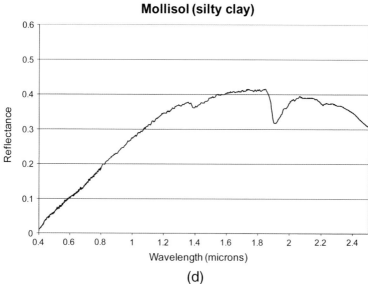

(d)

Figure 4.38   (*continued*)

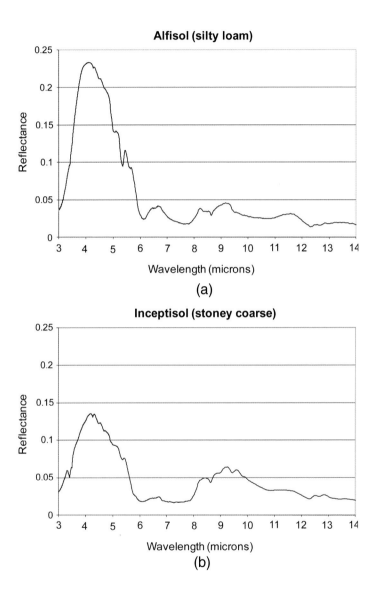

Figure 4.39 Spectral reflectance of four major soil types in the MWIR/LWIR spectral region: (a) alfisol, (b) inceptisol, (c) entisol, and (d) mollisol.

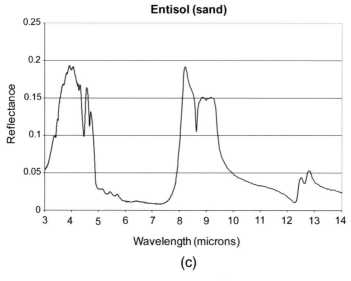

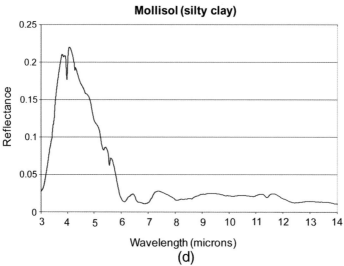

Figure 4.39   (*continued*)

Table 4.3 Major soil orders of the world.

Soil order	Primary characteristics
Alfisol	Tropical soil, easily cultivated, high supply of bases and water, nutrients concentrated in topsoil, low levels of organic matter.
Andisol	Soil from volcanic ash parent material, high organic content, highly erodible, high phosphorus retention and water capacity.
Aridisol	Common soil in arid areas (deserts), high salt content, low level of organic matter, lack of water, subject to wind erosion.
Entisol	Unconsolidated wind-blown or water-deposited sand, inherently infertile with few nutrients, inert parent material such as quartz.
Gelisol	Permafrost or soil associated with freezing and thawing, confined to higher latitudes and elevations.
Histosol	Highly organic or peat soil, degrades by shrinkage of high level of organic matter.
Inceptisol	Typical brown earth, increasing clay with depth, high resilience to erosion, high water content, few carbonates or amorphous silica.
Mollisol	Soil formed on limestone parent, erosion resistant, very dark brown or black surface, high calcium and crystalline clay content.
Oxisol	Red soil due to iron content, low supply of nutrients, strong acidity, loamy texture.
Spodosol	High humus, aluminum, and iron content, coarse texture.
Ultisol	Acidic soil, low base saturation, strongly leached, all nutrients except Al decreased substantially, high water capacity.
Vertisol	Soil with 30% or more clay, pronounced changes in volume with moisture.

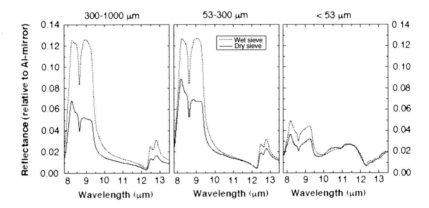

Figure 4.40 Effect of grain size on the LWIR spectral reflectance of silicate soils as characterized by wet- and dry-sieving of soil samples from Camp Lejuene, North Carolina [reprinted with permission from Johnson (1998) © 1998 Elsevier].

as depicted by the emissivity spectra in Fig. 4.41. This occurs because disturbing the soil by overturning, for example, brings a larger fraction of finer grain particles to the surface, which is eventually eroded away by weathering. Note that tamping the disturbed soil, in this case, was insufficient to eliminate the effect of the soil disturbance.

Organic matter is highly absorbing in the LWIR spectral region and reduces the depth of the other mineral features. Figure 4.42 illustrates the reduction of the silicate restrahlen feature for mollisols and ultisols with increasing organic content. Such content is quantified by the relative carbon composition. A small amount of organic content causes a more significant reduction in the short-wavelength lobe of the quartz reststrahlen doublet than the long-wavelength lobe, resulting in an overall reflectance peak near 9.5 μm. As organic content increases, the restrahlen feature becomes completely annihilated. Soil moisture has a similar impact in terms of reducing the spectral contrast of the silicate reststrahlen feature; however, the reduction is spectrally uniform in this case, such that the shape of the LWIR spectral reflectance remains somewhat unchanged. This is illustrated in Fig. 4.43.

4.2.6 Road materials

Most road materials consist of a combination of gravel, concrete, and asphalt. Concrete is a mixture of aggregates such as gravel (crushed rock), sand, water, and Portland cement, which is itself a mixture of calcium

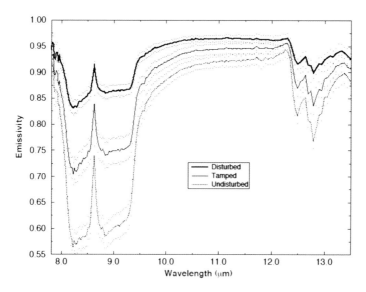

Figure 4.41 Comparison of LWIR spectral emissivity for undisturbed, disturbed, and tamped soil from Camp Lejeune, North Carolina [reprinted with permission from Johnson (1998) © 1998 Elsevier].

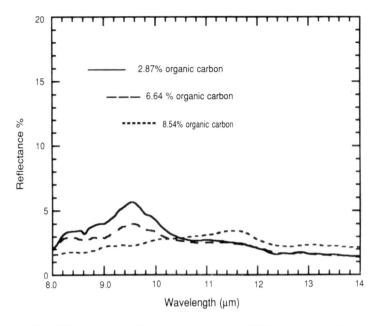

Figure 4.42 Effect of organic content on the LWIR spectral reflectance of silicate soils [reprinted with permission from Salisbury and D'Aria (1992b) © 1992 Elsevier].

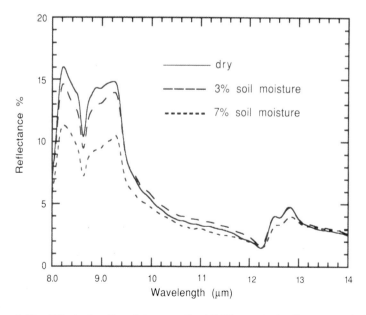

Figure 4.43 Effect of soil moisture on the LWIR spectral reflectance of silicate soils [reprinted with permission from Salisbury and D'Aria (1992b) © 1992 Elsevier].

silicates and calcium aluminate that hydrates in the presence of water. Therefore, concrete exhibits spectral features characteristic of the minerals composing the aggregate materials, water, and silicates. An example of a concrete reflectance spectrum is given in Fig. 4.44. The dominant spectral features include the quartz reststrahlen doublet, primary water absorption bands, and calcite features near 4, 6.5, and 11.3 μm. Asphalt is a mixture of aggregate along with a tar-like oil by-product called petroleum bitumen. This is a residue that remains after petroleum distillation and is a mixture of naturally occurring linear polymers dominated by hydrocarbons. Fresh

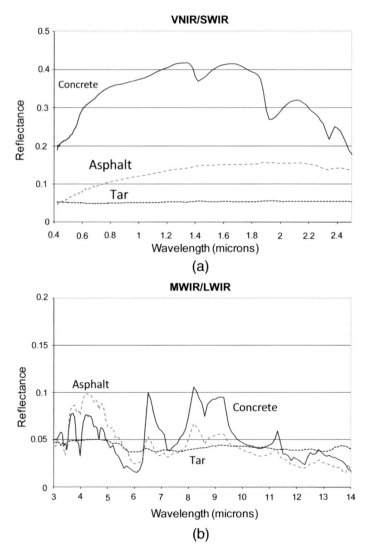

Figure 4.44 Typical spectral reflectance of road materials: (a) VNIR/SWIR and (b) MWIR/LWIR.

asphalt exhibits a spectral reflectance much like the tar example, which is strongly absorbent over the entire spectral range. Weathered asphalt exhibits the spectral characteristics of aggregate material because the tar breaks down. The example shown in Fig. 4.44 includes the silicate and calcite features observed in the concrete example.

4.2.7 Metals

At frequencies below the plasma frequency (including the entire range from the visible through the LWIR for most metals), metals are highly absorptive and thus also highly reflective. Figures 4.45 and 4.46 illustrate the optical constants for gold, silver, and aluminum [figures based on data from (Palik, 1985)]. The gold and aluminum examples exhibit absorption features in the visible and NIR spectral range, respectively. Otherwise, the real and imaginary parts are large and monotonically increasing with wavelength. The spectral reflectance calculated from the optical constants is illustrated for both the VNIR/SWIR and MWIR/LWIR spectral regions in Figs. 4.47 and 4.48. Except for the regions associated with the gold and aluminum absorption features in the VNIR, the metals are all highly reflective.

When the radiation frequency is much lower than the plasma frequency, the real and imaginary parts of the complex index of refraction become very large, and both approach the value

$$n \approx \kappa \approx \sqrt{\frac{\sigma}{2\omega\varepsilon_0}}. \qquad (4.74)$$

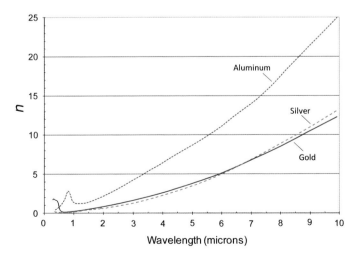

Figure 4.45 Real part of the index of refraction for gold, silver, and aluminum.

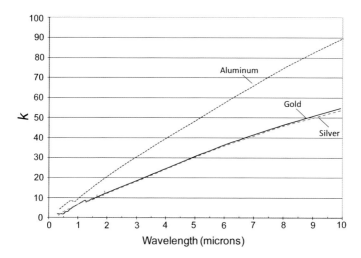

Figure 4.46 Imaginary part of the index of refraction for gold, silver, and aluminum.

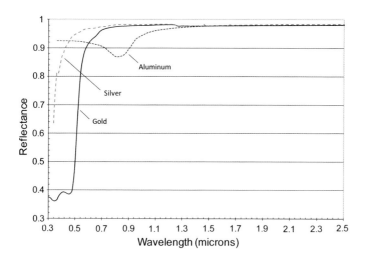

Figure 4.47 Surface reflectance for gold, silver, and aluminum in the VNIR/SWIR spectral region.

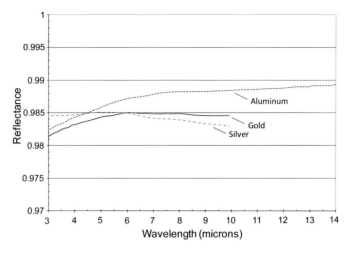

Figure 4.48 Surface reflectance for gold, silver, and aluminum in the MWIR/LWIR spectral region.

Under this approximation, the reflectance approaches

$$R \approx 1 - \frac{2}{n} \approx 1 - \sqrt{\frac{8\omega\varepsilon_0}{\sigma}}, \qquad (4.75)$$

a relationship known as a Hagen–Rubens formula. Figure 4.49 illustrates this approximation compared to the calculated aluminum spectral reflectance in the MWIR/LWIR spectral range.

4.2.8 Paints and coatings

The spectral reflectance properties of paints and coatings can be very diverse depending on their specific composition, but some common features can be understood by examining the basic makeup of paints and coatings. As illustrated in Fig. 4.50, they are generally composed of a binder material, such as polyurethane or alkyd, in which a large number of pigment and other filler particles are suspended. The pigments are selected to produce the desired visible color characteristics, while the binder supports adhesion and durability. Filler particles are often added to produce a desired sheen and other surface characteristics. In general, the VNIR spectral characteristics of paints and coatings are dictated by the pigment characteristics, while the SWIR, MWIR, and LWIR characteristics are influenced by the binder, filler particles, and substrate material. Modeling the spectral characteristics of real paints and coatings is extremely difficult because of the complex multiple scattering and interference effects that arise due to the high concentration of polydispersed particles. Effective modeling requires not only the optical

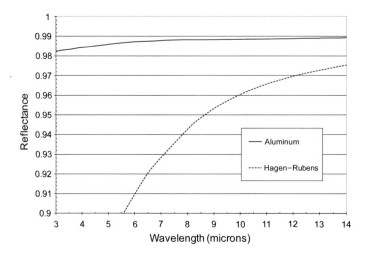

Figure 4.49 Surface reflectance for aluminum compared to Hagen–Rubens approximation.

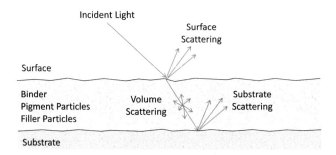

Figure 4.50   Basic composition of a paint or coating.

constants of the binder, pigments, filler particles, and substrates, but also the size and shape distribution of the pigment and filler scatterers. Furthermore, the two-stream scattering model that is typically employed is not accurate for modeling the high-density scattering situations characteristic of realistic paints.

To illustrate the effect of different pigments within a common coating type, Fig. 4.51 provides the VNIR/SWIR reflectance for coated paper of a variety of colors. Note that VNIR spectral properties vary significantly between samples due to different pigment types, while SWIR spectral properties are common for all samples. The latter are primarily indicative of the cellulose content of the paper substrate. Apparently, the coating binder is relatively transparent in this spectral region. Another example, shown in Fig. 4.52, is olive-green alkyd gloss paint on a pine-wood substrate. Again, the green pigment characteristics are evident in the visible region, while the cellulose features of the pine substrate material

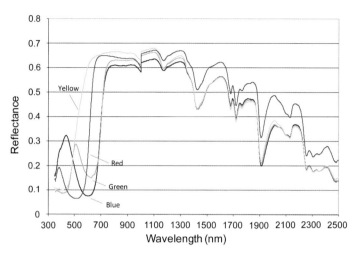

Figure 4.51 Spectral reflectance of file folders of different colors in the VNIR/SWIR spectral region.

show through the transparent binder in the SWIR spectral region. The MWIR spectral features include an overlap of a strong hydroxyl band near 2.7 µm, a water band near 2.9 µm, and strong C–H stretching bands near 3.45 µm that result in low reflectance over the 2.7- to 3.5-µm range. Also, a strong carbonyl band exists near 5.7 µm. The remainder of the MWIR spectral region exhibits fairly high reflectance, as the binder is again relatively transmissive. Reflectance is lower in the LWIR spectral region where characteristic features are typically associated with filler materials such as silicates. The reflectance increase near 9.4 µm is likely due to such silicate content.

4.3 Summary

A two-stream analysis method based on coupled differential equations for counterpropagating irradiance distributions can be used to relate the apparent spectral properties of layered media to the intrinsic spectral properties of the layers. This method accounts for both internal reflectance and transmission at the interfaces, as well as volume reflectance from distributed scatterers within the layers. To account for rough-surface effects, analytical methods typically involve semi-empirical models such as the Beard–Maxwell model, which provides a mathematical form of the bidirectional reflectance distribution function (BRDF) composed of specular surface reflectance, diffuse surface reflectance, and volume reflectance components, with parameters fit to measured characteristics. When a surface is sufficiently rough, a directional hemispherical reflectance (DHR) or hemispherical directional reflectance (HDR) can adequately describe the material for cases where the reflected or incident

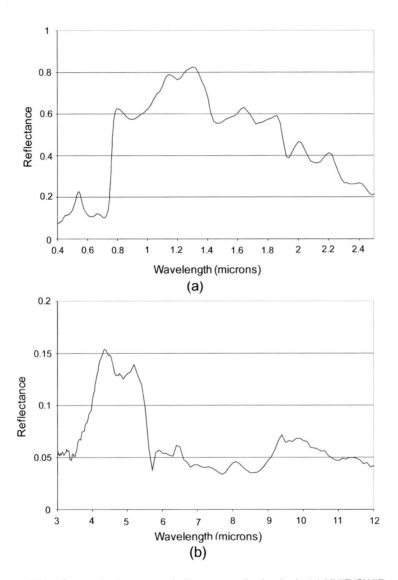

Figure 4.52 Spectral reflectance of olive-green alkyd paint in (a) VNIR/SWIR and (b) MWIR/LWIR spectral regions.

radiation is assumed to be diffuse. For a truly Lambertian surface, the DHR and HDR are equivalent and fully describe reflectance characteristics.

Especially in the MWIR and LWIR spectral region, thermal emission from observed materials must be considered along with reflectance and transmission. The emitted spectral radiance is the product of a blackbody spectral radiance distribution for the material temperature and the spectral emissivity, an apparent spectral property related to absorbance. In fact, the emissivity equals the absorbance according to Kirchoff's law under the strict conditions of thermal equilibrium; this approximation is commonly used in remote sensing.

The apparent spectral reflectance, transmission, and emissivity characteristics of common remote sensing materials can be understood by applying the fundamental principles established throughout the first three chapters of this book to the specific chemical and physical characteristics of the material. The examples described in this chapter elucidate the physical relationships between material composition and apparent spectral properties that form the foundation of hyperspectral remote sensing. A critical component of the application of this sensing technology toward remote sensing problems is a thorough phenomenological understanding of the materials of interest and their apparent spectral properties.

4.4 Further Reading

Two publicly accessible databases of apparent spectral properties of materials are useful for further understanding material characteristics and performing spectral signature analysis. The first is the *NIST Chemistry WebBook* (Lindstrom and Mallard, Eds.) from the National Institute of Standards and Technology (NIST) that provides chemical data and spectral properties for a variety of chemical species. The Advanced Spaceborne Thermal Emission Reflection Radiometer (ASTER) spectral library (Baldridge *et al.*, 2009) is a compilation of spectral reflectance measurements over the 0.4- to 15.4-μm spectral range of various remote sensing materials based on data from the National Aeronautics and Space Administration Jet Propulsion Laboratory (NASA/JPL), Johns Hopkins University Applied Physics Laboratory (JHU/APL), and United States Geological Survey (USGS). Optical constants of a variety of materials are tabulated in Palik (1985). More-detailed information on material signature pheonomenology associated with hyperspectral remote sensing applications can be found in books such as Anderson *et al.* (1994), Van der Meer and DeJong (2001), and Pieters and Englert (1993) but is mainly found in remote sensing journals and conference proceedings. Emissivity spectra of terrestrial materials have been measured and compiled into a database by Salisbury and D'Aria (1992a, 1993).

References

Anderson, R., Malila, W., Maxwell, R., and Reed L. (Eds.), *Military Utility of Multispectral and Hyperspectral Sensors*, Environmental Research Institute of Michigan, Ann Arbor, MI (1994).

Baldridge, A. M., Hook, S. J., Grove, C. I., and Rivera, G., "The ASTER spectral library – version 2.0," *Remote Sens. Environ.* **113**, 711–715 (2009), http://speclib.jpl.nasa.gov (last accessed Sept. 2011).

Boltzmann, L., "Ableitung des Stefan'schen gesetzes betreffend die Abhängigkeit der Wärmestrahlung von der Temperatur aus der elektromagnetischen Lichttheorie," *Annal. Phys. Chem.* **22**, 291–294 (1884).

Bransden, B. H. and Joachain, C. J., *Physics of Atoms and Molecules*, Second Edition, Prentice Hill, Harlow, UK (2003).

Dekker, A. G., Brando, V. E., Anstee, J. M., Pinnel, N., Kutser, T., Hoogenboom, E. J., Peters, S., Pasterkamp, R., Vos, R., Olbert, C., and Malthus, T. J. M., "Imaging spectrometry of water," in *Imaging Spectrometry: Basic Principles and Prospective Applications*, Van der Meer, F. D., and DeJong, S. M. (Eds.), Kluwer Academic, Dordrecht, Netherlands (2001).

Gates, D. M., "Physical and physiological properties of plants," in *Remote Sensing*, Ch. 5, pp. 224–252, National Academy of Sciences, Washington, DC (1970).

Gerhard, S. L., "The infrared absorption spectrum and the molecular structure of ozone," *Phys. Rev.* **42**, 622–631 (1932).

Heald, M. A., "Where is the Wien peak?" *Am. J. Phys.* **71**, 1322–1323 (2003).

Johnson, J. R., Lucey, P. G., Horton, K. A., and Winter, E. M., "Infrared measurements of pristine and disturbed soils 1: spectral contrast differences between field and laboratory data," *Remote Sens. Environ.* **64**, 34–46 (1998).

Karp, G., *Cell and Molecular Biology: Concepts and Experiments*, Sixth Edition, Wiley & Sons, Hoboken, NJ (2009).

Kirchhoff, G., "On the relation between the radiating and absorbing powers of different bodies for light and heat," *Philos. Mag. J. Science* **20**, p. 130 (1860).

Knipling, E. B., "Physical and physiological basis for the reflectance of visible and near-infrared radiation from vegetation," *Remote Sens. Environ.* **1**, 155–159 (1970).

Kubelka, P. and Munk, F., "An article on optics of paint layers," *Z. Tech. Phys.* **12**, 593–601 (1931).

Kumar, L., Schmidt, K., Dury, S., and Skidmore, A., "Imaging spectrometry and vegetation science," in *Imaging Spectrometry: Basic Principles and Prospective Applications*, Van der Meer, F. D., and DeJong, S. M. (Eds.), Kluwer Academic, Dordrecht, Netherlands (2001).

Lindstrom, P. J. and Mallard, W. G. (Eds.), *NIST Chemistry WebBook*, NIST Standard Reference Database Number 69 (1996), http://webbook. nist.gov/chemistry (last accessed Sep. 2011).

Malthus, T. J. and George, D. G., "Airborne remote sensing of macrophytes in Cefni Reservoir, Angelesy, UK," *Aquat. Bot.* **58**, 317–332 (1997).

Maxwell, J. R., Beard, J., Weiner, S., Ladd, D., and Ladd, S., *Bidirectional Reflectance Model Validation and Utilization*, Technical Report AFAL-73-303, Environmental Research Institute of Michigan, Ann Arbor, MI (1973).

Menken, K., Brezonik, P. L., and Bauer, M. E., "Influence of chlorophyll and colored dissolved organic matter (CDOM) on lake reflectance spectra: implications for measuring lake properties by remote sensing," *Lake Reservoir Manage.* **22**, 179–190 (2006).

Montanaro, M., Salvaggio, C., Brown, S., and Messinger, D., *NEFDS Beard-Maxwell BRDF Model Implementation in MATLAB*®, DIRS Technical Report 2007-83-174, Rochester Institute of Technology, Rochester, RI (2007).

Palik, E. D., *Handbook of Optical Constants of Solids*, Academic Press, Orlando, FL (1985).

Pieters, C. M. and Englert, P. A., *Remote Chemical Analysis*, Cambridge University Press, Cambridge, UK (1993).

Planck, M., *Theory of Heat*, Macmillan, New York (1957).

Sadava, D., Berenbaum, M. R., Orians, G. H., Purves, W. K., and Heller, H. C., *Life: The Science of Biology*, W.H. Freeman, New York (2009).

Salisbury, J. W and D'Aria, D. M., "Emissivity of terrestrial materials in the 8–12 μm atmospheric window," *Remote Sens. Environ.* **42**, 83–106 (1992a).

Salisbury, J. W and D'Aria, D. M., "Emissivity of terrestrial materials in the 3–5 μm atmospheric window," *Remote Sens. Environ.* **46**, 1–25 (1993).

Salisbury, J. W and D'Aria, D. M., "Infrared (8–14 μm) remote sensing of soil particle size," *Remote Sens. Environ.* **42**, 157–165 (1992b).

Sandford, B. and Robertson, L., "Infrared reflectance properties of aircraft paint," *Proc. IRIS Spec. Group Targets, Backgr. Discrim.*, p.111 (1985).

Stefan, J., "Über die Beziehung zwischen der Wärmestrahlung und der Temperatur," *Sitzungen der mathematisch-naturwissenschaftlichen Klasse der kaiserlichen Akademie der Wissenschaften* **79**, 391–428 (1879).

Torrance, K. E. and Sparrow, E. M., "Theory for off specular reflection from roughened surfaces," *J. Opt. Soc. Am.* 57, 1105–1114 (1967).

U.S. Department of Agriculture, *Soil Taxonomy: A Basic System of Soil Classification for Making and Interpreting Soil Surveys*, Agriculture Handbook Number 436 (1999).

Van der Meer, F. D. and DeJong, S. M. (Eds.), *Imaging Spectrometry: Basic Principles and Prospective Applications*, Kluwer Academic, Dordrecht, Netherlands (2001).

Chapter 5
Remotely Sensed Spectral Radiance

A fundamental goal of hyperspectral remote sensing is to extract information related to the intrinsic spectral properties of remotely imaged objects. In a laboratory environment, where sensing environments can be carefully controlled, the spectral reflectance and transmission properties discussed previously can be accurately measured, and such intrinsic properties inferred from them. In a remote sensing context, however, the situation can be quite complex due to the uncontrolled nature of illumination sources, atmospheric properties, and other environmental variables. To deal with these issues, it is of paramount importance to at least understand the impact of these variables on remotely sensed spectral radiance, and ultimately compensate the collected data for such influences.

This chapter describes the principles of radiative transfer modeling, which is the methodology for understanding how radiation from a scene propagates to a sensor and is influenced by the environment. The first section focuses explicitly on characteristics of the atmosphere and natural illumination sources. The spectral nature of these environmental influences is explored based on a sophisticated radiative transfer model, which is grounded in empirical data of the intrinsic spectral properties of atmospheric constituents. The second section provides observation models for the received spectral radiance at the sensor, based on scene and environmental characteristics for solid, gaseous, and liquid objects. The remotely sensed observable is called pupil-plane spectral radiance, as it relates to the measureable quantity at an entrance pupil of a hyperspectral sensor.

5.1 Radiative Transfer Modeling

Radiative transfer is the physical process by which radiation from various sources interacts with objects of interest, the atmosphere, and their local environments, ultimately resulting in measured spectral radiance at the sensor aperture. Figures 5.1 and 5.2 provide pictorials depicting the

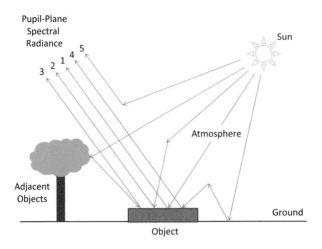

Figure 5.1   Radiative transfer processes for the solar reflective case.

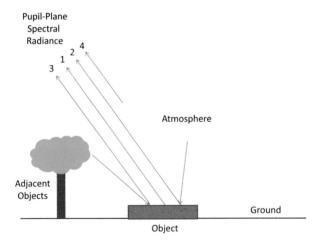

Figure 5.2   Radiative transfer processes for the thermally emissive case.

primary physical processes underlying a radiative transfer model for a flat object facet for both solar reflective and thermally emissive cases. In VNIR and SWIR spectral regions, optical radiation originating from the sun and reflecting from the object surface dominates the sensed spectral radiance. However, to give an accurate quantitative picture, indirect radiative transfer processes including atmospheric- and ground-scattered radiation must be included along with direct solar reflectance. In the LWIR, thermal emission is dominant, and solar radiation is negligible. This is not limited to direct emission from the object of interest, however, but includes emitted radiation from the atmosphere and local objects, including the surrounding ground plane that makes its way either directly or indirectly to the remote sensor. Because of dominant radiative-transfer processes, the VNIR/SWIR spectral region is often called the *reflective* spectral region, while the LWIR is called the *emissive* region. While often referred to as part of the emissive region, the MWIR contains comparable solar-reflected and thermally emitted components during the day.

In the solar-reflective case in Fig. 5.1, five primary components to spectral radiance are ultimately received by the sensor:

1. Direct solar irradiance reflected off of the object and transmitted through the atmosphere to the sensor.
2. Indirect solar irradiance scattered by the atmosphere, reflected off of the object, and transmitted through the atmosphere to the sensor.
3. Scattered irradiance from the ground and local objects reflected off of the object and transmitted through the atmosphere to the sensor.
4. Ground-scattered radiance scattered further by the atmosphere, reflected off of the object, and transmitted through the atmosphere to the sensor; this component is also known as the adjacency effect.
5. Upwelling path radiance scattered by the atmosphere directly to the sensor.

In this case, component 1 is called the *direct solar irradiance* component. Components 2, 3, and 4 are combined to make up the *indirect downwelling radiance* component. Component 5 is denoted as the *upwelling path radiance* component.

For the emissive case in Fig. 5.2, there are four primary components to spectral radiance that is ultimately received by the sensor:

1. Direct thermal emission of the object transmitted through the atmosphere to the sensor.
2. Downwelling thermal emission of the atmosphere reflected off of the object and transmitted through the atmosphere to the sensor.
3. Thermal emission from the ground and local objects reflected off of the object and transmitted through the atmosphere to the sensor.
4. Upwelling thermal emission of the atmospheric path directly to the sensor.

In this case, component 1 is called *direct thermal emission*, component 4 is denoted as the upwelling path radiance component, and components 2 and 3 make up the indirect downwelling component.

5.1.1 Atmospheric modeling

The earth's atmosphere clearly has a significant impact on all of the observed spectral radiance components; therefore, much of this chapter focuses on characterizing the nature of this impact. Spectral absorption characteristics of atmospheric gases have been previously discussed.

Spectral regions of atmospheric gas absorption exhibit a reduction in transmission over the remote imaging path but also an increase in upwelling and downwelling atmospheric radiation (i.e., radiation due to the imaging path itself) through thermal emission because emissivity of the atmosphere is nominally equal to its absorbance according to Kirchoff's law. Scattering in the atmosphere also has the impact of reducing overall transmission and increasing path radiance, particularly at shorter wavelengths.

Quantifying atmospheric impacts on radiative transfer is an extremely complex endeavor, from both measurement and modeling perspectives. In this chapter, characteristics are described based on detailed atmospheric modeling as opposed to empirical measurements. Atmospheric models numerically analyze radiative transfer through atmospheric paths in a manner similar to layered media, described in the previous chapter. Consider the situation depicted in Fig. 5.3 with an object at altitude h_0, a sensor at altitude h_N, a viewing path at an angle θ_r from the zenith direction, and solar source at a zenith angle of θ_s. The sensor is assumed to be oriented at (θ, ϕ) from the solar illumination direction, according to the spherical coordinate geometry described for scattering analysis in Chapter 2. Radiative transfer is modeled on a layer-by-layer basis, where the N layers are thin enough such that atmospheric parameters can be assumed to be constant within a layer. Each layer is characterized by absorption and scattering coefficients, and (unlike in the layered-media analysis) there are no interfaces between layers to cause reflection.

Extinction of radiation along the propagating path is caused by aggregate absorption and scattering of atmospheric constituents, both gases and aerosols, within a layer; radiation is added by both scattering into the path from the sun and thermal emission of the atmospheric constituents. Let $\beta_i(P, T)$ represent the absorption per unit concentration of

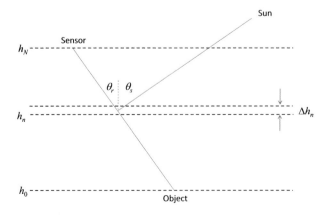

Figure 5.3   Geometry of layered atmospheric model.

the i'th atmospheric constituent for a specified pressure P and temperature T, and let C_i be the corresponding constituent concentration. Then the aggregate absorption coefficient is given by

$$\beta_a(P,T) = \sum_i C_i \beta_i(P,T). \qquad (5.1)$$

Similarly, an aggregate scattering coefficient β_s can be defined in this case, dependent only on constituent concentrations, or more specifically, size and shape distributions. Note that atmospheric pressure and temperature, along with the concentration of all of the atmospheric constituents, are altitude-dependent quantities, thus necessitating a layer-by-layer analysis. Typical pressure and temperature distributions for a standard midlatitude summer atmospheric model are given in Fig. 5.4, while Fig. 5.5 shows typical concentration distributions for water vapor, carbon dioxide, and ozone. The wavelength and altitude dependence of both β_a and β_s are implicit.

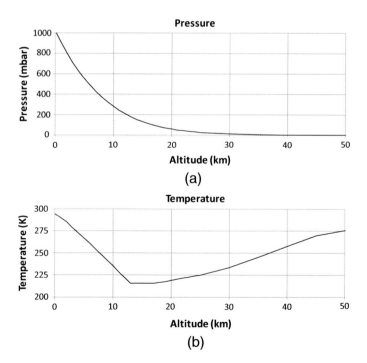

Figure 5.4　Typical altitude distribution of (a) pressure and (b) temperature.

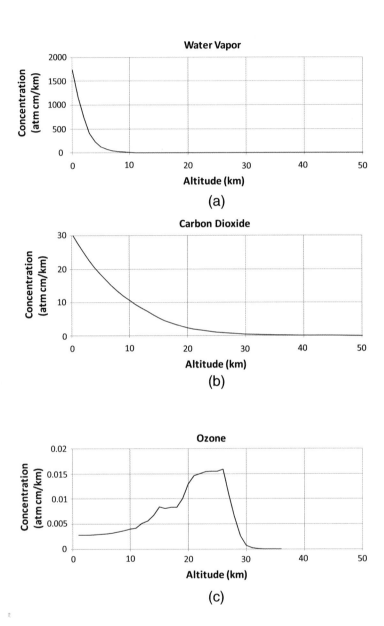

Figure 5.5 Typical concentration variation with altitude of (a) water vapor, (b) carbon dioxide, and (c) ozone.

Within a uniform layer, the differential equation characterizing radiative transfer (Smith, 1993) is

$$\cos\theta_r \frac{dL_\lambda(\lambda)}{dh} = -[\beta_a(P,T) + \beta_s]L_\lambda(\lambda)$$

$$+ \beta_a(P,T)B(\lambda,T) + \beta_s p(\theta,\phi)E_s(\lambda;h_n), \tag{5.2}$$

where $p(\theta,\phi)$ is the aggregate scattering phase function, $B(\lambda,T)$ is a blackbody radiance function, and $E_s(\lambda;h)$ is the solar spectral irradiance at layer altitude h_n. Considering propagation within the j'th layer from h_n to $h_n + \Delta h_n$, the propagating radiance is given from Eq. (5.2) by

$$L_\lambda(\lambda;h_n + \Delta h_n) = L_\lambda(\lambda;h_n)\, e^{-\frac{\beta_a(P,T)+\beta_s}{\cos\theta_r}\Delta h_n} + \beta_a(P,T)B(\lambda,T)\frac{\Delta h_n}{\cos\theta_r}$$

$$+ \beta_s p(\theta,\phi)E_s(\lambda;h_n)\frac{\Delta h_n}{\cos\theta_r}. \tag{5.3}$$

The exponential term in Eq. (5.3) represents atmospheric transmission through the layer. The second term is thermally emitted atmospheric path radiance, while the third term is atmospheric path radiance from scattered solar radiation.

The basic relationship in Eq. (5.3) can be used to propagate radiation from the object, either emitted or reflected, to the sensor aperture, given the characteristics of all atmospheric constituents as a function of altitude. To do so, the solar spectral irradiance $E_s(\lambda;h)$ must first be computed as a function of altitude. This is performed by considering a path from the top of the atmosphere, along the solar illumination path, down to the altitude of interest using the first term in Eq. (5.3), with exo-atmospheric solar spectral irradiance $E_{s,exo}(\lambda)$ as the originating source. If there are M total layers in the model atmosphere from the ground to space, then the solar spectral irradiance at the n'th layer is

$$E_s(\lambda;h_n) = E_{s,exo}(\lambda) \prod_{m=M}^{n} e^{-\frac{\beta_a(P,T)+\beta_s}{\cos\theta_r}\Delta h_n}. \tag{5.4}$$

The product term in Eq. (5.4) is the atmospheric transmission along the solar illumination path from the top of the atmosphere to the layer of interest. The two quantities of primary interest for remote sensing are the atmospheric path transmission from the object to the sensor,

$$\tau_a(\lambda) = \prod_{n=0}^{N} e^{-\frac{\beta_a(P,T)+\beta_s}{\cos\theta_r}\Delta h_n}, \tag{5.5}$$

and the atmospheric path radiance from the object to the sensor, assuming single scattering,

$$
L_a(\lambda) = \sum_{n=0}^{N} \left[\beta_a(P, T) B(\lambda, T) \frac{\Delta h_n}{\cos \theta_r} \right.
$$

$$
\left. + \beta_s p(\theta, \varphi) E_s(\lambda; h_n) \frac{\Delta h_n}{\cos \theta_r} \right] \left[\prod_{m=n}^{N} e^{-\frac{\beta_a(P,T)+\beta_s}{\cos \theta_r} \Delta h_m} \right]. \tag{5.6}
$$

The indirect downwelling radiance is similar in form to Eq. (5.6) but is integrated in the downward direction. The equations differ for more-sophisticated scattering models used in rigorous radiative-transfer codes outlined in the next section.

5.1.2 Moderate-resolution atmospheric transmission and radiation code

The results given in Eqs. (5.5) and (5.6) are based on a single-scattering radiative transfer model. In reality, light can be multiply scattered in the atmosphere, and more-sophisticated radiative transfer models account for this possibility. Furthermore, radiative transfer computations are typically performed in a spherical, earth-centered coordinate system, since viewing paths for remote sensors can be long enough such that the curvature of the earth and atmospheric layers must be considered. Finally, it is common to perform computations in wavenumber as opposed to wavelength spectral units because wavenumbers are better matched to gaseous spectral databases. For hyperspectral remote sensing, the most appropriate atmospheric modeling program is the moderate-resolution atmospheric transmission and radiance code (MODTRAN), which is capable of modeling atmospheric characteristics at a spectral resolution from 1 to 15 cm^{-1} using a correlated k-distribution band transmission model (Berk *et al.*, 1989; Anderson *et al.*, 1999). This model is based on the high-resolution transmission molecular absorption (HITRAN) database of measured atmospheric gas properties. It also utilizes a multiple-scattering discrete-ordinate radiative transfer program for a multilayered plane-parallel medium (DISORT) model to quantify atmospheric scattering processes.

As described in previous chapters, absorption characteristics of atmospheric gases exhibit very fine spectral structures that are spectrally resolved in the HITRAN database but not resolved at the coarser spectral resolution of MODTRAN. Performing radiative transfer calculations at HITRAN resolution and spectrally integrating the results to the MODTRAN resolution is an accurate modeling approach, but is computationally intractable for large spectral ranges associated with passive sensors. On the other hand, band averaging either the absorption or scattering coefficients prior to radiative transfer calculations is inaccurate because of the nonlinearity of the radiative transfer model. MODTRAN performs this accurately and efficiently using a numerical method called correlated k-distribution bands. To understand this numerical methodology, consider the problem of quantifying atmospheric transmission over a MODTRAN spectral band from σ_1 to σ_2 over a path length u for which the atmospheric parameters are constant. The layer transmission over this band is given by the integral

$$\tau = \frac{1}{\sigma_2 - \sigma_1} \int_{\sigma_1}^{\sigma_2} e^{-\beta_e(\sigma)u} d\sigma, \tag{5.7}$$

where $\beta_e(\sigma)$ is the aggregate extinction coefficient sampled at the HITRAN resolution. Numerically, this integral can be estimated by a summation of the form

$$\tau = \sum_{k=1}^{K} \alpha_k e^{-\beta_e(\sigma_k)u}, \tag{5.8}$$

where α_k are the quadrature weights at the $k = 1, 2, \ldots K$ spectral samples of $\beta_e(\sigma)$ associated with the numerical integration method used. Using the trapezoidal method, for example,

$$\alpha_k = \begin{cases} \Delta/2 & k = 1, N \\ \Delta & k \neq 1, N \end{cases}, \tag{5.9}$$

where

$$\Delta = \frac{\sigma_2 - \sigma_1}{N - 1}. \tag{5.10}$$

Because of the nonlinear nature of Eq. (5.8), the integral cannot be accurately estimated using a low-spectral-resolution form of $\beta_e(\sigma)$. To understand the k-distribution method, it is important to recognize that the order of the indices in Eq. (5.8) can be changed in any manner without impacting the result. Consider the mapping of $\beta_e(\sigma_k)$ into an order such that it is monotonically increasing or decreasing. The central point of the k-distribution method is that approximating Eq. (5.8) with a coarser set of samples for this remapped spectrum results in significantly less error than approximating it with the same number of samples for the original spectrum because of its inherent smoothness.

Mathematically, consider the mapping $\beta(g)$ that maps the smallest β to $g = 0$, the largest to $g = 1$, and all others such that they are linearly distributed between 0 and 1. Figure 5.6 compares the k-distribution to the original absorption spectrum for the carbon monoxide fundamental absorption band. The remapped absorption function, referred to as a k-distribution, resembles a cumulative distribution and is clearly a much smoother function. The transmission over the MODTRAN band is computed by

$$\tau = \sum_{k=1}^{K} \alpha_k e^{-\beta(g_k)\, u}, \tag{5.11}$$

where a smaller number of sample points can be used due to the well-behaved nature of the k-distribution. For a single computation, the computational complexity involved with determining and applying the mapping outweighs the benefit of a simpler numerical integration. However, since the atmospheric absorption line locations do not change with altitude and atmospheric conditions, by precomputing and storing the mapping, a substantial numerical benefit can be achieved in practice

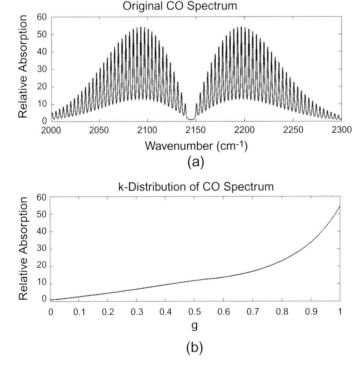

Figure 5.6 Application of k-distribution mapping to the carbon monoxide fundamental absorption band: (a) original spectrum and (b) corresponding k-distribution.

(Berk *et al.*, 1989). While the atmospheric absorption line locations do not change, their strengths and widths do change with concentration, pressure, and temperature. MODTRAN accommodates this by assuming that the mapping function is correlated over changes in these parameters, and extracting the mapping function from a lookup table based on the band-averaged line strength, the number of spectral lines within the band, the Lorentz bandwidth, and the Doppler bandwidth.

While MODTRAN is extensively used and referenced throughout this book as a standard radiative transfer code, it is not the only code used in the hyperspectral remote sensing field. Another rigorous radiative-transfer modeling capability, the Second Simulation of the Satellite Signal in the Solar Spectrum (6S), is also widely used in the field (Vermote *et al.*, 1997). The 6S code uses the same HITRAN 96 database as MODTRAN to model absorption by atmospheric gases, although it differs in computational method. Two other important differences include the modeling of aerosol scattering, where 6S uses a successive order of scattering method (Lenoble and Broyniez, 1984), and the incorporation of a bidirectional reflectance distribution function (BRDF) surface model to more accurately simulate pupil-plane spectral radiance. The BRDF model is not only designed to model simple opaque materials but also includes a fully developed vegetation canopy model based on user specification of tree-canopy parameters such as leaf area density and leaf orientation characteristics (Pinty and Verstraete, 1991). The spectral databases used in 6S support a spectral resolution of 2.5 nm, which is well suited for hyperspectral remote sensing analysis. The model compares favorably with MODTRAN, as described in a comparison study performed by Kotchenova *et al.* (2008).

5.1.3 Atmospheric path transmission

With a basic understanding of radiative transfer modeling in hand, we can examine spectral characteristics of the primary impacts of the atmosphere on remote sensing, namely atmospheric path transmission, atmospheric path radiance, and both direct and diffuse downwelling radiance. Multiple MODTRAN-generated results are discussed to understand the dependence of these quantities on viewing geometry and environmental conditions. The discussion begins in this section with an examination of atmospheric path transmission. As described in the previous section, molecular and atomic absorption properties of atmospheric gas constituents directly impact remote sensing by reducing the transmission over both object-to-sensor and source-to-object atmospheric paths. Figure 5.7 illustrates the nominal atmospheric transmission from ground to space over a zenith path for primary gas species as well as a typical composite atmosphere. The locations of gaseous absorption bands have been previously discussed and are not repeated here; however, it is important to recognize that even

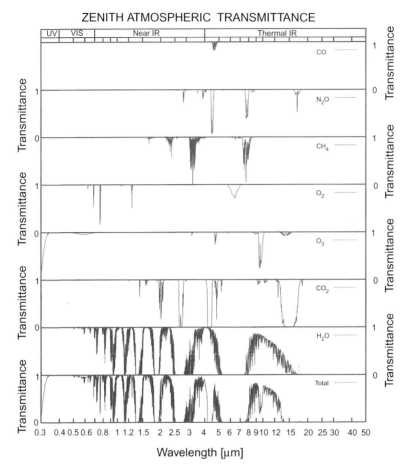

ZENITH ATMOSPHERIC TRANSMITTANCE

Wavelength [μm]

Figure 5.7 Typical ground-to-space atmospheric path transmission for a zenith path separated into constituent individual contributions [reprinted with permission from Petty (2004) © 2004 Sundog Publishing].

fairly weak absorption lines in atmospheric constituents can result in low transmission due to the long paths involved.

Atmospheric transmission is also impacted by extinction due to scattering from both atmospheric gas molecules and aerosols (dust, water droplets, etc.) that can exist in the atmosphere. Because of the small size of the scattering centers, molecular scattering can be approximated by the Rayleigh theory, resulting in an extinction coefficient that is inversely proportional to the fourth power of the optical wavelength. The result is that extinction is substantial only in the VNIR spectral range. Scattered radiation is blue in color, while transmitted radiation is reddened, as can be seen especially during sunset. Figure 5.8 illustrates the impact of molecular scattering on the zenith atmospheric transmission. The dashed black trend line is a Rayleigh scattering fit to the MODTRAN model based on a zenith

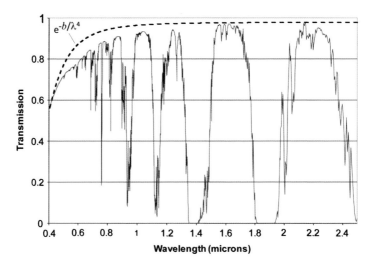

Figure 5.8 Impact of molecular scattering on zenith atmospheric transmission in the visible spectral region, depicted by comparing a desert atmospheric model with a Rayleigh scattering trend line.

atmospheric model with a desert model. The deviation of the actual trend in the transmission is due partly to the presence of larger scattering centers in the atmosphere, which result in an aggregate spectral dependence that is weaker than λ^4.

Atmospheric aerosols are often too large to be accurately modeled using the Rayleigh scattering theory. The Mie theory is appropriate under the approximation of spherical scattering centers. Figure 5.9 shows the range of particle sizes for atmospheric constituents and the spectral behavior of the modeled extinction efficiency, defined as the extinction cross section normalized to the aerosol physical cross section, for particles of various sizes (Petty, 2004). Molecular scattering, characterized by particles on the order of 0.1 nm, falls in the Rayleigh scattering regime in the visible spectral region and is negligible in the infrared. Hence, there is a λ^{-4} dependence of scattered light, causing its blue color. Atmospheric haze and other aerosols range in size from 0.1 to 10 μm, which begin to fall outside the Rayleigh regime at VNIR wavelengths. This causes scattered light to whiten. Cloud water droplets are on the order of 10 μm, such that scattering in the VNIR/SWIR spectral region is effectively white. Note that the locations of the ripples in the scattering distributions in Fig. 5.9(b) vary with particle size and are averaged out for typical particle size distributions. Note also that water droplet scattering at MWIR or LWIR wavelengths is overwhelmed by the high absorption.

Using MODTRAN, the primary influences on composite atmospheric-transmission profiles in both the VNIR/SWIR and MWIR/LWIR spectral regions are characterized in the remainder of this section. Three model

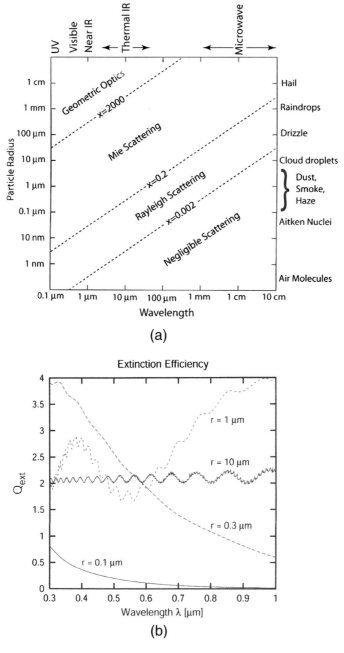

(a)

(b)

Figure 5.9 Characteristics of atmospheric scattering: (a) scattering regimes based on particle radius and spectral region, and (b) VNIR spectral behavior of the extinction coefficient for various particle sizes [reprinted with permission from Petty (2004) © 2004 Sundog Publishing].

atmospheres, summarized in Table 5.1, are used to investigate dependence on basic atmospheric conditions. These models are used to compare somewhat limiting atmospheric conditions, from dry desert to moist tropical conditions. VNIR/SWIR results are computed across the 0.4- to 2.5-μm spectral range at 10-cm^{-1} spectral resolution, while MWIR/LWIR results are computed over the 3- to 14-μm spectral range at 4-cm^{-1} spectral resolution.

The zenith ground-to-space transmission over the VNIR/SWIR spectral range is shown for the three model atmospheres in Fig. 5.10. There are two very apparent differences between the three models. The first is the difference in the absolute magnitude of the transmission, which is largely driven by absorption differences between the three atmospheric models. The second is a difference in the spectral shape of the transmission due to the varying extent of molecular and aerosol scattering over the dry-to-tropical conditions. Figure 5.11 illustrates the way in which VNIR/SWIR transmission varies with zenith angle over a ground-to-space path for the Ohio rural model. In this case, the spectral shape is the same, but absolute transmission varies as a function of path length. According to Eq. (5.5), zenith angle dependence should be roughly

$$\tau(\lambda;\theta) = [\tau(\lambda;0)]^{1/\cos\theta}. \qquad (5.12)$$

Table 5.1 Model atmosphere parameters for MODTRAN results.

Model	Characteristics
Baseline	Midlatitude summer model atmosphere, 23-km visibility, rural haze model. Rural area in midwest United States (Ohio), −84-deg longitude, 38-deg latitude, 4 July 2005 at noon local time, forest surface albedo.
Desert	Desert model atmosphere, 50-km visibility, desert haze model. Desert in southwest United States (California), −118-deg longitude, 35-deg latitude, 4 July 2005 at noon local time, desert surface albedo.
Tropical	Tropical model atmosphere, 5-km visibility, maritime haze model. Tropics in southeast United States (Florida), −80-deg longitude, 26-deg latitude, 4 July 2005 at noon local time, forest surface albedo.

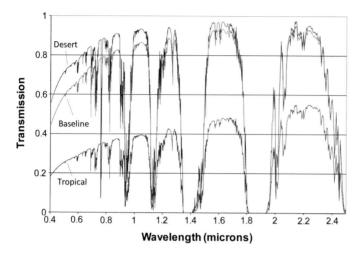

Figure 5.10   VNIR/SWIR atmospheric path transmission over a zenith path from ground to space.

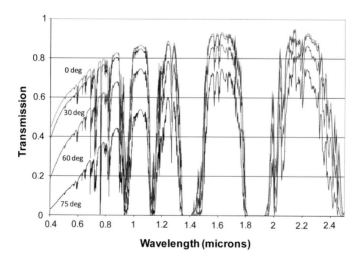

Figure 5.11   VNIR/SWIR atmospheric path transmission over a slant ground-to-space path for different zenith angles.

The impact of sensor altitude on transmission is shown in Fig. 5.12. Note that most of the loss in transmission occurs at altitudes below 5 km where the atmosphere is denser.

Figures 5.13, 5.14, and 5.15 illustrate the same atmospheric transmission results for the MWIR/LWIR spectral region. A few key differences between the MWIR/LWIR and VNIR/SWIR atmospheric transmission behavior should be noted. First, there is little difference for baseline and desert atmospheric conditions, while the tropical transmission is markedly worse due to water droplet scattering and the water absorption

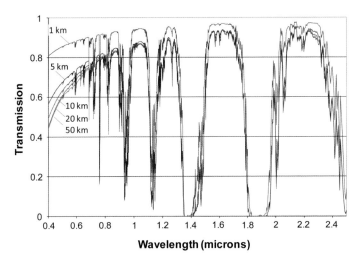

Figure 5.12 VNIR/SWIR atmospheric path transmission over a zenith path from ground to different altitudes.

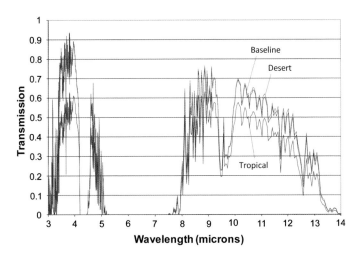

Figure 5.13 MWIR/LWIR atmospheric path transmission over a zenith path from ground to space.

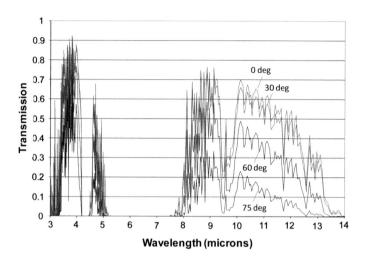

Figure 5.14   MWIR/LWIR atmospheric path transmission over a slant ground-to-space path for different zenith angles.

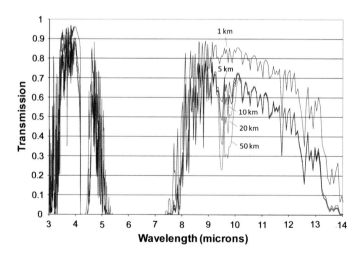

Figure 5.15   MWIR/LWIR atmospheric path transmission over a zenith path from ground to different altitudes.

continuum, the integrated tails of the multiple Lorentzian bands. The combination of these effects also results in a more considerable reduction in transmission as a function of zenith angle relative to the VNIR/SWIR case. Again, most of the transmission loss is attributed to altitudes below 5 km. The exception is the ozone absorption band from 9 to 10 μm, which dominates at higher altitudes.

5.1.4 Atmospheric path radiance

The same physical processes that are responsible for atmospheric transmission loss also result in the accumulated path radiance that occurs over a propagation path. Path radiance is defined as the component of radiance measured at the sensor end of an atmospheric path that does not originate from the source (e.g., ground object). That is, it originates along the propagation path itself. Atmospheric path radiance includes two parts. The first is thermal emission from atmospheric molecules. The second is scattering of radiation into the path from sources other than the object viewed along the sensor line of sight (LOS). One such source is solar radiation, either directly scattered by the atmosphere or indirectly scattered by multiple scattering from the atmosphere, ground, and other neighboring objects in the scene. Another source is thermal radiation that is scattered by the same mechanisms.

Figures 5.16, 5.17, and 5.18 illustrate VNIR/SWIR path radiance for an upwelling propagation path as a function of the same variables for which the transmission properties were characterized. A logarithmic scale is used for path radiance due to the large spectral dependence, attributed to the

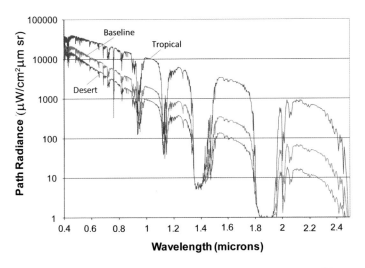

Figure 5.16 VNIR/SWIR atmospheric path radiance across a zenith ground-to-space path as a function of atmospheric conditions.

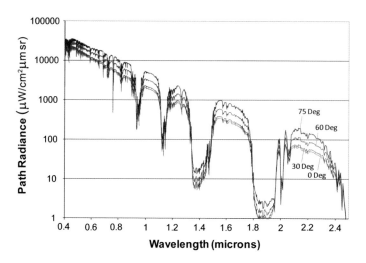

Figure 5.17   VNIR/SWIR atmospheric path radiance across a slant ground-to-space path as a function of zenith angle.

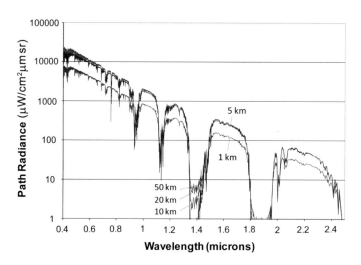

Figure 5.18   VNIR/SWIR atmospheric path radiance across a zenith path as a function of sensor altitude.

Rayleigh nature of the scattering processes that dominate path radiance in this spectral region, and resulting in its bluish color. Coincident with the nature of atmospheric transmission, the magnitude of VNIR/SWIR path radiance is substantially increased for moist atmospheric conditions and decreased for dry conditions. Note that the reduction in scattering with wavelength is less severe in the tropical case because of larger scattering centers, consistent with the Mie scattering analysis. The baseline, desert, and tropical models are based on specific assumptions of aerosol concentrations and size distributions; in a real atmosphere, these parameters can vary considerably, resulting in path radiance impacts that exhibit corresponding variability. Almost all of such path radiance is attributed to low altitudes and is moderately dependent on zenith angle.

Figures 5.19, 5.20, and 5.21 illustrate the corresponding results for the MWIR/LWIR spectral region, again for an entirely cloud-free propagation path. In this case, thermal radiative processes dominate scattering processes. To understand these results, it is helpful to bound them with the 230- and 280-K blackbody functions shown, as they exhibit a nominal spectral profile for thermally emitted radiation at high and low altitudes in the atmosphere, respectively. In 3- to 5-μm and 8- to 12-μm transmission windows, there is still considerable path radiance across the distributed path. Spectral regions of moderate transmission, such as the edges of atmospheric windows at 5 and 8 μm, correspond to areas of high path radiance because the received radiance originates from lower altitudes and is at a higher temperature, approaching the 280-K bounding blackbody function. Regions of high absorption, such as the 4.2-μm carbon dioxide absorption band, result in lower path radiance due to the

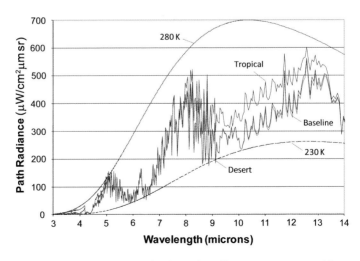

Figure 5.19 MWIR/LWIR atmospheric path radiance across a zenith ground-to-space path as a function of atmospheric conditions.

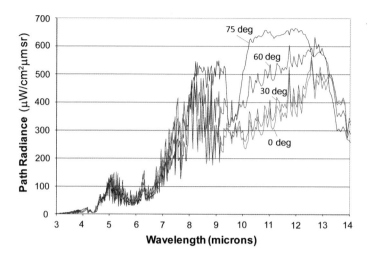

Figure 5.20 MWIR/LWIR atmospheric path radiance across a slant ground-to-space path as a function of zenith angle.

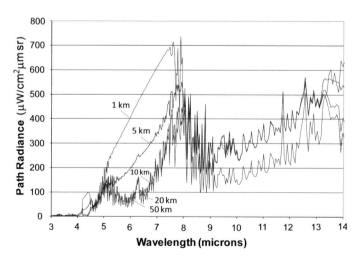

Figure 5.21 MWIR/LWIR atmospheric path radiance across a zenith path as a function of sensor altitude.

higher originating altitudes, approaching the 230-K bounding blackbody function. The moister tropical atmosphere produces significantly higher path radiance in both MWIR and LWIR atmospheric windows due to the higher atmospheric water content.

MWIR/LWIR path radiance only changes significantly as a function of zenith angle across the 8- to 12-μm transmission window. Elsewhere, increased radiation across the longer path lengths appears to be somewhat balanced by increased absorption of the radiated energy. In terms of altitude dependence, emitted atmospheric radiance in the carbon dioxide and water absorption bands is substantial at low altitudes, but the effective path radiance across longer paths to higher altitudes decreases because absorption in the colder high altitude dominates. In the atmospheric windows, low altitude emission dominates.

5.1.5 Downwelling radiance

In the VNIR/SWIR spectral region, almost all radiation collected by a remote sensor originates from the sun. At night, manmade and other celestial sources are appreciable, but their illumination levels are not generally adequate for hyperspectral remote sensing. Bright manmade sources or natural sources such as fires can produce significant daytime or nighttime radiance levels, but these are atypical for most applications. Thus, to understand VNIR/SWIR downwelling radiance, it is sufficient to understand direct and indirect solar illumination.

The sun is a G-class star with a mean radius of 695,000 km located at a mean distance from the earth of 149.68×10^6 km. It is essentially a heated plasma that radiates almost as a blackbody, with the exception that cooler solar atmospheric gases influence its spectral and spatial characteristics (Zissis, 1993). Figure 5.22 illustrates its exo-atmospheric solar irradiance compared to a 5770-K blackbody function based on the radiometry of earth–sun geometry, and a 5900-K blackbody normalized to the same integrated irradiance across the spectral range. From a standpoint of total integrated irradiance (1390 W/m^2), the 5770-K blackbody achieves the best approximation to the integrated exo-atmospheric solar irradiance. A normalized 5900-K blackbody function, however, is the best fit to solar spectral distribution.

Solar illumination at any other altitude within the atmosphere, including at ground level, is highly affected by absorption and scattering along the space-to-ground atmospheric path in the manner described by Eq. (5.4). Figure 5.23 illustrates direct solar irradiance at the ground for the three model atmospheres. While there is some difference due to a slight variation in solar zenith angles for local noon at the different geographical positions, the irradiance profiles are predominately equal to the product of the exo-atmospheric irradiance in Fig. 5.22 with the atmospheric transmission

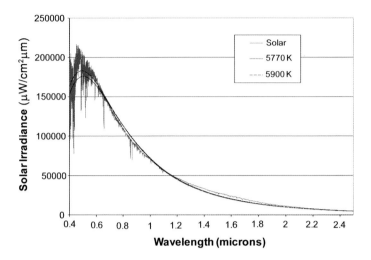

Figure 5.22   Exo-atmospheric solar spectral irradiance.

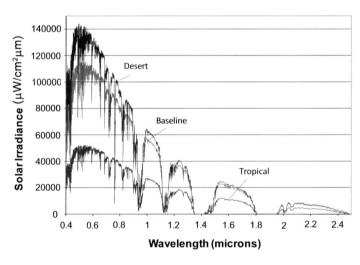

Figure 5.23   Direct solar irradiance at the ground as a function of atmospheric conditions.

profiles presented in Fig. 5.7. Diurnal variation with direct solar irradiance is illustrated in Fig. 5.24 for the baseline case. The change in time of day is due both to the change in space-to-ground transmission as a function of the solar zenith angle, as well as the cosine dependence of projecting the irradiance onto the ground plane. Finally, the impact of clouds on direct solar irradiance is shown in Fig. 5.25. Cirrus clouds represent a scattering loss, but direct irradiance transmitted through the clouds is still substantial. For cumulus and stratus clouds, however, solar irradiance is almost completely scattered, leaving essentially no direct component.

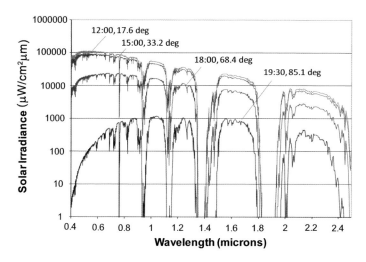

Figure 5.24 Direct solar irradiance at the ground as a function of time of day and solar zenith angle.

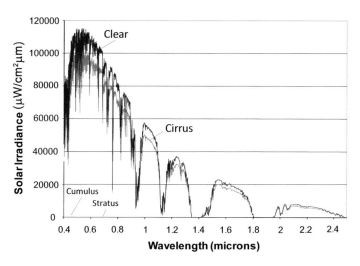

Figure 5.25 Direct solar irradiance at the ground as a function of cloud conditions.

Scattered solar irradiance, including multiply scattered radiation within the atmosphere and between the atmosphere and ground, makes up the diffuse solar downwelling radiance component. This is illustrated in Figs. 5.26, 5.27, and 5.28 for the same variables as the direct component. A logarithmic scale is used because of significant spectral and environmental variation. The diffuse solar irradiance component is highly correlated with the scattered path radiance previously investigated, and the trends with atmospheric conditions and solar zenith angle are the same. The impact of clouds illustrated in Fig. 5.28 depends on the size distribution of the water droplets that compose the clouds. The Rayleigh nature of cirrus cloud scattering indicates smaller scattering centers relative to cumulus and stratus clouds, which exhibit different spectral behavior. In particular, note the extremely low downwelling radiance in the SWIR for the cumulus cloud case. These cloud scattering results are based on solid cloud decks; scattered clouds are much more complex sources of scattering.

Unlike the VNIR/SWIR spectral region, thermally emitted radiation dominates solar radiation in the MWIR/LWIR spectral region. The exception to this is in the short-wavelength end of the MWIR spectral region during daytime hours. Consider a 300-K surface with a spectrally independent emissivity of 0.8 and reflectance of 0.2 that is illuminated according to the baseline atmospheric conditions. Figure 5.29 compares the emitted component to the reflected components in the 3- to 5-μm spectral region. The reflected component includes reflection of both the direct solar irradiance, which decreases with wavelength and dominates at the 3-μm end of the spectral region, and the emitted downwelling irradiance, which dominates at the 5-μm end. At wavelengths above 4 μm,

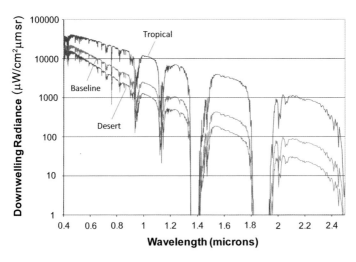

Figure 5.26 VNIR/SWIR diffuse downwelling radiance as a function of atmospheric conditions.

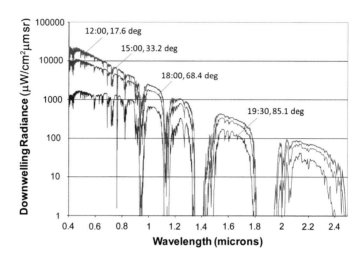

Figure 5.27 VNIR/SWIR diffuse downwelling radiance as a function of time of day and solar zenith angle.

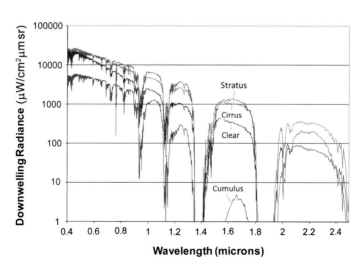

Figure 5.28 VNIR/SWIR diffuse downwelling radiance as a function of cloud conditions.

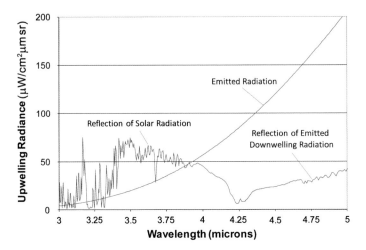

Figure 5.29 Comparison of emitted and reflected radiance components in the MWIR spectral region for daytime atmospheric conditions and a 20% reflective 300-K surface.

emitted radiance begins to dominate total reflected radiance. This trend continues in the LWIR spectral region where solar irradiance can be completely ignored. Recall that this example is for local noon, where solar radiation peaks. The crossover wavelength between reflected and emitted components shifts to shorter wavelengths as the solar zenith angle increases later in the day.

Figure 5.30 illustrates downwelling sky radiance in the MWIR/LWIR spectral region for the three atmospheric conditions. As in the

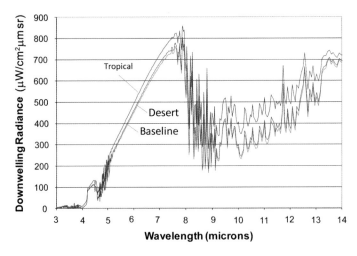

Figure 5.30 MWIR/LWIR diffuse downwelling radiance as a function of atmospheric conditions.

VNIR/SWIR case, these results correlate well with the zenith path radiance results previously shown. As expected, downwelling radiance is somewhat larger for the tropical case due to the increased atmospheric water concentration at low altitude. The impact of various cloud types is shown in Fig. 5.31. Cirrus clouds for the relatively thin cloud thickness modeled in this case have a negligible impact, although such clouds can become thick enough to impact radiometric properties. Cumulus and stratus clouds result in a downwelling radiance distribution resembling that of a perfect blackbody. This is consistent with the high emissivity of liquid water across the MWIR/LWIR spectral range. The variation in LWIR atmospheric downwelling emission from cirrus clouds is examined more closely in Fig. 5.32. These data, collected by a Fourier transform infrared (FTIR) spectrometer, are given in wavenumber units, which alter the shape of the spectral distribution (see Chapter 8 for more information on this instrument and mapping wavelength to wavenumber spectral radiance units). The results indicate the relative variation of cirrus emission between the limits of clear sky and liquid cloud emission. Again, the liquid cloud appears nominally as a blackbody. Further information on cloud emission is provided by Shaw *et al.* (2005) and Nugent *et al.* (2009).

5.2 Remote Sensing Models

Combining the radiative-transfer modeling capabilities described in the previous section with a characterization of the apparent spectral properties of remotely sensed materials of interest, we are now in a position to put the pieces together to form a complete model for pupil-plane spectral

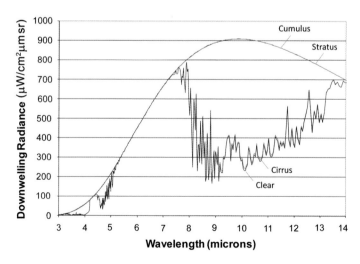

Figure 5.31 MWIR/LWIR diffuse downwelling radiance as a function of cloud conditions.

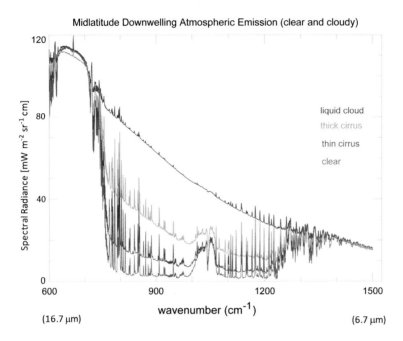

Figure 5.32   LWIR downwelling radiance measurements for different cirrus cloud thicknesses compared to clear sky and liquid clouds. [Graphic courtesy of Joseph Shaw, Montana State University.]

radiance for various remote sensing situations. Three such situations are addressed in this section. The first considers sensing of a solid material, where spectral signature information is captured within a combination of reflected and emitted radiation from the object. The second concerns hyperspectral sensing of gaseous effluents, where the spectral apparent property is the spectral transmission through the gas plume. Finally, the situation of spectral sensing in shallow water bodies is discussed. In this case, a water body is represented by an apparent spectral reflectance that relates all of the constituent properties of the materials within the water, as well as the bottom surface.

5.2.1 Facet model for a solid surface

We first consider remotely observed spectral radiance for a general flat, solid surface. Since complex objects can be decomposed (both theoretically and numerically) into a superposition of flat facets, this model serves as the basis for terrestrial remote imaging in a generalized sense. In fact, one can construct a sophisticated scene model from first principles by numerically assembling a complex scene from a large set of facets, and performing radiometrically accurate modeling of each facet in light of the atmosphere, environment, illumination, observation geometry, and interactions between facets. This is the basic approach taken in the Digital Image and Remote Sensing Image Generation (DIRSIG) code developed at the Rochester Institute of Technology for physics-based modeling for remote sensing applications (Schott, 2007).

Mathematical treatment of the general single-facet model is based on a spherical coordinate system defined with respect to the facet, such that the surface normal corresponds to $\theta = 0$, as shown in Fig. 5.33. The facet is characterized by its BRDF $\rho_{BRDF}(\theta_r, \phi_r, \theta, \phi, \lambda)$ and surface temperature T and is in a radiative environment characterized by direct solar irradiance $E_s(\theta_s, \phi_s, \lambda)$ and diffuse downwelling radiance $L_d(\theta, \phi, \lambda)$, where (θ_s, ϕ_s) corresponds to the solar position relative to spherical facet coordinates (θ, ϕ). The remote sensor is assumed to be in the direction of (θ_r, ϕ_r) in the facet-fixed coordinate system.

The upwelling radiance $L_u(\lambda)$ from the solid surface is the sum of thermally emitted and reflected components,

$$L_u(\theta_r, \phi_r, \lambda) = L_e(\theta_r, \phi_r, \lambda) + L_r(\theta_r, \phi_r, \lambda), \tag{5.13}$$

where the emitted component is given by

$$L_e(\theta_r, \phi_r, \lambda) = B(\lambda, T)\left[1 - \int_0^{2\pi}\int_0^{\pi/2} \rho_{BRDF}(\theta_r, \phi_r, \theta, \phi, \lambda)\cos\theta\sin\theta\, d\theta\, d\varphi\right], \tag{5.14}$$

the reflected component is given by

$$L_r(\theta_r, \phi_r, \lambda) = \rho_{BRDF}(\theta_r, \phi_r, \theta_s, \phi_s, \lambda)E_s(\theta_s, \phi_s, \lambda)$$
$$+ \int_0^{2\pi}\int_0^{\pi/2} \rho_{BRDF}(\theta_r, \phi_r, \theta, \phi, \lambda)L_d(\theta, \phi, \lambda)\cos\theta\sin\theta\, d\theta\, d\varphi, \tag{5.15}$$

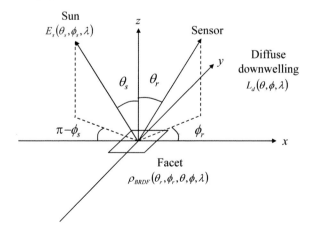

Figure 5.33 Single-facet model geometry.

and $B(\lambda, T)$ is a blackbody spectral radiance distribution for the surface temperature. Note that Eq. (5.14) is based on the assumptions governing Kirchoff's law, and the bracketed term represents the effective emissivity for the viewing direction. At the entrance aperture of a remote sensor at the indicated viewing orientation, spectral radiance is given by

$$L_p(\lambda) = \tau_a(\lambda) L_u(\theta_r, \phi_r, \lambda) + L_a(\lambda), \qquad (5.16)$$

where $\tau_a(\lambda)$ and $L_a(\lambda)$ correspond to the atmospheric path transmission and path radiance over the facet-to-sensor propagation path, respectively, that is, the quantities described in Sections 5.1.3 and 5.1.4. This radiance is often referred to as *pupil-plane radiance*, while that shown in Eq. (5.13) is often known as *ground-plane radiance*.

Under the Beard–Maxwell BRDF model, the BRDF is decomposed into a specular component ρ_s, a diffuse surface component ρ_d, and a volumetric scattering component ρ_v, according to

$$\rho_{BRDF}(\theta_r, \phi_r, \theta, \phi, \lambda) = \rho_s(\theta_r, \phi_r, \theta, \phi, \lambda) + \rho_d(\lambda) + \frac{2\rho_v(\lambda)}{\cos\theta + \cos\theta_r}.$$

$$(5.17)$$

With this decomposition, the emitted and reflected upwelling components in Eqs. (5.14) and (5.15) become

$$L_e(\theta_r, \phi_r, \lambda) = B(\lambda, T)\left[1 - \int_0^{2\pi}\int_0^{\pi/2}\rho_s(\theta_r, \phi_r, \theta, \phi, \lambda)\cos\theta\sin\theta\, d\theta\, d\varphi\right]$$

$$+ B(\lambda, T)[1 - \pi\rho_d(\lambda)]$$

$$+ B(\lambda, T)\left[1 - \rho_v(\lambda)\int_0^{2\pi}\int_0^{\pi/2}\frac{2\cos\theta\sin\theta}{\cos\theta + \cos\theta_r}d\theta\, d\varphi\right] \quad (5.18)$$

and

$$L_r(\theta_r, \phi_r, \lambda) = \rho_s(\theta_r, \phi_r, \theta_s, \phi_s, \lambda)E_s(\theta_s, \phi_s, \lambda)$$

$$+ \int_0^{2\pi}\int_0^{\pi/2}\rho_s(\theta_r, \phi_r, \theta, \phi, \lambda)L_d(\theta, \phi, \lambda)\cos\theta\sin\theta\, d\theta\, d\varphi$$

$$+ \rho_d(\lambda)\left[E_s(\theta_s, \phi_s, \lambda) + \int_0^{2\pi}\int_0^{\pi/2}L_d(\theta, \phi, \lambda)\cos\theta\sin\theta\, d\theta\, d\varphi\right]$$

$$+ \rho_v(\lambda)\left[\frac{2E_s(\theta_s, \phi_s, \lambda)}{\cos\theta_s + \cos\theta_r} + \int_0^{2\pi}\int_0^{\pi/2}L_d(\theta, \phi, \lambda)\frac{2\cos\theta\sin\theta}{\cos\theta + \cos\theta_r}d\theta\, d\varphi\right]$$

$$(5.19)$$

When the facet can be approximated as a diffuse Lambertian reflector $\rho_s = 0$,

$$\rho_d(\lambda) = \frac{\rho_{DHR}(\lambda)}{\pi} = \frac{\rho_{HDR}(\lambda)}{\pi}, \tag{5.20}$$

where DHR is directional hemispherical reflectance and HDR is hemispherical directional reflectance, as defined in Chapter 4, and $\rho_v = 0$, such that only the second lines of Eqs. (5.18) and (5.19) are required.

The relationship for pupil-plane radiance greatly simplifies under this diffuse reflector assumption to

$$L_p(\lambda) = \tau_a(\lambda)\,[1 - \rho_{DHR}(\lambda)]\,B(\lambda, T)$$
$$+ \frac{\tau_a(\lambda)\rho_{DHR}(\lambda)}{\pi}[E_s(\theta_s, \phi_s, \lambda) + E_d(\lambda)] + L_a(\lambda), \tag{5.21}$$

where $E_d(\lambda)$ is the integrated downwelling irradiance of the facet from indirect sources, given by

$$E_d(\lambda) = \int_0^{2\pi} \int_0^{\pi/2} L_d(\theta, \phi, \lambda)\cos\theta\sin\theta d\theta d\phi. \tag{5.22}$$

Often, angular dependence of the downwelling radiance is ignored and Eq. (5.22) simplifies further to

$$E_d(\lambda) = \pi L_d(\lambda). \tag{5.23}$$

For facets in shadow, an angularly independent approximation can be employed with the integral in Eq. (5.22) performed only over the effective solid angular region Ω_d of unobstructed viewing of the sky,

$$E_d(\lambda) = \Omega_d\,L_d(\lambda), \tag{5.24}$$

where Ω_d ranges from 0 to π in order to account for the cosine reduction in irradiance with angle θ. This assumption ignores the scattering of radiation onto the facet from obstructing material. We can characterize the term Ω_d/π as a scalar shadow coefficient.

In the reflective (VNIR/SWIR) spectral region, the first term in Eq. (5.21) can be ignored because solar irradiance dominates thermal emission. This leads to the diffuse facet model for the reflective spectral region:

$$L_p(\lambda) = \frac{\tau_a(\lambda)\rho_{DHR}(\lambda)}{\pi}[E_s(\theta_s, \phi_s, \lambda) + E_d(\lambda)] + L_a(\lambda). \tag{5.25}$$

In the emissive spectral region (LWIR and the long-wavelength end of the MWIR), solar irradiance can be ignored. Again employing the diffuse Lambertian assumption, the pupil-plane radiance becomes

$$L_p(\lambda) = \tau_a(\lambda)[1 - \rho_{DHR}(\lambda)]B(\lambda, T) + \frac{\tau_a(\lambda)\rho_{DHR}(\lambda)}{\pi}E_d(\lambda) + L_a(\lambda),$$

(5.26)

where $E_d(\lambda)$ is again defined in Eq. (5.22). Note that reflected radiation cannot be ignored in the emissive spectral region because downwelling radiance is generally at a level that is comparable to emitted radiance. Also, the approximation discussed for dealing with facets in shadow for the reflective case is not generally a good approximation in the emissive region due to the significant amount of emitted energy from local obstructing materials. In the MWIR spectral region, Eq. (5.26) can be used at night, but Eq. (5.21) must be used during the day to account for both solar and thermal influences.

The downwelling irradiance component of $E_d(\lambda)$ in Eqs. (5.25) and (5.26) can be very difficult to model due to potential effects of local obscuration, scattering, reflection, and emission. Variations of this model can be tailored to account more explicitly for these local or adjacency effects instead of implicitly incorporating them into angularly dependent downwelling radiance. In the reflective spectral region, the model can be modified as

$$L_p(\lambda) = \frac{A(\lambda)}{1 - \rho_e(\lambda)S(\lambda)}\rho_{DHR}(\lambda) + \frac{B(\lambda)}{1 - \rho_e(\lambda)S(\lambda)}\rho_e(\lambda) + L_a(\lambda),$$

(5.27)

where

$$A(\lambda) = \frac{\tau_a(\lambda)}{\pi}\Big[F_s E_{s,exo}(\lambda)\tau_s(\lambda)\cos\theta_s + F_d E_d(\lambda)\Big] + L_a(\lambda),$$ (5.28)

$\tau_s(\lambda)$ is the atmospheric transmission along the direct solar path to the object of interest, $B(\lambda)$ is the product of total ground irradiance and scattering from the local surround into the LOS, $S(\lambda)$ is spherical albedo accounting for ground-scattered radiance onto the object of interest, α is a scalar that accounts for whether the object is occluded from the direct solar irradiance or not ($\alpha = 0$ in shadow, $\alpha = 1$ otherwise), and β is a measure of the fraction of the hemisphere that is not occluded by local objects from the perspective of the object of interest ($\beta = \Omega_d/\pi$). In the emissive spectral region, the model can be modified to

$$L_p(\lambda) = \tau_a(\lambda)\left[1 - \rho_{DHR}(\lambda)\right]B(\lambda, T) + \beta\,\tau_a(\lambda)\frac{\rho_{DHR}(\lambda)}{\pi}E_d(\lambda)$$
$$+ (1 - \beta)\,\tau_a(\lambda)\rho_{DHR}(\lambda)B(\lambda, T_e) + L_a(\lambda), \tag{5.29}$$

where β is defined as before and T_e is the effective temperature of local occluding objects. Equations (5.27) and (5.29) are referred to as adjacency models.

5.2.2 Gaseous effluent model

Several hyperspectral remote sensing applications involve the sensing of gaseous effluents in the form of clouds or plumes above the ground's surface (Griffin *et al.*, 2005), according to the basic picture in Fig. 5.34. Since the interesting spectral features of such gaseous materials almost always reside in the MWIR/LWIR spectral region, a diffuse emissive model is used to describe the pupil-plane radiance for such a situation. The plume is characterized by its temperature T_p, thickness d, and the concentrations C_i and normalized absorption coefficients β_i, corresponding to the gaseous constituents that are assumed to be uniformly mixed in the plume. According to Beer's law, plume transmission is given by

$$\tau_p(\lambda) = e^{-\sum_i C_i \beta_i d}. \tag{5.30}$$

Ignoring scattering, emissivity equals unity minus the transmission at all wavelengths according to Kirchoff's law, which is ordinarily assumed to be valid.

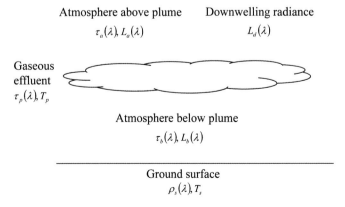

Atmosphere above plume
$$\tau_a(\lambda), L_a(\lambda)$$

Downwelling radiance
$$L_d(\lambda)$$

Gaseous effluent
$$\tau_p(\lambda), T_p$$

Atmosphere below plume
$$\tau_b(\lambda), L_b(\lambda)$$

Ground surface
$$\rho_s(\lambda), T_s$$

Figure 5.34 Gaseous effluent model definitions.

Ground surface is characterized by its directional hemispherical reflectance, denoted by $\rho_s(\lambda)$ and surface temperature T_s. Kirchoff's law is again assumed to apply, such that emissivity equals unity minus reflectance at all wavelengths. The downwelling radiance $L_d(\lambda)$ is defined just above the plume layer and is assumed to be angularly independent. The atmospheric path transmission and path radiance are separated into components above the plume, $\tau_a(\lambda)$ and $L_a(\lambda)$, and below the plume, $\tau_b(\lambda)$ and $L_b(\lambda)$.

Pupil-plane radiance consists of five basic components: (1) surface emission through the plume layer, (2) upwelling emission from the plume, (3) downwelling radiance reflected off of the surface through the plume layer, (4) downwelling plume emission reflected off of the surface back through the plume, and (5) path radiance. Pupil-plane radiance can be written mathematically in the corresponding order as

$$
\begin{aligned}
L_p(\lambda) = {} & \tau_a(\lambda)\,\tau_b(\lambda)\,\tau_p(\lambda)\,[1 - \rho_s(\lambda)]\,B(\lambda, T_s) \\
& + \tau_a(\lambda)\,[1 - \tau_p(\lambda)]\,B(\lambda, T_p) \\
& + \tau_a(\lambda)\,\tau_b(\lambda)\,\tau_p(\lambda)\,\rho_s(\lambda)\,[L_b(\lambda) + \tau_b(\lambda)\,\tau_p(\lambda)\,L_d(\lambda)] \\
& + \tau_a(\lambda)\,\tau_b^2(\lambda)\,\tau_p(\lambda)\,\rho_s(\lambda)\,[1 - \tau_p(\lambda)]\,B(\lambda, T_p) \\
& + L_a(\lambda) + \tau_a(\lambda)\,\tau_p(\lambda)\,L_b(\lambda).
\end{aligned}
\tag{5.31}
$$

In most applications, the plume layer is relatively close to the ground surface, such that it is reasonable to let $\tau_b(\lambda) = 1$ and $L_b(\lambda) = 0$, and to redefine $\tau_a(\lambda)$ and $L_a(\lambda)$ to represent the terrestrial imaging situation without the plume, that is, transmission and path radiance all the way to the ground. With these approximations, the pupil-plane radiance expression simplifies to

$$
\begin{aligned}
L_p(\lambda) = {} & \tau_a(\lambda)\,\tau_p(\lambda)\,[1 - \rho_s(\lambda)]\,B(\lambda, T_s) \\
& + \tau_a(\lambda)\,[1 - \tau_p(\lambda)]\,B(\lambda, T_p) \\
& + \tau_a(\lambda)\,\tau_p^2(\lambda)\,\rho_s(\lambda)\,L_d(\lambda) \\
& + \tau_a(\lambda)\,\tau_p(\lambda)\,[1 - \tau_p(\lambda)]\,\rho_s(\lambda)\,B(\lambda, T_p) \\
& + L_a(\lambda).
\end{aligned}
\tag{5.32}
$$

This can be written in the form

$$
L_p(\lambda) = a\tau_p^2(\lambda) + b\tau_p(\lambda) + c,
\tag{5.33}
$$

where

$$
a = \tau_a(\lambda)\,\rho_s(\lambda)\,[L_d(\lambda) - B(\lambda, T_p)],
\tag{5.34}
$$

$$
b = \tau_a(\lambda)\,[1 - \rho_s(\lambda)]\,[B(\lambda, T_s) - B(\lambda, T_p)],
\tag{5.35}
$$

and

$$c = \tau_a(\lambda) \, B(\lambda, T_p) + L_a(\lambda). \qquad (5.36)$$

In remote sensing of gaseous effluents, plume transmission $\tau_p(\lambda)$ represents the signal to be sensed; therefore, Eq. (5.33) characterizes pupil-plane radiance in terms of a quadratic function of the signal. The coefficient of the quadratic term is related to the contrast between downwelling radiance and blackbody emission associated with plume temperature. This term is often small because of similarity in apparent temperature between the plume layer and the atmosphere above it. In general, however, it cannot be completely ignored. The linear coefficient is related to the product of surface emissivity and blackbody contrast between the plume and surface temperatures. This is usually the primary signal term, although it becomes small when surface reflectance increases or the plume and surface temperature difference becomes small. Therefore, the magnitude of the remotely sensed signature of a gaseous effluent is extremely dependent on the reflectance and temperature characteristics of the surface beneath it.

5.2.3 Shallow-water model

Scattering processes within natural bodies of water have been touched on in previous sections, especially in the context of understanding the volume-scattered reflectance spectrum. In a shallow body of water, the situation is a bit more complex than in deep bodies of water because in shallow water spectral characteristics of the bottom surface also come into play. Theoretically, such a situation could be approached in a fashion similar to a gaseous effluent. However, there are some substantial differences, including: (1) hyperspectral sensing of water bodies is performed at visible wavelengths where water is transparent, (2) scattering processes play a vital role in radiative transfer, and thermal emission can be ignored, and (3) assumptions of uniformity of the optical properties with depth are generally invalid.

Due to the complexity of scattering processes involved in radiative transfer within a natural body of water, a completely analytical approach to modeling is not typically used. A detailed numerical scattering model called HydroLight (Mobley and Lundman, 2000) has been developed for this application. HydroLight is the analog to the MODTRAN atmospheric model for characterizing radiative transfer in natural bodies of water. Semi-analytical models based on HydroLight and empirical measurements have also been developed to provide more-transparent insight into the relationships between observed characteristics such as apparent top surface reflectance and physical characteristics of a body of water such as clarity, turbidity, depth, and bottom reflectance (Mobley *et al.*, 1993; Mobley, 1994). One such semi-analytical model (Lee *et al.*, 1998) is described in this section as an example.

Based on the physical picture depicted in Fig. 5.35 for light interaction within a body of water in the visible spectral region, subsurface volume reflectance $\rho_s(h, \lambda)$ for a water depth h is related to that for an infinitely deep body of water, the albedo $\rho_s(\infty, \lambda)$, according to

$$\rho_s(h, \lambda) = \rho_s(\infty, \lambda)[1 - A_0 e^{-[D_d(\theta_s)+D_{uc}(\lambda)]\alpha(\lambda)h}] + A_1 \rho_b(\lambda) e^{-[D_d(\theta_s)+D_{ub}(\lambda)]\alpha(\lambda)h}. \tag{5.37}$$

The first term accounts for reduction in volume reflectance as depth decreases, in that it equals an infinitely deep reflectance for large h and equals zero for $h = 0$. The second term accounts for reflected light from the bottom and becomes zero for large h. The attenuation coefficient $\alpha(\lambda)$ is related to the absorption and backscattering coefficients $a(\lambda)$ and $b(\lambda)$ according to

$$\alpha(\lambda) = a(\lambda) + b(\lambda), \tag{5.38}$$

where spectral dependence is explicitly shown for clarity. The bottom is assumed to be Lambertian and is characterized by DHR $\rho_b(\lambda)$. Subsurface reflectance is defined as the ratio of the reflected radiance to the incident solar irradiance at a subsurface zenith angle θ_w. It is, therefore, a BRDF with sr^{-1} units that is assumed to be independent of the receiver angle; that is, it is also Lambertian and must be multiplied by π to represent DHR.

The remaining terms in Eq. (5.37) are coefficients used to match the physically based equation with empirical measurements. Ideally, $A_0 = 1$ and $A_1 = 1/\pi$. The best-fit values to empirical data, however, are $A_0 = 1.03$ and $A_1 = 0.31$. $D_d(\theta_w), D_{uc}(\lambda),$ and $D_{ub}(\lambda)$ represent the angular

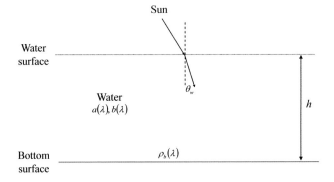

Figure 5.35 Shallow-water model definitions.

and directional characteristics of the vertically integrated downwelling column, upwelling column, and upwelling bottom extinction, respectively, relative to $\alpha(\lambda)$. Ideally, they are all constants.

According to the two-stream model discussed in Chapter 4, deep subsurface reflectance is given by

$$\rho_s(\infty, \lambda) = 0.5u(\lambda), \tag{5.39}$$

where

$$u(\lambda) = \frac{b(\lambda)}{a(\lambda) + b(\lambda)}, \tag{5.40}$$

and the downwelling and upwelling backscatter coefficients are assumed to be equal. The parameter $u(\lambda)$ is related to water clarity, with $u = 0$ representing clear water, and higher values representing decreased clarity. The semi-analytical model for deep-water subsurface reflectance is

$$\rho_s(\infty, \lambda) = [g_0 + g_1 \, u(\lambda)^{g_2}] \, u(\lambda), \tag{5.41}$$

where best-fit coefficients are $g_0 = 0.07, g_1 = 0.155$, and $g_2 = 0.752$. Similarly, best-fit parametric forms of the relative extinction constants are

$$D_d(\theta_s) = \frac{1}{\cos \theta_w}, \tag{5.42}$$

$$D_{uc}(\lambda) = 1.2 \, [1 + 2.0u(\lambda) \,]^{1/2}, \tag{5.43}$$

and

$$D_{ub}(\lambda) = 1.1 \, [1 + 4.9u(\lambda)]^{1/2}. \tag{5.44}$$

Absorption coefficient $a(\lambda)$ is modeled as the sum of three parts: pure-water absorption, phytoplankton absorption, and gelbstoff absorption (Smith and Baker, 1981). Pure-water absorption is based on the optical constants of water and has been previously described. Phytoplankton absorption is modeled as

$$a_p(\lambda) = [a_0(\lambda) + a_1(\lambda) \ln a_p(440)] \, a_p(440), \tag{5.45}$$

where absorption is in m^{-1} units, wavelength is in nanometer units, and

$$a_p(440) = 0.06C_a^{0.65}, \tag{5.46}$$

where C_a is the concentration of chlorophyll a, the primary absorbing constituent, in mg/m^3 units. The terms $a_0(\lambda)$ and $a_1(\lambda)$ are fitting parameters displayed in Fig. 5.36. The resulting phytoplankton absorption for typical C_a values ranging from 0.4 to 5 mg/m^3 are shown in Fig. 5.37. Gelbstoff absorption is modeled as

$$a_g(\lambda) = a_g(440) \, e^{-0.014(\lambda-440)}, \tag{5.47}$$

where $a_g(440)$ ranges from 0.05 to 0.3 m^{-1}. Figure 5.38 compares the three absorption terms for $C_a = 1$ mg/m^3 and $a_g(440) = 0.1$ m^{-1}.

Backscattering coefficient $b(\lambda)$ is modeled as the sum of clear-water scattering and particle scattering; the latter is dependent on water turbidity. Particle scattering is related to phytoplankton content and water turbidity and is modeled as

$$b(\lambda) = BC_a^{0.62} \frac{550}{\lambda}, \tag{5.48}$$

where B is a turbidity parameter. A nominal value of B is 0.3, and values in the range of 5 reflect high turbidity. Figure 5.39 illustrates the particle backscattering coefficient for three turbidity levels. Molecular scattering in clear water is generally insignificant relative to particle scattering except in the clearest-water conditions. Figure 5.40 illustrates empirical data for molecular scattering in seawater and freshwater.

When the various components described in this section are inserted into Eq. (5.37), subsurface reflectance due to volume scattering and bottom reflectance can be estimated. For remote sensors, however, the apparent

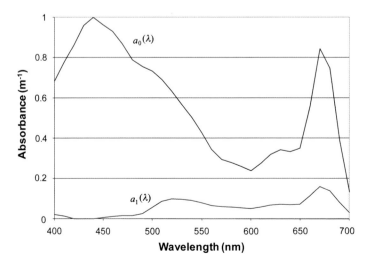

Figure 5.36 Phytoplankton absorption fitting parameters.

reflectance from above the water's surface is the quantity of interest. The relationship between the apparent and subsurface reflectance spectra is not as simple as one might expect, due to the total internal reflection that occurs upon upwelling through the air–water interface. Recall that the analysis is based on a diffuse model, and the reflectance values must account for such off-axis effects. The apparent reflectance in terms of DHR $\rho_{DHR}(\lambda)$ is related to subsurface reflectance through

$$\rho_{DHR}(\lambda) = \frac{0.518\rho_s(\lambda)}{1 - 1.562\rho_s(\lambda)}, \tag{5.49}$$

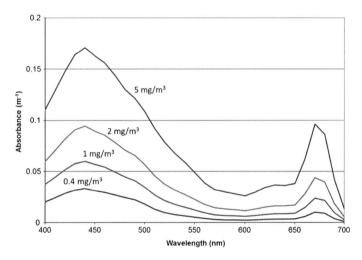

Figure 5.37 Phytoplankton absorption for typical chlorophyll a concentration levels.

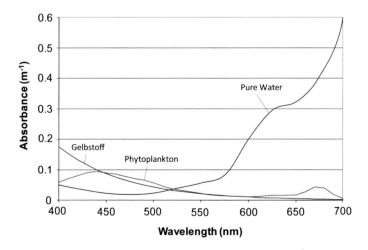

Figure 5.38 Comparison of absorption terms for natural water bodies.

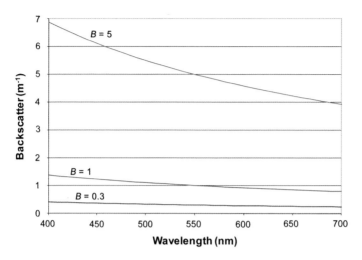

Figure 5.39 Particle backscattering coefficients for three turbidity levels.

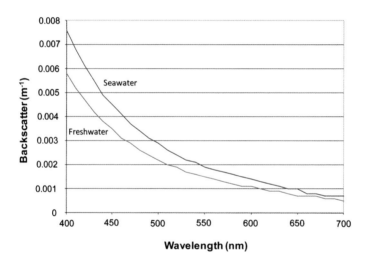

Figure 5.40 Molecular backscattering coefficients for clear water.

where the form of Eq. (5.49) is based on theory, and the coefficients are empirically determined.

Using the reflectance model based on Eqs. (5.37) and (5.49) and the supporting equations, a nonlinear analytical relationship is established between the apparent reflectance, a remotely observable quantity, and a small number of water body characteristics: chlorophyll *a* concentration, gelbstoff content, turbidity, depth, and bottom type or corresponding reflectance. In a forward-modeling sense, remotely sensed radiance can be derived from these parameters, along with corresponding solar and atmospheric characteristics. On the other hand, the analytical model also provides a basis for estimating underlying water characteristics from remotely sensed radiance.

5.3 Summary

Clearly, propagation of scene radiation through the atmosphere to a sensor has a profound effect on observed spectral radiance and must be considered in any hyperspectral remote sensing application. The radiative transfer process is very complicated and requires a sophisticated numerical model such as MODTRAN to characterize the expected effects. On the other hand, radiative transfer can be distilled down to three major atmospheric influences: atmospheric transmission, path radiance, and downwelling radiance. Each is strongly dependent on weather conditions, atmospheric conditions, and observation geometry, for which there is typically much uncertainty and variation.

If atmospheric properties can be established, it is possible to combine them with the apparent spectral properties of materials of interest to form an observation model for pupil-plane spectral radiance measured by a remote hyperspectral sensor system. Models vary depending on whether the imaged objects are solid materials, gaseous effluents, or parameters of water bodies, and simplifying assumptions such as Kirchoff's law and Lambertian surfaces are often made to make models more tractable. However, more-sophisticated computer models such as DIRSIG and HydroLight can treat more-generalized situations based on, as an example, a full bidirectional reflectance distribution function (BRDF) description of solid surfaces extracted from extensive materials databases. These methods use numerical processing techniques in a computationally efficient manner to estimate the integral equations at each point in a synthetic scene.

References

Anderson, G. P., Berk, A., Acharya, P. K., Matthew, M. W., Bernstein, L. S., Chetwynd, J. H., Dothe, H., Adler-Golden, S. M., Ratkowski, A. J., Felde, G. W., Gardner, J. A., Hoke, M. L., Richtsmeier, S. C., Pukall, B., Mello, J. B., and Jeong, L. S., "MODTRAN4: Radiative transfer modeling for remote sensing," *Proc. SPIE* **3866**, 2–10, 1999 [doi:10.1117/12.371318].

Berk, A., Bernstein, L. S., and Robertson, D. C., *MODTRAN: A Moderate Resolution Model for LOWTRAN 7*, Air Force Geophysics Laboratory Report GL-TR-89-0122, Spectral Sciences, Burlington, MA (1989).

Griffin, M. K., Czerwinski, R. N., Upham, C. A., Wack, E. C., and Burke, H. K., "A procedure for embedding plumes into LWIR imagery," *Proc. SPIE* **5806**, 78–87 (2005) [doi:10.1117/12.604895].

Kotchenova, S. Y., Vermote, E. F., Levy, R., and Lyapustin, A., "Radiative transfer codes for atmospheric correction and atmospheric retrieval: an intercomparison study," *Appl. Optics* **47**, 4455 –4464 (2008).

Lee, Z., Carder, K. L., Mobley, C. D., Steward, R. G., and Patch, J. S., "Hyperspectral remote sensing for shallow waters," *Appl. Optics* **37**, 6329–6338 (1998).

Lenoble, J. and Broyniez, C., "A comparative review of radiation aerosol models," *Beitr. Phys. Atmos.* **57**, 1–20 (1984).

Mobley, C. D., *Light and Water: Radiative Transfer in Natural Waters*, Academic Press, New York (1994).

Mobley, C. D., Gentili, B., Gordon, H. R., Jin, Z., Kattawar, G. W., Morel, A., Reinersman, P., Stamnes, K., and Starn, R. H., "Comparison of numerical models for computing underwater light fields," *Appl. Optics* **32**, 7484–7504 (1993).

Mobley, C. D. and Lundman, L. K., *HydroLight 4.1 Technical Documentation*, Sequoia Scientific Inc., Seattle, WA (2000).

Nugent, P. W., Shaw, J. A., and Piazzolla, S., "Infrared cloud imaging in support of Earth-space optical communication," *Opt. Express* **17**, 7862–7872 (2009).

Petty, G. W., *A First Course in Atmospheric Radiation*, Sundog, Madison, WI (2004).

Pinty, B. and Verstraete, M. M., "Extracting information on surface properties from directional reflectance measurements," *J. Geophys. Res.* **96**, 2865–2874 (1991).

Schott, J. R., *Remote Sensing: The Image Chain Approach*, Second Edition, Oxford University Press, Oxford, New York (2007).

Shaw, J., Nugent, P., Pust, N., Thurairajah, B., and Mizutani, K., "Radiometric cloud imaging with an uncooled microbolometer thermal infrared camera," *Opt. Express* **13**, 5807–5817 (2005).

Smith, F. G. (Ed.), *Atmospheric Propagation of Radiation*, Volume 2 of *The Infrared and Electro-Optical Systems Handbook*, SPIE Press, Bellingham, WA (1993).

Smith, R. C. and Baker, K. S., "Optical properties of the clearest natural waters (200–800 nm)," *Appl. Optics* **20**, 177–184 (1981).

Vermote, E. F., Tanre, D., Deuze, J. L., Herman, M., and Morcrette, J., "Second simulation of the satellite signal in the solar spectrum (6S): an overview," *IEEE T. Geosci. Remote* **35**, 675–685 (1997).

Zissis, G. J. (Ed.), *Sources of Radiation*, Volume 1 of *The Infrared and Electro-Optical Systems Handbook*, SPIE Press, Bellingham, WA (1993).

Chapter 6

Imaging System Design and Analysis

The preceding chapters of this book have taken the remote sensing data model introduced in Chapter 1 and explored the observable spectral information content for materials, and the way in which this material is transferred to the input of the sensor in the form of pupil-plane spectral radiance. At this point, we can turn our attention to sensor instrumentation, in this case, imaging spectrometry. Before doing so, however, we will take a slight diversion in this chapter to overview the principles of conventional EO/IR imaging sensors, major sensor components, and metrics and methods for characterizing sensor system performance. For the reader familiar with EO/IR imaging sensor technology, this chapter can serve as a useful refresher; for the unfamiliar reader, it can serve as a tutorial. For both, it will provide a launching point from which to address the design and performance of the imaging spectrometers described in the subsequent three chapters. Some familiarity with the basic principles of geometrical optics (Smith, 1990) and Fourier optics (Goodman, 1996) is assumed in the treatments provided. For more details on these topics, the reader is encouraged to consult references provided at the end of this chapter such as Pedrotti and Pedrotti (1987) and Goodman (1985).

6.1 Remote Imaging Systems

Generically, a remote imaging system consists of the six major subsystems depicted in Fig. 6.1: imaging optics, focal plane array (FPA), readout electronics, data processing, pointing and stabilization, and calibration system. The imaging optics capture incident radiation and focus it onto the FPA so that there is correspondence between FPA image position and incident field angle or object position. From a radiometric perspective, the optics map the incident radiance from a scene position into the focal plane irradiance at the corresponding focal plane position. The FPA usually consists of an array of detectors, each of which converts the incident irradiance into a detected electrical signal, such as current or voltage.

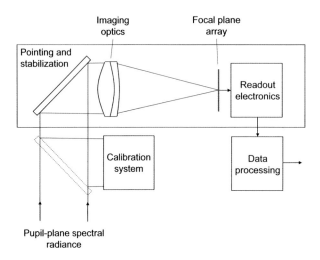

Figure 6.1   Generic schematic of a passive EO/IR imaging system.

The readout electronics multiplex analog signals from the entire array of detectors and perform the requisite amplification, filtering, and digitization to form a digital image for subsequent data processing. Most of this chapter focuses on the design and performance of these subsystems. Emphasis is given to conventional broadband imaging systems, while the extension to hyperspectral systems is left for subsequent chapters devoted to specific imaging spectrometer designs.

Airborne and space-based remote sensors generally reside on moving platforms; therefore, it is necessary to properly point and stabilize their LOS to form quality imagery. The term *pointing* is used in reference to moving the LOS of the imaging system in a prescribed manner with respect to the scene so that it forms an image of the scene, while *stabilization* refers to compensating for moving platform disturbances such as vibration and attitude variations to achieve fine image quality. Earlier generations of remote imaging sensors employed either single detectors or small detector arrays that needed to be scanned in two dimensions to form an image. The recent advent of large detector arrays, however, has enabled more-efficient approaches to obtaining area coverage.

Current-generation systems are typically based on either 1D (linear) or 2D (staring) detector arrays. When using 1D arrays, the LOS must be scanned in at least one direction to form an image. Two common pointing methods, called pushbroom and whiskbroom scanning, are depicted in Figs. 6.2 and 6.3. In pushbroom scanning, the linear FPA is oriented along the cross-track direction and is pushed along the flight direction at a fixed cross-track angle as the platform moves forward. In this mode, the linear FPA size directly dictates the swath or image width in the cross-track direction. In whiskbroom scanning, the FPA is oriented in the along-track

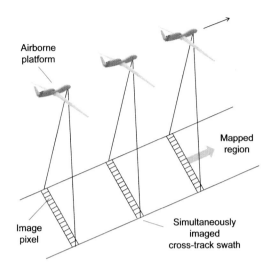

Figure 6.2 Pushbroom imaging geometry using linear FPAs.

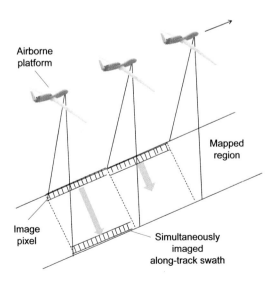

Figure 6.3 Whiskbroom imaging geometry using linear FPAs.

direction and is scanned cross-track as the platform moves forward. This allows the swath width to be somewhat decoupled from platform velocity but generally requires a more complex pointing and stabilization system using some combination of a scan mirror and mechanical gimbal. Note that there is a limiting case of a whiskbroom scanner, where a single detector element is scanned over the full image volume. For a given swath width, the use of more detectors allows for slower angular scanning and longer detector integration times to maintain the same coverage rate relative to this fully scanned approach, which has advantages in terms of sensor performance as described later in this chapter.

The use of 2D arrays provides the ability to capture a complete image without any line-of-sight scanning. When the platform is moving, the LOS must be stabilized during the frame integration time to avoid image smearing. This is performed by some combination of stabilized optics, such as a scan mirror, and mechanical assemblies, such as gimbals. Pointing the actual sensor platform, known as body pointing in space-based remote sensor design, can also be used for line-of-sight scanning and stabilization. Since typical remote imaging sensors require area coverage much greater than that provided with a single FPA, such systems are often operated with step-stare pointing, depicted in Fig. 6.4. This chapter addresses the influence of a basic pointing strategy on imaging optics and FPA design, as well as imaging system performance; however, it does not address the optomechanical design of pointing and stabilization systems in any further detail.

The nature of calibration and data processing subsystems is very application dependent. In situations where absolute radiometry is

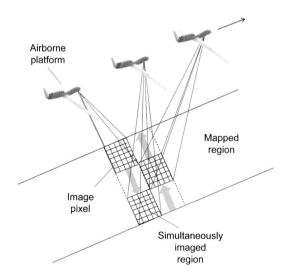

Figure 6.4   Step-stare imaging geometry using 2D FPAs.

demanded, calibration sources are used as references to provide traceability between raw collected data in digital units and a radiometric unit, such as pupil-plane spectral radiance. Even when absolute radiometry is not necessary for a particular application, a calibration strategy is usually needed to reduce image degradations due to FPA imperfections. Discussion of sensor calibration, even for conventional imaging systems, is deferred until Chapter 10. Similarly, data processing methods for various hyperspectral remote sensing applications are deferred to subsequent chapters. Instead, we focus on optical system design, FPAs, and system performance analysis in this chapter.

6.2 Optical System Design

The ultimate purpose of an optical system is to produce an irradiance image on the FPA that faithfully represents the radiance distribution of the scene with some magnification M. Defining (x, y) as the FPA coordinates and (x_g, y_g) as corresponding object (or scene) coordinates, FPA irradiance distribution $E_g(x, y)$ ideally equals

$$E_g(x, y) = K L_p\left(\frac{x}{M}, \frac{y}{M}\right),\qquad(6.1)$$

where $L_p(x_g, y_g)$ is the pupil-plane radiance received as a function of object coordinates and K is a scale factor. For the time being, any spectral dependence of these quantities is ignored. The subscript $E_g(x, y)$ denotes that this is the ideal geometric image. The basic parameters of an optical system are its focal length f and aperture diameter D (as depicted in Fig. 6.5), which lead to magnification

$$M = \frac{f}{R}\qquad(6.2)$$

for a common case of infinite conjugates, where the object range R is much greater than the focal length. Another quantity of interest is the f-number, $f/\# = f/D$.

6.2.1 Point spread function

In practice, all optical systems are limited in their ability to produce the ideal geometric image in Eq. (6.1). Such imperfections can be loosely placed into four categories: aberrations, distortion, absorption, and diffraction. An ideal wavefront for a point object will be spherical in nature as it exits an optical system and will converge to a perfect point on the focal plane. Aberrations refer to deformations of the real wavefront from the ideal that cause the rays to land at different locations on the focal plane

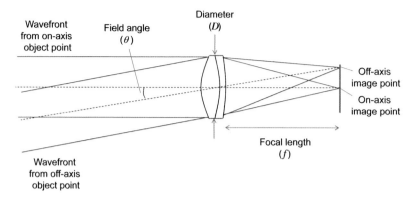

Figure 6.5 Basic parameters of an optical imaging system.

and broaden the image of a point object. This results in a blurring or loss of spatial resolution in the image, the extent of which is proportional to the magnitude of the aberration. Distortion is a particular type of aberration that causes the magnification to change as a function of image position, meaning that the image will not be a perfectly scaled rendition of the object, as shown in Eq. (6.1). Absorption decreases the transmission of pupil-plane radiance through the optical system to the FPA, reducing image contrast relative to system noise level. Finally, diffraction from the finite aperture size provides the ultimate limit to the achievable resolution. According to Fourier optics analysis (Goodman, 1985), the actual focal plane image for a linear, shift-invariant optical system is given by

$$E(x, y) = E_g(x, y) * h(x, y), \tag{6.3}$$

where $*$ represents a 2D spatial convolution operation, and $h(x, y)$ is the system point spread function (PSF). For a perfect, circularly symmetric optical system with the characteristics given here, the diffraction-limited PSF is given by the Airy function,

$$h_d(x, y) = \frac{4J_1^2\left(\frac{\pi\sqrt{x^2+y^2}}{\lambda(f/\#)}\right)}{\left(\frac{\pi\sqrt{x^2+y^2}}{\lambda(f/\#)}\right)^2}, \tag{6.4}$$

one dimension of which is displayed in Fig. 6.6, where $J_1(x)$ is a Bessel function of the first kind (Lebedev, 1972). The peak-to-null width of this function,

$$\varepsilon_0 = 1.22\,\lambda\,(f/\#), \tag{6.5}$$

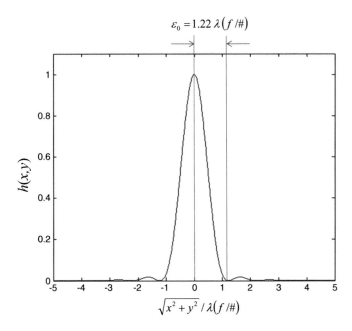

$$\varepsilon_0 = 1.22\,\lambda\left(f\,/\#\right)$$

The x-axis is labeled $\sqrt{x^2 + y^2}\,/\,\lambda\left(f\,/\#\right)$ and the y-axis is labeled $h(x,y)$.

Figure 6.6 PSF for a diffraction-limited optical system with circular aperture.

also known as the Airy radius, represents the fundamental limit to image resolution. However, geometrical aberrations of the optical system broaden the PSF; therefore, a fuller understanding of optical system design and performance must include the control and characterization of potential geometrical aberrations.

To maximize image quality, an optical system must limit aberrations over the entire extent of the image, control image distortion, maximize optical transmission, reduce any stray radiation that might impinge on the focal plane, and minimize the f-number, the latter of which both increases focal plane irradiance and decreases the Airy radius. Of course, optimization of image quality invariably becomes a trade-off between these competing design issues. Minimization of aberrations at one point in the field of view (FOV) tends to increase aberrations elsewhere. Also, aberrations become exceedingly difficult to correct when the f-number of the optical system is reduced to increase light efficiency and decrease the effects of diffraction. Effective aberration control requires many degrees of freedom in the design, allowing more optical surfaces to be optimized. However, this tends to lessen optical transmission due to both increased absorption within the optical elements and reflection losses at the interfaces between them. The primary goal of an optical designer is to effectively balance these competing design requirements in some manner that is optimal with respect to the desired imaging characteristics.

6.2.2 Optical aberrations

While a detailed understanding of optical aberrations is beyond the scope of this chapter, a brief examination of the nature of aberrations is helpful to further understand these performance limitations. Aberrations are often characterized as the optical path difference $W(\xi, \eta)$ between the real wavefront exiting the optical system for some incident field angle, and the corresponding ideal spherical wavefront that converges to a perfect point on the focal plane at the paraxial location (Smith, 1990). Normalized coordinates are used such that (ξ, η) represents the Cartesian pupil coordinates, ρ represents the radial pupil coordinate and ranges from zero to one across the aperture extent, h represents the normalized field coordinate and ranges from zero to one across the defined FOV, and $W(\xi, \eta)$ is normalized to some nominal wavelength, usually the center wavelength of the imaging spectral band. One wave corresponds to a 2π phase difference for the center wavelength between a real and ideal wavefront.

Third-order aberrations of a radially symmetric optical system are limited by symmetry to the form

$$W(x, y) = A_0 + A_1\xi + A_2\eta + A_3\rho^2 + A_4(\xi^2 - \eta^2)$$
$$+ A_5(2\xi\eta) + A_6\rho^2\xi + A_7\rho^2\eta + A_8\rho^4. \tag{6.6}$$

The coefficients A_i depend on characteristics of the optical system and can be a function of both field angle and wavelength. They can be computed by tracing paraxial rays through the optical system; thus, a change in the way marginal and chief rays traverse the optical system has an impact on aberrations. Third-order aberrations of primary interest, including field angle dependence, are summarized in Table 6.1. Spherical aberration is the only on-axis aberration and is due to marginal rays being refracted either too shallowly or too steeply. Astigmatism represents a cylindrical off-axis power such that the effective focal length for the image point differs in tangential (radial) and sagittal (orthogonal) directions. Field curvature is the tendency of the optical system to exhibit best focus on a curved rather than flat image plane and is closely related to astigmatism. Distortion does not impact the shape of the PSF but does impact the position of the PSF relative to the ideal optical system. It is often increased in favor of reducing other aberrations to achieve better spatial resolution at the expense of a distorted image.

A final type of aberration, chromatic aberration, is the result of the dispersion of optical components. Third-order analysis includes both axial chromatic aberration, which is a change in optical power with wavelength, and lateral chromatic aberration, which is a change in magnification with wavelength. Dispersion of optical glass is characterized by the Abbe

Table 6.1 Primary third-order optical aberrations.

Aberration	$W(\xi, \eta)$
Spherical	$W_{040}(\xi^2 + \eta^2)^2$
Coma	$W_{131}h(\xi^2 + \eta^2)\eta$
Astigmatism	$-0.5\, W_{222}h^2(\xi^2 - \eta^2)$
Field curvature	$(W_{220} + 0.5W_{222})\, h^2(\xi^2 - \eta^2)$
Distortion	$W_{311}h^3\eta$

number

$$V = \frac{n-1}{\Delta n}, \tag{6.7}$$

where n is the refractive index at the center of the spectral range of interest, and Δn is a measure of the change in index across the spectral range. For the visible region, the center wavelength is usually specified at $\lambda_d = 587.56$ nm, with the difference computed between $\lambda_c = 656.27$ nm and $\lambda_F = 486.13$ nm. Low-dispersion glasses with a high Abbe number are called crown glasses, while high-dispersion glasses are called flint glasses.

Each of the wavefront aberrations discussed can be correspondingly described in terms of a ray aberration, or the deviation of a ray from an (ξ, η) location in the exit pupil of the imaging system relative to the ideal paraxial image point. If one were to trace a number of rays from the object point through different pupil locations and measure the ray aberration for each, a fairly direct assessment of image blur would be provided. To examine this further, consider an example: an optical design of an $f/2$ camera lens with a 50-mm focal length designed for a 35-mm format image. The basic design depicted in Fig. 6.7 is a double-Gaussian arrangement based on two achromatic doublets placed about a central stop. The stop limits the effective aperture diameter to 25 mm. Achromatic doublets consist of a positive focal length (or converging) crown-glass lens element and a negative focal length, or diverging flint-glass lens element, that balance chromatic aberration over the visible spectral range. Symmetry around the stop attempts to make surface refractions closer to aplanatic, a condition for which spherical aberration and coma for all orders are both zero. It also aids in reducing field curvature and astigmatism. The initial and final lens elements provide additional degrees of freedom to reduce off-axis aberrations over the full FOV. A complete description of the optical design, including lens materials, curvatures, and distances, is given in Table 6.2. This specification was developed by performing a least-squares optimization of lens curvatures and distances from a nominal double-Gaussian starting point to minimize ray aberrations

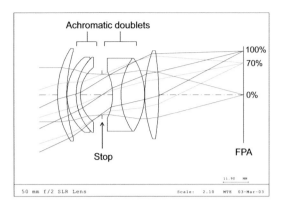

Figure 6.7   Optical ray trace of a 50-mm, $f/2$ camera lens.

Table 6.2   Optical specification of a 50-mm, $f/2$ camera lens.

SURFACE	RADIUS (cm)	THICKNESS (cm)	MATERIAL
OBJ:	INFINITY	INFINITY	
1:	38.75910	4.500000	LAF2_SCHOTT
2:	66.74342	0.100000	
3:	22.41910	4.335897	LAF2_SCHOTT
4:	23.86190	2.000000	K5_SCHOTT
5:	15.89393	10.243310	
STO:	INFINITY	5.000000	
7:	−24.66139	4.000000	SF4_SCHOTT
8:	29.21277	11.000000	LAF2_SCHOTT
9:	−30.51781	0.100000	
10:	58.23509	6.547165	LAF2_SCHOTT
> 11:	−130.04011	40.271813	
IMG:	INFINITY	−0.271813	

across three field locations (on-axis, 70% full field, and 100% full field) and three wavelengths (λ_d, λ_F, and λ_C).

Ray aberrations for this optical system are depicted in Fig. 6.8. Each plot illustrates the ray aberration in millimeters as a function of position of the analyzed fan of rays in the pupil for the three design wavelengths. The plots on the left are for the tangential direction, meaning the pupil direction parallel to the vector between the optical axis and the off-axis object point, and the plots on the right are for the sagittal direction, meaning the pupil direction orthogonal to this object height vector. Because of the circular symmetry of the optical system, these two directions fully specify the aberration, and the sagittal aberration becomes symmetric about the optical axis. Therefore, the x axes in the plots in Fig. 6.8 extend from one edge of the pupil to the other in the tangential case, and from the optical axis to the edge of the pupil in the sagittal case. From top to bottom, the plots

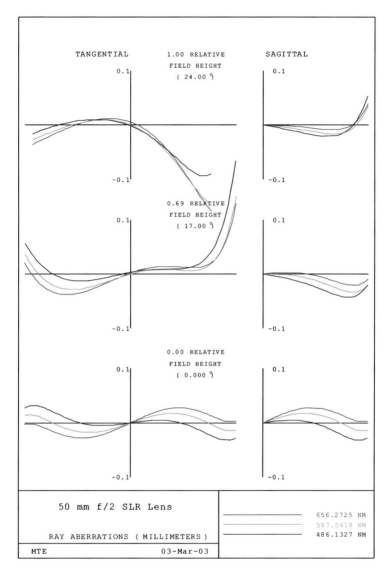

Figure 6.8 Ray aberrations for a 50-mm, $f/2$ camera lens.

show performance at full field, 70% full field, and on-axis. The difference in the plots for the three colors is indicative of chromatic aberration. In this case, monochromatic aberrations exceed chromatic aberrations, especially off-axis. Note the difficulty in controlling the off-axis monochromatic aberrations at the edges of the pupil. This is particularly difficult for systems with a low f-number.

Further insight can be gained by examining the field curves shown in Fig. 6.9. The longitudinal spherical aberration plots illustrate on-axis performance. In this case, however, aberration is measured in terms of

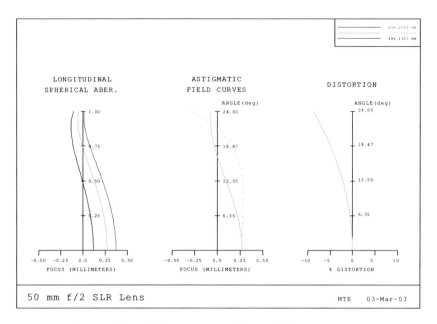

Figure 6.9 Field curves for a 50-mm, $f/2$ camera lens.

longitudinal error relative to the ideal paraxial focus position as a function of ray height in the pupil plane. The difference for the three wavelengths is a direct measure of axial chromatic aberration. The astigmatic field curves illustrate the best transverse and sagittal focus contours. The average of the two curves is indicative of field curvature, and the difference between the two contours is indicative of astigmatism. The last curve illustrates distortion as a function of field angle, which was left largely unconstrained in this design. To further control distortion would require an increase in other aberrations.

The composite PSF of an imaging system is a combination of a geometrical optics PSF with the diffraction-limited PSF previously discussed. The effect of geometric aberrations is generally space variant but can be assessed through spot diagrams illustrating where a fan of rays from particular field angles intercept the image plane. The fan is selected to fill the exit pupil such that the distribution of image spots, or spot density, is representative of the geometrical PSF. A spot diagram for the lens design in our example is illustrated in Fig. 6.10. This optical design provides a very tight on-axis spot and somewhat broader off-axis distributions. A useful measure of geometrical aberration is a root-mean-square (RMS) spot diameter, defined by

$$2\varepsilon = \left[\sum_{i} (x_i - x_o)^2 + (y_i - y_o)^2 \right]^{1/2}, \tag{6.8}$$

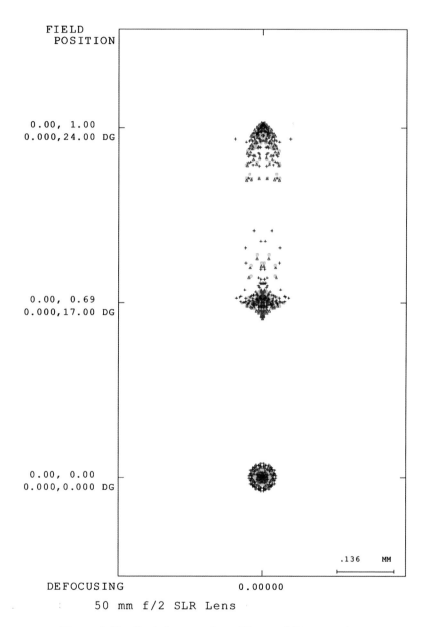

Figure 6.10 Spot diagrams for a 50-mm, $f/2$ camera lens.

where (x_i, y_i) correspond to the image plane intersection of the i'th ray in the ray fan, and (x_o, y_o) corresponds to the centroid of the ray fan. If one were able to describe the spot in terms of a normalized density function $h_g(x, y)$, then the total PSF would be a convolution of the diffraction-limited PSF with the geometric distribution

$$h(x, y) = h_g(x, y) * h_d(x, y). \tag{6.9}$$

The magnitude of ε relative to ε_0 is indicative of the limiting factor on resolution, geometric or diffraction. For this particular lens, the RMS spot diameter is about 35 µm on-axis and 75 µm off-axis, while the Airy radius corresponding to an $f/2$ system at a nominal 500-nm wavelength is about 1.2 µm. This system is far from diffraction limited, which is typical for low f-number, refractive optical systems in the visible spectral region. Diffraction-limited optical systems are more common in the infrared spectral regions, and for visible optical systems at higher f-numbers.

6.2.3 Modulation transfer function

The modulation transfer function (MTF) is widely used to characterize the overall imaging performance of an optical system and completely captures the limiting effects of aberrations and diffraction. The MTF is the magnitude of the optical transfer function $H(u, v)$, which is the Fourier transform of the PSF,

$$H(u, v) = \int_{-\infty}^{\infty} \int_{-\infty}^{\infty} h(x, y) \, e^{i2\pi(ux+vy)} dx dy. \tag{6.10}$$

Physically, the MTF represents the image contrast relative to that of the object of a sinusoidal pattern with a spatial frequency given by (u, v). For a diffraction-limited optical system with a circular pupil, the MTF is

$$\text{MTF}(\rho) = \begin{cases} \dfrac{2}{\pi} \left[\cos^{-1}\left(\dfrac{\rho}{\rho_0}\right) - \dfrac{\rho}{\rho_0} \sqrt{1 - \left(\dfrac{\rho}{\rho_0}\right)^2} \right] & \rho \leq \rho_0 \\ 0 & \rho > \rho_0 \end{cases}, \tag{6.11}$$

where ρ is a radial spatial frequency and

$$\rho_0 = \frac{1}{\lambda (f/\#)} \tag{6.12}$$

is the cutoff spatial frequency. This function is illustrated in Fig. 6.11 and monotonically decreases to zero at the cutoff frequency, beyond

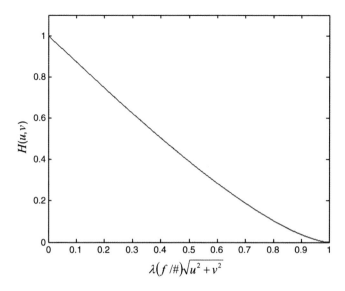

Figure 6.11 Modulation transfer function for a diffraction-limited optical system with circular aperture.

which the imaging system is not able to provide any resolution. In this aberration-free case, the MTF is determined by the area of overlap $A(\xi, \eta)$ between two renditions of the pupil function, shifted relative to each other by $(\lambda f u, \lambda f v)$. Another perhaps more intuitive interpretation is that the MTF represents the complex autocorrelation of the pupil function, where the aberrations represent the phase of the pupil function. Given a wavefront aberration in pupil plane $W(\xi, \eta)$, the monochromatic MTF for the aberrated optical system is given by

$$\text{MTF}(u, v) = \frac{\iint_{A(\xi,\eta)} e^{i\frac{2\pi}{\lambda}\left[W\left(\xi+\frac{\lambda f u}{2},\eta+\frac{\lambda f v}{2}\right)-W\left(\xi-\frac{\lambda f u}{2},\eta-\frac{\lambda f v}{2}\right)\right]}d\xi d\eta}{\iint_{A(0,0)} dxdy}. \quad (6.13)$$

Figure 6.12 illustrates the MTF of the camera lens as a function of field angle and wavelength and compares it to the diffraction-limited MTF. Again, it is seen that performance is better on-axis than off-axis but is far from diffraction limited in all cases.

6.2.4 Lens design

For remote sensing, optical systems often require long focal lengths to achieve adequate spatial resolution at long object ranges and require very large apertures to maintain adequate light-gathering power. Extending the basic design shown in Fig. 6.7 to very long focal lengths can result in impractically large systems. To minimize the overall length of the optical

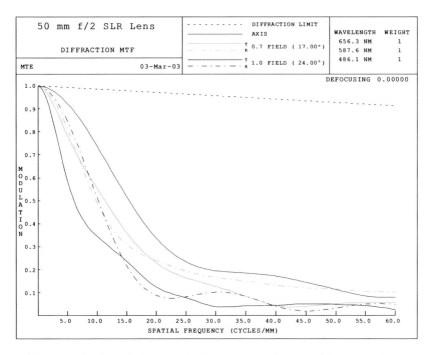

Figure 6.12   Modulation transfer function for a 50-mm, $f/2$ camera lens.

system, telephoto designs are typically used. These lens designs employ a shorter focal length front-lens group, called a primary or objective lens followed by a secondary lens group, an approach that provides the requisite real or virtual magnification and the desired effective focal length in a more compact package (Kidger, 2002). An example is depicted in Fig. 6.13. The challenge of the telephoto design is that the f-number of the primary lens is lower than the effective f-number of the entire optical system. This becomes particularly problematic for long focal length systems, as the primary optic can become massive due to its large physical diameter. In these cases, a more practical design approach is to use reflective optics, or mirrors, to minimize the weight of the optical system. This has the added advantages of supporting higher optical transmission over a very broad spectral range due to the high reflectance of metals, and also reducing chromatic aberrations. Most reflective telescopes are based on the two-mirror, circularly obscured form shown in Fig. 6.14, where the secondary element blocks the central portion of the entrance pupil. Differences in the particular optical design relate to the specific shapes of the primary and secondary mirrors, as outlined in Table 6.3. The use of aspheric surfaces supports better optical performance over a large FOV and has become practical with the recent emergence of precision diamond-turning manufacturing methods.

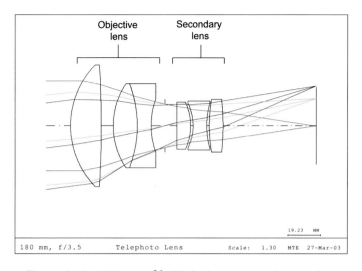

Figure 6.13 180-mm, $f/3.5$ telephoto camera lens design.

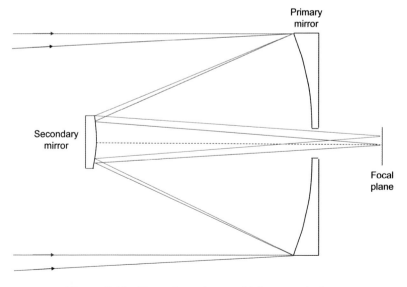

Figure 6.14 Two-mirror obscured telescope design.

Table 6.3 Common reflective telescope designs.

Optical design	Primary mirror	Secondary mirror	Optical performance
Parabolic	Concave parabaloid	Flat	Perfect on-axis, but poor over an appreciable FOV.
Dall–Kirkham	Concave parabaloid	Convex sphere	Moderately good performance over a limited FOV.
Cassegrain	Concave parabaloid	Convex hyperboloid	Good performance over a limited FOV; limited by coma and astigmatism.
Richey–Chretien	Concave hyperboloid	Convex hyperboloid	Good performance over a broad FOV; aplanatic; no coma or astigmatism.

Reflective two-mirror telescopes are often followed by downstream optics, which perform additional aberration correction, split out optical paths for multiple detectors, and relay the image plane onto FPAs. These issues can become important when considering hyperspectral remote sensing systems. Also, the central obscuration that is inherent in these designs can often be problematic when it comes to sensor calibration and stray-light control. One alternative to an obscured telescope is to use an off-axis design, such as the three-mirror anastigmatic (TMA) design shown in Fig. 6.15. This design can provide excellent imaging performance over a rectangular FOV but is difficult to align due to lack of any central symmetry.

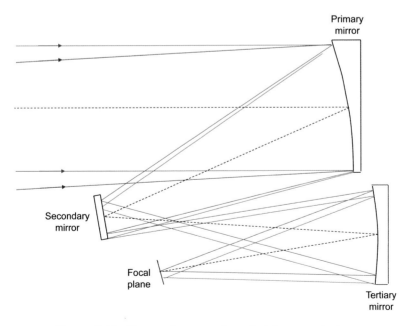

Figure 6.15 Three-mirror anastigmatic telescope design.

The basic optical design and analysis methods outlined can be applied to any part of the electro-optical and infrared spectral region. An important distinguishing characteristic between spectral regions, however, is the availability of high-transmission materials that can be manufactured into the smooth, precise surfaces required of optical systems. Figure 6.16 provides typical spectral transmission for a variety of optical and infrared materials. Optical transmission of common optical glasses, such as the borosilicate glass shown, for visible imaging systems typically cuts off in the SWIR spectral region. Fused silica maintains visible transmission while extending the range throughout the SWIR. Silicon and sapphire support good transmission across the MWIR spectral region, but only sapphire maintains transmission down to the visible region. Germanium, zinc selenide, and zinc sulfide all provide good LWIR transmission. Of these, only zinc sulfide can adequately cover the full spectral range from the visible through the LWIR.

The optical transmission characteristics shown in Fig. 6.16 are for uncoated windows, meaning that they include the Fresnel reflection losses at the front and back surfaces. These reflection losses are higher for the case of infrared optics due to their higher index of refraction. Refractive optics are almost always coated with thin-film AR coatings to minimize such reflection losses.

All of the infrared optical materials described are crystalline in nature, which means that they are fairly expensive, especially in large sizes. Since metals exhibit extremely high infrared reflectance, infrared imaging system designs are sometimes based on reflective instead of refractive optical elements. The obscured imaging systems are simpler to design and fabricate but exhibit a significant disadvantage in infrared systems due to central obscuration, which presents a radiating element in the pupil that can limit both system sensitivity and calibration. The latter is particularly troublesome for hyperspectral remote sensors, where absolute radiometric calibration is desired. For these reasons, off-axis reflective designs such as the TMA are of particular interest in the infrared spectral region. With the advancement of precision machining capabilities, complete telescopes can be made as diamond-turned, bolt-together systems with adequate tolerances for infrared wavelengths, significantly reducing manufacturing costs because they alleviate optical alignment challenges.

Another extremely important issue concerning MWIR and LWIR imaging system design is the issue of cold filtering and cold shielding. In these spectral regions, blackbody radiation from ambient sources can be an appreciable component of the energy collected by a detector. Therefore, it is essential to limit both the spectral range and the solid angle viewed by the infrared detector to those containing radiation from the scene, not the thermally emitting elements within the optical train itself. This is

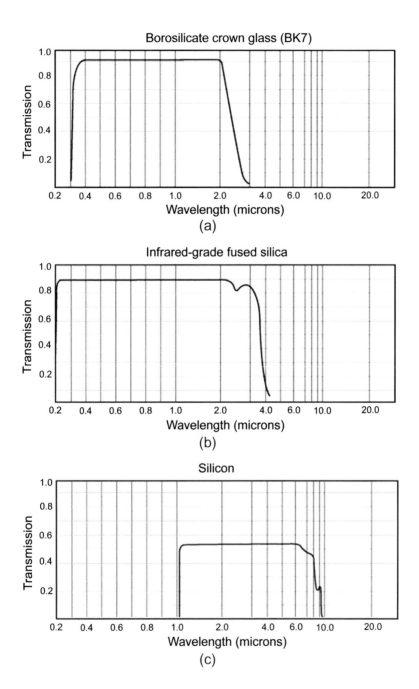

Figure 6.16 Spectral transmission of common optical and infrared materials: (a) borosilicate glass, (b) fused silica, (c) silicon, (d) sapphire, (e) germanium, (f) zinc selenide, and (g) zinc sulfide (courtesy of Newport Corporation).

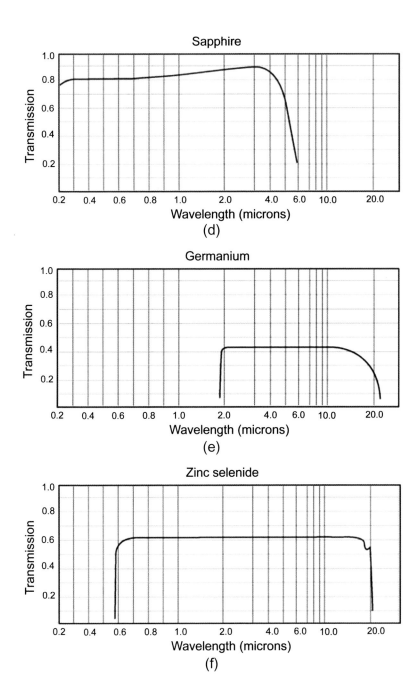

Figure 6.16 (*continued*)

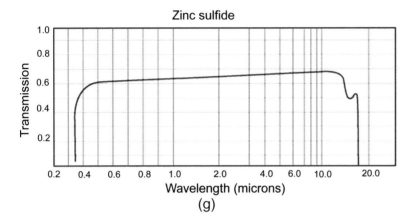

Figure 6.16   (*continued*)

accomplished by placing the FPA behind a spectral bandpass filter and cold shield, as shown in Fig. 6.17. Both of these elements are cooled in a cryogenic dewar to an acceptably low temperature, so that they not only cut out the incident radiation outside the passband and solid angle of the optical system but also emit a negligible amount of infrared radiation on their own. Dewar temperatures on the order of 80 K are achieved either with a cryogen such as liquid nitrogen or a closed-cycle mechanical cooler.

To quantitatively investigate this dewar arrangement, consider a situation where the optical system is illuminated from the imaged scene by a pupil-plane spectral radiance $L_p(\lambda)$ from on-axis. Optical transmission of the warm optical system and cold filter are represented as $\tau_o(\lambda)$ and $\tau_c(\lambda)$, respectively. Similarly, let the temperature of the warm optics, cold filter/shield, and detector be $T_o, T_c,$ and T_d, respectively. Finally, let Ω_c represent the solid angle of the cold shield opening from the perspective of the detector, and $\bar{\Omega}_c$ be the solid angle occluded by the cold shield. Spectral irradiance on the detector due to incident radiance from the scene, referred to as signal irradiance, is given by

$$E_s(\lambda) = \tau_o(\lambda)\tau_c(\lambda) \iint_{\Omega_c} L_p(\lambda) \cos\theta \sin\theta\, d\theta\, d\phi, \qquad (6.14)$$

where θ and ϕ are the standard spherical coordinates centered on the detector. Total spectral irradiance on the detector, referred to as background irradiance, is a combination of signal irradiance with the instrument self-emission. Based on these assumptions, this can be

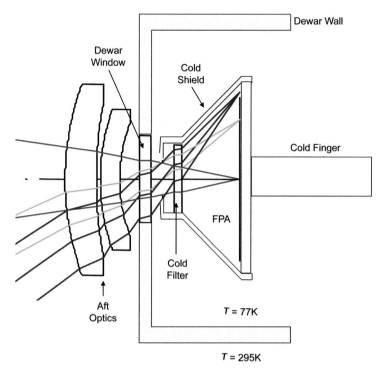

Figure 6.17   Cold shield and cold filtering arrangement for an infrared FPA.

computed by

$$E_b(\lambda) = E_s(\lambda) + \iint_{\bar{\Omega}_c} B(\lambda, T_c) \cos\theta \sin\theta d\theta d\phi$$

$$+ [1 - \tau_c(\lambda)] \iint_{\Omega_c} B(\lambda, T_c) \cos\theta \sin\theta d\theta d\phi$$

$$+ \tau_c(\lambda)[1 - \tau_o(\lambda)] \iint_{\Omega_c} B(\lambda, T_o) \cos\theta \sin\theta d\theta d\phi. \quad (6.15)$$

As will be discussed later, background irradiance increases the system noise level and reduces radiometric sensitivity; therefore, it should be minimized. If the dewar temperature is on an order similar to the ambient scene temperature, the second and third terms in Eq. (6.15) could easily swamp the signal term and result in a substantial degradation in performance. Proper cold shielding at cryogenic temperatures on the order of 80 K can make the second and third terms negligible. Matching the cold-filter passband to the spectral band of interest can minimize the fourth term.

Extending Eqs. (6.14) and (6.15) to the off-axis case is somewhat more difficult, not only due to the off-axis angles involved but also because background irradiance is greatly impacted by the location of the imaging system's exit pupil. Figure 6.17 depicts an optimal case where the exit pupil is located at the cold shield such that off-axis rays do not vignette at all on the cold shield. In this case, Eqs. (6.14) and (6.15) hold both on- and off-axis, and cold filter efficiency is said to be 100%. In the general case, however, vignetting can occur, and signal irradiance can fall off more rapidly with field angle than can background irradiance since signal irradiance is partially blocked by the cold shield. In discussing the camera lens design example, it was mentioned that aberrations are often controlled by placing the stop centrally within the optical system. The requirement of placing the stop near the FPA to achieve high cold-shield efficiency, as shown in Fig. 6.17, further limits the optical design for an infrared optical system. In reflective infrared telescope designs, refractive relay optics are often used to force the exit pupil onto the cold shield in the FPA dewar.

6.3 FPA Materials and Devices

The FPA converts irradiance distribution produced by imaging optics into an electrical signal for image display, recording, or processing by a computer. As such, it has a tremendous impact on image quality and can be the most costly component of an imaging system. A great diversity of detectors has been developed for EO/IR imaging systems, many of which can be fabricated into large, high-density linear and 2D arrays for scanned and staring remote imaging. The diversity in detectors is driven by two aims: (1) achieving sensitivity in different spectral regions (an issue of particular relevance to hyperspectral imaging systems) and (2) achieving sensitivity to small radiometric differences in the image. The first effort is typically an issue for detector materials, while the second can be highly influenced by optoelectronic device design. This section summarizes the most prevalent detector materials used to cover the VNIR through LWIR spectral region with excellent radiometric sensitivity and focuses on standard photoconductive and photovoltaic device designs.

The FPA consists of two primary components: an array of detectors and a multiplexer. The detector array produces an electrical response to incident radiation, typically a current, voltage, accumulated charge, or resistance charge. A multiplexer is an electronic component that conditions and integrates an electrical response at the unit-cell (or single-detector) level and produces either an analog or digital signal containing a serial representation of the detected image for downstream processing, recording, and display. Multiplexers, or readout integrated circuits (ROICs), are almost always silicon integrated circuits and are usually either charge-coupled device (CCD) or complementary metal-oxide semiconductor (CMOS) switching circuits. CCDs read out photo-integrated charges through a sequence of capacitive gates along rows of the array that are phased such that they act as shift registers for the charge packets. The phased operation for a single CCD pixel is shown in Fig. 6.18, where V_1, V_2, and V_3 are the phased voltage signals that control charge shifting. To read out an entire 2D FPA, such CCD shift registers are organized in rows, with a column shift register at the edge of the array, as depicted in Fig. 6.19. After each row shift, the entire column shift register is emptied into the output amplifier, thus forming an output analog signal that is a complete serialized representation of the image. There are many variations on this basic architecture, including those with multiple outputs to support higher image frame rates. The variation depicted in Fig. 6.19 is called a full-frame transfer architecture, which initially shifts the integrated charge packets for the entire image rapidly into a 2D CCD, which is blocked from any incident light to avoid degradations from image integration during the readout process.

A CMOS readout is an alternate type of ROIC design that uses switching circuits within each unit cell that sequentially connect the unit cells directly to the output amplifier. Such devices require more circuitry in the unit cell and can be a bit noisier than CCDs but offer greater flexibility in image readout, including the ability to perform electronic windowing, zooming, and tracking. In some cases, detector arrays can be formed monolithically on the silicon ROICs, particularly in the case of visible detector arrays employing CCD readouts. In the infrared, however, detector arrays and ROICs are often separately fabricated and then

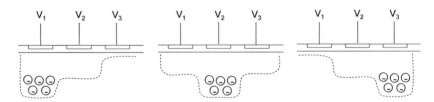

Figure 6.18 Depiction of charge-packet shifting in a single CCD pixel.

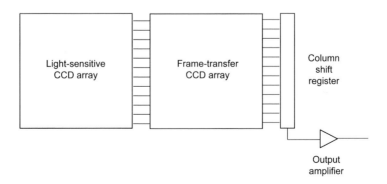

Figure 6.19 Full-frame transfer CCD architecture.

combined, or hybridized, by precisely aligning the individual detectors to the ROIC unit cells and establishing electrical connectivity through an array of indium bumps, as depicted in Fig. 6.20.

6.3.1 Quantum detectors

There are two fundamentally different types of EO/IR detectors: quantum detectors and thermal detectors. Quantum detectors are based on photo-induced electronic events; that is, incident photons produce electronic state changes, usually in the form of a transition from the valence band to the conduction band, that produce an observable electrical response. Thermal detectors, on the other hand, are based on radiative heating, and their measurable characteristic is temperature. The remainder of this section focuses on the operation and performance of the two most common types of quantum detectors from which FPAs are composed— photoconductors and photodiodes— with little attention given to ROIC design and performance.

Semiconductor quantum detectors have a band structure, or energy-versus-momentum diagram (depicted in Fig. 6.21), that consists of a valence band where electrons are bound to host semiconductor atoms in the crystalline lattice, and a conduction band where electrons are free to roam

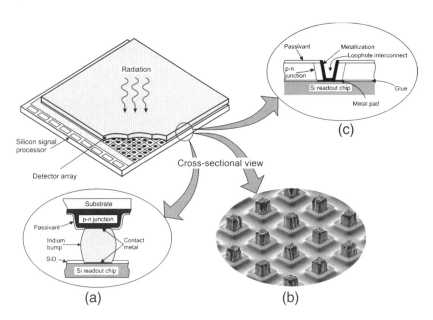

Figure 6.20 Hybrid FPA using bump bonding to connect the detector array to the ROIC: (a) detail schematic of indium bump interconnect approach, (b) microscopic image of indium bump array, and (c) detailed schematic of metalized via hole interconnect approach [reprinted with permission from Rogalski (2002) © 2002 Elsevier].

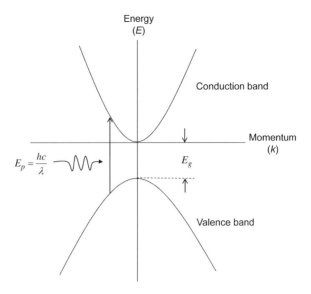

Figure 6.21   Band structure of a semiconductor quantum detector.

throughout the lattice with a given momentum. The conduction band is separated from the valence band by a bandgap energy E_g. If the wavelength of incident light is less than the cutoff wavelength, defined by

$$\lambda_c = \frac{hc}{E_g},\tag{6.16}$$

then it can be absorbed in the material and can elevate an electron from the valence band to the conduction band in the process. Concurrently, it creates an occupancy, or hole, in the valence band from the host ion that it left behind. If such photogenerated electrons and holes are sensed in some manner before they recombine with other electrons and holes in the lattice, then this forms the basis of a quantum detector. The probability that an incident photon will actually be absorbed by the semiconductor depends on all of the factors described in Chapter 3, including the transition moment integral between states and the density of states within both the valence and conduction bands.

Spectral absorption characteristics of a quantum detector can be characterized by the detector quantum efficiency $\eta(\lambda)$, which is a dimensionless quantity representing the probability that an incident photon with wavelength λ will be absorbed in the detector and produce a corresponding photoelectron in the conduction band. The quantum efficiency is identically zero above the cutoff wavelength and can theoretically approach a value of 1 at all wavelengths below it. In practice, some incident light is absorbed in the detector substrate, the crystalline material on which the detectors are grown, or in a region of the detector where it does not provide a measurable electrical response. Some light passes through the detector unabsorbed. Near the cutoff wavelength, the quantum efficiency typically rolls off gradually from its peak value to zero due to a limited number of unoccupied states at the bottom of the conduction band. Substrate absorption typically limits the shortwave response of detectors because a crystalline substrate has a larger bandgap and strongly absorbs light at wavelengths below its cutoff wavelength. In a hybrid FPA, substrates are sometimes thinned or even removed by machining after hybridization to enhance the short-wavelength response of the detector.

A quantity related to quantum efficiency that characterizes the light sensitivity of a quantum detector is its responsivity $R(\lambda)$, defined as the ratio of the photon-induced current I_d relative to the incident spectral flux, or

$$R(\lambda) = \frac{I_d}{E_d(\lambda)A_d}, \tag{6.17}$$

where $E_d(\lambda)$—the signal and background components of which are given in Eqs. (6.14) and (6.15)—is the focal-plane spectral irradiance, and A_d is the illuminated detector area. For a quantum detector, suppose that each photon at wavelength λ results in a photon event producing a free electron that is ultimately sensed by the readout circuitry with an overall efficiency $\eta(\lambda)$. Since the energy of such a photon equals hc/λ according to quantum theory, the responsivity of a quantum detector becomes

$$R(\lambda) = \frac{e\lambda}{hc}\eta(\lambda), \tag{6.18}$$

where e is the electron charge (1.6×10^{-19} C). For an ideal detector where quantum efficiency is unity up to the cutoff wavelength, responsivity is linearly proportional to wavelength. This is due to the fact that more long-wavelength photons are needed to carry a certain amount of energy than short-wavelength photons are needed to carry that same energy. Figure 6.22 depicts the basic characteristics of a real responsivity curve compared to the ideal, where the difference is contained entirely in the reduction in quantum efficiency near the cutoff wavelength and at shorter wavelengths due to the physical processes previously discussed.

6.3.2 Photoconductors

From a conceptual perspective, the simplest quantum detector is perhaps the photoconductive device depicted in Fig. 6.23. When there is no incident radiation, the device exhibits a dark conductivity, given by

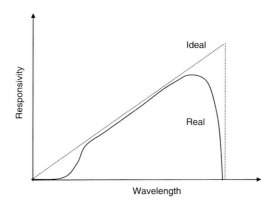

Figure 6.22 Shape of responsivity curve for ideal and real quantum detectors.

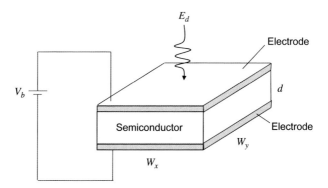

Figure 6.23 Layout of a simple photoconductive detector.

$$\sigma_0 = e\left(\mu_n n_0 + \mu_p p_0\right), \tag{6.19}$$

where μ_n and μ_p are the electron and hole mobility in $cm^2 V^{-1} s^{-1}$, and n_0 and p_0 are the electron and hole number densities under dark conditions (Dereniak and Boreman, 1996). Suppose that incident irradiance results in an electron–hole pair generation rate G and the generated carriers exhibit a lifetime of τ before recombining. Then conductivity would increase by an amount given by

$$\Delta\sigma = e\left(\mu_n + \mu_p\right)G\tau. \tag{6.20}$$

The generation rate is related to quantum efficiency according to the relationship

$$G = \frac{A_d}{hc} \int \eta(\lambda)E_d(\lambda)\lambda \, d\lambda, \tag{6.21}$$

based on the FPA spectral irradiance and detector area definitions given earlier. If the device were attached to an electric circuit such that there was a bias voltage V_b applied across it, the device current would be the combination of a dark current and a photocurrent

$$I = I_0 + I_p = \frac{\sigma_0 A_d V_b}{d} + \frac{\Delta\sigma A_d V_b}{d}. \tag{6.22}$$

This is what ultimately is sampled and converted to digital data through an electronic circuit. The lifetime of the electron–hole pairs can be greater than the transit time of the carriers across a distance d, given by

$$t_{transit} = \frac{d^2}{\mu V_b}, \tag{6.23}$$

such that one photon can result in multiple electron transits across the device and the circuit to which it is connected. This is the basis for the concept of photoconductive gain, which is the effective number of carrier transits across a device per absorbed photon. Compensating for differing electron and hole mobilities, photoconductive gain can be written as the product

$$\text{gain} = \frac{\tau}{t_{transit}}\left(1 + \frac{\mu_p}{\mu_n}\right). \tag{6.24}$$

This increases the measurable electrical signal for each absorbed photon.

6.3.3 Photodiodes

A common detector device used for remote imaging systems is the photovoltaic detector, or photodiode, the cross section for which is shown in Fig. 6.24. The diode is formed by the junction of a p- and n-type semiconductor with an energy gap appropriate for the spectral region for which the detector is intended to operate. The difference in carrier densities at the p–n junction causes a built-in voltage that bends the valence and conduction bands (as illustrated in Fig. 6.25), forming a depletion region in which all of the holes flow to the p-type material and electrons flow to the n-type material. This built-in voltage stops any current flow in the n-to-p direction while allowing current flow in the p-to-n direction, leading to the standard diode I–V characteristic

$$I = I_s(e^{eV_b/kT_d} - 1), \tag{6.25}$$

where I_s is the reverse saturation current and T_d is the diode temperature.

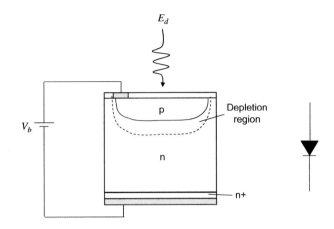

Figure 6.24 Cross-sectional diagram of a photodiode.

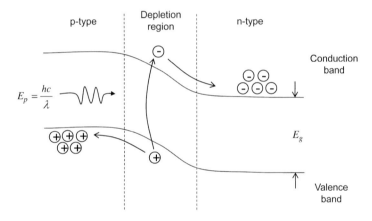

Figure 6.25 Band diagram across a diode junction.

When such a photodiode is illuminated by optical radiation at wavelengths where the photon energy exceeds the semiconductor bandgap, electron–hole pairs are generated in the p-layer, n-layer, and depletion region. Especially when produced within the depletion region, photogenerated electrons and holes can be forced in opposite directions by the built-in voltage and then flow through a circuit to which the diode is connected. This results in a photocurrent. There is some probability, however, that a photogenerated electron or hole can recombine with another hole or electron in the diode before reaching the other side of the depletion region, thereby not contributing to the photocurrent. Carriers generated in the depletion region have a much lower probability of recombination than those in the n- and p-regions, thus making the depletion region most responsible for the photocurrent. To enhance the effective size of the depletion region and thus increase the quantum efficiency, a variation called a p–i–n photodiode employs an insulator between the n- and p-layers.

Given the generation rate G of electron–hole pairs contributing to the photocurrent expressed in Eq. (6.21), the current–voltage (I–V) characteristic of the illuminated photodiode becomes

$$I = I_s(e^{eV_b/kT_d} - 1) + e\,G, \tag{6.26}$$

where the first term is the dark current and the second the photocurrent. The I–V characteristic curves for increasing detector irradiance are illustrated in Fig. 6.26. For an open-circuited diode, the voltage V_{oc} changes with detector irradiance in a nonlinear manner. However, when the diode is short circuited, the diode current I_{sc} increases more linearly with light level.

Photodiodes are usually operated using a circuit with one of the basic forms shown in Fig. 6.27. In a load-resistance circuit, output voltage is the product of the diode current with load resistance R_L and amplifier gain. In a transimpedance-amplifier circuit, effective load resistance is the feedback resistance R_f divided by the open-loop amplifier gain, and the output current closely resembles the short-circuit current, making the device more linear. Sometimes photodiodes are operated with a small reverse bias, which can increase saturation irradiance as well as the bandwidth of the diode by more quickly forcing the photogenerated charges out of the depletion region. The penalty is an increase in dark current, which increases the detector noise level, as will be explained soon. Operation

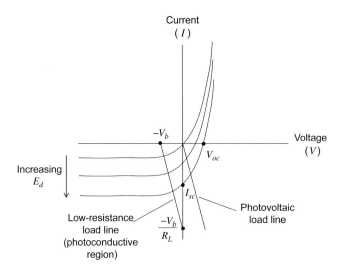

Figure 6.26 Current–voltage characteristic curve of a photodiode under increasing illumination conditions.

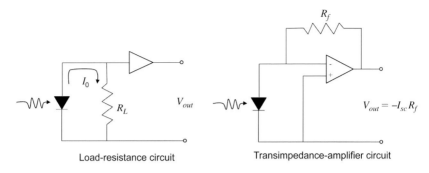

Figure 6.27 Standard photodiode operational circuits.

with a reverse bias is sometimes called the photoconductive mode, while operation with forward bias is termed the photovoltaic mode. The specific circuit design is typically the result of a trade-off between noise, linearity, and bandwidth. There is also a limit to the complexity of the circuit that can be realistically integrated into a single pixel unit cell of the ROIC.

6.3.4 Detector materials

A variety of semiconductor materials has been developed for use as EO/IR detectors. Material type is extremely important, since in most cases it dictates the wavelength cutoff and therefore the spectral range where the device can be optimally used. Obviously, a detector is not useful for sensing radiation at wavelengths longer than the cutoff wavelength, but it can provide inferior performance when used in a spectral region well below the cutoff wavelength. This occurs because the bandgap is smaller than necessary, resulting in increased dark noise, as there are more thermally excited electrons that have enough energy to transition into the conduction band. Figure 6.28 illustrates the bandgap, corresponding cutoff wavelength, and lattice constant for several semiconductor materials. Many optical detectors are made from semiconductor alloys, represented by lines between the primary crystalline materials whose quality is impacted heavily by the lattice constant matching of alloyed materials.

Silicon (Si) is the most prevalent electronic semiconductor and is the standard detector material used in the visible and NIR spectral region. Silicon has a bandgap of 1.12 eV, which corresponds to a cutoff wavelength of 1.1 μm. A typical quantum efficiency profile for a silicon detector is shown in Fig. 6.29. By thinning the substrate in semiconductor processing and illuminating the diodes through it, responsivity in the blue region can be enhanced as shown in this example. 2D detector arrays of more than 5000 $\times$ 5000 elements with less than 10-μm detector spacing, and linear arrays with more than 12,000 elements have been applied to remote sensing systems. Such arrays can be grown monolithically on

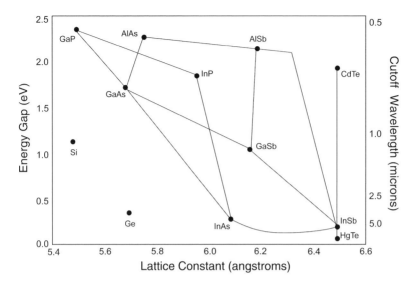

Figure 6.28   Bandgap, cutoff wavelength, and lattice constant for semiconductor detector materials.

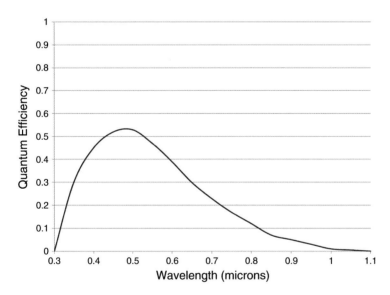

Figure 6.29   Typical quantum efficiency for silicon photodiode arrays (Eastman Kodak Company, 2010).

silicon CCD readouts, and hybrid devices with CMOS ROICs are also routinely manufactured.

Indium gallium arsenide (InGaAs) detectors, developed originally for fiber optic communications, are applicable for SWIR imaging applications. The cutoff wavelength can be varied by adjusting the relative composition of In and Ga in the $In_xGa_{1-x}As$ alloy, although longer-

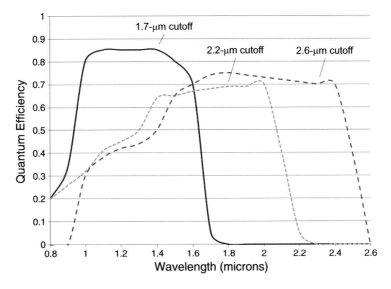

Figure 6.30 Typical quantum efficiency for InGaAs photodiode arrays (Goodrich Sensors Unlimited, 2011).

wavelength detectors suffer from higher dark current and defects due to the greater lattice mismatch with GaAs substrates. A typical cutoff wavelength is 1.7 μm, but devices with up to a 2.6-μm cutoff wavelength are possible with greater lattice-mismatch problems. Figure 6.30 illustrates typical quantum efficiency profiles for a range of these devices. Detector cooling is used to reduce dark current for longer-wavelength cutoff arrays, but the detector operating temperature can be higher for shorter-wavelength cutoff devices, as is described in Section 6.3.5. 2D arrays on the order of 320×240 and linear arrays on the order of 1024 elements have been developed, both with nominal 30-μm pixel pitch. Significant effort is underway to make larger and higher-density arrays from this material.

Indium antimonide (InSb) is typically the detector material of choice for the MWIR spectral region. InSb has a bandgap of 0.23 eV at 77 K, which corresponds to a cutoff wavelength of 5.5 μm. Quantum efficiency, as shown in Fig. 6.31 [figure based on data from Dereniak and Boreman (1996)], typically falls off considerably at short wavelengths, although substrate-thinned detector arrays have been employed with good response throughout the visible spectral range. Devices are always cooled to cryogenic temperatures of about 80 K to reduce dark current associated with the small bandgap. As the bandgap of InSb becomes smaller with increased temperature, operation at higher detector temperatures is not possible. InSb FPAs are made as hybrids with silicon ROICs and have been produced in 2D-array sizes on the order of 1000×1000 and in linear arrays of 6000 detectors, both with nominal 20-μm detector spacing.

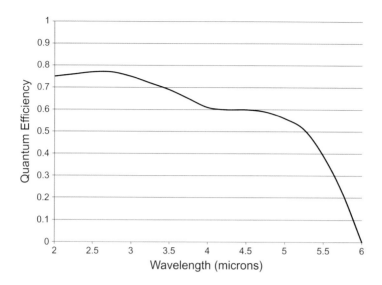

Figure 6.31   Typical quantum efficiency for InSb photodiode arrays.

The wavelength cutoff of mercury cadmium telluride (HgCdTe) can be tailored from 0.83 μm to beyond 15 μm by adjusting the relative Hg and Cd composition in the $Hg_xCd_{1-x}Te$ alloy. Therefore, application of HgCdTe detectors encompasses SWIR, MWIR, and LWIR spectral regions. Figure 6.32 illustrates some typical quantum efficiency profiles. All require cryogenic cooling to some level, with lower temperatures required for longer-wavelength devices. Typical operating temperatures are 180 K in the SWIR, 100 K in the MWIR, 80 K with a cutoff wavelength near the end of the LWIR (e.g., 9 μm), and down to 40 K for LWIR operation out to 12 μm. Area arrays on the order of 1000 × 1000 have been produced with roughly 20-μm detector spacing. Such FPAs are made as hybrids on silicon ROICs. SWIR HgCdTe detector arrays have recently been produced using a substrate removal technique to achieve response throughout the visible-to-SWIR spectral range by eliminating the CdTe substrate material that otherwise absorbs it before the light reaches the diodes (Chuh, 2004). A typical spectral quantum efficiency profile is shown in Fig. 6.33. While they are certainly versatile detector materials, HgCdTe detectors are generally expensive, especially for large-array formats and long-wavelength cutoffs, due to a relatively low manufacturing yield and problems with crystalline defects.

Extrinsic photoconductors have been used as alternatives to HgCdTe photodiodes for detection in the LWIR spectral range. The detection mechanism in such devices is based on an increase in the majority carrier density due to photo-ionization of impurities in the semiconductor lattice. Silicon and germanium are the most common host semiconductors,

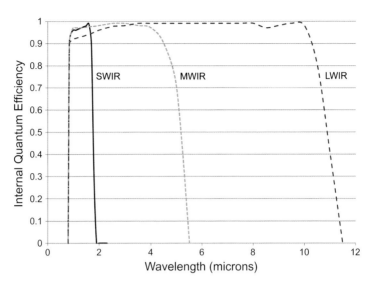

Figure 6.32 Typical quantum efficiency for HgCdTe photodiode arrays (Raytheon Vision Systems, 2009).

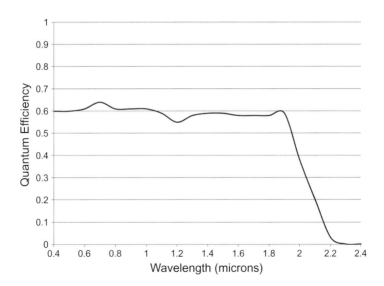

Figure 6.33 Spectral quantum efficiency for substrate-removed SWIR HgCdTe photodiode arrays (Simi *et al.*, 2001).

with the best performance demonstrated from the silicon-based devices. Table 6.4 provides ionization energy, cutoff wavelength, and highest operating temperature for a range of silicon impurities. Extremely low operating temperatures are required due to small ionization energies. One variant of extrinsic photoconductor development that has been applicable to infrared hyperspectral imaging sensors is the Si:As impurity band conduction device. This detector achieves higher quantum efficiency, on the order of 0.2, relative to conventional extrinsic photoconductors using a heavily doped donor band for high absorption in a small area. Also, a blocking layer is used to reduce the detector dark current. Such detectors have been developed in arrays on the order of 256×256 in size and operated with liquid helium cooling at 10 K.

Recent research in infrared detectors has focused on the use of bandgap-engineered quantum structures to achieve improved detection performance, better uniformity, and improved producibility using more-standard III–V semiconductor processes. Such devices include multiple quantum wells, superlattices, and quantum dots. The most mature design thus far is the GaAs/AlGaAs multiple-quantum-well detector. This device is composed of a number of GaAs quantum wells formed with $Al_xGa_{1-x}As$ barriers, where the band structure of the wells is a function of both well width and Al composition. Photodetection is based on intersubband absorption and tunneling of excited electrons into the continuum band, thereby increasing device conductivity. Detectors on the order of 512×512 in size have been demonstrated in the LWIR with nominally 60-K cooling. One limitation of such detectors is that intersubband absorption is both polarization and wavelength sensitive, a result of which is relatively low quantum efficiency on the order of 0.1. Diffractive structures employed for improving light coupling into the material can increase quantum efficiency somewhat, but generally at the expense of narrower spectral bandwidth (Gunupala and Bandara, 2000).

Table 6.4 Silicon-based extrinsic photoconductors.

Detector material	Ionization energy (eV)	Cutoff wavelength (μm)	Operating temperature (K)
Si:In	0.155	8.0	45
Si:Mg	0.103	12.0	27
Si:Ga	0.074	16.8	15
Si:Bi	0.071	17.6	18
Si:Al	0.069	18.1	15
Si:As	0.054	23.0	12
Si:B	0.045	27.6	10
Si:P	0.045	27.6	4
Si:Sb	0.043	28.8	4
Si:Li	0.033	37.5	4

6.3.5 Detector noise

Noise sources in a detector can vary depending on the specific device design and the electrical circuitry in the ROIC used to read out the detected photoelectrons. Some noise sources, however, are fundamental to all detectors and provide a sound basis for understanding detector performance capabilities and limitations. The most fundamental noise source is called detector shot noise, which is a result of the inherent statistical nature of light and the quantum detection process, and which exists even for an ideal optical detector. Consider a detector for which the photoelectron generation rate is G, given in Eq. (6.21). If spectral quantum efficiency were removed from this equation, the equation would represent the photon arrival rate onto the detector. If this rate were perfectly constant, then the expected number of generated photoelectrons within an integration time t_d would be the product $N = t_d G$, which would correspond to the photocurrent $I_p = eG$. However, corresponding ultimately to the Heisenberg uncertainty principle, photon arrival and absorption rates are not perfectly constant and exhibit some random variance from the mean rate expressed by G. Specifically, photon absorption is characterized by the Poisson random process and described by the probability mass function

$$P(n) = \frac{N^n}{n!} e^{-N},$$
(6.27)

where n is the measured number of photon-to-electron conversions, or photoelectrons. The mean value for the number of collected photoelectrons within integration time N is given by

$$N = \frac{A_d t_d}{hc} \int \eta(\lambda) E_d(\lambda) \lambda \, d\lambda.$$
(6.28)

The quantity in Eq. (6.27) provides the probability that n photoelectrons will actually be generated for any particular mean value N. Examples of this discrete probability density function are shown in Fig. 6.34. It is important to recognize that this noise source is signal dependent, increasing in variance as detector irradiance increases.

The Poisson statistical nature of the quantum detection process means that, if detector illumination is held perfectly constant in time, detector output would vary between integration times with a standard deviation given by that of the Poisson distribution. Similarly, if one were to measure the collected photoelectrons for many identical detectors across an FPA, each would provide a different output characterized by this inherent randomness. In both cases, the inherent Poisson variance represents a fundamental noise in the measurement; in an imaging system, the captured image would be a degraded version of the ideal image by this random

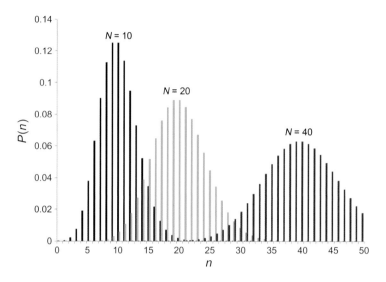

Figure 6.34   Example of Poisson probability density functions.

noise, which is called signal shot noise. For a Poisson random process, variance is identically equal to mean value, meaning that it is directly dependent on the signal level. This variance can be expressed in terms of the standard deviation of the number of collected photoelectrons,

$$\sigma_{n,\,shot} = \sqrt{N}, \tag{6.29}$$

or alternatively, by the corresponding standard deviation of the photocurrent,

$$\sigma_{i,\,shot} = \frac{e}{t_d} \sqrt{N}. \tag{6.30}$$

If the detector has a dark noise source, then shot noise is also associated with the resulting dark current due to the random statistical nature of discrete charge carriers. The dark current is commonly expressed in terms of a current density J_d, with units of $\mu A/cm^2$, where

$$J_d = \frac{e\,A_d\,(\mu_n n_0 + \mu_p p_0)\,V_b}{d} \tag{6.31}$$

for a photoconductor, and

$$J_d = \left| \frac{\beta\,kT_d}{e\,R_0 A_d} (e^{eV_b/\beta kT_d} - 1) \right| \tag{6.32}$$

for a photodiode (Dereniak and Boreman, 1996). In the latter equation, a diode nonideality factor β is included, and dark-current density is expressed in terms of the detector shunt resistance at zero bias R_0. The nonideality factor ranges from 1 for the case where the diffusion current dominates, and 2 for the case where the recombination current dominates. The zero bias shunt current is defined as

$$R_0 = \frac{1}{(\partial I/\partial V)|_{V=0}} = \frac{e\,I_s}{\beta\,kT_d}.$$

(6.33)

The product R_0A_d is an important figure of merit for the dark-noise behavior of photodiodes. Some typical R_0A_d values for InSb and HgCdTe photodiodes are given as a function of temperature and cutoff wavelength in Table 6.5. As the bandgap is decreased to extend wavelength cutoff, shunt resistance decreases, and dark current and associated noise increases. Dark noise can again be expressed in terms of the standard deviation of the number of collected photoelectrons,

$$\sigma_{n,dark} = \sqrt{\frac{J_d A_d t_d}{e}},$$

(6.34)

or that of the photocurrent,

$$\sigma_{i,dark} = \sqrt{\frac{e J_d A_d}{t_d}}.$$

(6.35)

Dark noise is also Poisson distributed with the mean value given by the term within the square root in Eq. (6.34). To counteract dark noise, infrared detectors are cooled to cryogenic temperatures. This, even more so than the need for limiting instrument self-emission, is the reason that infrared FPAs are placed within evacuated, cryogenic dewars, a requirement that significantly increases the cost and complexity of infrared systems.

Table 6.5 Typical R_0A_d values for InSb and HgCdTe infrared detectors.

Detector material	Cutoff wavelength (μm)	Operating temperature (K)	R_0A_d (ohm cm^2)
InSb	5.5	77	2×10^6
		60	2×10^7
HgCdTe	2.6	200	1×10^4
		100	5×10^7
	8.5	77	2×10^4
		60	5×10^4
	11.0	60	3×10^2
		50	1×10^4

An additional noise source common to all detectors is Johnson noise, which is associated with thermal carrier motion in resistive circuit elements. Detector shunt resistance provides the fundamental limit to how small Johnson noise can be, but it certainly can be increased due to other resistive elements in a detector readout circuit. Johnson noise is a signal-independent, zero-mean, additive random process that can be modeled with normal distribution. Its standard deviation can be expressed in terms of collected photoelectrons,

$$\sigma_{n, Johnson} = \sqrt{\frac{2kT_d t_d}{e^2 R_{eff}}}, \tag{6.36}$$

or that of the photocurrent,

$$\sigma_{i, Johnson} = \sqrt{\frac{2kT_d}{t_d R_{eff}}}. \tag{6.37}$$

Both Eqs. (6.36) and (6.37) are expressed in terms of an effective resistance R_{eff}, which can be derived by a detector readout circuit analysis. Replacing R_{eff} with R_0 provides the fundamental Johnson noise limit.

Since the three fundamental noise sources described are statistically independent, the noise variances simply add, leading to the following expressions for total detector noise:

$$\sigma_n = \sqrt{\sigma_{n, shot}^2 + \sigma_{n, dark}^2 + \sigma_{n, Johnson}^2} \tag{6.38}$$

and

$$\sigma_i = \sqrt{\sigma_{i, shot}^2 + \sigma_{i, dark}^2 + \sigma_{i, Johnson}^2}. \tag{6.39}$$

Other noise mechanisms can be germane to a particular detector material or device structure that add further to the total detector noise. Additional noise can be added in the ROIC, preamplifiers, and other readout electronics as well. Ultimately, this noise model is extended to include other system noise sources that, while added in downstream electronics or processing, are referred to in the same quantities of effective detector integrated photoelectrons σ_n or photocurrent σ_i. Expressing the signal level by the mean photoelectron level N or corresponding photocurrent I_p, the signal-to-noise ratio (SNR) of the detected signal is

$$\text{SNR} = \frac{N}{\sigma_n} = \frac{I_p}{\sigma_i}. \tag{6.40}$$

6.3.6 Detector performance

It is customary to characterize detector performance in terms of its noise-equivalent irradiance (NEI) or noise-equivalent power (NEP), that is, the difference in incident irradiance or radiative power (either in time or between detectors on an array) required to produce an SNR of 1. Such characterization is typically based on the assumption that detector spectral irradiance $E_d(\lambda)$ and spectral quantum efficiency $\eta(\lambda)$ are both constant with wavelength and are represented by spectrally integrated irradiance E_d and spectrally averaged quantum efficiency η. Under such assumptions, NEI can be computed based on the detector noise model already described by setting the SNR in Eq. (6.40) to 1 and inserting all of the relevant preceding equations, resulting in the expression

$$\text{NEI} = \frac{hc}{\lambda} \frac{\sigma_n}{A_d t_d \eta}$$

$$= \frac{hc}{\lambda} \frac{1}{A_d t_d \eta} \sqrt{\frac{\lambda}{hc} A_d t_d \eta \, E_d + \frac{J_d A_d t_d}{e} + \frac{2kT_d t_d}{e^2 R_{\text{eff}}}}. \tag{6.41}$$

Because of the shot noise component [the first term in the square root of Eq. (6.41)], the NEI is a function of the incident detector irradiance signal. The other two terms are signal independent and represent a noise floor for the detector, with the behavior illustrated in Fig. 6.35. At low irradiance, the first term can be ignored, and the NEI approaches the constant

$$\text{NEI}_{dark} = \frac{hc}{\lambda} \frac{1}{A_d t_d \eta} \sqrt{\frac{J_d A_d t_d}{e} + \frac{2kT_d t_d}{e^2 R_{\text{eff}}}}, \tag{6.42}$$

known as the dark-current limit. At high irradiance, the first term dominates, and NEI approaches

$$\text{NEI}_{BLIP} = \sqrt{\frac{hc}{\lambda} \frac{E_d}{A_d t_d \eta}}, \tag{6.43}$$

known as background-limited performance (BLIP). Ideally, the detector dark current will be minimized such that the detector operates in the BLIP regime, which allows for the best achievable performance for a given quantum efficiency. Setting the quantum efficiency in Eq. (6.43) to 1 provides an expression for the theoretical NEI limit for an ideal detector.

The NEP is simply related to the NEI by the product of the NEI and the detector area A_d. A related parameter, known as the specific detectivity D^*,

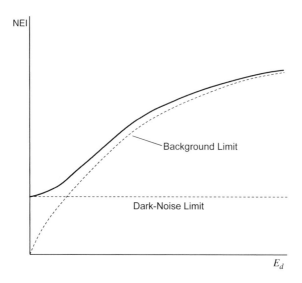

Figure 6.35   Typical variation of the NEI with detector irradiance.

is defined as

$$D^* = \frac{\sqrt{A_d B}}{\text{NEP}},$$
(6.44)

where B is the detector bandwidth ($B = 1/t_d$). The units of D^* are cm s$^{-1/2}$ W^{-1}, known as Jones. In the BLIP limit,

$$D^*_{BLIP} = \sqrt{\frac{\lambda\eta}{hc}\frac{1}{E_d}}.$$
(6.45)

A specific detectivity metric is an attempt to provide a detector figure of merit that is invariant to detector area and bandwidth, both of which strongly influence the NEI and NEP, to compare the fundamental detection capabilities of various detector implementations. Care must be taken in interpreting D^* characteristics, however, because they are inherently a function of the illumination conditions during D^* measurement. Often, the capabilities of a detector design are communicated in terms of how close the design comes to achieving the BLIP limit shown in Eq. (6.45).

6.4 Radiometric Sensitivity

The quality of imagery captured by an EO/IR remote sensor is primarily driven by four attributes: sharpness of the effective optical system PSF; the granularity in which the FPA is able to sample the irradiance image; the ratio of the image contrast to the noise level; and the presence of artifacts due to stray light, optical and focal plane imperfections, and electronic interference. Of these, the fourth is often the most difficult to characterize and tends to be system specific; however, there is a well-established methodology for relating image quality to the first three attributes and the system parameters that influence them. These three attributes are called spatial resolution, spatial sampling, and radiometric sensitivity.

Radiometric sensitivity is a characterization of how well an imaging system can preserve subtle spatial or temporal radiance differences in a scene. It is essentially a measure of image contrast relative to the system noise level. In this section, a systematic process is outlined for evaluating the radiometric sensitivity of a conventional imaging system based on its optical and FPA parameters. The process is described in a step-by-step manner beginning with the pupil-plane spectral radiance and ending with the SNR of the detected and quantized digital image. The analysis is performed with respect to the definition of basic scene characteristics by using a tri-bar pattern, which is a common object used to assess image quality and is depicted in Fig. 6.36. The tri-bar target is composed of three uniform, rectangular bars of width W and height $5W$ separated by their width W. When considering a reflective imaging situation, the bars are assumed to exhibit a Lambertian spectral reflectance $\rho_t(\lambda)$ and are placed in a uniform background with reflectance $\rho_b(\lambda)$. Both reflectance characteristics correspond to DHR. When considering an emissive imaging situation, both regions are usually assumed to be blackbody emitters with a target temperature T_t and background temperature T_b. In each case, image contrast is due to the target-to-background difference as opposed to the absolute magnitude of either. That is, the signal is either the reflectance difference,

$$\Delta\rho = \rho_t(\lambda) - \rho_b(\lambda), \tag{6.46}$$

or temperature difference,

$$\Delta T = T_t - T_b. \tag{6.47}$$

The difference signal could be due to a spatial difference, as in the case of the tri-bar pattern, or a temporal change in reflectance or temperature. There are other remote sensing applications where the extraction of absolute radiometric information from the scene (data such as temperature, reflectance, or irradiance) is of primary interest, as opposed to image contrast or radiometric differences. In these applications, the signal is composed by absolute radiometric levels instead of a temperature or reflectance difference.

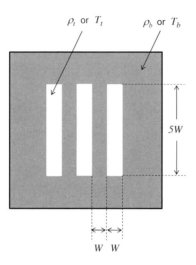

Figure 6.36   Tri-bar target pattern frequently used for assessing image quality.

6.4.1 Signal and background radiance

Radiometric sensitivity analysis, therefore, must begin with a definition of signal and background for a particular application. This is done in the pupil-plane spectral radiance domain, that is, the incident radiance displayed in Fig. 6.1. Background spectral radiance $L_b(\lambda)$ always refers to the total spectral radiance that enters the system regardless of its origin. Signal spectral radiance $L_s(\lambda)$ refers to the component of the background spectral radiance of interest to the application. For absolute radiometric remote sensing applications, $L_s(\lambda)$ is considered to be the component of pupil-plane radiance that is attributable to the scene, or

$$L_s(\lambda) = \tau_a(\lambda)L_u(\lambda), \tag{6.48}$$

where $\tau_a(\lambda)$ corresponds to the atmospheric path transmission and $L_u(\lambda)$ corresponds to the upwelling scene radiance of the object of interest. Background radiance includes atmospheric path radiance $L_a(\lambda)$ as well, such that

$$L_b(\lambda) = \tau_a(\lambda)L_u(\lambda) + L_a(\lambda). \tag{6.49}$$

For imaging applications in the reflective spectral region, the pupil-plane spectral radiance difference corresponding to $\Delta\rho$ is the signal

$$L_s(\lambda) = \frac{\Delta\rho}{\rho_b(\lambda)}\tau_a(\lambda)L_u(\lambda), \tag{6.50}$$

assuming that upwelling radiance is associated with the background region and not the target. Background radiance is again given by Eq. (6.49). According to the diffuse facet model from Chapter 5,

$$L_u(\lambda) = \frac{\rho_b(\lambda)}{\pi}[E_s(\theta_s, \phi_s, \lambda) + E_d(\lambda)]. \qquad (6.51)$$

For the emissive spectral region,

$$L_s(\lambda) = \tau_a(\lambda)[B(\lambda, T_t) - B(\lambda, T_b)], \qquad (6.52)$$

and

$$L_b(\lambda) = \tau_a(\lambda)B(\lambda, T_b). \qquad (6.53)$$

By differentiating the blackbody spectral radiance function with respect to temperature, and employing a first-order Taylor series approximation, Eq. (6.52) can be written in terms of ΔT as

$$\begin{aligned} L_s(\lambda) &= \frac{hc}{\lambda kT}\frac{\Delta T}{T}\frac{e^{hc/\lambda kT}}{e^{hc/\lambda kT} - 1}\tau_a(\lambda)B(\lambda, T_b) \\ &\approx \frac{hc}{\lambda kT}\frac{\Delta T}{T}\tau_a(\lambda)B(\lambda, T_b). \end{aligned} \qquad (6.54)$$

6.4.2 Focal plane irradiance

Using the signal and background pupil spectral radiance definitions most appropriate to the application, the focal plane spectral irradiance corresponding to each is given respectively by

$$\begin{aligned} E_s(\lambda, \theta) &= \frac{\pi K(\theta)}{4(f/\#)^2 + 1}\tau_c(\lambda)\tau_o(\lambda)L_s(\lambda) \\ E_b(\lambda, \theta) &= \frac{\pi K(\theta)}{4(f/\#)^2 + 1}\tau_c(\lambda)\tau_o(\lambda)L_b(\lambda) + E_o(\lambda, \theta), \end{aligned} \qquad (6.55)$$

where $(f/\#)$ is the effective f-number of the optical system, $K(\theta)$ represents the irradiance roll-off with field angle for the particular optical system, $\tau_c(\lambda)$ is the cold filter transmission, $\tau_o(\lambda)$ is the warm optics transmission, and $E_o(\lambda, \theta)$ represents the irradiance due to veiling glare, stray radiance, and instrument self-emission. The form of Eq. (6.55) is based largely on the radiometry of circularly symmetric imaging systems, illustrated in Figs. 6.37 and 6.38. On-axis irradiance can be computed by considering the exit pupil to be a circular source with a uniform pupil-plane

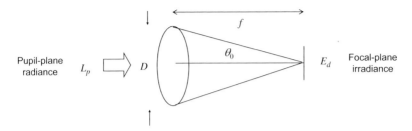

Figure 6.37   On-axis radiometry of a circularly symmetric imaging system.

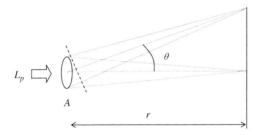

Figure 6.38   Irradiance reduction with field angle for an ideal optical system.

spectral radiance. The effective solid angle Ω is related to the numerical aperture (NA), which is $\sin \theta_0$, where θ_0 is the marginal ray angle shown in Fig. 6.37. The solid angle is

$$\Omega = \int_0^{\theta_0} \int_0^{2\pi} \cos \theta \, \sin \theta \, d\theta \, d\phi = \pi \sin^2 \theta_0 = \frac{\pi}{4 \, (f/\#)^2 + 1} \quad (6.56)$$

and relates pupil-plane spectral radiance and on-axis focal plane irradiance for perfectly transmissive optics. The product of the effective solid angle with the detector area is known as the optical throughput $A\Omega$ and is an invariant of the optical system and a key metric of its light-gathering power.

Off-axis, the relationship changes due to obliquity of the radiation as it passes through the exit pupil to the focal plane. For the simple single-lens geometry shown in Fig. 6.38, the differential irradiance is related by

$$dE = L_p d\Omega \cos \theta = L_p \frac{dA_\theta}{r_\theta^2} \cos \theta = L_p \frac{dA \cos \theta}{(r/\cos \theta)^2} \cos \theta, \quad (6.57)$$

resulting in the $\cos^4$ law for irradiance reduction,

$$K(\theta) = \cos^4 \theta. \quad (6.58)$$

In a real imaging system, especially one that is telecentric, the principal ray for an off-axis angle can land on the image plane at a shallower angle than the incident field angle can, resulting in an off-axis irradiance reduction that is less severe than that predicted by Eq. (6.58). Thus, the function $K(\theta)$ is a unique characteristic of optical design and is therefore kept generic in Eq. (6.55). Instrument irradiance includes the previously discussed instrument self-emission, which is an important component for emissive imaging systems. By approximating the cold shield as a uniform blackbody radiator in the pupil plane that fills the area outside the pupil shown in Fig. 6.37, an effective cold-shield solid angle can be computed as

$$\bar{\Omega} = \int_{\theta_0}^{\pi/2} \int_0^{2\pi} \cos\theta \, \sin\theta \, d\theta \, d\phi = \pi \cos^2\theta_0 = \frac{4\pi(f/\#)^2}{4\,(f/\#)^2 + 1}, \quad (6.59)$$

resulting in

$$E_0(\lambda) = \frac{4\pi(f/\#)^2}{4(f/\#)^2 + 1} B(\lambda, T_c)$$

$$+ \left[1 - \tau_c(\lambda)\right] \frac{\pi}{4(f/\#)^2 + 1} B(\lambda, T_c)$$

$$+ \left[1 - \tau_o(\lambda)\right] \tau_c(\lambda) \frac{\pi}{4(f/\#)^2 + 1} B(\lambda, T_o). \quad (6.60)$$

The first term in Eq. (6.60) represents cold shield emission, the second term represents cold filter emission, and the third represents warm optics emission.

6.4.3 Photoelectronic conversion

Conversion of focal plane irradiance into a detected signal is dependent on detector type. Assuming the case of a photodiode detector array with a detector area A_d and quantum efficiency $\eta(\lambda)$, the detected signal and background can be expressed either as a photocurrent,

$$I_s = eA_d \int \frac{\lambda}{hc} \eta(\lambda) E_s(\lambda) d\lambda$$

$$I_b = eA_d \int \frac{\lambda}{hc} \eta(\lambda) E_b(\lambda) d\lambda, \quad (6.61)$$

or in terms of the integrated photoelectrons within the frame integration time t_d,

$$N_s = A_d t_d \int \frac{\lambda}{hc} \eta(\lambda) E_s(\lambda) d\lambda$$

$$N_b = A_d t_d \int \frac{\lambda}{hc} \eta(\lambda) E_b(\lambda) d\lambda.$$

(6.62)

The use of integrated photoelectrons to represent signal and background makes comparisons to shot noise and unit cell charge capacity very direct and simple; therefore, signal, background, and noise characteristics are often specified in these terms.

6.4.4 Total system noise

Total system noise includes signal shot noise, dark-current noise, Johnson noise, and any other specific detector noise levels, along with additional noise contributions from downstream electronics and processing. Three such noise sources are added to the previously described detector noise sources, resulting in the aggregate noise model,

$$\sigma_n = \sqrt{\sigma_{n,shot}^2 + \sigma_{n,dark}^2 + \sigma_{n,Johnson}^2 + \sigma_{n,read}^2 + \sigma_{n,spatial}^2 + \sigma_{n,ADC}^2},$$

(6.63)

which assumes that all six noise sources are statistically independent. From earlier discussions of detector noise sources, it is evident that both shot noise and dark noise are Poisson distributed. At higher signal levels, the Poisson distribution can be approximated by a normal distribution with signal-dependent variance. Since the other noise sources are typically modeled by normal distributions as well, it is common to assume normal distribution for the total system noise.

The readout noise term $\sigma_{n,read}$ is included to represent the effective noise due to ROIC and other readout electronics that can result from multiple noise mechanisms. Its magnitude is either computed based on a complete circuit analysis or measured. In either case, it is often specified in terms of RMS photoelectrons, corresponding directly to $\sigma_{n,read}$. The spatial noise term $\sigma_{n,spatial}$ is included to represent the effect of residual nonuniformity and fixed pattern noise on the FPA. This effect is due to a small variation in detector responsivity (gain) and dark current (offset) among detectors across an FPA. The variation acts as noise spatially across the image but not temporally from frame to frame. The uncorrected gain and offset variation is typically on the order of a few to several percentage points of the mean level, indicating a dominant noise source. Fortunately, such differences are fairly stable over time periods that are long enough that they can be measured using uniform calibration sources and corrected in digital processing. Chapter 10 shows the limit of this nonuniformity correction (for a two-point calibration process) to be a result of FPA nonlinearity according to

$$\sigma^2_{n,spatial} = \frac{16R^2_{NU}R^2_{NL}}{N^2_{max}}[L(L_1 + L_2 - L) - L_1L_2]^2 + \frac{L^2_2\sigma^2_{n,1} - L^2_1\sigma^2_{n,2}}{(L_2 - L_1)^2},$$

(6.64)

where R_{NU} is the relative nonuniformity expressed as a fraction of the mean, R_{NL} is the relative detector response nonlinearity, N_{max} is the saturation level in integrated photoelectrons, $\sigma_{n,1}$ is the total noise level during collection of calibration data from a uniform source with pupil-plane radiance L_1, $\sigma_{n,2}$ is the total noise level during collection of calibration data from a uniform source with pupil-plane radiance L_2, and L is the pupil-plane radiance level at which the spatial noise is being characterized. This residual spatial noise is signal dependent with the characteristic relationship with illumination given in Fig. 6.39. It is minimized at two calibration points and increases away from the calibration points. Further details are provided in Chapter 10.

The final system noise model in Eq. (6.63) includes a noise term $\sigma_{n,ADC}$ due to the analog-to-digital conversion (ADC) process. The ADC of the output signal is a quantization process where the continuous current or voltage signal is represented by a fixed number of digital symbols. An analog value that falls between quantization levels must be rounded or truncated to the nearest level, resulting in an error in the representation on the order of the difference in quantization levels. In a binary system there

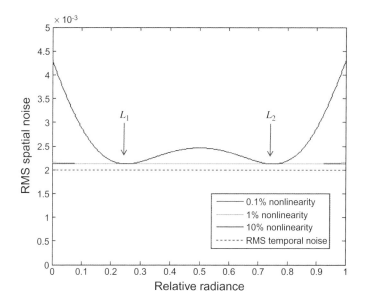

Figure 6.39 Characteristic spatial noise variation with illumination.

are 2^b levels, where b is the number of bits used to represent a data sample. If the integrated-photoelectron level ranges from zero to N_{max}, the standard deviation of the quantization noise can be expressed as (Oppenheim and Schafer, 1975)

$$\sigma_{n,ADC} = \frac{2^{-b}N_{max}}{\sqrt{12}}.$$

(6.65)

The numerator is the difference in quantization levels, which denotes twice the maximum error in the representation. The factor in the denominator is based on a uniform distribution of this error term.

The SNR based on the previous signal and noise models is given by

$$\text{SNR} = \frac{N_s}{\sigma_n} = \frac{I_s}{\sigma_i},$$

(6.66)

where N_s and I_s correspond to the signal integrated-photoelectron count and signal photocurrent, respectively. The corresponding background levels N_b and I_b increase the measurement level and thereby increase the shot noise term. In this way, components added to the background, such as path radiance and instrument internal emission, merely drive up system noise through shot and spatial noise components without adding to the signal, thus resulting in a decrease in SNR.

6.4.5 Total system performance

System performance is commonly characterized in noise-equivalent signal quantities specific to an application. For an absolute measurement case, a relevant quantity is the noise-equivalent spectral radiance (NESR) or band-integrated, noise-equivalent radiance (NEL), defined as the radiance signal differences that equal the noise standard deviation. After computing the SNR based on the signal and background spectral radiance expressions in Eqs. (6.48) and (6.49), the NEL is

$$\text{NE}L = \frac{L_u}{\text{SNR}},$$

(6.67)

where L_u is the radiance integrated over the spectral band of the imaging system. For reflective band imaging, the quantity of interest is the noise-equivalent reflectance difference NE$\Delta\rho$:

$$\text{NE}\Delta\rho = \frac{|\Delta\rho|}{\text{SNR}},$$

(6.68)

where the SNR is based on the pupil-plane spectral radiance signal in Eq. (6.50) and background in Eq. (6.51). The noise-equivalent reflectance difference corresponds to a scene reflectance difference that causes a signal change equal to the system noise standard deviation. A noise-equivalent temperature difference (NEΔT) is similarly defined for imaging in the emissive band as

$$\mathrm{NE}\Delta T = \frac{|\Delta T|}{\mathrm{SNR}}, \tag{6.69}$$

where the SNR is based on the signal in Eq. (6.52) or Eq. (6.54) and the background in Eq. (6.53). The NE$\Delta \rho$ and NEΔT are approximately related to the NEL by

$$\mathrm{NE}\Delta\rho \approx \frac{\mathrm{NE}L}{L_u}\rho_b, \tag{6.70}$$

and

$$\mathrm{NE}\Delta T \approx \frac{\mathrm{NE}L}{L_u}\frac{\lambda k T^2}{hc}, \tag{6.71}$$

where ρ_b is the average background reflectance over the spectral band and λ is the center wavelength.

Noise-equivalent radiometric sensitivity metrics are often measured in a laboratory environment using controlled radiometric sources that uniformly fill the entrance pupil of the sensor and also provide uniform illumination across the FOV. In this case, both $L_s(\lambda)$ and $L_b(\lambda)$ are represented by the calibration source radiance $L(\lambda)$, and any impact of an atmospheric path is not included. The system noise level σ_n is estimated directly from the captured image data and can be calculated by the standard deviation of an ensemble of measurements across pixels within a frame, or for a single pixel in time from frame to frame. The former case includes the spatial noise term, and the latter case does not. Therefore, contribution from spatial noise can be isolated from the difference between spatial and temporal variance measurements. NEL can be calculated directly from the known radiance and the SNR and is computed by the ratio of the ensemble mean to its standard deviation. The NE$\Delta\rho$ or NEΔT can be inferred based on the NEL through Eq. (6.70) or Eq. (6.71), or alternatively, these metrics can be measured more directly using a collimator with a calibrated tri-bar target projected into it. In that case, the signal ($\Delta\rho$ or ΔT) is known, the SNR is estimated by the ratio of the mean difference to the background standard deviation for the collected image data, and these metrics are calculated by either Eq. (6.68) or Eq. (6.69).

6.5 Spatial Sampling

Modern imaging systems employing detector arrays provide a spatially sampled representation of the continuous irradiance distribution produced by imaging optics on the focal plane. We have already discussed the impact of optics and diffraction on the spatial fidelity of an imaging system through the PSF or MTF, but we must further consider the impact of this spatial sampling by the FPA. Similar to ADC in the radiometric dimension, spatial sampling limits the spatial fidelity of the imagery produced. In some respects, sampling is a very simple process; however, it is inherently nonlinear and therefore cannot be completely described by linear system characteristics such as the MTF. When sampling frequency is sufficiently high relative to fine details in an image, the impact of sampling can be largely ignored. When it is not, however, then aliasing can occur, and the imagery can become corrupted. Let p_x and p_y be spatial sampling periods at the detector array, and w_x and w_y be corresponding detector widths. By considering magnification of an optical system, $M = f/R$ as expressed in Eq. (6.2), a simple imaging performance metric directly related to the sampling period is the ground-sampled distance (GSD),

$$
\begin{aligned}
\text{GSD}_x &= \frac{R}{f} p_x \\
\text{GSD}_y &= \frac{R}{f} p_y,
\end{aligned}
\tag{6.72}
$$

where R is the scene-to-sensor range or object distance and a small angle approximation is used, since $R \gg f$. GSD gives some sense of the resolution of an imaging system but by itself provides an incomplete and optimistic characterization of resolution.

To understand the limitations of sampled systems, it is helpful to consider the angular spectrum of an image, which is its Fourier transform representation. For example, Fig. 6.40 illustrates the tri-bar target image and its corresponding angular spectrum. The angular spectrum is provided in spatial frequency coordinates u and v (typically, cycles per millimeter for the image and cycles per meter for the object). In the u direction, the angular spectrum for a tri-bar target consists of a peak-at-zero frequency, fundamental peaks at frequencies of $-1/2W$ and $+1/2W$ on either side of the zero-frequency peak corresponding to the bar period of $2W$, and second-harmonic frequencies at $-3/2W$ and $+3/2W$ on either side of the zero-frequency peak. There is no periodicity in the v direction, so all peaks line up at zero frequency. According to the Shannon sampling theorem, a bandlimited signal can be perfectly reconstructed using bandlimited interpolation from a periodically sampled representation if the sampling frequency is greater than twice the maximum frequency in the sampled

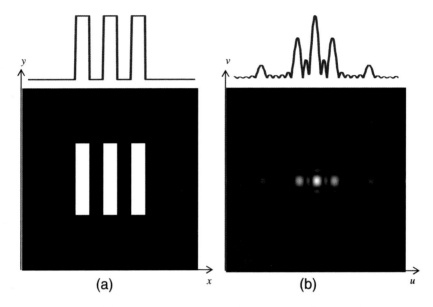

Figure 6.40 (a) Tri-bar target image and (b) corresponding angular spectrum. The plots above the images indicate distribution through the center.

signal. Stated another way, a Nyquist frequency can be defined that is half the sampling frequency. The Shannon sampling theorem states that spatial frequency content below the Nyquist frequency is properly represented in the sampled image data, while spatial frequency content above the Nyquist frequency is aliased, or nonlinearly converted to lower frequencies. With regard to a tri-bar target, the Nyquist frequency must be beyond the highest harmonic in order to perfectly reproduce the image, which is difficult because the square bar pattern contains significant energy in many harmonics. However, an optical imaging system has the effect of low-pass filtering the image before it is sampled by the detector array, removing higher harmonics and blurring sharp edges. Therefore, to avoid aliasing, it is only necessary for the Nyquist frequency to exceed the maximum spatial frequency passed by the imaging system.

Let (u, v) represent spatial frequency coordinates corresponding to the image and (u_G, v_G) be corresponding coordinates for the object. The two sets of coordinates are related according to

$$
\begin{aligned}
u_G &= \frac{f}{R} u \\
v_G &= \frac{f}{R} v.
\end{aligned}
\tag{6.73}
$$

Nyquist frequencies in the image domain, which is where sampling analysis is typically performed, are given by

$$u_N = \frac{1}{2p_x}$$

$$v_N = \frac{1}{2p_y}.$$
$$\text{(6.74)}$$

If a spatial frequency is sampled at Nyquist, then there are two samples across the period. For example, if the fundamental frequency of the tri-bar target is sampled at Nyquist, then there would be one sample on the bar and one between the bars if phasing was optimum. Even in this case, nonoptimal phasing can cause contrast reduction because the samples would straddle the edges of the bars. For a finer sampling, the fundamental frequency would not be aliased, and one would have a chance to resolve the pattern. For a coarser sampling, the pattern might be seen but at a lower frequency. Of course, even if the fundamental frequency is below Nyquist, the harmonics can be aliased. However, even a diffraction-limited optical system has an optical cutoff frequency of $\rho = \lambda(f/\#)$, where ρ is a radial spatial frequency coordinate. In this case, aliasing can be completely avoided if

$$p_x \leq \frac{\lambda(f/\#)}{2}, \tag{6.75}$$

and

$$p_y \leq \frac{\lambda(f/\#)}{2}. \tag{6.76}$$

In this case, the higher spatial-frequency content of the object is eliminated by the optical system via low-pass filtering before being sampled by the detector array. Figure 6.41 illustrates an irradiance image and an angular spectrum corresponding to a tri-bar target for three cases of diffraction-limited resolution, specifically with peak-to-null widths $\varepsilon_0 = W/2, W$, and $2W$. In the first case, the MTF passes both fundamental frequency peaks and second harmonics, providing a very sharply resolved bar pattern [Fig. 6.41(a)]. In the second case, the fundamental peak is preserved, but harmonics are eliminated, rounding off the sharp edges of the tri-bar image [Fig. 6.41(b)]. In the third case, which corresponds to the Rayleigh resolution criterion defined by the Airy radius in Eq. (6.5), fundamental peaks are only partially passed, and the pattern is marginally resolved [Fig. 6.41(a)]. Irradiance plots through the bar patterns above the images in Fig. 6.41 illustrate the impact of spatial blurring on modulation depth.

The combination of optical blurring and sampling is examined in Fig. 6.42. In this case, a tri-bar target is blurred by a diffraction-limited optical system with a peak-to-null width $\varepsilon_0 = W$. This image is then delta-sampled at different sampling frequencies, specifically at Nyquist, twice Nyquist, and five times Nyquist, relative to the bar period. In the top row of images, samples are perfectly aligned with the bar pattern, thus maximizing contrast. In the bottom row, samples are shifted by one-half the sampling period, representing worst-case phasing of the samples. In each case, samples are replicated into blocks to clearly indicate the sample period. For the Nyquist case, the bar pattern is clearly preserved with perfect phasing, but is not preserved when the samples are not properly phased [Fig. 6.42(a)]. At twice Nyquist, the pattern is preserved with both phasings, even though the sampled rendition varies in appearance with phasing [Fig. 6.42(b)]. At five times Nyquist [Fig. 6.42(c)], the sampled version more closely resembles the image prior to sampling in the middle column of Fig. 6.41, and sample phasing has little effect on the result other than a slight apparent image shift.

In summary, FPA sampling can be characterized as GSD that fundamentally limits spatial resolution. Specifically, image content at spatial frequencies above the Nyquist frequency is aliased and therefore not resolved. Instead, it is nonlinearly mapped into low-frequency content

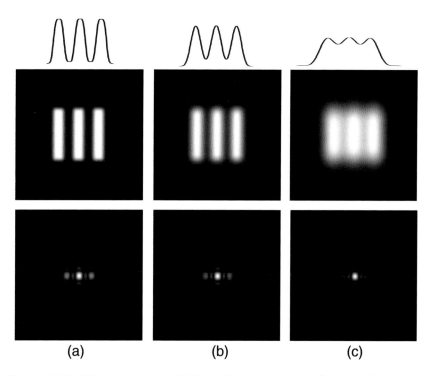

Figure 6.41 Tri-bar images (middle row) and corresponding angular spectra (bottom row) for diffraction-limited optical systems with peak-to-null widths equal to (a) $W/2$, (b) W, and (c) $2W$. The top row provides distributions through the center of the tri-bar images to show loss of resolution.

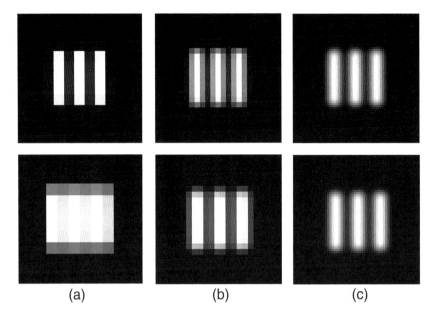

(a)   (b)   (c)

Figure 6.42 Tri-bar images for a diffraction-limited optical system with a peak-to-null width equal to W and delta-sampling at (a) Nyquist, (b) twice Nyquist, and (c) five times Nyquist. In the top row, samples are perfectly aligned with the bar pattern, while in the bottom row, sampling is shifted by one half the sampling period.

that corrupts the image. Even image content at spatial frequencies below Nyquist can be reduced in contrast, depending on sample phasing. It is necessary, however, to consider the combined effect of sampling and PSF of the optics, as well as other contributors for a more complete assessment of spatial resolution.

6.6 Spatial Resolution

Spatial resolution of an EO/IR imaging system is impacted by detector array sampling, the PSF of the imaging system, and radiometric sensitivity. A resolution metric characterizes the ability to distinguish features of a particular spatial scale in a scene. One standard definition for resolution is the Rayleigh criterion, which states that two point sources can be marginally resolved if their spacing in the image is greater than $1.22\lambda(f/\#)$, which is the peak-to-null width of the Airy function. This metric is based on the ability to recognize a dip in irradiance in the sum of the two diffraction-limited PSFs between two point source locations. As such, it does not account for the deleterious effect of sampling on resolution as outlined in the previous section, it ignores other degradations of the PSF in the optical system other than diffraction, and it assumes that there is no noise in the observed image that could detract from one's ability to recognize the dip in irradiance to distinguish the points.

6.6.1 Ground-resolved distance

A more comprehensive spatial resolution metric is the ground-resolved distance (GRD) that incorporates these other influences. GRD is defined as the minimum period $2W$ of a tri-bar target for which the bars are distinguishable by the optical imaging system. Resolving bar patterns requires two conditions: (1) the fundamental frequency of the bar pattern in the image domain must be at or below the Nyquist frequency corresponding to the detector array to avoid aliasing, and (2) the amplitude of the fundamental frequency component of the bar pattern in the image must be above the system noise level to be distinguished as a spatial pattern. The first condition is easily met when the GSD is less than or equal to W; that is, there are two samples within a period of the tri-bar pattern. Since GRD is specified with regard to the bar period as opposed to the bar width, it is at best twice the GSD. To satisfy the second condition, the fundamental frequency must be less than the MTF cutoff frequency, which is similar to the Rayleigh criterion. However, this can be insufficient if the SNR is so low that the frequency component falls below the noise floor, as seen in Fig. 6.43, which illustrates a bar-pattern image with three diffraction-limited PSFs at SNRs of $1, 5$, and 10. In all cases, images are sampled at 20 times Nyquist. The product of SNR and $\text{MTF}(u, v)$ characterizes the SNR for a particular spatial frequency (u, v). Therefore, this product for the bar frequency must be above some minimal level for the bar pattern to be visible in the image.

GRD is defined as a bar pattern that meets the Nyquist sampling criterion and supports an adequate SNR at the fundamental frequency for the pattern to be recognized. Mathematically, GRD can be expressed as

$$
\begin{aligned}
\text{GRD}_x &= \frac{R}{f} max \left[\frac{1}{u_R}, \frac{1}{u_N} \right] \\
\text{GRD}_y &= \frac{R}{f} max \left[\frac{1}{v_R}, \frac{1}{v_N} \right],
\end{aligned}
\tag{6.77}
$$

where u_N and v_N are the Nyquist spatial frequencies defined in the image plane as expressed in Eq. (6.71), and u_R and v_R represent the spatial frequency at which the bar pattern amplitude is three times the noise level. The factor of 3 is a rather arbitrary metric but is somewhat representative of the recognition capabilities of image analysts. These spatial frequencies are mathematically defined as

$$
\begin{aligned}
\text{MTF}(u_R, 0) &= \frac{3}{\text{SNR}} \\
\text{MTF}(0, v_R) &= \frac{3}{\text{SNR}}.
\end{aligned}
\tag{6.78}
$$

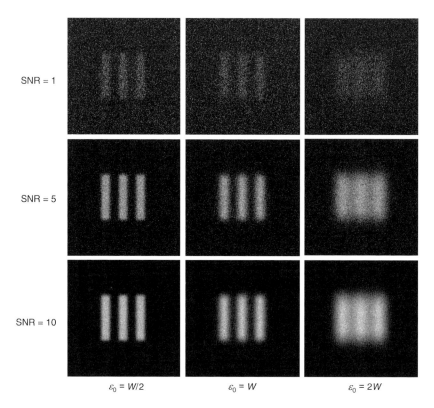

Figure 6.43 Variation in the imaged tri-bar pattern as a function of both diffraction-limited MTF cutoff frequency and SNR.

When $u_N < u_R$ or $v_N < v_R$, spatial resolution is sampling limited for the respective direction. Otherwise, it is optics limited.

6.6.2 System modulation transfer function

The MTF in Eq. (6.78) is a full-system MTF that should account for all impacts on the PSF of an imaging system. These impacts can involve the aggregate effect of optical aberrations, diffraction, wavefront irregularities, optical blur, finite sampling width, stabilization system jitter, pointing system drift, atmospheric turbulence, and other system-specific MTF limiters. Due to the linearity assumption, a system MTF can be expressed as the following product:

$$\begin{aligned}
\text{MTF}(u, v) = {} & \text{MTF}_{diffraction}(u, v)\, \text{MTF}_{optics}(u, v)\, \text{MTF}_{wavefront}(u, v) \\
& \text{MTF}_{blur}(u, v)\, \text{MTF}_{detector}(u, v)\, \text{MTF}_{jitter}(u, v) \\
& \text{MTF}_{drift}(u, v)\, \text{MTF}_{turbulence}(u, v).
\end{aligned} \qquad (6.79)$$

Accepted models for each of the effects are described next, but the reader should recognize that other MTF models could be more appropriate for a particular system design, and an optical system can have additional influences on the MTF that are not captured here. Nevertheless, the model in Eq. (6.79) is fairly comprehensive and widely applicable. The top line in Eq. (6.79) characterizes optical MTF effects, and the bottom line captures other system impacts. The diffraction-limited MTF for a circular aperture was previously given in Eq. (6.11); however, this MTF can be extended to model the MTF loss that occurs from the central obscuration in obscured, reflective telescopes that are common for remote sensing (O'Neill, 1956). Letting η represent the ratio of the obscuration diameter to the aperture diameter,

$$\text{MTF}(u, v) = \frac{A + B + C}{1 - \eta^2}, \tag{6.80}$$

where

$$A = \begin{cases} \frac{2}{\pi}\left[\cos^{-1}\left(\frac{\rho}{\rho_0}\right) - \frac{\rho}{\rho_0}\sqrt{1 - \left(\frac{\rho}{\rho_0}\right)^2}\right] & \rho \leq \rho_0 \\ 0 & \rho > \rho_0 \end{cases}, \tag{6.81}$$

$$B = \begin{cases} \frac{2\eta^2}{\pi}\left[\cos^{-1}\left(\frac{\rho}{\eta\rho_0}\right) - \frac{\rho}{\eta\rho_0}\sqrt{1 - \left(\frac{\rho}{\eta\rho_0}\right)^2}\right] & \rho \leq \eta\rho_0 \\ 0 & \rho > \eta\rho_0 \end{cases}, \tag{6.82}$$

$$C = \begin{cases} -2\eta^2 & \rho < \frac{1-\eta}{2}\rho_0 \\ \frac{2\eta}{\pi}\sin\phi + \left(\frac{1+\eta^2}{\pi}\right)\phi - \frac{2(1-\eta^2)}{\pi}\tan^{-1} & \\ \quad \times\left[\left(\frac{1+\eta}{1-\eta}\right)\tan\left(\frac{\phi}{2}\right)\right] - 2\eta^2 & \frac{1-\eta}{2}\rho_0 \leq \rho \leq \frac{1+\eta}{2}\rho_0 \\ 0 & \rho > \frac{1+\eta}{2}\rho_0 \end{cases}, \tag{6.83}$$

$$\phi = \cos^{-1}\left[\frac{1 + \eta^2 - (2\rho/\rho_0)^2}{2\eta}\right], \tag{6.84}$$

$$\rho = \sqrt{u^2 + v^2}, \tag{6.85}$$

and $\rho_0 = D/\lambda$. For an unobscured aperture, $\text{MTF}(u, v) = A$, which is consistent with Eq. (6.11).

An optics MTF is determined through geometrical optics analysis, as described previously. For most optical systems, the MTF is space variant, which means that it changes across the sensor FOV and is orientation dependent. It is common to characterize it in terms of tangential and sagittal directions, which would need to be transformed to rectilinear spatial frequency coordinates. Based on statistical optics analysis, the MTF corresponding to random wavefront irregularity due to optics manufacturing limitations can be expressed as

$$\text{MTF}_{wavefront}(u, v) = e^{-4\pi^2\sigma_w^2[1-e^{-\lambda^2 f^2(u^2+v^2)/L^2}]}, \tag{6.86}$$

where σ_w is the RMS wavefront error in waves, and L is the correlation length of the wavefront error in the exit pupil. Optical blur due to defocus or other aberrations can be modeled as a Gaussian function,

$$\text{MTF}_{blur}(u, v) = e^{-\pi^2(\delta_x^2 u^2+\delta_y^2 v^2)/4}, \tag{6.87}$$

where δ_x and δ_y are blur spot radii. When caused by an axial defocus Δz in the image domain,

$$\delta_x = \delta_y = \frac{0.62 \, D \, \Delta z}{f^2}. \tag{6.88}$$

The effect of spatial integration of finite detector elements can be expressed as a rectangular PSF with widths w_x and w_y. The corresponding MTF is given by

$$\text{MTF}_{detector}(u, v) = \left| \frac{\sin(\pi w_x u)}{\pi w_x u} \frac{\sin(\pi w_y v)}{\pi w_y v} \right|. \tag{6.89}$$

When the argument of absolute value is negative, there is a contrast reversal of image content. This is known as spurious resolution and occurs in a spatial frequency region beyond Nyquist, making it aliased as well. The combination of sampling and detector MTF provides the total effect of detector array sampling. Figure 6.44 illustrates impacts on a tri-bar target for the same parameters as in Fig. 6.42 for delta sampling. The difference in these figures is entirely due to spatial integration but is barely perceptible. The results in Fig. 6.44 are for the 100% fill factor case, where detector width equals detector pitch. When the fill factor, defined as the ratio of detector width to detector pitch, is less than 100%, the first null of the MTF moves to a higher frequency relative to Nyquist, and the resolution is slightly enhanced at the expense of radiometric sensitivity due to the smaller detector area.

Limitations in pointing and stabilization can be separated into two major components: random jitter within the frame integration time, and linear line-of-sight (LOS) drift during integration time. The jitter MTF assumes normally distributed, random LOS motion with an RMS angular deviation, typically in μrad units of σ_{LOS} within frame integration time t_d. With jitter power spectral density $S(v)$, given in units of $\mu rad/Hz^{1/2}$ as a function of temporal frequency v, the RMS jitter is

$$\sigma_{LOS}^2 = \int_0^\infty \left[1 - \frac{\sin^2(\pi t_d v)}{(\pi t_d v)^2} \right] |S(v)|^2 dv, \tag{6.90}$$

and the resulting MTF is

$$\mathrm{MTF}_{jitter}(u, v) = e^{-2\pi^2 f^2 \sigma_{LOS}^2 (u^2 + v^2)}. \tag{6.91}$$

The bracketed term in Eq. (6.90) is a high-pass filter that eliminates the low-frequency components of jitter power spectral density that can cause frame-to-frame motion but not within-frame motion. The effect of low-frequency motion on resolution can be modeled as linear drift components with angular drift rates ω_x and ω_y, typically expressed in milliradians per

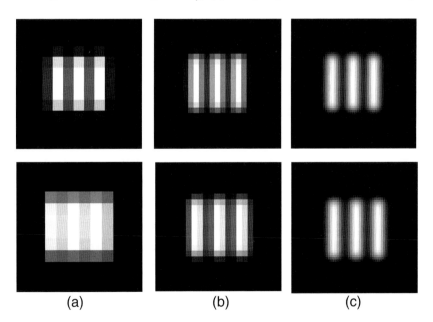

(a)   (b)   (c)

Figure 6.44 Tri-bar images for a diffraction-limited optical system with a peak-to-null width equal to W and 100% fill factor sampling at (a) Nyquist, (b) twice Nyquist, and (c) five times Nyquist. In the top row, samples are perfectly aligned with the bar pattern, while in the bottom row, sampling is shifted by one-half the sampling period.

seconds. The resulting MTF has the same basic form as that of spatial integration:

$$\text{MTF}_{drift}(u, v) = \left| \frac{\sin(\pi \omega_x t_d f u)}{\pi \omega_x t_d f u} \frac{\sin(\pi \omega_y t_d f v)}{\pi \omega_y t_d f v} \right|. \tag{6.92}$$

Atmospheric turbulence has an impact similar to wavefront aberrations on optical performance, except in this case the aberrations are distributed across the imaging path from the object to the sensor and are also temporally nonstationary. Analysis of atmospheric turbulence is quite complex (Fried, 1996), but an accepted form of turbulence MTF based on a statistical optics analysis is

$$\text{MTF}_{turb}(u, v) = e^{-3.44(\lambda f \rho / r_o)^{5/3} [1 - \alpha (\lambda f \rho / D)^{1/3}]}, \tag{6.93}$$

where ρ is defined in Eq. (6.85); r_0 is the correlation diameter of atmospheric path, which represents the aperture diameter beyond whose size the spatial resolution becomes turbulence limited, and $\alpha = 0$ under long exposure conditions and $\alpha = 0.5$ for short exposure conditions. The correlation diameter can be modeled as

$$r_0 = 2.1 \left[\frac{5.84\pi^2}{\lambda^2 \cos\theta} \int_{h_1}^{h_2} C_n^2(h) \left(\frac{h - h_1}{h_2 - h_1} \right)^{5/3} dh \right]^{-3/5}, \tag{6.94}$$

where h_0 is the object altitude, h_N is the sensor altitude, θ is the sensor zenith angle, and $C_n^2(h)$ is the index structure parameter distribution in $m^{-2/3}$ units that characterizes turbulence strength with altitude. The index structure parameter ranges from 10^{-14} $m^{-2/3}$ for benign conditions to 10^{-12} $m^{-2/3}$ for turbulent conditions at 1-m altitude, and decreases with altitude at a nominal h^{-1} rate, although there are more-sophisticated models of altitude dependence. In general, the MTF can be expressed as a linear combination of short- and long-exposure MTFs according to

$$\text{MTF}_{turb}(u, v) = e^{-V t_d / r_o} e^{-3.44(\lambda f \rho / r_o)^{5/3} [1 - 0.5(\lambda f \rho / D)^{1/3}]}$$
$$+ \left(1 - e^{-V t_d / r_o}\right) e^{-3.44(\lambda f \rho / r_o)^{5/3}}, \tag{6.95}$$

where V is the effective lateral velocity of the atmosphere relative to the entrance pupil of the sensor.

The composite MTF in Eq. (6.79) varies with both field angle and wavelength. Field angle dependence is accounted for in the analysis of an optical system by characterizing resolution at various locations in the sensor FOV. For hyperspectral imaging systems, spatial resolution can be

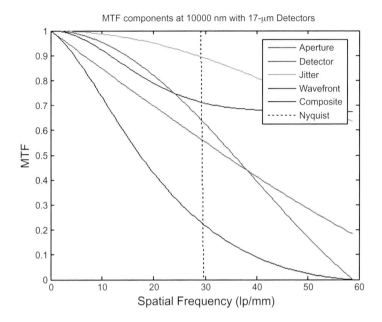

Figure 6.45 Example MTFs for a typical LWIR imaging system.

characterized as a function of spectral band in an analogous manner. For a broadband system, however, it is necessary to define a polychromatic MTF as a weighted average over the spectral band,

$$\text{MTF}_{poly}(u, v) = \frac{\int R(\lambda)E_d(\lambda)MTF(u, v; \lambda)d\lambda}{\int R(\lambda)E_d(\lambda)d\lambda}, \qquad (6.96)$$

where $R(\lambda)$ is detector responsivity and $E_d(\lambda)$ is a representative focal plane irradiance spectrum. Several examples of MTFs for an LWIR imaging system are illustrated in Fig. 6.45.

6.7 Image Quality

Several approaches have been developed to quantify overall image quality. This section focuses on one approach called the generalized image-quality equation (GIQE), a semiempirical equation relating primary imaging system performance metrics to image analyst tasks (Leachtenauer *et al.*, 1997). The GIQE is based on extensive regression analyses of the ability of image analysts to interpret imagery of differing fundamental attributes. Image quality is defined in terms of the Normalized Image Interpretability Rating Scale (NIIRS), a ten-level scale that characterizes image analysis tasks that can be supported with imagery of increasing quality. A subset of the image rating levels is given in Table 6.6. Each increasing level corresponds roughly to an improvement factor of 2 in GRD. The GIQE predicts the rating level Q based on imaging system characteristics, according to

$$\text{NIIRS} = 11.81 + 3.32 \log_{10} \left[\frac{\text{RER}_{GM}}{\text{GSD}_{GM}} \right] - 1.48 H_{GM} - \frac{G}{\text{SNR}}, \quad (6.97)$$

where RER is a relative edge response, H is an edge height overshoot, G is a noise gain factor, the subscript GM corresponds to a geometric mean, and GSD is measured in inches. For example,

$$\text{GSD}_{GM} = \sqrt{\text{GSD}_x \text{GSD}_y}. \quad (6.98)$$

As just mentioned, GSD is given in units of inches; the other parameters are all dimensionless.

The relative edge response and edge height overshoot are both determined from the MTF according to

$$\text{ER}_x(\xi) = 0.5 + \frac{1}{\pi} \int_0^{D/\lambda_{\min}} \frac{\text{MTF}(u, 0)}{u} \sin(2\pi w p_x \xi) \, du$$

$$\text{ER}_y(\eta) = 0.5 + \frac{1}{\pi} \int_0^{D/\lambda_{\min}} \frac{\text{MTF}(0, v)}{v} \sin(2\pi v p_y \eta) \, dv, \quad (6.99)$$

$$\text{RER}_x = \text{ER}_x(0.5) - \text{ER}_x(-0.5)$$

$$\text{RER}_y = \text{ER}_y(0.5) - \text{ER}_y(-0.5), \quad (6.100)$$

and

$$H_x = \begin{cases} \max \text{ER}_x(\xi = 1.0 \text{ to } 3.0) & \text{ER}_x(\xi) \text{ is not monotonically increasing} \\ \text{ER}_x(\xi = 1.25) & \text{ER}_x(\xi) \text{ is monotonically increasing} \end{cases}$$

$$H_y = \begin{cases} \max \text{ER}_y(\eta = 1.0 \text{ to } 3.0) & \text{ER}_y(\eta) \text{ is not monotonically increasing} \\ \text{ER}_y(\eta = 1.25) & \text{ER}_y(\eta) \text{ is monotonically increasing.} \end{cases}$$

$$(6.101)$$

In Eqs. (6.99)–(6.101), ξ and η are spatial coordinates in the image plane normalized to the detector sampling pitch $\xi = x/p_x$ and $\eta = y/p_y$, and are centered at the edge. The function ER is a normalized edge-response function in the specified direction, RER is the relative rise in edge response over one detector spacing, and H is the peak overshoot as depicted in Fig. 6.46. Noise gain G is a measure of the relative decrease in SNR due to the use of edge-enhancement filters often used by image analysts to accentuate image detail. SNR is related to the differential signal between a 7 and 15% target reflectance in the VNIR/SWIR spectral region, and 300- and 302-K blackbody temperature in the MWIR/LWIR spectral region.

Table 6.6   Examples of NIIRS image-quality levels.

Image quality	Image interpretability
0	None
1	Detect a medium-sized port facility.
	Distinguish between taxiways and runways at a large airfield.
2	Detect large hangars at airfields.
	Detect large buildings such as hospitals and factories.
3	Identify the wing configuration of a large aircraft.
	Detect a large surface ship in port by type.
	Detect a train on railroad tracks.
4	Identify large fighter aircraft by type.
	Identify large tracked and wheeled vehicles by type.
	Identify railroad tracks, control towers, and switching points.
5	Identify radar as vehicle mounted or trailer mounted.
	Identify individual rail cars and locomotives by type.
6	Identify radar antennas as parabolic or rectangular.
	Identify the spare tire on a medium-sized truck.
	Identify automobiles as sedans or station wagons.
7	Identify ports, ladders, and vents on electronics vans.
	Detect vehicle mounts for antitank guided missiles.
	Identify individual rail ties.
8	Identify rivet lines on bomber aircraft.
	Identify a handheld surface-to-air missile.
	Identify windshield wipers on a vehicle.
9	Identify vehicle registration numbers on trucks.
	Identify screws and bolts on missile compartments.
	Detect individual spikes on railroad ties.

Table 6.7 provides image quality values as a function of GSD for an infinite SNR, $H = 1$, and RER values of $0.25, 0.5$, and 1.0. An RER near unity corresponds to detector-limited resolution.

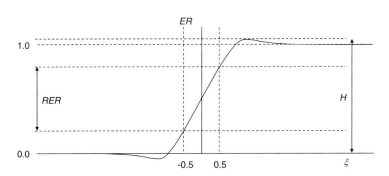

Figure 6.46   Example edge response function indicating definitions for RER and H.

Table 6.7 Image quality as a function of GSD for several RER levels with infinite SNR and $H = 1$.

GSD (m)	NIIRS		
	RER = 0.25	RER = 0.5	RER = 1.0
0.1	6.4	7.4	8.4
0.3	4.8	5.8	6.8
0.6	3.8	4.8	5.8
1.0	3.0	4.0	5.0
3.0	1.5	2.5	3.5
6.0	0.5	1.5	2.5
10.0	−0.3	0.7	1.7

6.8 Summary

Performance of an electro-optical (EO/IR) system is dependent on the capability of its optical system to produce an undistorted, unaberrated, and intense image of a scene on a focal plane, as well as the ability of its focal plane array (FPA) to convert an image into an analog or digital form with low noise and fine spatial sampling. The performance of other system components such as system stabilization and readout electronics can have significant impacts as well. Optical system performance is largely dictated by optical throughput $A\Omega$, effective optical transmission as a function of wavelength, and optical system modulation transfer function (MTF). FPAs are primarily characterized by quantum efficiency as a function of wavelength and noise for given operating parameters.

A full understanding of total imaging system performance includes radiometric sensitivity, spatial sampling, and the MTF system. These characteristics can be combined in terms of ground-resolved distance (GRD) or image quality as a complete metric, related to the ability of the system to preserve scene content. Some of these metrics are directly extensible to hyperspectral imaging systems, although analysis in that case would require additional attention to preserving spectral features of the collected spectra. Nevertheless, the fundamental principles of design and performance of broadband imaging systems provide a sound foundation from which to make this extension.

6.9 Further Reading

A basic introduction to geometrical optics can be found in texts by Pedrotti and Pedrotti (1987), Fowles (1989), or Dereniak and Boreman (1996). Further information on optical aberration theory and optical design can be found by Smith (1990) or Kidger (2002). An excellent reference on Fourier optics is given by Goodman (1985). Dereniak and Boreman (1996) and Rogalski (2011) provide a complete treatment of FPAs and detector performance analysis with an emphasis on infrared technology.

References

Chuh, T. Y., "Recent developments in infrared and visible imaging for astronomy, defense, and homeland security," *Proc. SPIE* **5563**, 19–34, 2004 [doi:10.1117/12.565661].

Dereniak, E. L. and Boreman, G. D., *Infrared Detectors and Systems*, Wiley, New York (1996).

Eastman Kodak Company, "Kodak KAI-0340 image sensor: Device performance," Rev. 5.0 MTD/PS-0714, July 2010, http://www.kodak.com.

Fried, D. L., "Limiting resolution looking down through the atmosphere," *J. Opt. Soc. Am.* **56**, 1380–1384 (1996).

Fowles, G. R., *Introduction to Modern Optics*, Dover, Mineola, NY (1989).

Goodman, J. W., *Introduction to Fourier Optics*, Second Edition, McGraw-Hill, Boston (1996).

Goodman, J. W., *Statistical Optics*, Wiley, New York (1985).

Goodrich Sensors Unlimited, "InGaAs products: Focal plane arrays," http://www.sensorsinc.com (last accessed Sep 2011).

Gunupala, S. D. and Bandara, S. V., "Quantum well infrared photodetector (QWIP) focal plane arrays," *Semiconduct. Semimet.* **62**, 197–282 (2000).

Kidger, M. J., *Fundamental Optical Design*, SPIE Press, Bellingham, WA (2002) [doi:10.1117/3.397107].

Leachtenauer, J. C., Malila, W., Irvine, J., Colburn, L., and Salvaggio, N., "General image quality equation: GIQE," *Appl. Optics* **36**, 8322–8328 (1997).

Lebedev, N. N., *Special Functions and Their Applications*, Dover, New York (1972).

Newport Corporation, "Optical materials," http://www.newport.com (last accessed Sep 2011).

Oppenheim, A. V. and Schafer, R. W., *Digital Signal Processing*, Prentice-Hall, Englewood Cliffs, NJ (1975).

O'Neill, E. L., "Transfer function for an annular aperture," *J. Opt. Soc. Am.* **46**, 285–288 (1956).

Pedrotti, F. L. and Pedrotti, L. S., *Introduction to Optics*, Prentice-Hall, Englewood Cliffs, NJ (1987).

Raytheon Vision Systems, "The infrared wall chart," 2009, http://www.raytheon.com (last accessed Sep 2011).

Rogalski, A., "Infrared detectors: an overview," *Infrared Phys. Techn.* **43**, 187–210 (2002).

Rogalski, A., *Infrared Detectors*, Second Edition, CRC Press, Boca Raton, FL (2011).

Simi, C., Winter, E., Williams, M., and Driscoll, D., "Compact Airborne Spectral Sensor (COMPASS)," *Proc. SPIE* **4381**, pp. 129–136 (2001) [doi: 10.1117/12.437000].

Smith, W. J., *Modern Optical Engineering*, Second Edition, McGraw-Hill, New York (1990).

Chapter 7
Dispersive Spectrometer Design and Analysis

Imaging spectrometers are essential instruments for hyperspectral remote sensing systems that capture pupil-plane spectral radiance from a scene and ultimately form it into a hyperspectral data cube for processing, analysis, and information extraction. An imaging spectrometer is an EO/IR system, thus it contains all of the basic imaging system components outlined in the previous chapter, exhibits similar design challenges, and is characterized by some of the same performance metrics. In addition to imaging, however, an imaging spectrometer must also capture radiation spectra for each image pixel with adequate fidelity; this places additional demands on system design and adds another dimension to performance characterization.

The primary design challenge is that a hyperspectral data cube is essentially a 3D construct, with two spatial dimensions and one spectral dimension, while FPAs are 2D. Therefore, some means are necessary to capture this additional dimension. Often this is done using the time dimension. Various designs that perform this function in different ways are explored in this chapter and subsequent two chapters. In this chapter, we focus on arguably the most common type, the dispersive imaging spectrometer.

Dispersive spectrometers use either a prism or grating to spatially spread a spectrum of incident radiation across a detector array. In a dispersive *imaging* spectrometer, depicted conceptually in Fig. 7.1, the spectrum is dispersed in one direction, while the image is relayed in an orthogonal direction. In the direction of dispersion, it is necessary to limit the field of view to one spatial element; this is performed using a slit. If the field of view is not thusly limited, irradiance distribution on the detector array will resemble an image with large lateral chromatic aberration. In that way, a back-end spectrometer only captures a single line of the 2D field produced by front-end optics. A scanning mechanism, such as the pushbroom or whiskbroom method discussed in Chapter 6, is required to image from frame to frame in a direction orthogonal to the slit. As this occurs, a

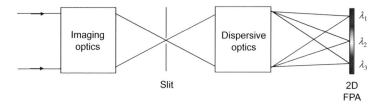

Figure 7.1 Dispersive imaging spectrometer concept.

hyperspectral data product is formed as a sequence of spatial–spectral frames, as depicted in Fig. 7.2. The composite set forms a hyperspectral data cube.

The imaging spectrometer itself is basically an imaging relay from the slit to the FPA, with lateral dispersion to map the spectrum across the detector array in the cross-slit direction. This dispersion is generally performed in a collimated space by a prism or grating to minimize aberrations. The dispersion is precisely matched to the focal length of the reimaging optics and detector spacing such that desired spectral sampling is achieved. The primary performance parameters that characterize a dispersive spectrometer are its resolving power, spectral range, optical throughput, spatial–spectral distortion, and radiometric sensitivity, all of which are explored in this chapter.

7.1 Prism Spectrometers

There are two primary ways to produce dispersion: *prisms* and *gratings*. Prisms are the oldest spectrometer component, producing angular

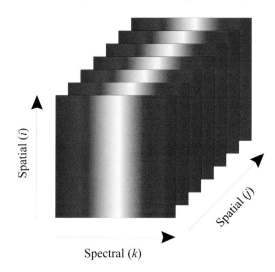

Figure 7.2 Sequence of frames from a dispersive imaging spectrometer, producing a hyperspectral data cube.

dispersion of light due to a combination of material dispersion and nonparallel optical surfaces. The amount of achievable dispersion is limited by the material characteristics available and optical aberrations introduced by the prism elements. These limitations are addressed by using curved prism surfaces in the spectrometer design. The use of prisms as dispersive elements can be advantageous due to the excellent spectral purity that prism spectrometers can achieve. Spectral purity is an attribute of a spectrometer that characterizes its ability to confine energy from a monochromatic source to a confined band in the spectrum and resist scattering that spreads energy throughout the spectral band. An instrument without good spectral purity can produce anomalous spectral features due to ghosting and scattering artifacts. Prism spectrometers are well suited for hyperspectral remote sensing applications requiring moderate spectral resolution.

7.1.1 Prism dispersion

Spectral dispersion by a prism is a well-known phenomenon that occurs due to the change in refractive index of the prism material with wavelength (Smith, 1990). Since a refractive index decreases with wavelength in regions of low absorption, prisms refract short-wavelength radiation at greater deviations than they refract long wavelength radiation. This is quantitatively analyzed in Fig. 7.3, which shows the path of an incident ray at angle θ_1 to the first surface normal of a prism with apex angle α. The deviation caused by the prism is given by

$$\phi = \phi_1 + \phi_2 = \theta_1 - \theta_1' + \theta_2 - \theta_2', \tag{7.1}$$

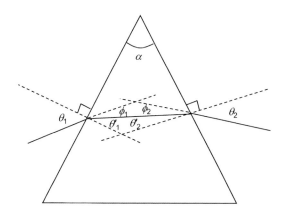

Figure 7.3 Geometry for analyzing prism dispersion.

where the angles are defined in the figure. By Snell's law,

$$\sin \theta_1 = n \sin \theta_1' \tag{7.2}$$

and

$$\sin \theta_2 = n \sin \theta_2'. \tag{7.3}$$

The index n is wavelength dependent, which is what causes the desired angular dispersion. From basic geometrical properties, it is easy to show that

$$\alpha = \theta_1' + \theta_2'. \tag{7.4}$$

By combining Eqs. (7.1) through (7.4), the total angular deviation of the ray entering the prism is given by

$$\phi = \theta_1 - \alpha + \sin^{-1}\left[n \sin\left\{ \alpha - \sin^{-1}\left(\frac{\sin \theta_1}{n} \right) \right\} \right]. \tag{7.5}$$

The angular dispersion of the prism is represented by $d\phi/d\lambda$, which is a function of the material dispersion $dn/d\lambda$ through the nonlinear relationship in Eq. (7.5). This angular dispersion depends on the apex angle of the prism as well as the incident angle. Unfortunately, the nature of these relationships is obscured by the complexity of this dispersion equation.

To gain further insight into prism dispersion, it is common to assess prisms at or near the angle of minimum deviation (Wolfe, 1997), which can be defined as the incident angle, where

$$\frac{d\phi}{d\theta_1} = 0. \tag{7.6}$$

By either differentiating Eq. (7.1) using the parts given in Eqs. (7.2), (7.3), and (7.4), or directly differentiating Eq. (7.5), it can be shown that the minimum deviation condition is satisfied when

$$\frac{\cos \theta_1 \cos \theta_2'}{\cos \theta_2 \cos \theta_1'} = 1. \tag{7.7}$$

This condition is satisfied for a symmetric case where the external and internal angles are equal:

$$\theta_1' = \theta_2' = \frac{\alpha}{2} \tag{7.8}$$

and

$$\theta_1 = \theta_2 = \sin^{-1}\left[n\sin\left(\frac{\alpha}{2}\right)\right]. \tag{7.9}$$

The corresponding deviation angle is then given by

$$\phi = 2\sin^{-1}\left[n\sin\left(\frac{\alpha}{2}\right)\right] - \alpha. \tag{7.10}$$

Suppose that a prism is used at minimum deviation for some nominal wavelength λ_0. Since the index of refraction is wavelength dependent, the deviation angle will also be wavelength dependent. From Eq. (7.10), angular dispersion is linearly related to index dispersion by

$$\frac{d\phi}{d\lambda} = 2\frac{\sin(\alpha/2)}{\cos\theta_1}\frac{dn}{d\lambda}. \tag{7.11}$$

Based on the geometry depicted in Fig. 7.4, this can be further simplified to

$$\frac{d\phi}{d\lambda} = \frac{B}{D}\frac{dn}{d\lambda}, \tag{7.12}$$

where D is the diameter of the incident beam and B is the prism base. Retaining only the first term of the Taylor series expansion of the angular dispersion about the nominal wavelength, spectral dependence of the total ray deviation can be approximated as

$$\phi(\lambda) = \phi(\lambda_0) + \frac{B}{D}\frac{dn}{d\lambda}(\lambda - \lambda_0). \tag{7.13}$$

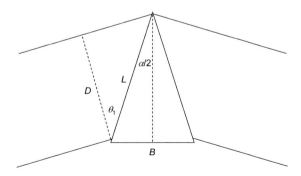

Figure 7.4 Prism geometrical definitions.

The resolving power R of a dispersive material is defined as the ratio of nominal wavelength λ_0 to minimum resolvable wavelength difference $\Delta\lambda_{min}$, or

$$R = \frac{\lambda_0}{\Delta\lambda_{min}}. \tag{7.14}$$

According to the Rayleigh criterion, the minimum resolvable angular difference of rays exiting the prism is $1.22\lambda_0/D$; therefore, it follows from Eq. (7.13) that

$$\Delta\lambda_{min} = \frac{1.22\,\lambda_0}{B\left(\frac{dn}{d\lambda}\right)}. \tag{7.15}$$

Thus, the resolving power of a prism is given by

$$R = 0.82\,B\frac{dn}{d\lambda}. \tag{7.16}$$

Both the resolving power and spectral range of a prism spectrometer are determined by the optical properties of the prism material; specifically, resolving power is defined by material dispersion, and spectral range is defined by the optical transmission spectrum. Figure 7.5 illustrates the dispersion characteristics of several optical glasses across the visible spectral range. These materials are transmissive throughout at least part of the SWIR spectral range to about 1.5 μm and exhibit highly nonlinear

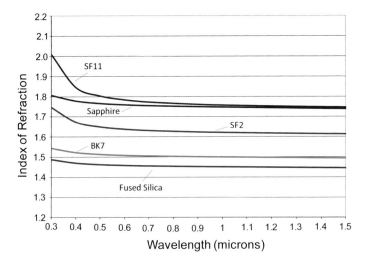

Figure 7.5   Dispersion of optical glasses in the visible and NIR spectral region.

dispersion. Infrared materials, such as those depicted in Fig. 7.6, are typically used for prism spectrometers in the SWIR to LWIR. Alkali halides (NaCl, KBr, and CaF$_2$) are generally hygroscopic, thus their environment must be controlled to ensure that they do not degrade from exposure to moisture. Zinc selenide has a high index of refraction and therefore exhibits high surface-reflection losses. Prisms can be AR coated to reduce such reflection losses, but it is difficult to achieve low losses over both a wide spectral range and a large angular field of view. In most cases, index dispersion is not linear; therefore, angular dispersion characteristics of prism spectrometers are typically nonlinear as well.

7.1.2 Prism spectrometer design

The classical design of a prism spectrometer shown in Fig. 7.7 places a conventional flat-sided prism in the collimated space of a one-to-one, or unit magnification, refractive relay. Let the spatial direction at the focal plane be represented by x and the spectral direction by y. The relay shown in the figure has unity magnification because the lens focal lengths are identical. In general, dissimilar focal lengths can be used, and the spectrometer relay will have a magnification $M_{spec} = f_2/f_1$. In this case, the overall system has an effective imaging focal length

$$f = M_{spec}f_{obj} = \frac{f_2}{f_1}f_{obj}, \qquad (7.17)$$

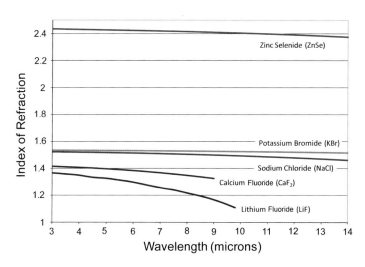

Figure 7.6 Dispersion of optical glasses in the MWIR to LWIR spectral region.

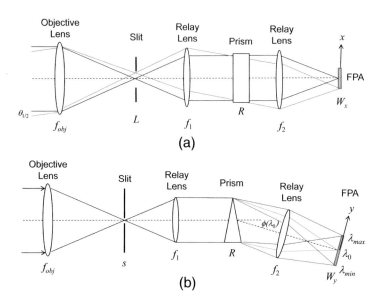

Figure 7.7 Traditional prism spectrometer design: (a) imaging direction and (b) dispersion direction.

where f_{obj} is the focal length of the objective lens or fore-optics preceding the slit. Dispersion at the center of the focal plane surface is given by

$$\frac{dy}{d\lambda} = f_2 \frac{d\varphi}{d\lambda} = 1.22 R (f / \#) . \tag{7.18}$$

The total system f-number is used in Eq. (7.18). To design a prism spectrometer, it is first necessary to relate the degrees of freedom in an optical design to the desired imaging characteristics. First, the length of slit L must match the desired field of view. For a simple optical system, the image height is given by $f \tan \theta$, such that

$$L = 2 f_{obj} \tan \theta_{1/2}, \tag{7.19}$$

where $\theta_{1/2}$ is the half-angle field of view. This is related to the FPA width W_x in the spatial direction by $W_x = M_{spec} L$, given the FPA coordinates in Fig. 7.8.

Spectral direction can be defined by three wavelengths: the minimum wavelength λ_{min}, maximum wavelength λ_{max}, and central wavelength λ_0. The interval $[\lambda_{min}, \lambda_{max}]$ is the spectrometer spectral range and is assumed to bound the spectral dimension of a detector array. The central wavelength is defined as the wavelength that strikes the center of the FPA at the optical axis. The optical axis of the final reimaging lens is deviated by an amount $\phi(\lambda_0)$, such that the following relationships hold for spectral

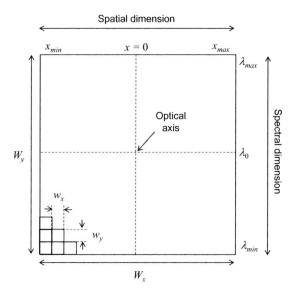

Figure 7.8 FPA coordinates in a dispersive imaging spectrometer.

mapping based on a simple lens mapping:

$$f_2 \tan[\phi(\lambda_0) - \phi(\lambda_{max})] = \frac{W_y}{2},$$

(7.20)

and

$$f_2 \tan[\phi(\lambda_{min}) - \phi(\lambda_0)] = \frac{W_y}{2}.$$

(7.21)

Given spectral range $[\lambda_{min}, \lambda_{max}]$ and FPA width W_y, Eqs. (7.20) and (7.21) can be used to determine f_2 and λ_0. The solution is complicated by the nonlinear nature of angular dispersion $\phi(\lambda)$. Using the approximate linear dispersion relation in Eq. (7.13), the equations simplify to

$$f_2 \tan\left[\frac{B}{D}\frac{dn}{d\lambda}(\lambda_{max} - \lambda_0)\right] = \frac{W_y}{2}$$

(7.22)

and

$$f_2 \tan\left[\frac{B}{D}\frac{dn}{d\lambda}(\lambda_0 - \lambda_{min})\right] = \frac{W_y}{2}.$$

(7.23)

If $dn/d\lambda$ is fixed over the spectral range,

$$\lambda_0 = (\lambda_{max} + \lambda_{min})/2$$

(7.24)

and

$$f_2 = \frac{W_y}{2\tan\left[\frac{B}{D}\frac{dn}{d\lambda}\frac{(\lambda_{max}+\lambda_{min})}{2}\right]}. \tag{7.25}$$

Alternatively, if one selects a lens f_2, Eq. (7.25) can be solved for the matching FPA width W_y. If more-complicated spectrometer relay optics are used that have a mapping function described by $g(\theta; f_2)$ as opposed to $f_2\tan\theta$ (as assumed in the preceding analysis), tangent function $\tan\theta$ is simply replaced by $g(\theta; f_2)/f_2$ in Eqs. (7.20) through (7.25).

While useful for nonimaging spectrometer instruments, the traditional design concept depicted in Fig. 7.7 is generally impractical for hyperspectral imaging applications due to significant aberrations introduced by the thick, flat-sided prism element over the imaging field. To minimize aberrations over the field, imaging spectrometers tend to employ some combination of prisms with curved surfaces, wavefront correction optics, and double-pass spectrometer arrangements. The Littrow arrangement illustrated in Fig. 7.9 is a common design form that exhibits some basic features of a practical imaging spectrometer (Sidran *et al.*, 1966). The reflective parabolic mirror eliminates the limited optical transmission, chromatic aberrations, and spherical aberrations due to refractive relay lenses. The prism is located in a collimated region and operates in a double-pass arrangement, doubling the angular dispersion for a given prism thickness. The parabolic mirror, however, introduces significant off-axis aberrations that are compounded by the prism, limiting the achievable field of view of the instrument.

Replacing the parabolic mirror in the Littrow design with a different mirror form can reduce off-axis aberrations at the expense of on-axis

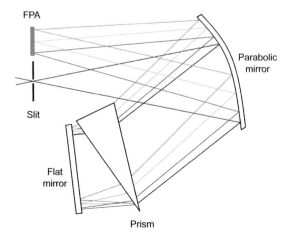

Figure 7.9 Littrow prism spectrometer design.

performance. To further reduce aberrations, a Schmidt corrector can be introduced near the prism, either as an additional refractive element or by introducing curvature into the flat mirror. One such reflecting Schmidt imaging spectrometer design uses an elliptical relay mirror and an aspheric mirror to provide the Schmidt aberration correction (Breckinridge *et al.*, 1983). This approach supports fields of view on the order of 15 deg and maintains good spectral resolution across a broad 0.4- to 2.5-μm spectral range with 50-μm detectors and $f/4$ optics.

The use of aplanatic refractions is another method to control aberrations in an imaging spectrometer. The aplanatic condition is

$$\frac{u}{n} = \frac{u'}{n'}, \tag{7.26}$$

where u and u' are the incident and refracted ray angles relative to the optical axis, and n and n' are the indices of refraction on the respective sides of the interface. Figure 7.10 illustrates the loci of conjugates that satisfy the aplanatic condition for positive and negative refraction from a spherical surface. When this condition is satisfied, surface refraction is free of spherical aberration and coma of all orders.

Prism spectrometers with low optical aberrations and high throughput (i.e., low f-number) can also be designed by using a Littrow-like configuration with prisms and mirrors that both exhibit optimized spherical surfaces. The double aplanatic prism spectrometer design approach depicted in Fig. 7.11 is based on spherical prism surfaces, which refract light nearly at aplanatic conjugates (Warren *et al.*, 1997). The mirror operates near its center of curvature for 1:1 imaging, with a small lateral displacement to allow separation of the slit (located at O) from the FPA (located at I). This double aplanatic design can achieve low aberrations

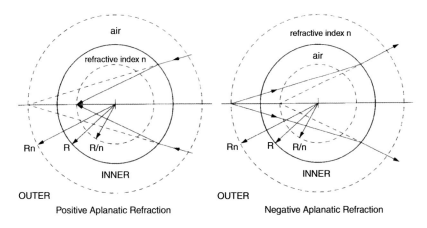

Figure 7.10   Loci of aplanatic conjugates (Warren *et al.*, 1997).

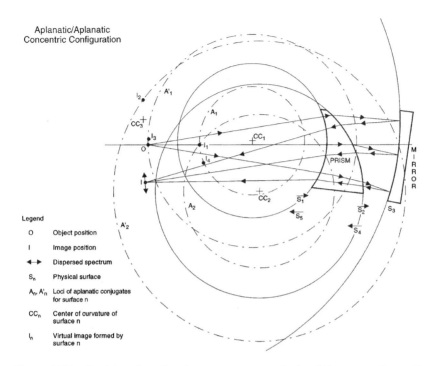

Figure 7.11 Double aplanatic prism spectrometer design (Warren *et al.*, 1997).

at low *f*-numbers and is used as a starting point for numerical ray-trace optimization of performance across the full angular field and spectral range. Variations of this design include using an aspheric mirror to reduce residual spherical aberration, and an additional lens near the focal plane to flatten field and control distortion.

7.2 Grating Spectrometers

Gratings are another common optical element used to produce dispersion in imaging spectrometers. Gratings are optical surfaces that contain a periodic structure with features on the order of an optical wavelength that diffract incident radiation into possibly multiple diffracted orders, each of which is spectrally dispersed. By controlling the dimensions and shape of a periodic structure, one can control the amount of dispersion and the amount of energy diffracted into a particular order. In this way, it is possible to achieve more control over dispersive properties compared to prism materials. Specifically, it is possible to achieve higher resolving powers and more readily operate in the infrared spectral range, where prism materials are more limited. Because of imperfections in the production of fine grating structures, gratings tend to exhibit more scattering than prisms, reducing the spectral purity of grating spectrometers. Gratings also exhibit other limitations due to multiple orders and a wavelength-dependent grating efficiency, issues discussed further in this section.

7.2.1 Grating diffraction

Gratings are optical surfaces or volumes that produce a periodic modulation of an incident wavefront through either index-of-refraction or surface-relief variations (Pedrotti and Pedrotti, 1987). Such gratings disperse light by a combination of diffraction and interference. Consider the geometry depicted in Fig. 7.12, where a grating with periodicity of Λ is illuminated by a plane wave at an incident angle θ_i. The grating equation,

$$\sin \theta_d = \sin \theta_i + m\frac{\lambda}{\Lambda}, \tag{7.27}$$

defines the allowable diffracted angles that satisfy the Floquet phase-matching condition, where m is an integer defining the diffracted order, and λ is the optical wavelength. From Eq. (7.27), angular dispersion is found to be

$$\frac{d\theta_d}{d\lambda} = \frac{m}{\Lambda \cos \theta_d}. \tag{7.28}$$

In this case, dispersion is closely linear (not exactly, due to the cosine term in the denominator), and angular dispersion can be approximated by

$$\phi(\lambda) = \theta_d(\lambda) - \theta_i = \theta_d(\lambda_0) - \theta_i + \frac{m}{\Lambda \cos \theta_d(\lambda_0)}(\lambda - \lambda_0), \tag{7.29}$$

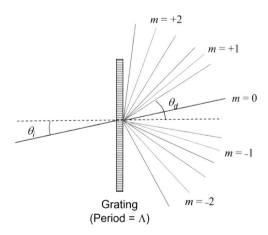

Figure 7.12 Diffraction and dispersion from a transmission grating.

where $\phi(\lambda)$ represents the ray deviation after passing through the grating to the m'th diffracted order. Note that longer wavelengths are diffracted to larger angles than are shorter wavelengths, which is the opposite of prism dispersion. Again using the Rayleigh criterion concerning the minimum-resolvable angular separation, it is easily shown that the minimum-resolvable spectral separation is

$$\Delta\lambda_{min} = 1.22\frac{\Lambda}{D}\frac{\cos\theta_d}{m}\lambda, \qquad (7.30)$$

and the resolving power is

$$R = 0.82\frac{m}{\cos\theta_d}\frac{D}{\Lambda}. \qquad (7.31)$$

Resolving power is nominally the product of the diffracted order and the number of rulings across the aperture diameter.

One fundamental issue that must be contended with in a grating spectrometer design is multi-order diffraction. As shown in Fig. 7.13, it is possible for short-wavelength radiation from higher diffraction orders to overlap with long-wavelength radiation from lower orders if the spectral range of the spectrometer is large. Assuming that the spectral range is limited by some combination of the detector response and an optical filter to an interval $[\lambda_{min}, \lambda_{max}]$, the condition for avoiding order overlap is

$$(m + 1)\,\lambda_{min} > m\lambda_{max}. \qquad (7.32)$$

This results in a spectral range restricted by

$$\lambda_{max} - \lambda_{min} < \frac{\lambda_{min}}{m} \qquad (7.33)$$

to avoid overlap of the diffracted orders. As is discussed in the next section, order-sorting filters can be used to extend the spectral range beyond this limitation.

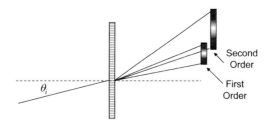

Figure 7.13   Order overlap in a diffraction grating.

The grating equation only provides directions of diffraction, not the energy distribution across the diffracted orders. The directions of diffraction are completely independent of the shape or profile of a periodic amplitude or phase disturbance that the grating produces. The grating profile determines the diffracted energy distribution and can be determined using Fourier optics analysis. Let $t(y)$ represent the grating amplitude transmittance function over a single grating period, where y is the spectral direction but is now defined at the grating surface. By definition, $t(y)$ must be periodic; that is, $t(y + i\Lambda) = t(y)$ for all integer values of i. Due to this periodicity, the transmittance can be expanded into a Fourier series representation,

$$t(y) = \sum_{m=-\infty}^{\infty} A_m e^{i2\pi my/\Lambda}, \tag{7.34}$$

where the integer m corresponds to the grating order, and each order represents a single plane-wave component of the angular spectrum. Fourier series coefficients A_m are given by

$$A_m = \frac{1}{\Lambda} \int_0^{\Lambda} t(y) e^{-i2\pi my/\Lambda} dy. \tag{7.35}$$

Diffraction efficiency is defined as irradiance in a particular order relative to incident irradiance and corresponds to the squared magnitude of a Fourier series coefficient corresponding to a particular grating order. That is, the diffraction efficiency in the m'th order is given by

$$\eta_m = |A_m|^2. \tag{7.36}$$

In the case of phase gratings, $t(y)$ is generally wavelength dependent, and the diffraction efficiency also shares this wavelength dependence.

Examination of the Fourier series coefficient relationship in Eq. (7.35) quickly leads to an optimal grating profile for an imaging spectrometer. First, absorption gratings are clearly undesirable due to the losses they introduce, making diffraction efficiency low in all orders. The preferred approach is to use a phase grating,

$$t(y) = e^{i2\pi W(y)}, \tag{7.37}$$

where $W(y)$ is the phase profile in waves for a particular wavelength of interest. The phase profile is selected to maximize diffraction efficiency in a single diffracted order. Typically, the $m = +1$ or -1 order is used in a grating spectrometer, with the major exception being Echelle grating spectrometers that exploit higher dispersion and spectral resolution in higher orders. Theoretically, diffraction efficiency is maximized using the blazed phase profile,

$$W(y) = y/\Lambda, \tag{7.38}$$

for which diffraction efficiency is identically one in the +1 order and zero in all others. Under this condition, the refracted deviation from a single grating facet exactly matches the diffracted direction of the +1 order.

In practice, it is not possible to produce a blazed grating for which Eq. (7.38) is satisfied, independent of wavelength. To understand this, consider the blazed surface profile depicted in Fig. 7.14 with a blaze angle β, such that

$$\tan\beta = \Delta/\Lambda \tag{7.39}$$

for a given amplitude Δ of a grating surface structure. The phase disturbance for such a surface relief transmission grating is given by

$$W(\lambda) = \frac{[n(\lambda) - 1]\,\Delta}{\lambda}\frac{y}{\Lambda} = \frac{[n(\lambda) - 1]\,\tan\beta}{\lambda}\,y. \tag{7.40}$$

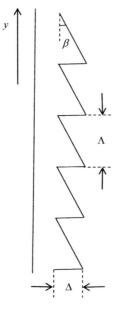

Figure 7.14 Blazed surface relief profile for a phase transmission grating.

Defining $W_0(\lambda)$ as the amplitude of the blazed phase disturbance

$$W_0(\lambda) = \frac{\Lambda}{\lambda}[n(\lambda) - 1]\tan\beta, \qquad (7.41)$$

it follows from Eqs. (7.35)–(7.37) that the diffraction efficiency of the m'th order is

$$\eta_m(\lambda) = \frac{\sin^2\{2\pi[m - W_0(\lambda)]/2\}}{\{2\pi[m - W_0(\lambda)]/2\}^2}. \qquad (7.42)$$

Maximum +1 order diffraction efficiency is achieved by setting $W_0(\lambda) = 1$, giving the blaze condition

$$\tan\beta = \frac{\lambda}{[n(\lambda) - 1]\Lambda}. \qquad (7.43)$$

Unfortunately, this can only be achieved for a single wavelength, for example, central wavelength λ_0, and diffraction efficiency descends across the full spectral range, according to Eq. (7.42), due to the inherent wavelength dependence of Eq. (7.41). Since $n(\lambda)$ decreases with wavelength, it is not possible to compensate wavelength dependence by material dispersion and satisfy the blazed condition in Eq. (7.43) across a larger spectral range.

Relative to the surface relief grating previously described, an alternate approach to producing a phase grating is to produce a periodic index profile in the grating material. The analog to a blazed surface relief grating is a blazed index grating for which the index of refraction varies linearly over the grating period with a peak index difference Δn. In this case, Eq. (7.41) changes to

$$W_0(\lambda) = \frac{\Delta n\, d}{\lambda}, \qquad (7.44)$$

where d is the grating thickness, and the blaze condition becomes

$$\Delta n = \frac{\lambda}{d}. \qquad (7.45)$$

Again, Eq. (7.44) is inherently wavelength dependent, resulting in the wavelength-dependent diffraction efficiency through Eq. (7.42). If the thickness of the phase grating is sufficiently large, the preceding thin phase grating analysis must be replaced by a full coupled wave analysis, and the gratings operate in the Bragg diffraction regime. Bragg gratings can also be made to provide near-unity diffraction efficiency in a single order at a specific wavelength, but the diffraction efficiency again becomes wavelength dependent, albeit by a different set of theoretical relationships.

While transmission gratings are sometimes used for imaging spectrometers, reflection gratings are more commonly used. The primary reason for this is that gratings made from metallic materials can achieve lower losses across broad spectral ranges, since they avoid surface reflection losses that occur with transmission gratings. Such gratings have traditionally been manufactured using ruling machines but more recently have been produced by precision diamond turning or a combination of electron-beam lithography and an etching process. When transmission gratings are used, Bragg gratings produced by holographic methods are fairly common.

To understand the performance of a blazed reflection grating, it is necessary to determine an effective phase profile when illumination occurs at an arbitrary angle of incidence. The geometry is shown in Fig. 7.15, where the optical path difference (OPD) is computed between two reflected rays off of adjacent facets. The important measure is the OPD between the two rays from a planar incident wavefront to a planar exiting wavefront in incident and diffracted directions, respectively. These wavefront surfaces are denoted by the dashed line segments in Fig. 7.15. Based on this geometry,

$$\Delta = \Delta_2 - \Delta_1 = \Lambda \sin(\theta_i + 2\beta) - \Lambda \sin\theta_i. \qquad (7.46)$$

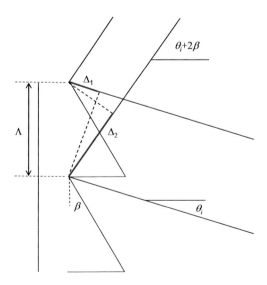

Figure 7.15 Blazed surface relief profile for a reflection grating.

Therefore,

$$W_0(\lambda) = \frac{\Delta}{\lambda} = \frac{\Lambda}{\lambda}[\sin(\theta_i + 2\beta) - \sin \theta_i], \qquad (7.47)$$

and diffraction efficiency is again computed from Eq. (7.42). The condition for unity diffraction efficiency in the +1 order is

$$\sin(\theta_i + 2\beta) = \sin \theta_i + \frac{\lambda}{\Lambda}. \qquad (7.48)$$

This condition is satisfied when the angle of reflection from a single blaze corresponds to an angle of the first diffracted order. Due to the wavelength dependence in Eq. (7.47), this condition is only satisfied for a single wavelength, as in the case of transmission gratings, and diffraction efficiency decreases across the spectral range. This reduction in diffraction efficiency with wavelength is a fundamental characteristic of grating spectrometers that limits the achievable spectral range in practice.

7.2.2 Grating spectrometer design

A traditional transmission grating spectrometer design, illustrated in Fig. 7.16, operates analogously to the prism spectrometer in Fig. 7.7 with some modification, due to different angular dispersion relationships as well as the common presence of an order-sorting filter directly in front of the FPA to compensate for the order overlap problem. Again, given spectrometer magnification $M_{spec} = f_2/f_1$, imaging detector width W_x is matched to the angular field of view according to

$$W_x = 2M_{spec}f_{obj} \tan \theta_{1/2}. \qquad (7.49)$$

Spectral dispersion is in the opposite direction (as is that of the prism spectrometer), and is matched to the detector array size in the spectral dimension by the following equations:

$$f_2 \tan[\phi(\lambda_0) - \phi(\lambda_{min})] = \frac{W_y}{2} \qquad (7.50)$$

and

$$f_2 \tan[\phi(\lambda_{max}) - \phi(\lambda_0)] = \frac{W_y}{2}. \qquad (7.51)$$

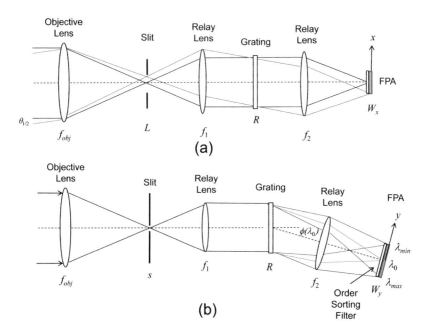

Figure 7.16 Traditional transmissive grating spectrometer design: (a) imaging direction and (b) dispersion direction.

For linear dispersion, the equations simplify to

$$f_2 \tan\left[\frac{m}{\Lambda \cos\theta_d(\lambda_0)}(\lambda_0 - \lambda_{min})\right] = \frac{W_y}{2} \tag{7.52}$$

and

$$f_2 \tan\left[\frac{m}{\Lambda \cos\theta_d(\lambda_0)}(\lambda_{max} - \lambda_0)\right] = \frac{W_y}{2}, \tag{7.53}$$

from which

$$\lambda_0 = (\lambda_{max} + \lambda_{min})/2 \tag{7.54}$$

and

$$f_2 = \frac{W_y}{2\tan\left[\frac{m}{\Lambda \cos\theta_d(\lambda_0)}\left(\frac{\lambda_{max} - \lambda_{min}}{2}\right)\right]}. \tag{7.55}$$

If the dispersion is not assumed to be linear, then

$$\phi(\lambda) = \sin^{-1}\left[\sin\theta_i + \frac{m\lambda}{\Lambda}\right] - \theta_i, \tag{7.56}$$

and the design equations become more complex. In this case, let $\Delta\phi$ be the angular dispersion across the full spectral range $[\lambda_{min}, \lambda_{max}]$. To match the FPA width in the spectral direction,

$$f_2 \tan\Delta\phi = \frac{W_y}{2}. \tag{7.57}$$

In the spatial direction, the equations are unchanged from the linear dispersion approximation. From Eq. (7.56), the design equations become

$$\Delta\phi = \sin^{-1}\left[\sin\theta_i + \frac{m\lambda_0}{\Lambda}\right] - \sin^{-1}\left[\sin\theta_i + \frac{m\lambda_{min}}{\Lambda}\right] \tag{7.58}$$

and

$$\Delta\phi = \sin^{-1}\left[\sin\theta_i + \frac{m\lambda_{max}}{\Lambda}\right] - \sin^{-1}\left[\sin\theta_i + \frac{m\lambda_0}{\Lambda}\right]. \tag{7.59}$$

If f_2 is known, Eq. (7.57) can be used to find the required total angular dispersion $\Delta\phi$ to match FPA width W_y, and then Eqs. (7.58) and (7.59) can be solved for grating period Λ and center wavelength λ_0 to match spectral range $[\lambda_{min}, \lambda_{max}]$ for a given incident angle θ_i and grating order m.

As mentioned, multi-order diffraction is a unique design issue that must be addressed with grating spectrometers. If the spectral range of a system falls within the limits imposed by Eq. (7.33), it can be easily accommodated by filtering all of the optical energy within these limits somewhere in the system before the FPA, possibly including the FPA spectral response itself. However, it is often desirable to achieve a larger spectral range; this can be obtained by employing order-sorting filters. Figure 7.17 illustrates the concept of an order-sorting filter for a VNIR spectrometer covering the 0.4- to 1.1-μm spectral range. Without any such filtering, the second order corresponding to the 0.4- to 0.55-μm spectral region overlaps the first order of the 0.8- to 1.1-μm region. However, the second order can be rejected over this part of the FPA by placing a high-pass filter that begins to transmit radiation at 0.7 μm, for example, over the FPA region associated with the 0.75- to 1.1-μm band. Filtering tolerance, based on the width of the transition region and filter–FPA alignment, is fairly loose (on the order of 0.5 μm) in this particular case. The order-sorting concept can be extended to larger relative spectral ranges, but this can involve more filters with tighter tolerances, since multiple diffracted

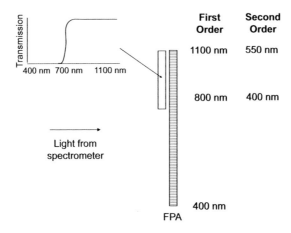

Figure 7.17 Order sorting in a VNIR grating spectrometer.

orders can be involved. It is common to design the filter so that the edges between sections fall within water bands or other spectral locations of less importance to the intended remote sensing applications for the spectrometer.

In some cases, the multi-order nature of gratings can be a benefit to spectrometer design. For example, consider the case where an imaging spectrometer is to be designed across the entire VNIR/SWIR range from 0.4 to 2.5 μm. The optimal detector for VNIR is silicon, while SWIR requires another material such as HgCdTe. Clearly, diffraction efficiency in the first order will fall off considerably across this extensive spectral range. However, the second order, associated with the 0.4- to 1.25-μm band, aligns perfectly with the first order of the 0.8- to 2.5-μm band and achieves similar diffraction efficiency and optical aberrations. Therefore, it is possible to use a single spectrometer for this entire spectral range with two optimized FPAs by using both orders and separating VNIR and SWIR energy between the spectrometer and FPAs with a low-pass spectral filter that transmits VNIR and reflects SWIR. This is known as a dichroic filter.

Grating spectrometers are designed using considerations similar to prism spectrometer design considerations. While transmission gratings are sometimes employed, reflection gratings are more common due to their increased light efficiency. The spectrometer designs discussed in this section are all based on reflection gratings. A Littrow configuration (shown in Fig. 7.18) can be used for grating spectrometers by tilting the grating such that the blazed facets are orthogonal to the incident rays, so $\theta_d(\lambda_0) = -\theta_i(\lambda_0)$ for the $m = -1$ order. In this fashion, first-order diffraction is retroreflective for the center wavelength. Separation between the slit and the FPA is achieved by operating off-axis in the along-slit direction, as shown in Fig. 7.18, or by deviating slightly from the exact Littrow in the

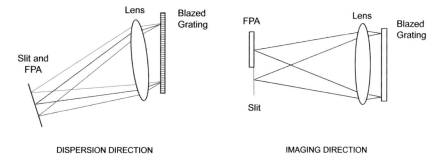

Figure 7.18 Littrow grating spectrometer design.

cross-slit direction. As with the Littrow prism spectrometer design, this can result in increased aberrations, limiting spectral and spatial resolutions for a low f-number system. To eliminate the refractive optic, a concave grating can be used with the slit and focal plane situated on the circle, matching the radius of curvature of the grating, as depicted in Fig. 7.19. This is known as the Paschen–Runge design. The Czerny–Turner design, illustrated in Fig. 7.20, places a flat blazed reflection grating in the collimated space of a two-mirror reimaging system using parabolic or spherical mirrors. All of these designs have been widely used in nonimaging spectrometers but are limited in achievable f-numbers for imaging spectrometers due to off-axis aberrations, including distortion.

One successful approach to achieving excellent grating imaging spectrometer performance in a compact optical system has been to employ concentric imaging design forms (Lobb, 1994). Concentric imaging systems, such as the Offner imaging design shown in Fig. 7.21, have a rotational symmetry about any line through a common center of curvature.

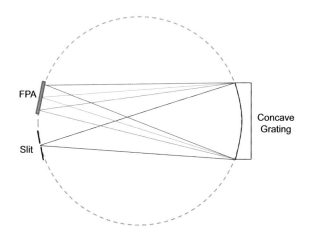

Figure 7.19 Paschen–Runge grating spectrometer design.

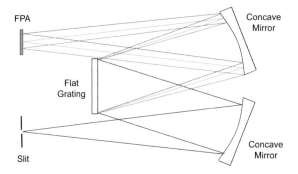

Figure 7.20   Czerny–Turner grating spectrometer design.

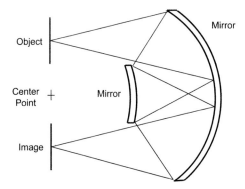

Figure 7.21   Offner concentric imaging design.

This results in perfect imaging for any radial line from the optical axis, with no axial aberrations, coma, or sagittal field curvature. Additionally, as a result of using only reflective elements, there are no chromatic aberrations in an Offner implementation. There is tangential field curvature (i.e., curvature of the best focal surface for tangential lines) that can be controlled at the expense of other aberrations by deviation from perfect concentricity. Figure 7.22 illustrates the way in which the Offner design can be applied to a grating imaging spectrometer (Mouroulis and Thomas, 1998). This requires the use of a convex blazed grating that can be produced by ruling or diamond turning but has most successfully been produced by electron-beam lithography. Because of the asymmetry introduced by grating diffraction, a split-Offner design is employed, where orientation of the two mirrors is slightly asymmetric. The use of this design has resulted in imaging spectrometers with extremely low values of spatial–spectral distortion, a topic addressed in the next section.

Another concentric imaging design is the Dyson variety, shown in Fig. 7.23, based on a spherical mirror and thick lens. Object and image planes are located near the center of curvature of the mirror against the

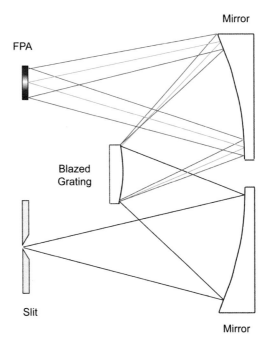

Figure 7.22 Offner grating spectrometer design.

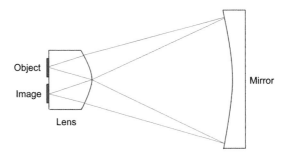

Figure 7.23 Dyson concentric imaging design.

flat surface of the lens. The other lens surface is concentric with the mirror and optimally placed to compensate aberrations. Using a multi-element lens with concentric, cemented interfaces provides further correction of spherical aberration and axial chromatic aberration. The Dyson concept can be translated into a grating spectrometer by placing the grating surface on the primary mirror, as shown in Fig. 7.24 (Mouroulis *et al.*, 2008). This once again requires the manufacture of a blazed reflection grating on a spherical surface, which is concave in this case.

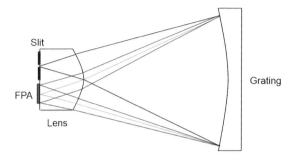

Figure 7.24 Dyson grating spectrometer design.

7.3 Imaging Spectrometer Performance

Characterization and analysis of imaging spectrometer performance follows the same methodology as described for broadband imaging systems, with some extensions to accommodate the spectral-sensing dimension. First, it is essential to precisely understand the mapping of spectral information on the FPA. While linear dispersion approximations can be adequate for initial spectrometer layout, the actual mapping is impacted by nonlinear dispersion and distortion of the imaging optics. In addition to defining spatial resolution based on the PSF, the spectral resolution of an imaging spectrometer is characterized by a spectral response function (SRF), which is a function of the resolving power of the dispersive element as well as the PSF of the optical system. Radiometric sensitivity is typically characterized by the SNR or NESR as a function of the spectral band.

7.3.1 Spatial and spectral mapping

Equations (7.5) and (7.29) provide the exact dispersion characteristics of prisms or gratings, respectively, that make up dispersive spectrometers. In the spectral direction, an optical system allows a wavelength-dependent angle exiting the dispersive element to transform to a spectral dimension on the FPA. The nomenclature uses the term *row* to represent a spatial direction on the FPA (i.e., a row is along the slit direction and contains a spatial line for a single wavelength) and the term *column* to represent a spectral direction on the FPA. That is, a column is across the slit direction and contains a spectrum for a single spatial position. The geometry is depicted in Fig. 7.8. Thus far, the relationship between the FPA position y in the column direction from the center and the exit angle ϕ from the dispersive element relative to the optical axis has been assumed to be

$$y(\lambda) = f_2 \tan[\phi(\lambda) - \phi(\lambda_0)], \qquad (7.60)$$

where the final lens is oriented such that the central wavelength λ_0 falls on the center of the FPA at $y = 0$. Given detector spacing w, this can be put in terms of FPA row index k according to

$$k = \text{int}\left[\frac{y(\lambda)}{w}\right] + k_0, \tag{7.61}$$

where k_0 corresponds to the row index at the optical axis and $\text{int}(x)$ is a nearest integer rounding function. A real imaging system exhibits some generally nonlinear mapping,

$$y(\lambda) = g[f_2, \phi(\lambda) - \phi(\lambda_0)], \tag{7.62}$$

as opposed to the distortion-free mapping in Eq. (7.60), with a functional form that can be determined by ray-trace analysis. Two potential nonlinearities between FPA position and wavelength occur, the first due to the dispersion relationship and the second due to lens distortion. Placing Eq. (7.62) into Eq. (7.61) provides the nonlinear relationship between FPA row k and wavelength λ, as is illustrated in Fig. 7.25. This relationship can be inverted to determine the spectral band center λ_k for each FPA row. Inversion of the nonlinear mapping to determine the spectral-band center for each FPA element must be accurately performed to extract quantitative spectral data from the instrument.

Unfortunately, the distortion characteristics of an imaging dispersive spectrometer are not always limited to spectral direction or even separable along spectral and spatial directions. Instead, there can be spatial–spectral distortion, which is a deviation from the ideal situation where spectral information is mapped precisely along columns and spatial information along rows. Consider Fig. 7.26, which illustrates pin-cushion distortion in a conventional imaging system. In a spectrometer, the distorted rows represent mapping of the slit for single spectral bands, and the columns represent mapping of the spectrum for a single position in the slit. By

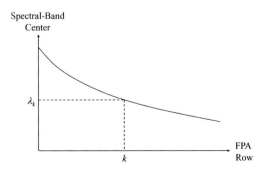

Figure 7.25 Nonlinear dispersion characteristic of a dispersive spectrometer.

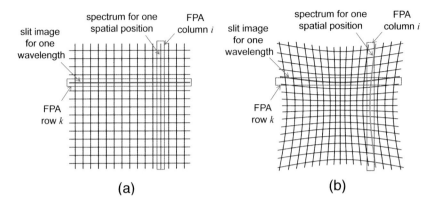

Figure 7.26 Spatial–spectral distortion in a dispersive imaging spectrometer: (a) undistorted mapping and (b) distorted mapping.

representing the slit position from the optical axis as x, it then becomes apparent that spatial–spectral distortion can be characterized as a general mapping into the (column, row) space (i, k) of the FPA as $\lambda_{i,k}$ and $x_{i,k}$. Instead of having the same spectral-band center across an FPA row, the band center now varies across the row. Likewise, the imaged spatial location from the slit varies across a column.

Like nonlinear dispersion, which is actually part of spatial–spectral distortion, the aforementioned mappings must be determined either through an exact ray trace or empirical measurements. The latter is referred to as spectral calibration and is discussed in Chapter 10. Characterization of spatial–spectral distortion is represented by Fig. 7.27, where *smile* (or *frown*) refers to the maximum deviation of a slit image for a single wavelength from linear, relative to the detector width; and *keystone* refers to the maximum deviation of the image of a slit point across the spectral range from linear, relative to the detector width. Typically, both smile and keystone must be controlled at subpixel levels. Smile and keystone are controlled by minimizing distortion of the reimaging system composing the spectrometer. Smile can also be compensated for to some extent by using a curved slit opposing the smile direction whose image forms a line on the FPA. This introduces a purely spatial distortion of the imaging system relative to the object but is often more tolerable for system applications for which spectral fidelity is more important. Since the magnitude of the smile can be wavelength dependent, curving the slit can only remove it at a single wavelength; there will still be some residual smile across the spectral range.

7.3.2 Spatial and spectral response functions

Spectral calibration is essential for extracting quantitative spectral information from a dispersive imaging spectrometer. Often it is also

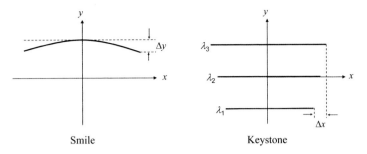

Smile Keystone

Figure 7.27 Characterization of spatial–spectral distortion in a dispersive imaging spectrometer.

important to characterize spatial and spectral response functions of an instrument. The spatial response function, alternatively known as the PSF as described in Chapter 6, describes the relative distribution in the spatial direction of the FPA from a point source object. In the case of an imaging spectrometer, the PSF can vary with both spatial position in the slit and the spectral-band center. Analogously, the SRF describes the distribution of wavelengths that fall onto any particular FPA element. For any particular element, designated by the (column, row) pair (i, k), the SRF has a width characterizing the spectral resolution of the instrument and is centered about the spectral-band center λ_k. For a variety of reasons yet to be seen, the SRF can also vary in both the spectral and spatial directions of the FPA. The SRF is theoretically limited to the diffraction-limited spectral resolution of the prism or grating instrument through the resolving power discussed earlier in this chapter but is often coarser due to finite geometric sizes of the slit and detectors. To understand these geometric impacts, consider the generic imaging spectrometer depicted in Fig. 7.28 composed of a dispersive element with a resolving power R and an imaging relay with lenses or mirrors of magnification M_{spec}, slit width s, detector width w, and an f-number at the exit pupil plane specified by $(f/\#)$. Let λ_k be the wavelength corresponding to the central ray through the slit to a particular FPA element centered at position y_k. Over small displacements about these central values, the relationship between FPA spatial position and wavelength is

$$y - y_k = 1.22\, R(f/\#)(\lambda - \lambda_k). \tag{7.63}$$

Because of general nonlinear-dispersion characteristics, the resolving power actually is a function of λ_k, but that dependence is merely implied for now.

By limiting the analysis to only geometrical considerations and ignoring diffraction and optical aberrations, the SRF can be determined simply by quantifying the fraction of the image of the slit that overlaps the

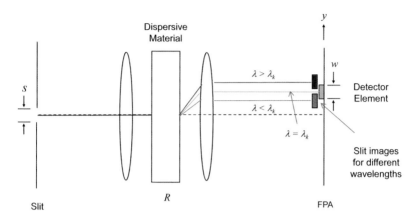

Figure 7.28 Characterization of the spatial and spectral response functions in a dispersive imaging spectrometer.

detector element of interest as a function of wavelength. Similarly, the PSF can be characterized by the extent of the slit positions that overlap the detector element in the image domain. It is apparent that these functions are somewhat coupled, and the analysis can be performed by considering a geometric support function $f(\lambda, y)$ that defines the range of wavelengths and detector positions for which incident light rays can pass through the slit and onto a particular detector. By definition, the support function is equal to 1 when both of these conditions are met, and 0 elsewhere. Even though it is common design practice to match the imaged slit width s, magnified by M_{spec}, to detector width w, we consider the general case where they are not perfectly matched.

First, consider the case where $M_{spec} s > w$, the support function depicted in Fig. 7.29. The horizontal dotted lines represent the edges of the detector, so the support function can only be nonzero between them; otherwise, the rays miss the detector element. The diagonal lines illustrate the condition for which rays pass through the slit. The slope of these lines is due to dispersion of the optics. To illustrate this further, three vertical colored line segments represent the extent of imaged slit width on the detector for three different wavelengths, one at the spectral-band center, one below it, and one above it. In this case, the magnified slit width overfills the detector width such that the detector is fully illuminated across a spectral range about the band center. As wavelength is further decreased or increased, however, the image of the slit width extends partially and then fully off the detector until it reaches the point where no light from the slit lands on the detector of interest. The union of regions between the horizontal and diagonal lines defines the set of wavelengths and spatial positions that meet the two conditions. This is indicated by the shaded region in the figure, for which $f(\lambda, y) = 1$.

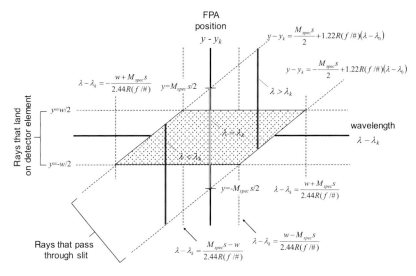

Figure 7.29 Spatial–spectral geometric support function for $M_{spec}s > w$. The vertical colored bars represent the images of the slit width on the detector at three different wavelengths. The support function is 1 in the shaded region and 0 elsewhere.

Figure 7.30 illustrates the case where $M_{spec}s < w$, for which similar logic follows. In this case, the imaged slit width underfills the detector, so the dimensions of support region $f(\lambda, y)$ are slightly different. This has implications concerning the radiometric sensitivity of the system. Ordinarily, the light-collecting power of the optical system is defined by the combination of the detector area and f-number of the optics. In this case, however, using the detector area overestimates the light-collecting power because some of the light is blocked by the slit before reaching it. Therefore, the ratio $M_{spec}s/w$ represents the effective transmission factor in radiometric sensitivity analysis. In the derivation of the spectral–spatial support function, it should be recognized that the situations depicted in Figs. 7.29 and 7.30 become identical when $M_{spec}s = w$. Also, Figs. 7.29 and 7.30 illustrate a grating spectrometer situation where longer wavelengths are dispersed more toward larger angles than toward shorter wavelengths. In the prism spectrometer case, the diagonal dispersion lines have a negative slope, which can be represented by using a negative resolving power R. The result concerning the SRF and PSF, however, is completely unchanged due to symmetry.

The geometric spatial PSF in the y direction for an imaging spectrometer is given by the projection of the support function in the spectral direction,

$$h_g(y) = \frac{\int f(\lambda, y)\, d\lambda}{\int f(\lambda, \acute{y})\, d\lambda} = \text{rect}\left(\frac{y}{M_{spec}s}\right). \tag{7.64}$$

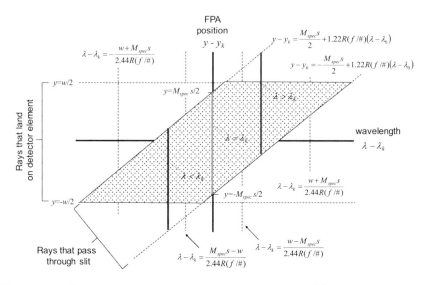

Figure 7.30   Spatial–spectral geometric support function for $M_{spec}s < w$. Vertical colored bars represent the images of the slit width on the detector at three different wavelengths. The support function is 1 in the shaded region and 0 elsewhere.

This PSF incorporates, or replaces, the detector PSF in MTF analysis. In the along-slit direction, the PSF is of similar form, with the detector width taking the place of the slit width (that is, the detector PSF in a conventional imaging system). Representing the optical PSF of a spectrometer imaging system as $h_o(x, y)$, including diffraction and all system aberrations, the total spectrometer PSF is then given by the convolution of the geometric PSF with the optical PSF,

$$h(x, y) = \text{rect}\left(\frac{x}{w}, \frac{y}{M_{spec}s}\right) * h_o(x, y). \tag{7.65}$$

This can also be modeled by multiplying the corresponding MTFs.

The geometric SRF is given by the projection of the support function in the spatial direction,

$$g_z(\lambda) = \frac{\int f(\lambda, y)\,dy}{\int f(\lambda_0, y)\,dy} = \text{rect}\left[\frac{1.22R(f/\#)(\lambda - \lambda_k)}{M_{spec}s}\right] * \text{rect}\left[\frac{1.22R(f/\#)(\lambda - \lambda_k)}{w}\right]. \tag{7.66}$$

This is depicted in Fig. 7.31. Given the same optical PSF as was previously discussed, accounting for diffraction and system aberrations, the total SRF

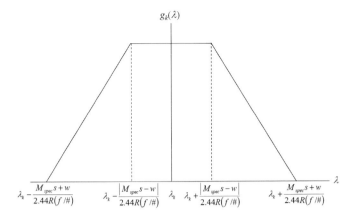

Figure 7.31 Geometric spectral response function for a dispersive spectrometer.

for the k'th spectral band is given by

$$g_k(\lambda) = \text{rect}\left[\frac{1.22R(f/\#)(\lambda - \lambda_k)}{M_{spec}s}\right]$$

$$* \text{rect}\left[\frac{1.22R(f/\#)(\lambda - \lambda_k)}{w}\right] * h_0\left[0, 1.22R(f/\#)\lambda\right]. \qquad (7.67)$$

where spectral dependence of the resolving power is now explicitly shown. One very important consequence of Eqs. (7.65) and (7.67) is the inherent coupling between spatial and spectral resolution in a dispersive imaging spectrometer. Spatial resolution is defined based on the PSF, or corresponding MTF, in a manner identical to the way it was described in Chapter 6 for a conventional imaging system. Geometrically, spectral resolution can be defined by the FWHM of the SRF, which can be seen from Fig. 7.31 to be

$$\delta\lambda = \frac{\max(M_{spec}s, w)}{1.22R(f/\#)}. \qquad (7.68)$$

Smaller slit and detector widths support finer spectral resolution, but with degraded radiometric sensitivity due to reduced light throughput, as is seen in the next section. As discussed previously, the implicit wavelength dependence of resolving power implies that spectral resolution be wavelength dependent. From a radiometric perspective, it is preferable to define spectral resolution, radiometric as

$$\delta\lambda_k = \int g_k(\lambda)\, d\lambda, \qquad (7.69)$$

which turns out to be identical to the FWHM in the geometric case, where the optical PSF is ignored.

7.3.3 Radiometric sensitivity

Characterization of radiometric sensitivity for a dispersive spectrometer follows closely to that previously discussed for a conventional imaging system. Assume that the entrance pupil of such an imaging spectrometer system, including all preceding fore-optics, is illuminated with signal and background spectral radiance $L_s(\lambda)$ and $L_b(\lambda)$, respectively, where they are defined for this particular situation in Chapter 6. At the spectrometer entrance slit, this results in the corresponding spectral irradiance distributions

$$
\begin{aligned}
E_s(\lambda) &= \frac{\pi K(x)}{4(f/\#)^2 + 1}\tau_o(\lambda)L_s(\lambda) \\
E_b(\lambda) &= \frac{\pi K(x)}{4(f/\#)^2 + 1}\tau_o(\lambda)L_b(\lambda) + E_o(\lambda),
\end{aligned}
\tag{7.70}
$$

where the along-slit dependence in the irradiance roll-off function along slit direction $K(x)$ is explicitly shown, $\tau_o(\lambda)$ is the (warm) optics transmission, and $E_o(\lambda)$ represents irradiance due to instrument self-emission and stray radiation. For a particular detector corresponding to a center wavelength λ_k and SRF $g_k(\lambda)$, the corresponding focal plane spectral irradiance distribution is

$$
\begin{aligned}
E_s(\lambda) &= \frac{\pi K(x, \lambda)}{4(f/\#)^2 + 1}\tau_o(\lambda)\,\tau_{spec}(\lambda)\,\tau_c(\lambda)\,L_s(\lambda)\,g_k(\lambda) \\
E_b(\lambda) &= \frac{\pi K(x, \lambda)}{4(f/\#)^2 + 1}\tau_o(\lambda)\,\tau_{spec}(\lambda)\,\tau_c(\lambda)\,L_b(\lambda)\,g_k(\lambda) + E_o(\lambda),
\end{aligned}
\tag{7.71}
$$

where $\tau_{spec}(\lambda)$ is the spectrometer transmission, $\tau_c(\lambda)$ is the cold filter transmission, the self-emission and stray radiation term now includes all sources up to the FPA, and roll-off is now defined as a function of both spatial and spectral FPA dimensions. Given the focal plane spectral irradiance, the remainder of this analysis proceeds similarly to that of a conventional imaging system by first computing the detected photoelectron counts,

$$
\begin{aligned}
N_s &= A_d t_d \frac{\min(M_{spec}s, w)}{w}\int \frac{\lambda}{hc}\eta(\lambda)\,E_s(\lambda)\,d\lambda \\
N_b &= A_d t_d \frac{\min(M_{spec}s, w)}{w}\int \frac{\lambda}{hc}\eta(\lambda)\,E_b(\lambda)\,d\lambda,
\end{aligned}
\tag{7.72}
$$

incorporating the effective transmission loss from the undersized slit width. Since spectral bandwidth $\delta\lambda_k$ is small, the integrals in Eq. (7.72) can be approximated by

$$N_s = A_d t_d \frac{\min(M_{spec} s, w) \lambda_k}{w} \frac{\lambda_k}{hc} \eta(\lambda_k) E_s(\lambda_k) \delta\lambda_k$$

$$N_b = A_d t_d \frac{\min(M_{spec} s, w) \lambda_k}{w} \frac{\lambda_k}{hc} \eta(\lambda_k) E_b(\lambda_k) \delta\lambda_k. \tag{7.73}$$

From this, the SNR for a band centered at λ_k with bandwidth $\delta\lambda_k$ is

$$\text{SNR}(\lambda_k) = \frac{N_s(\lambda_k)}{\sqrt{N_b(\lambda_k) + \sigma^2_{n,dark} + \sigma^2_{n,Johnson} + \sigma^2_{n,read} + \sigma^2_{n,spatial} + \sigma^2_{n,ADC}}}, \tag{7.74}$$

incorporating the noise model described in Chapter 6, where shot-noise variance [or the first term in the denominator of Eq. (7.74)], is simply $N_b(\lambda_k)$. The fractional term preceding the integrals in Eqs. (7.72) and (7.73) is required to account for the radiometric loss that occurs when the magnified slit width is less than the detector width in the cross-slit direction, as previously discussed.

A widely used specification for imaging spectrometers is the NESR, defined as

$$\text{NESR}(\lambda_k) = \frac{L_s(\lambda_k)}{\text{SNR}(\lambda_k)}, \tag{7.75}$$

which is typically quantified as a function of the band center over the spectral range, as indicated. This quantity can be computed directly from the SNR but can also be written in a simpler parametric form based on a few assumptions and definitions. First, the optical throughput of the imaging spectrometer is defined as

$$A\Omega = \frac{\min(M_{spec} s, w)}{w} \frac{\pi A_d}{4(f/\#)^2 + 1}. \tag{7.76}$$

Next, effective signal and background transmission profiles are computed as

$$\tau_s(\lambda_k) = \tau_o(\lambda_k) \tau_{spec}(\lambda_k) \tau_c(\lambda_k) \eta(\lambda_k) \tag{7.77}$$

and

$$\tau_b(\lambda_k) = \frac{\tau_o(\lambda_k)\,\tau_{spec}(\lambda_k)\,\tau_c(\lambda_k)\,E_b(\lambda_k) + E_o(\lambda_k)}{E_s(\lambda_k)}\eta(\lambda_k). \qquad (7.78)$$

Also, the total additive, signal-independent noise is defined as

$$\sigma_{n,add} = \sqrt{\sigma_{n,dark}^2 + \sigma_{n,Johnson}^2 + \sigma_{n,read}^2 + \sigma_{n,spatial}^2 + \sigma_{n,ADC}^2}. \qquad (7.79)$$

Based on these assumptions and definitions, the NESR at a center wavelength for a particular spectral band becomes

$$NESR(\lambda_k) = \frac{\sqrt{\sigma_{n,add}^2 + A\Omega\frac{t_d}{hc}\tau_b(\lambda_k)L_s(\lambda_k)\lambda_k\delta\lambda_k}}{A\Omega\frac{t_d}{hc}\tau_s(\lambda_k)\lambda_k\delta\lambda_k}. \qquad (7.80)$$

In a background-limited regime, the shot-noise term (second term in the square root) dominates the additive noise term, and the equation reduces to

$$NESR(\lambda_k) = \frac{1}{\tau_s(\lambda_k)}\sqrt{\frac{\tau_b(\lambda_k)L_s(\lambda_k)}{A\Omega\frac{t_d}{hc}\lambda_k\delta\lambda_k}}. \qquad (7.81)$$

In a noise-limited regime, the additive noise term dominates, and the equation reduces to

$$NESR(\lambda_k) = \frac{\sigma_{n,add}}{A\Omega\frac{t_d}{hc}\tau_s(\lambda_k)\lambda_k\delta\lambda_k}. \qquad (7.82)$$

In the latter case, the NESR is independent of illumination and more strongly dependent on the optical throughput, dwell time, signal transmission, and spectral bandwidth.

7.4 System Examples

Dispersive imaging spectrometers have been designed, built, and deployed from ground, airborne, and space-based platforms for a variety of remote sensing applications. The designs employed exhibit a fairly large diversity based on the different requirements for each specific application, the advancement of optical and FPA component technology over time, and personal preferences of the designers. A small number of widely used systems are discussed here to illuminate how the design issues discussed in this chapter have been put into practice, and the overall performance that has been achieved.

7.4.1 Airborne Visible/Infrared Imaging Spectrometer

The Airborne Visible/Infrared Imaging Spectrometer (AVIRIS) was developed by the Jet Propulsion Laboratory as a scientific research instrument and was the first airborne hyperspectral imaging sensor to cover the entire spectral range from 0.4 to 2.5 μm (Green *et al.*, 1998). The imaging spectrometer was first integrated into the Q-bay of a high-altitude NASA ER-2 aircraft, where it achieves about 20-m GSD and 11-km swath width down-looking from a 20-km altitude. It has also been operated from a Twin Otter aircraft at a much lower altitude of 4 km, with GSD and swath-width scales to 4 m and 2 km, respectively. Unlike the dispersive imaging spectrometer designs described in this chapter, AVIRIS uses single-pixel dispersive spectrometers with a whiskbroom, cross-track scanning mirror that performs spatial imaging in a fully scanned manner. That is, there is no spatial dimension to the FPAs. The scanner covers a 30-deg cross-track field of view, producing a 677-pixel swath at a scan rate of 12 Hz.

Collected light from the scanner is focused using reflective parabolic and elliptical fore-optics onto an array of fibers that transmit spectral irradiance to four spectrometers, each of which covers a portion of the full spectral range. Silica glass fibers are used for spectrometers A and B, covering the 0.4- to 0.7-μm and 0.7- to 1.8-μm spectral regions; zirconium fluoride glass fibers couple into spectrometers C and D, covering the 1.3- to 1.9-μm and 1.9- to 2.5-μm spectral regions. Due to a high NA of the fibers, the spectrometers operate at $f/1$ using an off-axis Schmidt design. Spectrometer A uses a 32-element array of 200×200-μm Si detectors at ambient temperature, while the other three spectrometers use 64-element InSb detector arrays cooled to 77 K. The detector element sizes for spectrometers B and C match those of spectrometer A, while larger 214×500-μm detectors are used in spectrometer D due to lower radiance levels. The large fiber core and detector element sizes are required to compensate for the short 87-μsec integration time that results from the fully scanned nature of the imaging system. This further results in a fairly large instrument, depicted in Fig. 7.32, relative to the imaging spectrometers described later in this chapter that utilize 2D FPAs. The reason for the size difference is that 2D arrays allow a spectrometer to eliminate the bulky scanner and capture spectra from multiple spatial locations simultaneously, supporting longer integration times with the same effective coverage rate.

The nominal specifications of the AVIRIS sensor are summarized in Table 7.1. One major advantage of the fully scanned design is that light for a particular spectral band passes through the same optical path and is sensed by the same detector element, regardless of spatial position. This supports excellent spatial uniformity relative to imaging spectrometers

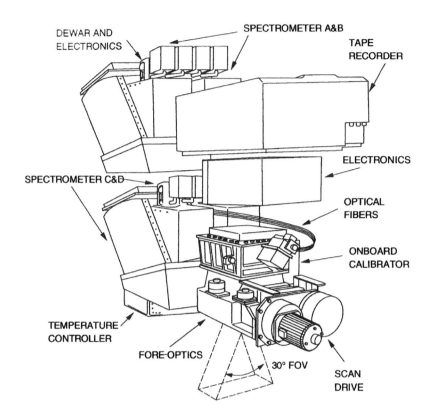

Figure 7.32 Layout of the AVIRIS VNIR/SWIR imaging spectrometer [reprinted with permission from Green *et al.* (1998)].

comprised of 2D FPAs, whose performance is often limited by the relative nonuniformity and nonlinearity of the array as well as changes in optical characteristics with field angle. Furthermore, the AVIRIS instrument includes a mechanical shutter that closes at the end of each scan line, providing a dark reference to eliminate the impact of detector dark-current drift. It also has an onboard calibrator composed of a stable, calibrated quartz halogen lamp and narrowband spectral filter that are measured before and after each flight line. The result is a hyperspectral sensor system that can be well calibrated in both a relative and an absolute sense, a well-recognized attribute of the AVIRIS sensor by hyperspectral analysts, for whom radiometric calibration is very important. Hyperspectral sensor calibration is further explored in Chapter 10.

Achieving adequate radiometric sensitivity for quantitative imaging spectroscopy is a challenge for fully scanned hyperspectral instruments such as AVIRIS. This is achieved not only through the use of large detector elements, but also with $f/1$ spectrometer designs that use convex blazed gratings with a correction surface unique to each spectrometer to maintain

Table 7.1 AVIRIS sensor parameters.

Characteristic	Specification
Aperture diameter	2.54 cm
Effective focal length	20 cm
f-number	8.0
Fiber aperture	200 μm
Detector width	200 μm
Spectrometer magnification	1:1
Angular sampling (IFOV[*])	1 mrad
FOV	30 deg
Spectral range	400 to 2500 nm
Spectral resolution	10 nm
Number of spectral channels	244
Scan rate	12 Hz
Digitization	12 bits

[*] instantaneous field of view

high optical throughput. SNR performance of the AVIRIS sensor for object reflectance of 0.5, 23.5 solar zenith angle and midlatitude summer atmospheric conditions is illustrated in Fig. 7.33. By means of a variety of sensor improvements during the first ten years of AVIRIS's existence, SNR has increased from an initial level much less than 100, to a final level approaching 1000 across the full spectral range. Coupled with its narrow spectral resolution and excellent absolute calibration, radiometric performance has made this sensor an extremely valuable remote sensing tool for a variety of applications, including ecology, geology, vegetation science, hydrology, and atmospheric science. AVIRIS has also served as a reference instrument for a wide range of airborne and satellite remote imaging systems.

7.4.2 Hyperspectral Digital Imagery Collection Experiment

A widely used instrument for advancing defense applications of hyperspectral remote sensing in the VNIR/SWIR spectral range has been the Hyperspectral Digital Imagery Collection Experiment (HYDICE) imaging spectrometer (Rickard *et al.*, 1993). HYDICE was developed for pushbroom hyperspectral imaging across the 400- to 2500-nm spectral range from fixed-wing aircraft at a nominal 10,000-ft altitude. The optical design depicted in Fig. 7.34 consists of an unobscured Paul–Baker telescope with a 2.7-cm aperture and a Schmidt double-pass spectrometer that uses an Infrasil-SF1 prism. A curved slit is used to minimize optical distortion to a level of about 0.2-pixel smile and keystone. HYDICE uses a 320×210-element InSb FPA cooled to 65 K to reduce dark current to acceptable levels. The FPA design includes special passivation and wedge-shaped AR coating to provide good response down to 400 nm. As shown in Fig. 7.35, the FPA is subdivided into three sections that use a different

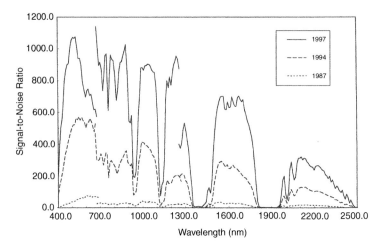

Figure 7.33 Modeled AVIRIS radiometric sensitivity [reprinted with permission from Green *et al.* (1998)].

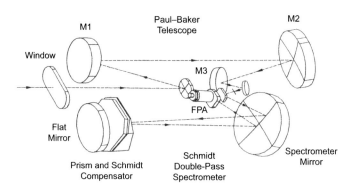

Figure 7.34 Design of the HYDICE VNIR/SWIR imaging spectrometer (Rickard *et al.*, 1993).

integrating capacitance (i.e., different effective gain and charge capacity) to optimize the FPA for varying mean FPA irradiance levels across a large spectral range. The effective charge capacity is nominally 9.8×10^6 electrons in the VNIR spectral region, 4.8×10^6 electrons in the short-wavelength end of the SWIR spectral region (or SWIR1), and 1.2×10^6 electrons in the long-wavelength end of the SWIR (SWIR2). The entire instrument is mounted in the aircraft in a nadir-viewing orientation to a Zeiss T-AS three-axis stabilization mount to minimize LOS jitter. The performance characteristics of the HYDICE instrument are summarized in Table 7.2. Radiometric sensitivity shown in Fig. 7.36 was modeled for the following sensor and environmental parameters: 60-Hz frame rate, 6-km altitude, 5% target reflectance, midlatitude summer atmosphere, and

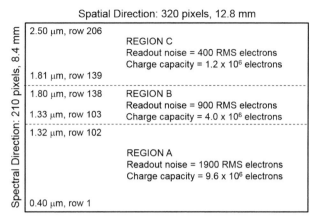

Figure 7.35 HYDICE FPA layout.

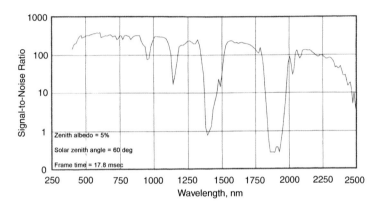

Figure 7.36 Modeled HYDICE radiometric sensitivity (Rickard *et al.*, 1993).

60-deg solar zenith angle. The modeled SNR is representative of flight performance.

7.4.3 Hyperion

Hyperion is a space-based hyperspectral imaging payload that is part of the Earth Observer 1 satellite deployed by NASA (Pearlman *et al.*, 2003). It was designed for earth remote sensing over the VNIR/SWIR spectral region and achieves a 3-m GSD and 7.5-km swath width by pushbroom imaging from a 705-km orbit. The overall Hyperion payload is depicted in Fig. 7.37. Hyperion fore-optics consists of a reflective, three-mirror anastigmatic telescope with a 12-cm aperture and 132-cm focal length that images onto a 56-µm-wide slit. A dichroic filter behind the slit with a nominally 950-nm transition wavelength relays the light toward separate VNIR and SWIR spectrometers. Each spectrometer is an Offner-like relay with a convex blazed grating and magnification of 1.38. The

Table 7.2 HYDICE sensor parameters.

Characteristic	Specification
Aperture diameter	2.7 cm
Effective focal length	8.1 cm
f-number	3.0
Slit width	40 µm
Detector width	40 µm
Spectrometer magnification	1:1
Angular sampling (IFOV)	0.51 mrad
FOV	8.94 deg
Spectral range	400 to 2500 nm
Spectral resolution	8 to 20 nm
Number of spectral channels	210
Maximum frame rate	120 Hz
Digitization	12 bits

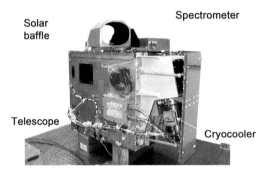

Figure 7.37   Picture of the Hyperion VNIR/SWIR imaging spectrometer (Lee *et al.*, 2001).

VNIR spectrometer uses a 768×384 element silicon FPA with 20-µm detectors and a 283-K operating temperature. The FPA includes an order-sorting filter, and the signal from 3×3 pixel groups are binned to provide an effective 256×128 element array, of which only 70 spectral channels are used. The SWIR spectrometer uses a 256×256 HgCdTe FPA with 60-µm detectors and a 110-K operating temperature, of which 172 channels are used. Performance characteristics of the Hyperion instrument are summarized in Table 7.3. Radiometric sensitivity shown in Fig. 7.38 was modeled for the following sensor and environmental parameters: 224-Hz frame rate, 30% target reflectance, midlatitude summer atmosphere, and 60-deg solar zenith angle. The circles indicate flight SNR measurements.

7.4.4 Compact Airborne Spectral Sensor

The Compact Airborne Spectral Sensor (COMPASS) was designed to support whiskbroom imaging from low-altitude aircraft, particularly military unmanned aerial vehicles (Simi *et al.*, 2001). The overall design is

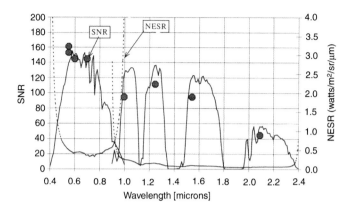

Figure 7.38 Modeled (line) and measured (points) Hyperion radiometric sensitivity (Lee *et al.*, 2001).

Table 7.3 Hyperion sensor parameters.

Characteristic	Specification
Aperture diameter	12 cm
Effective focal length	132 cm (telescope)
f-number	11.0 (telescope)
Slit width	56 μm
Detector width	60 μm
Spectrometer magnification	1.38:1
Angular sampling (IFOV)	0.043 mrad
FOV	0.62 deg
Spectral range	400 to 2500 nm
Spectral resolution	10 nm
Number of spectral channels	242
Maximum frame rate	224 Hz
Digitization	12 bits

illustrated in Fig. 7.39 and consists of a three-mirror anastigmatic telescope feeding an Offner grating spectrometer with a single HgCdTe FPA. The design includes two relatively unique features: a dual-zone blazed grating and a substrate-removed HgCdTe FPA. The grating was designed such that the central region is optimally blazed for the VNIR spectral range and the outer annular region is blazed for the SWIR spectral region. This is performed by special electron-beam lithographic methods. The resulting aggregate grating diffraction efficiency is an area-weighted sum for the two grating regions and results in more uniform efficiency over a large spectral range, as shown in Fig. 7.40. This design supports a broader spectral range than that achievable with a single blaze grating and specifically enhances the optical throughput in the SWIR spectral region, where pupil-plane spectral radiance is typically much lower compared to that in the VNIR spectral region. The FPA consists of 256×256 HgCdTe

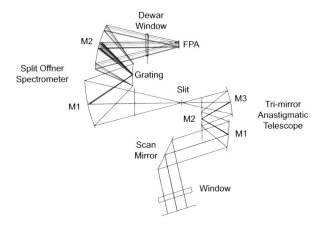

Figure 7.39 Design of the COMPASS VNIR/SWIR imaging spectrometer (Simi *et al.*, 2001).

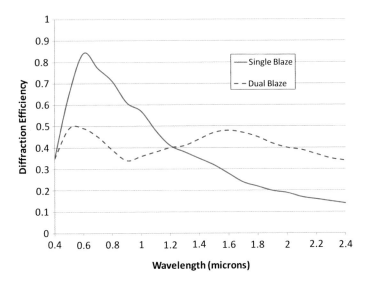

Figure 7.40 Comparison of COMPASS dual-blaze grating diffraction efficiency to that of a single blaze grating.

detector elements and is specially processed to remove the CdZnTe substrate after hybridization with the ROIC to provide good response in the visible spectral region. It exhibits 40-µm detectors and is operated near 180 K. Performance characteristics of the COMPASS instrument are summarized in Table 7.4. Radiometric sensitivity was modeled for 5% target reflectance, midlatitude summer atmosphere, 15-km visibility, 10,000-ft altitude, and 80-deg solar zenith angle. For these parameters, the sensor nominally achieves an SNR of 120 in the visible spectral range, 80 in the NIR, 60 in SWIR1, and 30 in SWIR2.

Table 7.4 COMPASS sensor parameters.

Characteristic	Specification
Aperture diameter	3.3 cm
Effective focal length	10 cm
f-number	3.0
Slit width	40 µm
Detector width	40 µm
Spectrometer magnification	1:1
Angular sampling (IFOV)	0.35 mrad
FOV	5.15 deg
Spectral range	400 to 2350 nm
Spectral resolution	7.6 nm
Number of spectral channels	256
Maximum frame rate	350 Hz
Digitization	14 bits

7.4.5 Spatially Enhanced Broadband Array Spectrograph System

The Spatially Enhanced Broadband Array Spectrograph System (SEBASS) was designed for pushbroom MWIR/LWIR hyperspectral imaging from low-altitude fixed-wing aircraft (Hackwell *et al.*, 1996). Its optical design is depicted in Fig. 7.41 and is based on the dual aplanatic prism spectrometer principles discussed earlier. A three-mirror anastigmatic telescope images onto a curved slit for minimizing spatial–spectral distortion; this is followed by a dichroic beamsplitter with a 6.5-µm transition wavelength that feeds separate MWIR and LWIR imaging spectrometers. Both spectrometers are double-pass, dual aplanatic spectrometers with an aplanatic lens near the FPAs to reduce the effective f-number and further minimize aberrations. An LiF prism is used in the MWIR, and an NaCl prism is used in the LWIR. Both focal planes are 128×128 arrays of Si:As impurity-band-conduction detectors that require cooling to 10 K. This is achieved by placing the entire spectrometer in a liquid helium cryostat, which has the added benefit of completely eliminating the instrument self-emission, except for that from the dewar window. Performance characteristics of the SEBASS instrument are summarized in Table 7.5. At a 122-Hz effective frame rate using two-frame co-adding, the instrument is able to achieve an NESR of roughly 0.8 µW/cm^2 µm sr for nominal 300-K blackbody radiance.

7.4.6 Airborne Hyperspectral Imager

The Airborne Hyperspectral Imager (AHI) was designed for pushbroom LWIR hyperspectral imaging from low-altitude aircraft (Lucey *et al.*, 2003). It was originally developed for mine detection applications and was specifically designed for tactical operation. As such, the design employs a

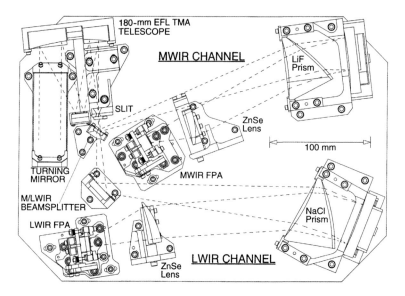

Figure 7.41 Design of the SEBASS MWIR/LWIR imaging spectrometer (Hackwell *et al.*, 1996).

Table 7.5 SEBASS sensor parameters.

Characteristic	Specification
Aperture diameter	2.25 cm
Effective focal length	7.5 cm
f-number	3.0
Slit width	180 μm
Detector width	75 μm
Spectrometer magnification	1:2.4
Angular sampling (IFOV)	1 mrad
FOV	7.3 deg
Spectral range	2.0 to 5.2 μm (MWIR)
	7.8 to 13.5 μm (LWIR)
Spectral resolution	25 nm (MWIR)
	50 nm (LWIR)
Number of spectral channels	128 (MWIR)
	128 (LWIR)
Maximum frame rate	244 Hz
Digitization	14 bits

256×256 HgCdTe FPA to eliminate the need for liquid-helium cooling (which can be a major logistical hindrance) and attempts to achieve good radiometric performance without cooling the entire spectrometer. The instrument layout is depicted in Fig. 7.42. The imaging spectrometer is roughly a Czery–Turner design using a flat-grating, gold-coated $f/4$ optics, and a curved slit to control spatial–spectral distortion. To control self-emission from the uncooled spectrometer, the system employs a linear

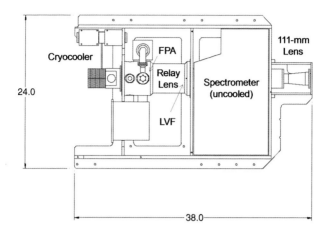

Figure 7.42 Layout of the AHI LWIR imaging spectrometer (Lucey *et al.*, 2003).

variable filter (LVF) at an intermediate image plane within a cryogenically cooled dewar that contains the LVF, FPA, and a one-to-one relay lens between them. An LVF, which is further described in Chapter 10, is a narrowband, thin-film dielectric filter whose passband gradually changes laterally across the surface. In the AHI system, the LVF is matched to the dispersion at the output spatial–spectral image of the spectrometer to eliminate radiation out of band to each spectral band. The efficacy of this approach is limited to how well this matching occurs in practice, since a mismatch between dispersion and spectral resolution of the spectrometer with the LVF produces a combination of reduced optical throughput and increased background radiance. Performance characteristics of the AHI instrument are summarized in Table 7.6. At a 150-Hz frame rate, the instrument is able to achieve an NESR of roughly 3 μW/cm^2 μm sr for a nominal 300-K blackbody radiance.

Table 7.6 The AHI sensor parameters.

Characteristic	Specification
Aperture diameter	2.78 cm
Effective focal length	11.1 cm
f-number	1.8
Slit width	300 μm
Detector width	750 μm
LVF bandpass	95 nm
Angular sampling (IFOV)	0.5 mrad
FOV	7.3 deg
Spectral range	7.9 to 11.5 μm
Spectral resolution	52 nm
Number of spectral channels	256
Maximum frame rate	150 Hz

7.5 Summary

Prism and grating spectrometers are the most common imaging spectrometer instruments used for hyperspectral remote sensing purposes. As seen by the provided system examples, they are capable of achieving moderate spatial and spectral resolution, and their imaging operation is well suited for pushbroom or whiskbroom imaging from airborne or space-based platforms. While prism instruments provide better spectral purity, grating instruments have become very common due to the additional design flexibility they provide. Particularly with the advent of manufacturing techniques for producing high-efficiency reflection gratings on convex surfaces, grating spectrometer designs can support fine spectral resolution, a broad spectral range, and large FOVs in a compact sensor package.

Critical performance characteristics of dispersive spectrometers parallel those of broadband imaging systems, with the additional consideration of a spectral dimension. Specifically, spatial–spectral distortion, spectral response function (SRF), and noise-equivalent spectral radiance (NESR) as a function of spectral band are important additional attributes that must be factored into a design. Because the slit in a dispersive imaging spectrometer impacts both spatial and spectral resolution, these characteristics are coupled in this particular hyperspectral sensor design. Furthermore, slit width represents a trade-off between spectral resolution, spatial resolution, and radiometric sensitivity. Because of limited irradiance within narrow spectral bands, it is not unusual for hyperspectral sensors to exhibit coarser spatial resolution compared to broadband imaging systems to achieve an acceptable NESR.

Practical hyperspectral remote sensor designs covering the VNIR/SWIR and MWIR/LWIR spectral regions have been developed and implemented from airborne and space-based platforms using both prism and grating designs. Spectral resolution for these systems is nominally 10 nm in the VNIR/SWIR and 50 nm in the LWIR. Spatial resolution is typically on the order of 1 m, depending on flight altitude. Compared to broadband imaging systems, spatial coverage is somewhat limited by FPA size, spectrometer FOV, and the integration time needed to achieve adequate radiometric sensitivity. In the LWIR, practical applications are also limited by the need for spectrometer cooling to achieve the excellent radiometric sensitivity required to capture subtle features associated with materials of interest. More-advanced spectrometer designs are addressing these current shortfalls.

References

Breckinridge, J. B., Page, N. A., Shannon, R. R., and Ridgers, J. M., "Reflecting Schmidt imaging spectrometers," *Appl. Optics* **22**, 1175–1180 (1983).

Green, R. O., Eastwood, M. L., Sarture, C. M., Chrien, T. G., Aronsson, M., Chippendale, B. J., Faust, J. A., Pavri, B. E., Chovit, C. J., Solis, M., Olah, M. R., and Williams, O., "Imaging Spectroscopy and the Airborne Visible/Infrared Imaging Spectrometer (AVIRIS)," *Remote Sens. Environ.* **65**, 227–248 (1998).

Hackwell, J. A., Warren, D. W., Bongiovi, R. P., Hansel, S. J., Hayhurst, T. L., Mabry, D. J., Sivjee, M. G., and Skinner, J. W., "LWIR/MWIR imaging hyperspectral sensor for airborne and ground-based remote sensing," *Proc. SPIE* **2819**, 102–107 (1996) [doi:10.1117/12.258057].

Lee, P., Carman, S., Chan, C.K., Flannery, M., Folkman, M., Iverson, P. J., Liao, L., Luong, K., McCuskey, J., Pearlman, J., Rasmussen, C. S., Viola, D., Watson, W., Wolpert, P., Gauther, V., Mastandrea, A., Perron, G., and Szarek, G., "Hyperion: a 0.4 mm to 2.5 mm hyperspectral imager for the NASA Earth Observing-1 mission," TRW Space and Technology Division, Redondo Beach, CA, 2001, http://www.dtic.mil (last accessed Sept. 2011).

Lobb, D. R., "Theory of concentric designs for grating spectrometers," *Appl. Optics* **33**, 2648–2658, 1994.

Lucey, P. G., Williams, T., and Winter, M., "Recent results from AHI, an LWIR hyperspectral imager," *Proc. SPIE* **5159**, 361–369 (2003) [doi:10.1117/12.505096].

Mouroulis, P., Green, R. O., and Wilson, D. W., "Optical design of a coastal ocean imaging spectrometer," *Opt. Express* **16**, 1087–1096 (2008).

Mouroulis, P. and Thomas, D. A., "Compact, low-distortion imaging spectrometer for remote sensing," *Proc. SPIE* **3438**, 31–37 (1998) [doi:10.1117/12.328120].

Pearlman, J. S., Barry, P. S., Segal, C. C., Shepanski, J., Basio, D., and Carman, S. L., "Hyperion, a space-based imaging spectrometer," *IEEE T. Geosci. Remote Sens.* **41**, 1160–1173 (2003).

Pedrotti, F. L. and Pedrotti, L. S., *Introduction to Optics*, Prentice-Hall, Englewood Cliffs, NJ (1987).

Rickard, L. J., Basedow, R., Zalewski, E., and Silverglate, P., "HYDICE: An airborne system for hyperspectral imaging," *Proc. SPIE* **1937**, 173–179 (1993) [doi:10.1117/12.157055].

Sidran, M., Stalzer, H. J., and Hauptmann, M. H., "Drum dispersion equation for Littrow-type prism spectrometers," *Appl. Optics* **5**, 1133–1138 (1966).

Simi, C., Winter, E., Williams, M., and Driscoll, D., "Compact Airborne Spectral Sensor (COMPASS)," *Proc. SPIE* **4381**, 129–136 (2001) [doi:10.1117/12.437000].

Smith, W. J., *Modern Optical Engineering*, Second Edition, McGraw-Hill, New York (1990).

Warren, D. W., Hackwell, J. A., and Gutierrez, D. J., "Compact prism spectrographs based on aplanatic principles," *Opt. Eng.* **36**, 1174–1182 (1997) [doi:10.1117/1.601237].

Wolfe, W. L., *Introduction to Imaging Spectrometers*, SPIE Press, Bellingham, WA (1997) [doi:10.1117/3.263530].

Chapter 8
Fourier Transform Spectrometer Design and Analysis

While dispersive spectrometers are a fairly direct way to separate optical radiation into its constituent irradiance spectrum, there are a few applications for which alternate imaging spectrometer designs are advantageous. In particular, this chapter describes a class of instrument designs that employs interferometry to produce signal modulation on a per-pixel basis related to the irradiance spectrum by a Fourier transform. Such an instrument is called a Fourier transform spectrometer (FTS) and requires digital processing to compute a representation of the irradiance spectrum from measured data. These spectrometers are also called Fourier transform infrared (FTIR) spectrometers, because they have historically been used in MWIR and LWIR spectral regions to capture extremely high-resolution spectra of gaseous materials. They are able to do so because their spectral resolution is limited only by mechanical parameters and not fundamental component characteristics such as prism material dispersion and grating periods. Laboratory instruments have been developed that achieve stability over very large interferometer baselines, producing the excellent resolution needed to resolve the fine rotational features of these materials. Traditionally, such instruments have been nonimaging; however, they have recently extended to imaging operation to support specific hyperspectral remote sensing applications, where their high achievable spectral resolution and natural staring imaging format are advantageous (Wolfe, 1997). The theoretical treatment of FTS operation is initiated with a discussion of traditional, nonimaging systems but is then extended to particular issues concerning imaging operation, along with a description of some examples of systems.

8.1 Fourier Transform Spectrometers

There are two basic FTS types discussed in this chapter: temporal FTS instruments, which are described in this section as well as Section 8.2, and spatial FTS instruments, described in Section 8.3. The most widely used FTS approach is based on a Michelson interferometer (Pedrotti and Pedrotti, 1987), as depicted in Fig. 8.1, and is called a temporal FTS for reasons that will become apparent. In a Michelson interferometer, incident light from a source is split into two paths by a beamsplitter, reflected off mirrors in both paths, and recombined onto a detector. One path includes a moving or controllable mirror, for which the optical path difference (OPD) between the interfering wavefronts can be precisely controlled. Assume that total transmission through one arm of an interferometer is τ_1 and the other is τ_2. Theoretically, transmission is constrained such that $\tau_1 + \tau_2 < 1$. Ideally, $\tau_1 = \tau_2 = \frac{1}{2}$. To cancel the effect of any dispersion in the beamsplitter material, a transparent substrate of an identical material with the same thickness but maximum transmission is located in one arm of the interferometer. This element is called a *compensator*.

At a certain position of the moving mirror, the OPD between interfering paths becomes identically zero. This position is called the *zero path difference* (ZPD) position and is indicated by the dotted line in Fig. 8.1. Considering the source as an incident, monochromatic plane wave with irradiance E_0 and wavelength λ, irradiance at the detector for an arbitrary OPD Δ is given by

$$E(\Delta) = \left\langle \left| \sqrt{\tau_1 E_0} \cos(\omega t - kz) + \sqrt{\tau_2 E_0} \cos\{\omega t - k(z + \Delta)\} \right|^2 \right\rangle$$

$$= \frac{\tau_1 + \tau_2}{2} E_0 \left[1 + \frac{2\sqrt{\tau_1 \tau_2}}{\tau_1 + \tau_2} \cos\left(\frac{2\pi\Delta}{\lambda}\right) \right], \qquad (8.1)$$

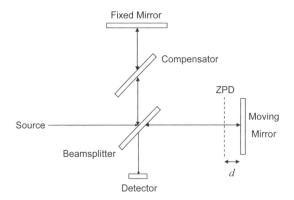

Figure 8.1 Michelson interferometer setup.

where $k = 2\pi/\lambda$ and $\omega = 2\pi v$. In the ideal case, Eq. (8.1) simplifies to

$$E(\Delta) = \frac{1}{2}E_0\left[1 + \cos\left(\frac{2\pi\Delta}{\lambda}\right)\right]. \tag{8.2}$$

If d is the displacement of the mirror from the ZPD position and θ is the incident angle of the plane wave relative to the optical axis, then the OPD can be shown to be

$$\Delta = 2d/\cos\theta. \tag{8.3}$$

In a nonimaging FTS, the incident angle is assumed to be zero. The analysis that follows is clearer if wavenumber $\sigma = 1/\lambda$ is used instead of the wavelength. With this substitution, Eq. (8.2) becomes

$$E(\Delta) = \frac{1}{2}E_0[1 + \cos(2\pi\sigma\Delta)]. \tag{8.4}$$

The behavior of the Michelson interferometer with monochromatic light at two different wavelengths is depicted in Fig. 8.2. As the OPD varies, a sinusoidal modulation occurs where the period of oscillation is related to the wavelength of the light because of the spectral dependence of the cosine argument in Eq. (8.4). At the ZPD position, light of all wavelengths constructively interferes so that sinusoidal patterns for differing wavelengths are all phased up at this corresponding mirror position. If transmission of the two interferometer arms is not exactly one-half, the same modulation occurs, but the overall transmission of the bias component (the mean value of the interference pattern in Fig. 8.2) is

$$\tau_b = \frac{\tau_1 + \tau_2}{2}, \tag{8.5}$$

and the modulation depth (the ratio of the sinusoidal amplitude to the mean value) is reduced to

$$m = \frac{2\sqrt{\tau_1\tau_2}}{\tau_1 + \tau_2}. \tag{8.6}$$

The transmission and modulation depth are assumed for now to equal 1 for notational simplicity but are factored in when radiometric sensitivity is examined later in this chapter.

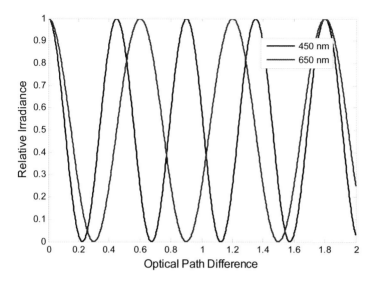

Figure 8.2 Sinusoidal modulation of a Michelson interferometer for light at two different wavelengths.

8.1.1 Interferograms

Now consider the response of the Michelson interferometer to a polychromatic source with spectral irradiance $E_0(\sigma)$. Detector irradiance as a function of OPD is then given by the integral

$$E(\Delta) = \int_0^\infty \frac{1}{2} E_0(\sigma) \left[1 + \cos(2\pi\sigma\Delta) \right] d\sigma, \qquad (8.7)$$

which is referred to as an *interferogram*. As an example, consider the irradiance spectra given in Fig. 8.3. These spectra are representative of a 320-K blackbody radiator viewed through 100 ppm m of methane gas at 280 K, captured by an ideal $f/3$ imaging system with a cold filter that truncates the spectrum to the 3- to 12-μm spectral band. Note the strong methane feature in the 7.5-μm region. This example is chosen because it includes both an extended spectrum and some fine spectral features with which to explore FTS operation. Two versions of the irradiance spectrum are displayed: $E_\lambda(\lambda)$ in μW/cm²μm units in Fig. 8.3(a), and $E_\sigma(\sigma)$ in μW/cm² cm^{-1} units in Fig. 8.3(b). The spectra are related by

$$E_\sigma(\sigma) = \frac{\lambda}{\sigma} E_\lambda(\lambda). \qquad (8.8)$$

Because it is more natural to use wavenumber units when dealing with an FTS, the produced spectrum is in the form of $E_\sigma(\sigma)$ and not $E_\lambda(\lambda)$.

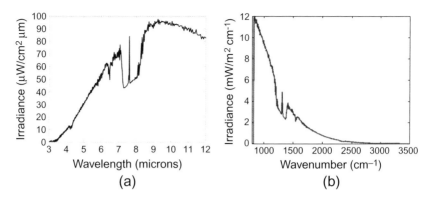

Figure 8.3 Example methane spectra in both (a) wavelength and (b) wavenumber units.

The interferogram corresponding to the spectrum example is depicted in Fig. 8.4 for a maximum OPD of both 0.125 [Fig. 8.4(a)] and 0.02 cm [Fig. 8.4(b)]. The interferogram in Fig. 8.4(b) is just the central part of the interferogram in Fig. 8.4(a). In both cases, the mirror moves symmetrically about the ZPD position, leading to both positive and negative OPDs and making what is known as a two-sided interferogram. Since all wavelengths interfere constructively at the ZPD position, there is a large central peak to the distribution. As the magnitude of an OPD increases, sinusoidal patterns for constituent wavelengths dephase, and the modulation depth of the pattern about the bias level decreases. Note that this interferogram is produced by varying the mirror displacement in time, hence the designation as a temporal FTS.

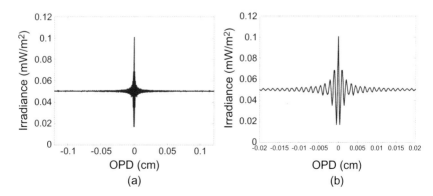

Figure 8.4 Interferogram of the spectrum example for (a) 0.125- and (b) 0.02-cm maximum OPD.

8.1.2 Spectrum reconstruction

Recognizing Eq. (8.7) as a discrete cosine transform, one would expect to be able to reconstruct a spectrum from the interferogram by a Fourier transform. Consider the Fourier transform of the interferogram, given by

$$S(\sigma) = \int_{-\infty}^{\infty} E(\Delta) e^{i2\pi\sigma\Delta} d\Delta. \tag{8.9}$$

Inserting Eq. (8.7) into Eq. (8.9) and separating the two terms leads to

$$S(\sigma) = \frac{1}{2} \int_{-\infty}^{\infty} \left[\int_{0}^{\infty} E(\sigma') d\sigma' \right] e^{i2\pi\sigma\Delta} d\Delta$$
$$+ \frac{1}{2} \int_{-\infty}^{\infty} \int_{0}^{\infty} E(\sigma') \cos(2\pi\sigma'\Delta) e^{i2\pi\sigma\Delta} d\sigma' d\Delta. \tag{8.10}$$

The first term represents the Fourier transform of the bias term in Eq. (8.7) and results in a very large peak-at-zero frequency proportional to the total irradiance over the spectral range of the instrument. This term can easily be removed by subtracting the mean or bias level from the interferogram prior to performing the Fourier transform. Removal is not necessary from a theoretical standpoint but is necessary from a practical, numerical perspective to ensure that side lobes of the very large zero-frequency peak do not interfere with the reconstructed spectrum. If the interferogram bias level is removed prior to computing the Fourier transform, the first term in Eq. (8.10) is zero, and the cosine in the second term can be expressed as the sum of two exponentials, leading to the two terms

$$S(\sigma) = \frac{1}{4} \int_{-\infty}^{\infty} \int_{0}^{\infty} E_0(\sigma') e^{i2\pi(\sigma+\sigma')\Delta} d\sigma' d\Delta$$
$$+ \frac{1}{4} \int_{-\infty}^{\infty} \int_{0}^{\infty} E_0(\sigma') e^{i2\pi(\sigma-\sigma')\Delta} d\sigma' d\Delta. \tag{8.11}$$

Integration over Δ results in Dirac delta functions $\delta(x)$, such that

$$S(\sigma) = \frac{1}{4} \int_{0}^{\infty} E_0(\sigma') \delta(\sigma + \sigma') d\sigma' + \frac{1}{4} \int_{0}^{\infty} E_0(\sigma') \delta(\sigma - \sigma') d\sigma'. \tag{8.12}$$

This leads to the final result

$$S(\sigma) = \frac{1}{4} E_0(\sigma) + \frac{1}{4} E_0(-\sigma). \tag{8.13}$$

The Fourier transform of an interferogram after bias removal leads to two terms: the irradiance spectrum and a mirror image of the irradiance spectrum, as is depicted in Fig. 8.5 for the irradiance spectrum example. The spectrum of the source, therefore, can be produced by capturing an interferogram with a Michelson interferometer and computing the Fourier transform over the spectral range of the source. Since the result in Eq. (8.9) is real-valued due to the symmetry of the interferogram, the same result can be achieved by using the cosine transform:

$$S(\sigma) = \int_{-\infty}^{\infty} E(\Delta)\,\cos(2\pi\sigma\Delta)\,d\Delta. \tag{8.14}$$

8.1.3 Spectral resolution

In practice, movement of the mirror is restricted to a particular physical range, so that the OPD limits of integration in Eq. (8.7) or Eq. (8.9) are finite. Consider on-axis radiation ($\theta = 0$), and let the total displacement of the mirror from the ZPD in either direction be given by distance d. In this case, Eq. (8.5) becomes

$$S(\sigma) = \int_{-2d}^{2d} E(\Delta)\,e^{i2\pi\sigma\Delta}d\Delta, \tag{8.15}$$

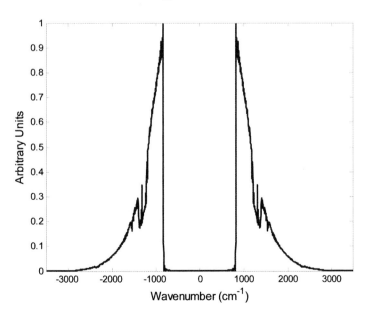

Figure 8.5 Fourier transform of the interferogram example to reconstruct the spectrum.

based on the relationship between Δ and d in Eq. (8.3). By recognizing that Eq. (8.15) is the product of Eq. (8.7) and a rect function of width $2d$, the Fourier transform represented in the ideal case by Eq. (8.13) becomes

$$S(\sigma) = \left\{ \frac{1}{4}E_0(\sigma) + \frac{1}{4}E_0(-\sigma) \right\} * \frac{\sin(4\pi\sigma d)}{4\pi\sigma d}. \tag{8.16}$$

That is, the ideal spectrum is convolved with the sinc function of a width that is inversely proportional to total mirror displacement. It is common to define spectral resolution $\delta\sigma$ of an FTS as the peak-to-null width of this sinc function, such that

$$\delta\sigma = \frac{1}{4d}. \tag{8.17}$$

Using the definition of resolving power given previously for dispersive spectrometers,

$$R = \frac{\lambda}{\delta\lambda} = \frac{\sigma}{\delta\sigma} = \frac{4d}{\lambda}. \tag{8.18}$$

Resolving power is directly proportional to the total OPD $4d$ from end to end over the double-sided interferogram. The effect of mirror displacement on spectral resolution is illustrated in Fig. 8.6, where an example spectrum is reconstructed based on truncated interferograms with maximum OPDs of $0.125, 0.0625$, and 0.03125 cm, corresponding to decreasing spectral resolution of $4, 8$, and 16 cm^{-1}. In this way, it is mechanical parameter d (as opposed to material characteristics) that completely defines spectral resolution.

8.1.4 Spectral range

Equation (8.15) uses a continuous Fourier transform to describe the transformation of the interferogram to the spectrum. In reality, the interferogram is sampled, and a discrete form of Eq. (8.11) is used:

$$S(\sigma) = \frac{1}{N} \sum_{j=1}^{N} E(\Delta_j) \, e^{i2\pi\sigma\Delta_j}, \tag{8.19}$$

where N is the number of interferogram samples and Δ_j are their corresponding OPDs. The expression in Eq. (8.19) is called the discrete Fourier transform (DFT) and is usually processed in a computationally efficient manner called a fast Fourier transform (FFT), where samples are uniformly spaced and N is a power of 2 (Oppenheim and Schafer, 1975).

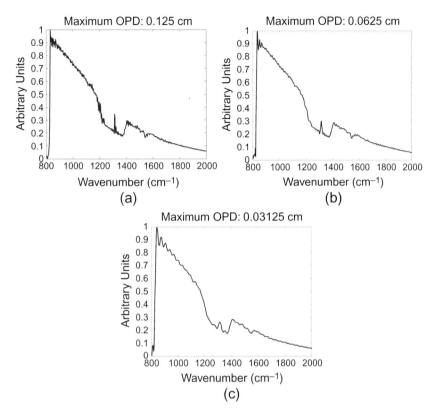

Figure 8.6 Reconstructed spectra at different spectral resolutions: (a) 4-cm^{-1} spectral resolution, (b) 8-cm^{-1} spectral resolution, and (c) 16-cm^{-1} spectral resolution.

Consider a uniformly distributed set of samples with a differential mirror displacement between samples of δ. To achieve Nyquist sampling of the irradiance spectrum, the spectrum must be bandlimited to a maximum wavenumber of

$$\sigma_{max} = \frac{1}{2(2\delta)},$$
(8.20)

or a minimum wavelength of

$$\lambda_{min} = 4\delta.$$
(8.21)

This defines the spectral range of an FTS. Maximum wavelength λ_{max} is independent of interferogram sampling but is practically limited by optics transmission and detector response.

8.1.5 Apodization

As expressed in Eq. (8.16), the SRF of an FTS is of the form of a sinc function, which exhibits high side lobes that limit the spectral purity of the instrument. To address this problem, an apodization function $W(\Delta)$ is often employed in the computation of the DFT. That is, Eq. (8.19) is replaced with

$$S(\sigma) = \frac{1}{N} \sum_{j=1}^{N} W(\Delta_j)\, E(\Delta_j)\, e^{i2\pi\sigma\Delta_j}. \tag{8.22}$$

The SRF associated with Eq. (8.22) then becomes

$$g(\sigma) = w(\sigma) * \frac{\sin(4\pi\sigma d)}{4\pi\sigma d}, \tag{8.23}$$

where $w(\sigma)$ represents the Fourier transform of the apodization function. Apodization produces a trade-off between spectral purity and spectral resolution, and selection of the apodization function employed is based on the relative importance of these attributes. This is illustrated in Fig. 8.7, where an interferogram with 0.0625-cm maximum OPD is reconstructed with cosine apodization and compared to a nonapodized spectrum. Cosine apodization is tailored in this case so that the zeroes of the cosine function coincide exactly with the endpoints of the interferogram. In the apodized spectrum, spectral resolution is slightly degraded, but most of the ripples in the nonapodized spectrum are removed. These artifacts can be mistaken for subtle spectral features, and the use of apodization can help avoid such misinterpretation.

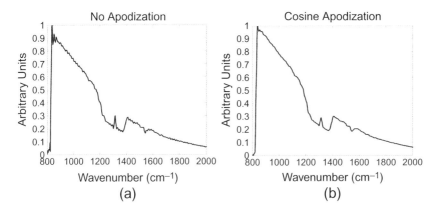

Figure 8.7 Reconstructed spectra (a) without apodization and (b) with cosine apodization.

8.1.6 Uncompensated interferograms

Some implementations of an FTS intentionally do not employ the compensator depicted in Fig. 8.1. In this case, the interferogram is of the form

$$E(\Delta) = \int_0^\infty \frac{1}{2} E_0(\sigma) \left[1 + \cos\{2\pi\sigma\Delta + \phi(\sigma)\} \right] d\sigma, \qquad (8.24)$$

where $\phi(\sigma)$ represents a spectrally dependent phase error due to the dispersion of the beamsplitter material. To effectively compute the spectrum in the presence of this phase error, the DFT in Eq. (8.15) must be modified to the form

$$S(\sigma) = \frac{1}{N} \sum_{j=1}^{N} E(\Delta_j) \, e^{i[2\pi\sigma\Delta_j + \phi(\sigma)]}. \qquad (8.25)$$

This precludes the use of efficient FFT computational methods and requires that dispersion be precisely known. Because of the presence of this phase error, irradiance at the ZPD position is given by

$$E(0) = \int_0^\infty \frac{1}{2} E_0(\sigma) \left[1 + \cos\{\phi(\sigma)\} \right] d\sigma. \qquad (8.26)$$

In this case, dephasing occurs such that total constructive interference does not result. Therefore, an uncompensated interferogram will not result in the same large peak at the ZPD position as a compensated ZPD will have. This can be beneficial in practice because it alleviates the dynamic range requirements of a detector. That is, the available dynamic range is not used up to capture the large ZPD peak and can be optimized to capture the interference signal across a full range of OPDs.

8.2 Imaging Temporal Fourier Transform Spectrometers

An FTS can be extended to accommodate imaging by incorporating an FPA and imaging optics, as depicted in Fig. 8.8. In this case, a remote scene is imaged by fore-optics onto an intermediate image plane, and this intermediate image is relayed onto an FPA through a Michelson interferometer. Ideally, the Michelson interferometer exists in collimated space, where the wavefront associated with a specific image location is planar. This planarity is not necessary for the spectrometer to work but simplifies the analysis and results in optimal performance. The spectrometer operates by capturing a set of frames of the FPA as one mirror is precisely moved across a certain range. The trace of a single FPA

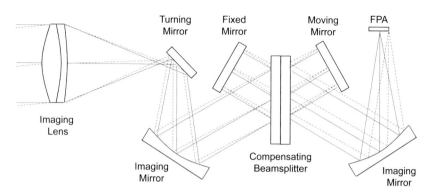

Figure 8.8 Imaging FTS based on a Michelson interferometer.

pixel over the set of frames represents an interferogram associated with a particular location. Combining the interferograms for all detector pixels will produce an interferogram dataset organized as depicted in Fig. 8.9. By Fourier transforming all of the signals in a frame-to-frame direction, an irradiance spectrum can be formed for each FPA pixel, thus producing a hyperspectral image.

In an imaging FTS, the frame sequence can be expressed as

$$E(x, y, d) = \int_0^\infty \frac{1}{2} E_0(x, y, \sigma) \left[1 + \cos\{4\pi\sigma d / \cos\theta(x, y)\}\right] d\sigma, \quad (8.27)$$

where (x, y) represents the image-plane spatial coordinates of a particular FPA pixel, and $\theta(x, y)$ represents the inverse of the lens-mapping function. For example, $\theta(x, y)$ can be given by

$$\theta(x, y) = \tan^{-1}\left(\frac{\sqrt{x^2 + y^2}}{f} \right), \quad (8.28)$$

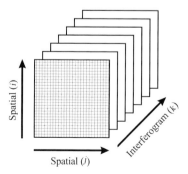

Figure 8.9 Frame sequence from a temporal imaging FTS.

where f is the focal length of the final reimaging lens. Because of the angular dependence of Eq. (8.27), the spatial pattern of a single frame with a uniform input radiance over the FOV represents a bulls-eye pattern, as depicted in Fig. 8.10, which is the resulting spatial distribution for a spatially uniform, monochromatic source at different mirror displacements. The specific results depicted are for a 100-mm focal length for spectrometer lenses, 50-mm full-image field, and 10-μm wavelength. Since detector elements integrate spatially over this pattern, the interferogram modulation depth decreases with increasing mirror displacement as the spatial frequency content of the spatial pattern increases. This essentially causes an apodization effect, which degrades the spectral resolution of the interferometer with field angle.

8.2.1 Off-axis effects

To quantify the apodization effect, it is necessary to compute the phase variation of the fringe pattern over the detector element size. Defining ϕ as the phase term in Eq. (8.27),

$$\phi = 4\pi\sigma d/\cos\theta. \tag{8.29}$$

The differential phase change with field angle is given by

$$d\phi = -\frac{4\pi\sigma d \sin\theta}{\cos^2\theta}d\theta. \tag{8.30}$$

Based on the mapping function in Eq. (8.28), angular variation corresponding to a detector of width w is nominally

$$d\theta = \frac{w}{f}\cos^2\theta. \tag{8.31}$$

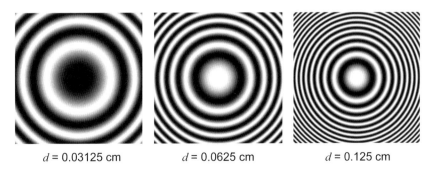

| d = 0.03125 cm | d = 0.0625 cm | d = 0.125 cm |

Figure 8.10 Spatial irradiance distribution for a uniform scene at different mirror displacements for a 100-mm focal length, 50-mm full-image field, and 10-μm wavelength.

Inserting Eq. (8.31) into Eq. (8.30) results in

$$|d\phi| = \frac{4\pi\,\sigma\,d\,w\sin\theta}{f} = \frac{2\pi\,R\,w\sin\theta}{2f}. \tag{8.32}$$

If the phase variation over detector element size equals 2π, the effect is similar to the cosine apodization effect illustrated in Fig. 8.7. If it exceeds this amount, spectral resolution is further degraded. Therefore, a reasonable metric is to require the maximum phase variation across the detector width to be less than 2π, which limits the maximum field angle to

$$\sin\theta < \frac{2f}{R\,w}. \tag{8.33}$$

As an example, consider 50-mm focal-length spectrometer lenses, a 50-µm detector element width, and a resolving power of 10,000, corresponding to 0.5-cm^{-1} spectral resolution at a minimum wavelength of 5000 cm^{-1} or 2 µm. The maximum field angle according to this criteria is roughly 11.5 deg, which nominally supports a 200×200 element FPA. Even at this fairly high spectral resolution and large detector element size, the FOV is not overly restrictive.

As long as the condition in Eq. (8.33) is met, a Michelson interferometer will theoretically produce interferogram modulation supportive of the desired resolving power. Even with low phase variation across the detector element width, the periodicity of the modulation for monochromatic light will be spatially dependent due to the presence of the field-angle-dependent term in the phase in Eq. (8.27). This can be compensated for in the computation of the spectrum by using a slightly modified DFT for each detector pixel of the form

$$S(x, y, \sigma) = \frac{1}{N} \sum_{j=1}^{N} I(x, y, d_j)\, e^{i[4\pi\sigma d_j/\cos\theta(x,y)]}. \tag{8.34}$$

If this is not performed, the off-axis spectra will be improperly scaled in the spectral direction by the factor $\cos\theta(x, y)$.

8.2.2 Additional design considerations

As with a nonimaging FTS, it is imperative that mirror orientation be precisely controlled as the mirror is moved to map out an interferogram. Relative tilts between the two mirrors produce a linear fringe pattern on the FPA that reduces modulation depth over the entire FOV. Furthermore, mirror positions d_j must be precisely controlled and measured relative to the integration times of the frame sequence. This is generally performed by using a mensuration system composed of a He–Ne or other laser source that is directed through the interferometer. Relative mirror displacement is precisely measured by measuring the phase of the laser beam. A white-light source is often used to locate the ZPD position, or the ZPD can be found from the measured interferogram data.

The relay optical system into which the Michelson interferometer is immersed is designed to maximize the planarity of the interfering wavefronts. If the wavefronts are not planar due to aberrations in the optical system, nearly perfect interference can still be achieved at a ZPD position; however, modulation depth can decrease as the OPD increases because the spatial shape of the interfering wavefronts may no longer match up. This has the effect of producing apodization of the inteferogram, which reduces spectral resolution in a manner that has already been discussed.

8.3 Spatial Fourier Transform Spectrometers

A spatial FTS is a class of interferometric imaging spectrometers that produces an interferogram across a spatial dimension of an FPA, as opposed to a frame-to-frame, or temporal, dimension of the more conventional Michelson design. To understand spatial FTSs, consider Young's double-slit experiment depicted in Fig. 8.11. A source illuminates two slits of width s and separation Δs, and diffracted light is collimated onto a screen. Considering the source to be an incident, monochromatic plane wave with irradiance E_0 and wavelength λ, irradiance at the detector for an arbitrary OPD Δ is given by

$$
E(x) = K \left\langle \left| \frac{1}{2} \sqrt{E_0} \cos(\omega t - kz \cos\theta + kx \sin\theta) \right. \right.
$$

$$
\left. \left. + \frac{1}{2} \sqrt{E_0} \cos(\omega t - kz \cos\theta - kx \sin\theta) \right|^2 \right\rangle
$$

$$
= \frac{1}{2} K E_0 \left[1 + \cos\left(\frac{4\pi x \sin\theta}{\lambda} \right) \right], \tag{8.35}
$$

where θ is the angle indicated in Fig. 8.11 and K is a constant that accounts for a fraction of the source irradiance that passes the slits onto the screen.

If the source irradiance is polychromatic with spectral irradiance $E_0(\sigma)$, then screen irradiance is in the form of an interferogram:

$$
E(x) = K \int_0^\infty \frac{1}{2} E_0(\sigma) \left[1 + \cos(4\pi\sigma x \sin\theta) \right] d\sigma. \tag{8.36}
$$

As in the case of a Michelson interferometer, the source spectrum can be computed to within a proportionality constant by a discrete Fourier

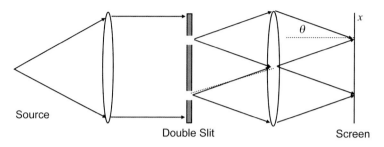

Figure 8.11   Young's double-slit experiment.

transform:

$$S(\sigma) = \frac{1}{N} \sum_{j=1}^{N} E(x_j)\, e^{i4\pi\sigma x \sin\theta}.$$ (8.37)

For an ideal lens,

$$\theta = \tan^{-1}\left(\frac{\Delta s}{2f}\right),$$ (8.38)

and

$$\sin\theta \approx \frac{\Delta s}{2f}.$$ (8.39)

The spectral resolution and range of a spatial FTS is dependent on the size and sample spacing of the detector array placed at the screen location given in Fig. 8.11 to capture the interferogram. Assume that a detector array is composed of N elements that are uniformly spaced by detector width w. If the detector array is symmetrically located to capture a double-sided interferogram, it directly follows under the approximation in Eq. (8.39) that the resolving power is

$$R = \frac{Nw\Delta s}{\lambda f}.$$ (8.40)

Again, by analogy to a Michelson interferometer, spectral range is dictated by detector sampling according to

$$\lambda_{min} = \frac{2w\Delta s}{f}.$$ (8.41)

The interferometer depicted in Fig. 8.11 is not a practical imaging spectrometer because the slits produce very small optical throughput; that is, the constant K is extremely small. The Sagnac interferometer design shown in Fig. 8.12 is a more practical approach. In this design, a slit is located at the image plane of the preceding fore-optics. Light exiting the slit is split by a 50/50 beamsplitter, with half (shown in red) propagating clockwise through the interferometer and the other half (shown in blue) propagating counterclockwise. When mirrors are located equidistant from the beamsplitter and the system is perfectly aligned, the two wavefronts constructively interfere on the output side of the beamsplitter before the lenses. However, a difference d in the length of the interferometer arms incurs a lateral shear Δs between the two wavefronts. When used in the Fourier transform configuration shown, this produces an interferogram in the cross-slit direction along the FPA in the same manner as described before. A cylindrical lens is added such that the slit is imaged in the along-slit dimension. Thus, the system performs imaging in the along-slit direction and produces an interferogram for each imaged pixel in the orthogonal, cross-slit direction.

As depicted in Fig. 8.13, the data format from a spatial FTS consists of a sequence of frames that contain along-slit spatial information in one direction and corresponding interferograms in the orthogonal direction. As the slit is scanned across the scene via aircraft motion or some other optical scanning mechanism, the sequence of frames captures a full set of interferograms for a 2D image area. After transforming all of the interferograms to spectra, a hyperspectral data cube is formed.

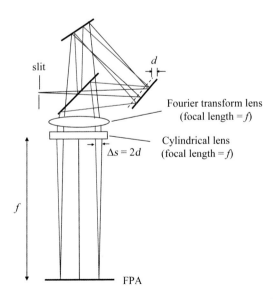

Figure 8.12 FTS using the Sagnac interferometer design.

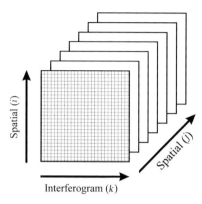

Figure 8.13   Frame sequence from a spatial FTS.

8.4 Radiometric Sensitivity

Characterization of radiometric sensitivity for an FTS follows closely that of a dispersive spectrometer. First, an imaging FTS of the Michelson interferometer variety is considered. Later, this is extended to the case of a spatial FTS.

8.4.1 Signal-to-noise ratio

To begin, assume that the front aperture of an imaging spectrometer system, including all preceding fore-optics, is illuminated with signal spectral radiance $L_s(\lambda)$ and background spectral radiance $L_b(\lambda)$, defined for the particular situation as previously discussed. Notwithstanding the interferometer, this corresponds to an image plane spectral irradiance distribution

$$E_s(\lambda) = \frac{\pi K(x, y)}{4(f/\#)^2 + 1} \tau_o(\lambda) L_s(\lambda)$$

$$E_b(\lambda) = \frac{\pi K(x, y)}{4(f/\#)^2 + 1} \tau_o(\lambda) L_b(\lambda) + E_o(\lambda),$$

$$(8.42)$$

where the 2D spatial dependence in irradiance roll-off function $K(x, y)$ is explicitly shown, $\tau_o(\lambda)$ is the warm optics transmission, and $E_o(\lambda)$ represents irradiance due to instrument self-emission and stray radiance. Spatial dependence of the irradiance is implied. With the interferometer inserted, the signal corresponds to a modulation term associated with pupil-plane signal radiance, while the background corresponds to a bias term associated with pupil-plane background radiance. Based on these definitions,

$$E_s(\lambda) = \frac{m}{2} \frac{\pi K(x,y)}{4(f/\#)^2 + 1} \tau_o(\lambda) \tau_{spec}(\lambda) \tau_c(\lambda) L_s(\lambda)$$

$$E_b(\lambda) = \frac{1}{2} \frac{\pi K(x,y)}{4(f/\#)^2 + 1} \tau_o(\lambda) \tau_{spec}(\lambda) \tau_c(\lambda) L_b(\lambda) + E_o(\lambda),$$

(8.43)

where $\tau_{spec}(\lambda)$ is the interferometer transmission that includes τ_b in Eq. (8.5), m is the interferometer modulation depth from Eq. (8.6) but is potentially reduced by other interferometer imperfections, $\tau_c(\lambda)$ is the cold filter transmission, and $E_o(\lambda)$ includes all self-emission and stray radiance sources up to the FPA.

Consider a signal corresponding to a particular wavelength band centered at λ_k, with effective SRF $g(\lambda - \lambda_k)$ given by Eq. (8.23) with $\sigma = 1/\lambda$. The associated signal on a per-frame basis is

$$N_{s,frame}(\lambda_k) = A_d t_d \int \frac{\lambda}{hc} \eta(\lambda) E_s(\lambda) g(\lambda - \lambda_k) d\lambda.$$

(8.44)

If N_f frames are collected and processed with a nonapodized DFT, the effective integrated signal in the spectral domain is

$$N_s(\lambda_k) = N_f A_d t_d \int \frac{\lambda}{hc} \eta(\lambda) E_s(\lambda) g(\lambda - \lambda_k) d\lambda.$$

(8.45)

Because of the narrowband nature of $g(\lambda - \lambda_k)$, the integral can be approximated as

$$N_s(\lambda_k) = N_f A_d t_d \frac{\lambda_k}{hc} \eta(\lambda_k) E_s(\lambda_k) \delta\lambda_k,$$

(8.46)

where

$$\delta\lambda_k = \int g(\lambda - \lambda_k) d\lambda.$$

(8.47)

Although the spectral resolution in wavenumber units is constant over the spectral range, the spectral resolution in wavelength units varies according to

$$\delta\lambda_k = \lambda_k^2 \delta\sigma.$$

(8.48)

Background is determined in a similar manner, but in this case, spectral integration is not weighted by an SRF. The corresponding integrated background in the spectral domain is

$$N_b = N_f N_{b,frame} = N_f A_d t_d \int \frac{\lambda}{hc} \eta(\lambda) E_b(\lambda) d\lambda. \qquad (8.49)$$

Note that the background is integrated over the entire spectral range (as opposed to a single band) and remains constant for all spectral bands.

Given the quantities in Eqs. (8.48) and (8.49), the SNR is given by

$$SNR(\lambda_k)$$

$$= \frac{\sqrt{N_f} N_{s,frame}(\lambda_k)}{\sqrt{N_{b,frame}(\lambda_k) + \sigma_{n,dark}^2 + \sigma_{n,Johnson}^2 + \sigma_{n,read}^2 + \sigma_{n,spatial}^2 + \sigma_{n,ADC}^2}}, \qquad (8.50)$$

where all of the noise terms have been previously defined for the dispersive spectrometer and conventional imaging system cases. The total noise variance is N_f times the noise variance per frame, assuming frame-to-frame independence of the noise processes. The equation is intentionally written using per-frame parameters to explicitly show SNR dependence on the number of interferogram samples. When apodization is considered, Eq. (8.50) becomes

$$SNR(\lambda_k)$$

$$= \frac{F \sqrt{N_f} N_{s,frame}(\lambda_k)}{\sqrt{N_{b,frame}(\lambda_k) + \sigma_{n,dark}^2 + \sigma_{n,Johnson}^2 + \sigma_{n,read}^2 + \sigma_{n,spatial}^2 + \sigma_{n,ADC}^2}}, \qquad (8.51)$$

where

$$F = \sqrt{\frac{1}{N} \sum_{j=1}^{N} W^2(\Delta_j)} \qquad (8.52)$$

based on apodization weights $W(\Delta_j)$ defined in Eq. (8.22).

8.4.2 Noise-equivalent spectral radiance

The noise-equivalent spectral radiance (NESR), defined as

$$NESR(\lambda_k) = \frac{L_s(\lambda_k)}{SNR(\lambda_k)}, \qquad (8.53)$$

is used as a radiometric sensitivity metric. This quantity can be computed directly from the SNR in Eq. (8.51). Like the formulation in a dispersive case, the NESR can also be written in a simpler parametric form based on similar assumptions and definitions. First, optical throughput of the imaging spectrometer is defined as

$$A\Omega = \frac{\pi A_d}{4(f/\#)^2 + 1}. \tag{8.54}$$

Next, effective signal and background transmission profiles are computed as

$$\tau_s(\lambda) = \tau_o(\lambda)\,\tau_{spec}(\lambda)\,\tau_c(\lambda)\,\eta(\lambda) \tag{8.55}$$

and

$$\tau_b(\lambda) = \frac{\tau_o(\lambda)\,\tau_{spec}(\lambda)\,\tau_c(\lambda)\,E_b(\lambda) + I_o(\lambda)}{E_s(\lambda)}\eta(\lambda). \tag{8.56}$$

Also, the total additive, signal-independent noise is defined as

$$\sigma_{n,add} = \sqrt{\sigma_{n,dark}^2 + \sigma_{n,Johnson}^2 + \sigma_{n,read}^2 + \sigma_{n,spatial}^2 + \sigma_{n,ADC}^2}. \tag{8.57}$$

Based on these definitions and the assumption that spectrally varying parameters are uniform over SRF $g(\lambda - \lambda_k)$, the NESR at a center wavelength λ_k for a particular spectral band becomes

$$\text{NESR}(\lambda_k) = \frac{1}{F\sqrt{N_f}} \frac{\sqrt{\sigma_{n,add}^2 + A\Omega \frac{t_d}{hc} \int \frac{\tau_b(\lambda)}{2} L_s(\lambda)\,\lambda\,d\lambda}}{A\Omega \frac{t_d}{hc} \frac{m\tau_s(\lambda_k)}{2}\lambda_k\delta\lambda_k}. \tag{8.58}$$

In a background-limited regime, the shot-noise term (second term in the quadratic) dominates the additive-noise term, and the equation reduces to

$$\text{NESR}(\lambda_k) = \frac{1}{F\sqrt{N_f}} \frac{1}{m\tau_s(\lambda_k)\,\lambda_k\,\delta\lambda_k} \sqrt{\frac{2}{A\Omega}\frac{hc}{t_d}\int \tau_b(\lambda)\,L_s(\lambda)\,\lambda\,d\lambda}. \tag{8.59}$$

In a noise-limited regime, the additive-noise term dominates, and the equation reduces to

$$\text{NESR}(\lambda_k) = \frac{2}{F\sqrt{N_f}} \frac{hc}{t_d} \frac{\sigma_{n,add}}{A\Omega\,m\,\tau_s(\lambda_k)\,\lambda_k\,\delta\lambda_k}. \tag{8.60}$$

In the latter case, the NESR is independent of illumination and more strongly dependent on optical throughput, dwell time, signal transmission, and spectral bandwidth.

Some modifications to Eq. (8.58) must be made to account for a spatial FTS. The first modification involves what is used for dwell time t_d. A temporal FTS is fully multiplexed both spatially and spectrally. Therefore, the total time to collect an image is given by $T_{total} = N_f t_d$ if contiguous frames are collected. A spatial FTS is only partially multiplexed, as it must scan over one spatial dimension. Therefore, $T_{total} = N_y t_d$, where N_y is the number of image lines to be captured in the cross-slit direction. Also, incident energy must be spread across N_x detector elements in an interferogram dimension. Adjusting for this second issue,

$$\text{NESR}(\lambda_k) = \frac{\sqrt{N_x}}{F} \frac{\sqrt{\sigma_{n,add}^2 + A\Omega \frac{t_d}{hc} \int \frac{\tau_b(\lambda)}{2N_x} L_s(\lambda)\,\lambda\,d\lambda}}{A\Omega \frac{t_d}{hc} \frac{m\tau_s(\lambda_k)}{2} \lambda_k\,\delta\lambda_k} \quad (8.61)$$

in the case of a spatial FTS. This becomes

$$\text{NESR}(\lambda_k) = \frac{1}{F} \frac{1}{m\,\tau_s(\lambda_k)\,\lambda_k\,\delta\lambda_k} \sqrt{\frac{2}{A\Omega} \frac{hc}{t_d} \int \tau_b(\lambda)\,L_s(\lambda)\lambda\,d\lambda} \quad (8.62)$$

in a background-limited regime, and

$$\text{NESR}(\lambda_k) = \frac{2}{F\sqrt{N_x}} \frac{hc}{\tau_d} \frac{\sigma_{n,add}}{A\Omega\,m\,\tau_s(\lambda_k)\,\lambda_k\,\delta\lambda_k} \quad (8.63)$$

in a noise-limited regime. That is, it becomes a factor of $\sqrt{N_x}$ higher in the background-limited case and N_x higher in the noise-limited case.

8.4.3 Imaging spectrometer sensitivity comparison

A theoretical NESR comparison of dispersive, temporal FTS, and spatial FTS instruments is shown in Fig. 8.14 for the somewhat idealized design example summarized in Table 8.1 (Eismann *et al.*, 1996). Conclusions are as follows. The fully multiplexed characteristic of a temporal FTS results in an NESR advantage but also a disadvantage, due to the fact that the spectrum is corrupted by shot noise associated with the entire spectral range, as opposed to that of only each spectral band. For background-limited operation (which this case illustrates at all but the lowest spectral bandwidths), the multiplex advantage and disadvantage effectively cancel. The spatial FTS exhibits the disadvantage due to spectral multiplexing but not the full spatial multiplex advantage. This results in poorer performance

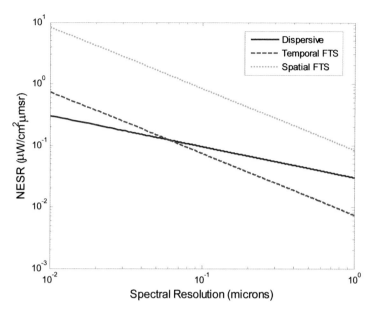

Figure 8.14 Theoretical NESR comparison of a dispersive spectrometer, a temporal FTS, and a spatial FTS.

Table 8.1 Parameters used for NESR performance comparison.

Characteristic	Specification
Spectral range	8 to 12 μm
Focal plane array size	128 × 128
Hyperspectral image size	128 × 128 × N (spectral)
Total integration time	1 sec
Detector quantum efficiency	0.6
Additive noise per frame	300 rms electrons
Detector element width	50 μm
Sensor f-number	2.0
Optical system transmission	1.0
Interferometer modulation depth	1.0
Stray radiance	None
Source object	300-K blackbody

than seen in all other designs. The dispersive design does not exhibit either the multiplex advantage or disadvantage. It performs better than a temporal FTS when the number of spectral bands is greater than the number of scanned spatial elements, but has poorer performance when spectral resolution is relaxed. In this particular example, the crossover point between the dispersive spectrometer and temporal FTS NESR curves is at about 70-nm spectral resolution. Recall, however, that the primary advantage of an FTS is the finer achievable spectral resolution. Therefore, one typically needs to resort to this design to achieve better spectral resolution, compared to a dispersive spectrometer in a region where it is not advantageous from a radiometric sensitivity standpoint.

Based on this performance comparison, it appears that the spatial FTS is a fundamentally inferior approach to hyperspectral imaging. Two additional considerations should be made, however, before arriving at such a conclusion. First, the multiplex advantage of the FTS approach, whether temporal or spatial, increases when the system falls into a noise-limited regime. Therefore, the spatial FTS can provide a means for achieving decent radiometric performance using low-cost FPAs, for example, using microbolometer detector arrays in the longwave infrared (Lucey *et al.*, 2008). In airborne applications, the spatial design can be advantageous over a Michelson approach in terms of being more compatible with scanning imaging geometry and not requiring the complexity of precisely controlled moving parts. Furthermore, the fundamental difference in NESR performance can be overcome by removing the slit and anamorphic optics and performing more sophisticated processing to extract spectral and spatial information.

Consider a situation where the intermediate image is relayed in both dimensions through an interferometer, the slit is widened to pass the entire FPA image, and the image moves at lateral velocity v in the direction of interferometer shear. With these modifications, the FPA image consists of the moving image modulated by the fixed interferogram pattern; that is,

$$E(x, y) = K \int_0^\infty \frac{1}{2} E_0(x - vt, y, \sigma) \left[1 + \cos(4\pi\sigma x \sin\theta)\right] d\sigma. \quad (8.64)$$

An individual frame at time t provides interferogram samples at different OPDs for different spatial locations in the scene. As the image shifts in time, however, a complete interferogram for every spatial position can be extracted if image motion is properly tracked through a succession of frames. This is illustrated using actual airborne data in Fig. 8.15, where the top image shows an uncorrected slice through the data, and the bottom image shows a slice through the motion-compensated data. For each, the horizontal axis corresponds to the x axis, and the vertical axis corresponds to the time axis. The end result of this spatial–temporal processing is an NESR performance commensurate with that of a temporal FTS design by virtue of eliminating the slit and making the instrument fully multiplexed. The spatial FTS approach is more easily implemented in an airborne imaging situation because it can build up the image one line at a time in a pushbroom manner, as opposed to requiring sophisticated pointing optics to support staring over long interferogram collection times. Therefore, using this more complicated processing system with a slitless spatial FTS

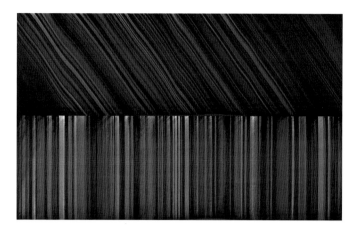

Figure 8.15 Motion compensation in a field-widened spatial FTS. Top shows an uncorrected slice through the data, and bottom shows a slice through the motion-compensated data [reprinted with permission from (Lucey *et al.*, 2008) © 2008 Optical Society of America].

design can make this approach advantageous over temporal FTS design for some airborne remote sensing applications.

8.5 System Examples

Several imaging FTSs have been designed, built, and deployed from ground, airborne, and space-based platforms for a variety of remote sensing applications. The designs employed exhibit some diversity based on the different requirements for each specific application, the advancement of optical and FPA component technology over time, and personal preferences of the designers. A small number of these systems are discussed to illuminate how design issues discussed in this section have been put into practice and to show the overall performance achieved.

8.5.1 Field-Portable Imaging Radiometric Spectrometer Technology

The Field-Portable Imaging Radiometric Spectrometer Technology (FIRST) spectrometer, depicted in Fig. 8.16, is a commercial LWIR imaging FTS developed by Telops, Inc. primarily for remote chemical-sensing applications (Chamberland *et al.*, 2005). The interferometer is basically of the Michelson interferometer design; however, it employs retroreflectors for mirrors to achieve robustness against external mechanical interferences during interferogram scanning. The design supports an OPD ranging from −0.6 to +2.5 cm. Over a nominal 8- to 12-μm spectral range, this supports 1-cm^{-1} spectral resolution in a double-sided interferogram mode and 0.24-cm^{-1} in a single-sided mode.

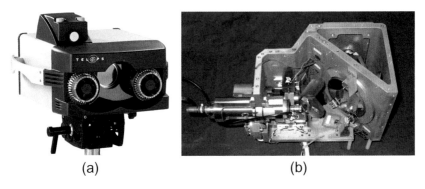

(a)   (b)

Figure 8.16 FIRST imaging spectrometer: (a) sensor unit and (b) interferometer module [images courtesy of Telops, Inc.].

Mirror scan rate can be adjusted across a large range, from 50 μm/sec to 20 cm/sec, trading off the speed of interferogram acquisition with radiometric sensitivity.

The FIRST sensor uses a 320×256 photovoltaic HgCdTe FPA with 30-μm detector pitch that can be electronically windowed to any subregion of the array. This windowing is used to exchange FOV for frame rate. By adjusting the extent of mirror motion, spectral resolution can be adjusted from 0.25 to 150 cm^{-1}, although performance is optimized at 4 cm^{-1}. The maximum frame rate is 300 frames per second (fps) for full field and 1400 fps for a 128×128 pixel subregion. The latter frame rate corresponds to the collection of a full interferogram for 4-cm^{-1} spectral resolution every 2 sec. A data acquisition system captures interferogram data in a 4-GB double-data-rate synchronous dynamic random access memory (DDR-SDRAM) and uses three field-programmable gate arrays to perform real-time DFTs to produce a corresponding hyperspectral data cube. Two internal calibration blackbody sources with a temperature range of -10 to 20 °C, a temperature stability of 5 mK, and a surface emissivity greater than 0.9996 are used for radiometric calibration. The complete unit weighs roughly 45 lbs and consumes 110 W of power. For typical operation at a 4-cm^{-1} spectral resolution and 0.5-Hz collection rate, the system achieves a single-scan NESR of roughly 4 μW/cm^{2}μm sr at 10 μm when viewing a 300-K blackbody source. Other sensor parameters are provided in Table 8.2.

8.5.2 Geosynchronous Imaging Fourier Transform Spectrometer

The Geosynchronous Imaging Fourier Transform Spectrometer (GIFTS) sensor is a developmental system designed to be placed on a geosynchronous satellite for atmospheric sounding in support of meteorological applications (Stobie *et al.*, 2002). At a 40,000-km orbit, the system is intended to achieve 4-km GSD and nominal 0.6-cm^{-1}

Table 8.2 FIRST sensor parameters.

Characteristic	Specification
Aperture diameter	4.3 cm
Effective focal length	8.6 cm
f-number	2.0
Detector array size	320×256
Detector width	30 μm
Angular sampling (IFOV)	0.35 mrad
FOV	5.1×6.4 deg
Spectral range	8 to 12 μm
Spectral resolution	0.25 to 150 cm^{-1}
Number of spectral channels	3 to 1600
Maximum frame rate	300 Hz (320×256)
	1400 Hz (128×128)
Digitization	14 bits

spectral resolution over both the MWIR (4.4 to 6.1 μm) and LWIR (8.8 to 14.5 μm) spectral regions. To achieve this performance, the system design is composed of a 0.6-m focal length, $f/2.25$ silicon carbide telescope, a cryogenic FTS unit, reflective relay optics, and advanced 128×128 MWIR and LWIR HgCdTe FPAs. A sensor schematic is provided in Fig. 8.17. The instrument includes two blackbody reference sources and a source selector for onboard radiometric calibration (the topic of Chapter 10), along with a visible imaging camera. A Lyot stop matches the stop in the intermediate image plane to minimize stray radiance from its thermal emission.

The design of the FTS is based on a cryogenically cooled design initially developed for a high-altitude balloon (Huppi *et al.*, 1979). The optical system is composed of a standard Michelson interferometer that uses a flex-pivot mirror mounting arrangement with a torque motor drive unit to precisely control the moving mirror. Mirror tilt is minimized by using a parallelogram configuration with flex pivots at the corners, and constant mirror motion is achieved via tachometer feedback of the torque motor. The entire assembly can be cryogenically cooled to reduce instrument self-emission, is roughly $10 \times 18 \times 30$ inches, and weighs less than 90 lbs. 128×128 imaging sensors are designed to be operated at 60 K to reduce dark current. The MWIR FPA is characterized by a 600-μsec frame time and 20-million-electron charge capacity, while the LWIR FPA is characterized by a 150-μsec frame time and 85- to 100-million-electron charge capacity. For a 10-sec total collection time, this supports collection of more than 15,000 interferogram samples, which can be coadded to 2048 samples while maintaining 0.6-cm^{-1} spectral resolution. Additional sensor characteristics are provided in Table 8.3. Examples of MWIR and LWIR atmospheric retrievals are given in Fig. 8.18 to illustrate system capabilities. These results are provided in spectral

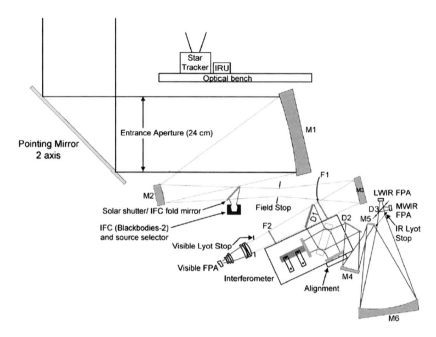

Figure 8.17 Layout of the GIFTS instrument [graphic courtesy of Bill Smith, NASA Langley Research Center].

Table 8.3 GIFTS sensor parameters.

Characteristic	Specification
Aperture diameter	27 cm
Effective focal length	60 cm
f-number	2.25
Detector array size	128×128
Detector width	60 µm
Angular sampling (IFOV)	0.1 mrad
FOV	0.7×0.7 deg
Spectral range	4.4 to 6.1 µm (MWIR)
	8.8 to 14.5 µm (LWIR)
Spectral resolution	0.6 cm^{-1}
Number of spectral channels	2048 (MWIR/LWIR)
Maximum frame rate	1500 Hz
Digitization	14 bits

apparent temperature units (described further in Chapter 11), in relation to atmospheric compensation from emissive hyperspectral data.

8.5.3 Spatially Modulated Imaging Fourier Transform Spectrometer

The Spatially Modulated Imaging Fourier Transform Spectrometer (SMIFTS) sensor is a cryogenically cooled, spatially modulated FTS that

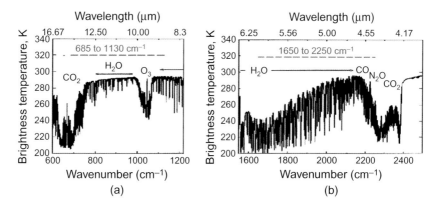

Figure 8.18 Atmospheric retrievals from the GIFTS instrument: (a) LWIR and (b) MWIR [graphic courtesy of Bill Smith, NASA Langley Research Center].

covers the NIR to MWIR spectral range, nominally 1 to 5.2 μm (Lucey *et al.*, 1993). The sensor design, depicted in Fig. 8.19, is composed of a Sagnac interferometer using a MgF_2 beamsplitter and compensator with two Zerodur® mirrors, and an anamorphic lens system composed of an off-axis parabolic mirror and an off-axis cylindrical mirror. A cold filter/slit wheel is used to support adjustment of the slit size and sensor passband. The entire system is cooled to roughly 65 K with liquid nitrogen and is mounted behind an imaging telescope and optional scan mirror for imaging from either airborne or fixed platforms. The baseline design uses a 128×128 InSb FPA and can achieve 100-cm^{-1} spectral resolution. Sensor parameters are summarized in Table 8.4, although some are dependent on the fore-optics used.

8.5.4 Fourier Transform Hyperspectral Imager

The Fourier Transform Hyperspectral Imager (FTHSI) sensor is a spatially modulated FTS of the Sagnac design that was developed as a NASA demonstration payload for earth observation as part of the MightySat II.1 program (Otten *et al.*, 1995). This sensor covers the VNIR spectral range, nominally 350 to 1050 nm, using a Sagnac interferometer with a 1024 × 1204-Si CCD detector array. To limit data size, the FPA readout performs 2×2 pixel binning to provide an effective array size of 512×512 with 28-μm detector spacing. The MightySat satellite is a 109-kg microsatellite with nominally a 500-km orbit, at which the FTHSI sensor achieves a 15.3-km swath width and 30-m GSD using a pushbroom imaging mode. The sensor unit itself is roughly $42 \times 25 \times 10$ cm in size and weighs about 20 lbs, not including data acquisition and control electronics. For a 20% reflective target during typical daylight conditions, the sensor can achieve an SNR of about 200 with 1.7-nm spectral resolution at 450-nm wavelength. Other sensor parameters are summarized in Table 8.5.

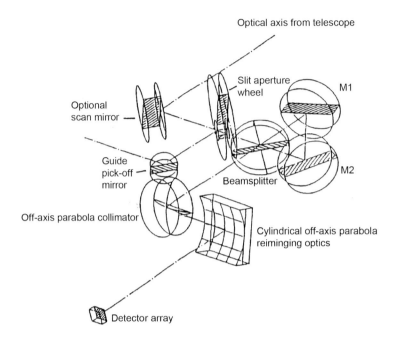

Figure 8.19   SMIFTS sensor unit (Lucey *et al.*, 1993).

Table 8.4   SMIFTS sensor parameters.

Characteristic	Specification
Aperture diameter	Dependent on fore-optics
Effective focal length	Dependent on fore-optics
f-number	20
Detector array size	128×128
Angular sampling (IFOV)	Dependent on fore-optics
FOV	Dependent on fore-optics
Spectral range	1.0 to 5.2 µm
Spectral resolution	100 cm^{-1}
Number of spectral channels	25

Table 8.5   FTHSI sensor parameters.

Characteristic	Specification
Aperture diameter	5 cm
Effective focal length	30 cm
f-number	6.0
Detector array size	512×512
Detector width	28 µm
Angular sampling (IFOV)	0.09 mrad
FOV	1.75 deg
Spectral range	350 to 1050 nm
Spectral resolution	1.7 nm (at 450 nm)
Number of spectral channels	256
Maximum frame rate	60 Hz

8.6 Summary

The Fourier transform spectrometer (FTS) is an alternate imaging-spectrometer approach for hyperspectral remote sensing that decouples spectral resolution from the properties of optical materials and spatial resolution of the instrument. Historically, FTSs have been used due to their multiplex advantage and ability to achieve very fine spectral resolution. As seen in this chapter, their multiplex advantage can be outweighed by the increased shot noise that occurs due to integrated background radiance over a spectral range. Fine spectral resolution is achievable, but long interferogram acquisition times are needed to concurrently achieve an adequate noise-equivalent spectral radiance (NESR).

The staring imaging format of a temporal FTS is advantageous for fixed imaging applications but can be problematic when instruments are on moving platforms due to scene motion during interferogram collection time that can corrupt the processed spectra. This can be compensated for by means of sophisticated pointing and stabilization systems and potentially through data processing, but these challenges hinder this type of hyperspectral design in situations where good spatial resolution is needed. The spatial FTS design is more amenable to imaging from moving platforms but exhibits an NESR disadvantage unless used in a slitless mode, with corrections of complicated scene motion accomplished through an interferogram.

References

Chamberland, M., Farley, V., Vallieres, A., Villemaire, A., and Legault, J.-F., "High-performance field-portable imaging radiometric spectrometer technology for hyperspectral imaging applications," *Proc. SPIE* **5994**, 5994ON-1–11 (2005) [doi:10.1117/12.632104].

Eismann, M. T., Schwartz, C. R., Cederquist, J. N., Hackwell, J. A., and Huppi, R. J., "Comparison of infrared imaging hyperspectral sensors for military target detection applications," *Proc. SPIE* **2819**, 91–101 (1996) [doi:10.1117/12.258056].

Huppi, R. J., Shipley, R. B., and Huppi, E. R., "Multiplex and high throughput spectroscopy," *Proc. SPIE* **191**, 26–32 (1979).

Lucey, P. G, Horton, K. A., and Williams, T., "Performance of a long-wave infrared hyperspectral imager using a Sagnac interferometer and an uncooled microbolometer array," *Appl. Optics* **47**, F107–F113 (2008).

Lucey, P. G., Horton, K. A., Williams, T. J., Hinck, K., Budney, C., Rafert, T., and Rusk, T. B., "SMIFTS: A cryogenically-cooled spatially-modulated imaging infrared interferometer spectrometer," *Proc. SPIE* **1937**, 130–141 (1993) [doi:10.1117/12.157050].

Oppenheim, A. V. and Schafer, R. W., *Digital Signal Processing*, Prentice-Hall, Englewood Cliffs, NJ (1975).

Otten, L. J., Sellar, R. G., and Rafert, J. B., "MightySat II.1 Fourier transform hyperspectral imager payload performance," *Proc. SPIE* **2583**, 566–575 (1995) [doi:10.1117/12.228602].

Pedrotti, F. L. and Pedrotti, L. S., *Introduction to Optics*, Prentice-Hall, Englewood Cliffs, NJ (1987).

Stobie, J., Hairston, A. W., Tobin, S. P., Huppi, R. J., and Huppi, R., "Imaging sensor for the Geosynchronous Imaging Fourier Transform Spectrometer (GIFTS)," *Proc. SPIE* **4818**, 213–218 (2002) [doi:10.1117/12.458133].

Wolfe, W. L., *Introduction to Imaging Spectrometers*, SPIE Press, Bellingham, WA (1997) [doi:10.1117/3.263530].

Chapter 9
Additional Imaging Spectrometer Designs

While dispersive and Fourier transform spectrometers are the predominate instruments used for hyperspectral remote sensing applications, there are a variety of other instruments that have also been developed for this purpose, and novel imaging spectrometer concepts continue to be explored. Some of these devices are developed for niche applications where specific characteristics of interest are not supported by more-conventional instrument designs. One such characteristic is the ability to perform hyperspectral imaging in a staring format in a stationary observation scenario without resorting to interferometric concepts or mechanical scanning components. The variety of innovative spectrometer approaches reported in the literature is quite extensive and diverse. This chapter merely outlines the principles of operation underlying a few of these hyperspectral imaging approaches.

9.1 Fabry–Pérot Imaging Spectrometer

A Fabry–Pérot interferometer, also known as an etalon, consists of two optically flat, parallel mirrors that are precisely spaced to support constructive interference for a specified optical wavelength (Pedrotti and Pedrotti, 1987). Consider the example shown in Fig. 9.1, where the mirrors are characterized by an amplitude transmissivity t, amplitude reflectivity r, and phase shift upon reflection of ϕ. The mirrors are assumed to be spaced by a distance d, and the material in the optical cavity between the mirrors is assumed to exhibit a dielectric index n that can be wavelength dependent but ideally is not. According to the coherent dielectric thin film analysis described in Chapter 3, a steady-state condition of the form

$$\mathcal{E} = (r^2 e^{i2\phi} e^{i2\pi n d\sigma \cos\theta} \mathcal{E} + t\mathcal{E}_0) e^{i2\pi n d\sigma \cos\theta} \tag{9.1}$$

can be written for an incident plane wave with an incident angle θ from the optical axis, where $\mathcal{E}$ is the incident electric field amplitude on the second

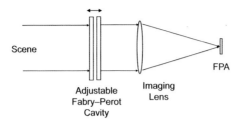

Figure 9.1 Fabry–Pérot interferometer.

mirror, $\mathcal{E}_0$ is the incident electric field amplitude on the interferometer, and σ is the wavenumber of the incident plane wave. Solving Eq. (9.1) for $\mathcal{E}$ leads to

$$\mathcal{E} = \frac{t\,\mathcal{E}_0}{e^{i2\pi nd\sigma\cos\theta} - r^2 e^{i2\phi} e^{i2\pi nd\sigma\cos\theta}}. \tag{9.2}$$

The transmitted electric field is then given by

$$\mathcal{E}_t = \frac{t^2\mathcal{E}_0}{e^{i2\pi nd\sigma\cos\theta} - r^2 e^{i2\phi} e^{i2\pi nd\sigma\cos\theta}}. \tag{9.3}$$

The irradiance transmission of the interferometer is consequently given by

$$\tau = \frac{|\mathcal{E}_t|^2}{|\mathcal{E}_0|^2}. \tag{9.4}$$

Substituting Eq. (9.3) into Eq. (9.4) yields

$$\tau = \left[\frac{t^2}{e^{i2\pi nd\sigma\cos\theta} - r^2 e^{i2\phi} e^{i2\pi nd\sigma\cos\theta}}\right]\left[\frac{t^2}{e^{-i2\pi nd\sigma\cos\theta} - r^2 e^{-i2\phi} e^{-i2\pi nd\sigma\cos\theta}}\right]. \tag{9.5}$$

With some algebraic manipulation and substitution of trigonometric functions for the complex exponentials, it can be shown that irradiance transmission is

$$\tau = \frac{t^4/(1 - r^2)^2}{1 + \frac{4r^2}{(1-r^2)^2}\sin^2(2\pi nd\sigma\cos\theta + \phi)}. \tag{9.6}$$

Defining

$$\tau_0 = t^4/(1 - r^2)^2 \tag{9.7}$$

as the peak transmission and

$$\rho = r^2 \qquad (9.8)$$

as the mirror irradiance reflectance, Eq. (9.6) can be written in the form

$$\tau = \frac{\tau_0}{1 + \frac{4\rho}{(1-\rho)^2} \sin^2(2\pi n d\sigma \cos\theta + \phi)}. \qquad (9.9)$$

The transmission expressed in Eq. (9.9) exhibits a series of peaks that occur when the argument of the sinusoid in the denominator is a multiple of π, which occurs for

$$\sigma_{peak} = \frac{1}{\lambda_{peak}} = \frac{m\pi - \phi}{2\pi n d \cos\theta}, \qquad (9.10)$$

where m is an integer. Peak transmission is τ_0, and the FWHM of the peaks is given by

$$\delta\sigma = \frac{1 - \rho}{\pi d \sqrt{\rho}}. \qquad (9.11)$$

This is depicted in Fig. 9.2 for the case of a 1-μm air-gap separation in the visible spectral region. To use a Fabry–Pérot etalon as a spectrometer, the separation is precisely controlled to capture spectrally resolved images successively in time. Equation (9.11) represents the spectral resolution of the instrument, and the corresponding resolving power is

$$R = \frac{\sqrt{\rho}}{1 - \rho} m\pi, \qquad (9.12)$$

where m is the index corresponding to the specific peak satisfying Eq. (9.10). Because of multiple transmission peaks, spectral range is limited to spectral separation. This condition is given by

$$\sigma_{max} - \sigma_{min} = \frac{1}{\lambda_{min}} - \frac{1}{\lambda_{max}} = \frac{1}{2nd}. \qquad (9.13)$$

Using Eq. (9.10) with $\phi = 0$ and $\theta = 0$, it can be shown that the free spectral range in wavelength units is

$$\lambda_{max} - \lambda_{min} \approx \frac{\lambda_{min}}{(m + 1)}. \qquad (9.14)$$

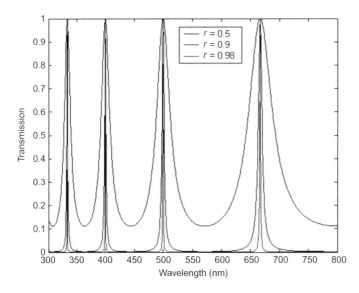

Figure 9.2   Example transmission of a Fabry–Pérot interferometer.

To achieve a large spectral range, it is necessary to operate an interferometer in a low order, meaning that mirrors are spaced on the order of an optical wavelength. When the interferometer is used as an imaging spectrometer, it is important to take into account the shift in peak wavelength that occurs with field angle through Eq. (9.10) during spectral calibration.

There are two primary ways that have been employed to engineer a Fabry–Pérot imaging spectrometer into a practical instrument: (1) mechanically controlling the air gap between two mirrors and (2) electrically controlling the index of refraction of an electro-optic material sandwiched between two fixed mirrors. Both methods can theoretically be employed at any part of the EO/IR spectrum, but the mechanical approach is more practical at longer wavelengths because optical and mechanical tolerances scale with wavelength. The electro-optical approach is more practical at visible wavelengths where materials are more conducive to this approach, particularly tunable liquid-crystal materials. Two LWIR instruments are described to illustrate these issues.

The Adaptive Infrared Imaging Spectroradiometer (AIRIS) is a Fabry–Pérot imaging spectrometer that operates in a low order (small m) to maximize the free spectral range, requiring very small separation between interferometer mirrors (Gittins and Marinelli, 1998). As depicted in Fig. 9.3, mirror separation is controlled by three piezoelectrically based inchworm drive motors that have a maximum linear velocity of 2 mm/sec. Mirror alignment is maintained via feedback from four capacitive sensors based on gold pads located on the outer surface of the mirrors. The system

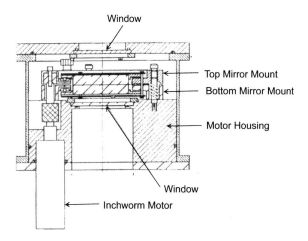

Figure 9.3 AIRIS Fabry–Pérot interferometer module (Gittins and Marinelli, 1998).

can achieve mirror spacing to an accuracy of about 2 nm and has exhibited a wavelength tuning repeatability of about 2 cm^{-1} with a minimum scan time of 1 to 10 msec. Devices have been demonstrated with 80-nm spectral resolution with a 2-μm spectral range in the LWIR spectral region. In addition to being able to directly perform staring hyperspectral imaging, this device has the advantage of being able to rapidly capture narrowband images at selected wavelengths of interest without needing to capture the complete spectrum. This can be a significant advantage for chemical sensing applications, where the measurement of a small set of narrowband images can be adequate for detection and characterization of particular species of interest.

An alternate approach that has been demonstrated and is depicted in Fig. 9.4 uses a fixed Fabry–Pérot interferometer filled with a liquid crystal material. Dielectric stack mirrors are deposited on flat ZnSe substrates to achieve greater than 90% reflectance, with a polymer alignment mirror for orienting the liquid crystal material. Transparent conducting layers between the dielectric stacks and mirror substrates are used to control the electric field across the liquid crystal material, varying the index of refraction for linearly polarized light in the direction of the extraordinary axis. Therefore, it is necessary to use a linear polarizer to operate this device as an imaging spectrometer, meaning that at least half of the incident light is lost by the polarizer. The advantage, however, is that it provides a staring hyperspectral imager with absolutely no moving parts.

One challenge in implementing a liquid crystal Fabry–Pérot interferometer is material absorption. Any such losses are amplified within the resonant optical cavity formed by the dielectric mirrors through the interferometric nature of the device, making it more sensitive to absorption

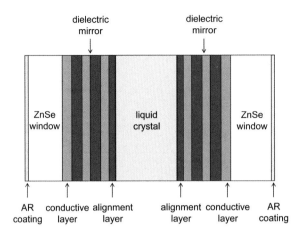

Figure 9.4   Liquid crystal Fabry–Pérot interferometer module.

than other liquid crystal devices. Good transmission is achievable in the VNIR spectral region where liquid crystals are very transparent, and imaging spectrometers based on this concept using silicon FPAs are commercially available. An attempt to extend this approach to the LWIR spectral region, however, uncovered material limitations (Daly *et al.*, 2000). In this particular instrument, index control from 1.52 to 1.86 was demonstrated, providing wavelength control from 8.5 to 9.4 μm with a 0- to 50-V applied voltage and 8.5-μm mirror spacing. Achieved spectral resolution was about 150 nm, but spectral transmission was poor due to absorption in the liquid crystal and transparent electrode materials. Transmittance of the uncoated ZnSe substrates with and without the liquid crystal is shown in Fig. 9.5, indicating several liquid-crystal absorption lines in this spectral region. When the highly reflective coating is added, the poor liquid-crystal transmission is amplified by the multipass nature of the interferometer, resulting in a very low overall transmission, shown in Fig. 9.6.

9.2 Acousto-optic Tunable Filter

Acousto-optic tunable filters (AOTFs) provide another means for producing an electronically tunable imaging spectrometer. Like a liquid crystal Fabry–Pérot spectrometer, such devices are inherently polarimetric in nature. A good example of an AOTF instrument is the Polaris-II instrument designed for a 42-inch telescope at the Lowell Observatory (Glenar *et al.*, 1992). The instrument is based on birefringent diffraction in a TeO$_2$ crystal in a noncollinear geometry, as illustrated in Fig. 9.7. Bragg diffraction occurs between ordinary and extraordinary states in the crystal, such that ordinary (out-of-plane) polarization is diffracted into extraordinary (in-plane) polarization, and vice versa. By orienting the

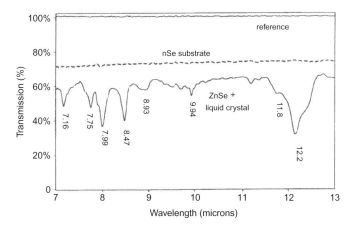

Figure 9.5 Transmission of uncoated ZnSe substrate with and without a liquid crystal (Daly *et al.*, 2000).

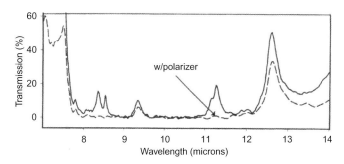

Figure 9.6 Overall transmission of the liquid crystal Fabry–Pérot interferometer module (Daly *et al.*, 2000).

optical axis 108 deg from the acoustic wave vector direction, the device symmetrically diffracts between ordinary and extraordinary polarizations (and vice versa).

By virtue of the spectral selectivity of Bragg diffraction, the acousto-optic device acts as a tunable spectral filter (Chang, 1978). Spectral transmission is given by

$$\tau(\lambda) = \tau_0 \frac{\sin^2[\pi(\lambda - \lambda_0)\,\delta\lambda]}{[\pi(\lambda - \lambda_0)\,\delta\lambda]^2}, \tag{9.15}$$

where the center wavelength is

$$\lambda_0 = \frac{\Delta n V_a}{v_a} C(\theta_i), \tag{9.16}$$

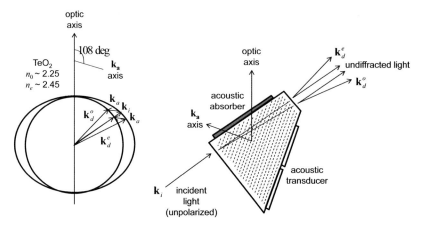

Figure 9.7 Bragg diffraction from an AOTF in noncollinear geometry.

spectral resolution is

$$\delta\lambda = \frac{0.9\lambda_0^2}{L\Delta nC(\theta_i)\tan(\theta_i - \alpha)}, \tag{9.17}$$

$$C(\theta_i) = \frac{\sin^2(\theta_i)}{\sin(\theta_i - \alpha)}, \tag{9.18}$$

α is the acoustic wave vector direction relative to the optic axis, V_a is the acoustic velocity, Δn is the birefringence, θ_i is the incident angle, and L is the interaction length. The spectral passband is electronically tuned by adjusting acoustic frequency v_a. Representative specifications of commercial TeO_2 AOTF devices are given in Table 9.1.

Table 9.1 Representative AOTF specifications.

Characteristic	Specification
Spectral range	0.4 to 5 μm
Birefringence	0.15
Entrance aperture	0.2 to 1.0 cm^2
Deflection angle	5 to 10 deg
Angular FOV	6 deg
Throughput	< 0.02 cm^{2}sr
Resolving power	10 to 2000
Tuning range	One octave
Optical efficiency	>80% across tuning range
Polarization isolation	>104
rf power	0.5 to 2 W
rf drive frequency	20 to 250 Mhz
Transducer material	LiNbO$_3$

Figure 9.8 illustrates how an AOTF can be used to produce a polarimetric imaging spectrometer. The device is placed near the image plane of a telescope and is reimaged by an elliptical mirror onto two CCD camera FPAs. Orthogonal linear polarizations are symmetrically diffracted by the AOTF and therefore form images of the field stop located just prior to the AOTF that are spatially separated on the CCD array. Each image corresponds to an orthogonal polarization state, and the spectral band is tuned successively by changing the acoustic frequency driving the AOTF. This type of instrument has been integrated into the Lowell Observatory and has acquired orthogonally polarized images of Saturn at 724 nm. While the AOTF is an interesting device on which to build an imaging spectrometer, it has some limitations that must be understood. First, optical throughput of the device is fairly small, limiting it to large $f/\#$ or small-aperture optical systems. Second, an AOTF cannot capture simultaneous spectral information. In fact, the time delay between spectral measurements becomes large if a large number of spectral bands is desired. These limitations inhibit the use of the approach for airborne or space-based, terrestrial imaging applications, especially when fine spatial resolution is needed.

9.3 Wedge Imaging Spectrometer

Interference filters provide another approach to spectral sensing. Based on the analysis described in Section 9.1 with regard to a Fabry–Pérot interferometer, a spectral bandpass filter can be created by placing a gap with an OPD of $\lambda/2$ between two high-reflectance mirrors. The implementation illustrated in Fig. 9.4 uses dielectric rather than metallic mirrors for this purpose. To achieve high reflectance with multilayer films, alternating high and low refractive index layers with quarter-wave OPD can be used, as depicted in Fig. 9.9. Letting n_H and n_L represent the

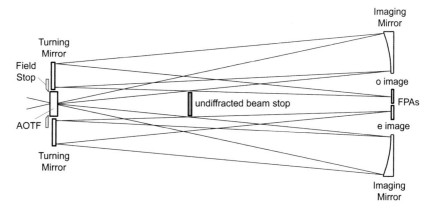

Figure 9.8 Polarimetric imaging spectrometer based on a TeO$_2$ AOTF.

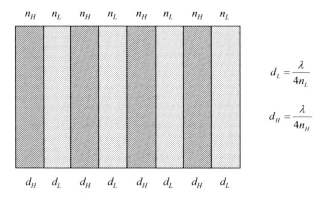

Figure 9.9 Multilayer dielectric thin film mirror.

refractive index of the respective layers, the transfer matrix for this design wavelength associated with N layers is given by

$$M = \begin{bmatrix} -\dfrac{n_H}{n_L} & 0 \\ 0 & -\dfrac{n_H}{n_L} \end{bmatrix}^N = \begin{bmatrix} \left(-\dfrac{n_H}{n_L}\right)^N & 0 \\ 0 & \left(-\dfrac{n_H}{n_L}\right)^N \end{bmatrix}. \tag{9.19}$$

Reflectance at the design wavelength is

$$\rho = \left[\frac{(n_H/n_L)^{2N} - 1}{(n_H/n_L)^{2N} + 1} \right]^2. \tag{9.20}$$

Peak reflectance increases with the number of layers, as shown in Fig. 9.10, where $n_H = 1.5$ and $n_L = 1.3$. As the wavelength deviates from the design wavelength, the film thickness no longer matches the quarter-wave criterion, and reflectance decreases. This is compensated for in a broadband dielectric mirror by varying the thickness of the layers slightly to exchange peak reflectance for spectral bandwidth. In a *linear variable filter* (LVF), however, the thickness of all layers is linearly graded across the filter such that the design wavelength varies linearly across it. A linearly varying dielectric bandpass spectral filter can be created by the structure shown in Fig. 9.11, which is simply a Fabry–Pérot cavity between two dielectric mirrors, with an optical transmission given by Eq. (9.9) using peak reflectance in Eq. (9.20). As the filter dimensions vary linearly across one dimension of the structure, the resonant wavelength and spectral resolution vary in proportion to them.

To create a wedge imaging spectrometer, an LVF is placed directly over a detector array, as depicted in Fig. 9.12, to create an FPA that

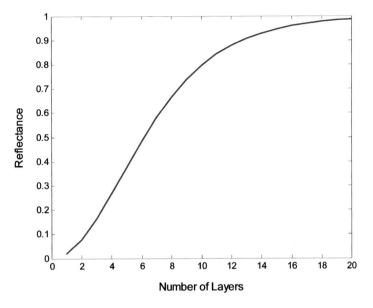

Figure 9.10 Peak reflectance as a function of the number of dielectric mirror layers.

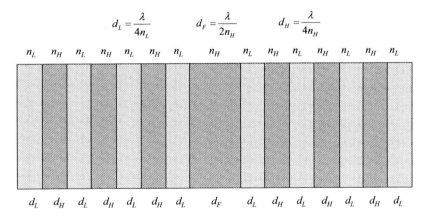

Figure 9.11 Multilayer dielectric thin film Fabry–Pérot interference filter.

provides spectral selectivity similar to that of a dispersive spectrometer (Elerding *et al.*, 1991). That is, it performs narrow spectral filtering with a center wavelength that increases linearly along one dimension of the FPA. When placed at the focal plane of a broadband imaging system, the resulting sensor captures spatial and spectral information in a fashion similar to a the way dispersive spectrometer captures this information. Unlike a dispersive imaging spectrometer, however, each row of a wedge spectrometer corresponds not only to a different spectral band, but also to a different field angle, and consequently, a different part of the scene.

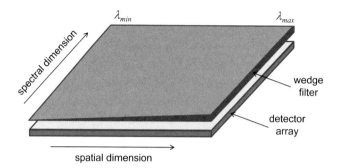

Figure 9.12 Linear variable filter layout on a FPA to create a wedge imaging spectrometer.

Figure 9.13 illustrates how such a wedge spectrometer can be employed in a pushbroom imaging mode to collect hyperspectral imagery. At any instant in time, the spectrometer captures a set of line images, each at a different position in the along-track direction, and a different spectral band, as indicated by the colors in the figure. These are linearly related according to the geometry of the LVF and detector array. As the aircraft moves forward, the image moves across the filtered detector array, and the frame-to-frame output of each row represents a single band of a hyperspectral image. However, these hyperspectral image bands are spatially shifted relative to each other with a shift that is linearly related to the difference in center wavelength. The time delay between adjacent spectral bands is GSD/v, where v is forward velocity. The relationship between the wedge spectrometer frame sequence and the ground-plane hyperspectral image coordinates is depicted in Fig. 9.14.

The intriguing features of a wedge spectrometer are the simplicity and compactness of its optical design. Theoretically, any scanning broadband imaging system could be retrofitted with a wedge FPA to turn it into an imaging spectrometer. LVFs with moderate spectral resolution and good linearity can be fabricated by using precisely controlled knife-edge masks in the filter deposition process. The primary disadvantage that has yet to be thoroughly addressed is the spatial misregistration between spectral bands that occurs due to a combination of time delay between spectral bands and unknown LOS disturbances. This misregistration can result in poor spectral fidelity and significantly degrade its utility as a remote sensor. The problem is exacerbated by spatial variations in the center wavelength due to LVF fabrication errors and off-axis wavelength shifts. Wedge imaging spectrometers have been developed and demonstrated from airborne platforms for both the VNIR and SWIR spectral ranges (Jeter and Blasius, 1999) and have been used as an atmospheric corrector for the NASA Earth Observer 1 (EO-1) satellite (Reuter *et al.*, 2001).

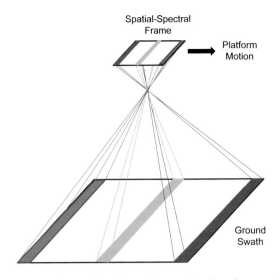

Figure 9.13 Pushbroom imaging with a wedge imaging spectrometer.

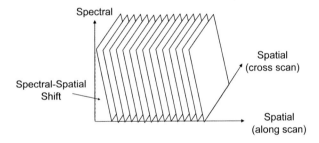

Figure 9.14 Data format of a wedge imaging spectrometer.

Nominal specifications for the airborne instruments are given in Tables 9.2 and 9.3.

9.4 Chromotomographic Imaging Spectrometer

The chromotomographic imaging spectrometer (CTIS) is designed to achieve the staring, fully multiplexed nature of an imaging FTS without resorting to interferometry. To do so, it captures spatial–spectral image content within either a succession of FPA frames or a single larger frame in an encoded form, and then reconstructs a hyperspectral data cube by means of digital processing to invert the spatial–spectral encoding. Two forms of the design approach are described in this section, the first based on a rotating direct-view prism (Mooney *et al.*, 1997), and the second on a set of gratings diffracting in multiple orders and directions (Descour and Dereniak, 1995). Discussion of design and operation begins with the former approach.

Table 9.2 Representative VNIR wedge imaging spectrometer specifications.

Characteristic	Specification
Detector material	Si
Effective focal length	2.74 cm
Detector width	18 μm
Angular sampling (IFOV)	0.66 mrad
FOV	19.4 deg
Spectral range	400 to 1000 nm
Spectral resolution	1.1 to 1.6 nm
Number of spectral channels	432
Digitization	12 bits

Table 9.3 Representative SWIR wedge imaging spectrometer specifications.

Characteristic	Specification
Detector material	InSb
Effective focal length	6.0 cm
Detector width	40 μm
Angular sampling (IFOV)	0.66 mrad
FOV	12.2 deg
Spectral range	1200 to 2400 nm
Spectral resolution	6.3 nm
Number of spectral channels	190
Digitization	14 bits

9.4.1 Rotating direct-view prism spectrometer

Consider the imaging system design depicted in Fig. 9.15, where an image is relayed through a direct-view prism onto a 2D FPA. A direct-view prism is composed of two or more prisms with opposing dispersion directions and precisely optimized material dispersion and apex angles, such that there is no deviation for some central wavelength λ_0 but some angular dispersion about this wavelength. It is similar to standard prisms used to construct dispersive spectrometers, except for this unique feature of not deviating the central wavelength. Unlike dispersive spectrometers, however, there is no slit in the intermediate image plane preceding the spectrometer; instead there is a field stop matched to the full FPA extent that passes the entire 2D field. Therefore, at a given instant in time for which the direct-view prism can be assumed stationary, the irradiance image at the FPA exhibits a large lateral chromatic aberration. That is, the image for each spectral band is laterally shifted in the dispersion direction proportional to the wavelength difference between the spectral band of interest and the center wavelength. Since the FPA spectrally integrates these shifted narrowband images, the result is an image that is heavily

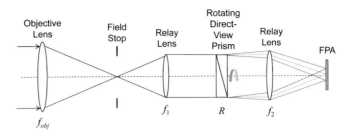

Figure 9.15 CTIS using a direct-view prism.

blurred in the dispersion direction by an amount dictated by the degree of lateral dispersion of the prism.

The CTIS operates by capturing a sequence of frames while a direct-view prism rotates about the optical axis through a 360-deg angular extent. Each captured frame exhibits a similar lateral chromatic aberration but in a different orientation, and the spatial–spectral content of a scene is encoded in the resulting frame sequence through this changing dispersion orientation. To understand how a hyperspectral image can be reconstructed from such data, an analogy is made to tomographic imaging processes used to reconstruct 3D objects from a sequence of 2D projections captured from different viewing directions. In the CTIS case, the object is a 3D hyperspectral data cube: two spatial dimensions and one spectral dimension. As depicted in Fig. 9.16, the effect of the direct-view prism is to produce an irradiance image on the FPA that is a projection of the hyperspectral data cube, where the projection angle is the inverse tangent of the product of the resolving power and f-number of the direct-view prism and the reimaging system. From the analysis in Chapter 7,

$$\frac{dx}{d\lambda} = R(f/\#). \tag{9.21}$$

In the figure, dispersion of the prism is oriented in a single spatial direction (x). As the direct-view prism rotates, the dispersion direction varies in orientation angle ϕ from the x axis, and the components of lateral dispersion dispersion, lateral in the two spatial axes become

$$\frac{dx}{d\lambda} = R(f/\#)\cos\phi \tag{9.22}$$

and

$$\frac{dy}{d\lambda} = R(f/\#)\sin\phi. \tag{9.23}$$

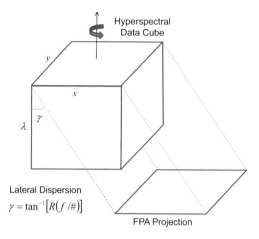

Figure 9.16 CTIS projection of hyperspectral data cube onto the FPA.

If $E_\lambda(x, y, \lambda)$ represents spectral irradiance of an image in the field stop, then integrated irradiance at the FPA for a prism orientation ϕ is given by

$$E(x, y, \phi) = \int_{\lambda_{min}}^{\lambda_{max}} E_\lambda(x, y, \lambda) * \delta\,[x - R(f/\#)\,(\lambda - \lambda_0)\,\cos\phi,\ y$$
$$- R(f/\#)\,(\lambda - \lambda_0)\,\sin\phi]\,d\lambda, \tag{9.24}$$

where $[\lambda_{min}, \lambda_{max}]$ is the spectral range of the combination of detector response and other limiting filters in the system, the asterisk represents a convolution operation, and $\delta(x, y)$ is a 2D Dirac delta function.

Instrument operation proceeds by rotating the direct-view prism completely about the optical axis and capturing a frame succession at a number of precisely measured angles ϕ, much like a tomographic imaging approach for 3D objects. According to the projection-slice theorem, the Fourier transform of a projection of a 3D object is a 2D slice through a 3D Fourier transform of the object. This is illustrated in Fig. 9.17. By virtue of this theorem, the 2D Fourier transform of each captured frame provides a set of data samples along a slice of the 3D Fourier transform $E_\lambda(u, v, w)$, where u and v correspond to standard spatial frequency axes and w represents the Fourier transform dimension corresponding to the spectral axis. Rotation of the direct-view prism is synonymous with rotation of $E_\lambda(u, v, w)$ about the w axis, causing the slice to sample different parts of the Fourier transform of the hyperspectral data cube. As the prism rotates through a full 360 deg, a solid region of the cube is sampled. Performing a 3D Fourier transform of this sampled region produces a hyperspectral data cube with spatial and spectral resolution proportionate to the extent of the sampled regions in the Fourier cube in respective directions.

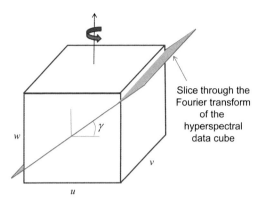

Figure 9.17 Representation of a CTIS projection as a slice in the Fourier transform domain.

The mathematical process for reconstructing a hyperspectral data cube from a sequence of image frames can be described starting with Eq. (9.24), which can be written in discrete form as

$$E(x, y, \phi) = \frac{\delta\lambda}{K} \sum_{k=1}^{K} E_\lambda(x, y, \lambda_k) * \delta[x - R(f/\#)(\lambda_k - \lambda_0)\cos\phi, y$$

$$- R(f/\#)(\lambda_k - \lambda_0)\sin\phi], \qquad (9.25)$$

where $\delta\lambda$ is the spectral bandwidth and k is the spectral index. The 2D Fourier transform of each frame is given by

$$E(u, v, \phi) = \Im_{2-D}\{E(x, y, \phi)\}$$

$$= \frac{\delta\lambda}{K} \sum_{k=1}^{K} E_\lambda(u, v, \lambda_k) e^{-i2\pi R(f/\#)(\lambda_k - \lambda_0)[\cos\phi\, u, + \sin\phi\, v]}. \qquad (9.26)$$

Given M frames at a set of prism orientations $\{\phi_m\}$, transformed data can be placed in vectors

$$\mathbf{y}(u, v) = \begin{bmatrix} E(u, v, \phi_1) & E(u, v, \phi_2) & \cdots & E(u, v, \phi_M) \end{bmatrix}^T. \qquad (9.27)$$

It is desired to reconstruct K bands corresponding to 2D spatial Fourier transforms of hyperspectral data cube bands, represented in vector form as

$$\mathbf{x}(u, v) = \begin{bmatrix} E(u, v, \lambda_1) & E(u, v, \lambda_2) & \cdots & E(u, v, \lambda_K) \end{bmatrix}^T. \qquad (9.28)$$

The relationship between the two vectors through Eq. (9.26) can be written as

$$\mathbf{y}(u, v) = \mathbf{A}(u, v)\,\mathbf{x}(u, v), \tag{9.29}$$

where $\mathbf{A}(u, v)$ is a matrix characterizing the orientation of Fourier samples. According to previously given relationships, $\mathbf{A}(u, v)$ is an $M \times K$ matrix with coefficients

$$A_{m,k}(u, v) = e^{-i2\pi R(f/\#)(\lambda_k - \lambda_0)[\cos \phi_m\, u,+\, \sin \phi_m\, v]}. \tag{9.30}$$

Reconstruction can be performed independently for each (u, v) sample by using either the pseudo-inverse,

$$\hat{\mathbf{x}}(u, v) = [\mathbf{A}^T(u, v)\,\mathbf{A}(u, v)]^{-1}\mathbf{A}^T(u, v)\,\mathbf{y}(u, v), \tag{9.31}$$

or some other numerical matrix inversion approach, such as singular value decomposition (SVD), that robustly handles rank deficiency issues with $\mathbf{A}(u, v)$. In summary, reconstruction of a hyperspectral data cube from a frame sequence involves three steps: (1) 2D Fourier transform of each frame, (2) estimation of the spectrum of each spatial frequency sample $\mathbf{x}(u, v)$ from the corresponding data vector $\mathbf{y}(u, v)$ using the system matrix $\mathbf{A}(u, v)$, and (3) 2D Fourier transform of each resulting spectral band back into the spatial domain. Each operation is a linear transformation.

When viewed in the Fourier domain as in Fig. 9.17, it is apparent that as ϕ cycles through 360 deg, there are conical sections oriented along the $+w$ and $-w$ directions, with a common apex at the origin for which no data samples are captured. This means that low-spatial-frequency scene information will be missing in a reconstructed hyperspectral data cube, or imagery will be high-pass filtered with a higher characteristic frequency for finer-scale spectral features. Compensation for this missing cone, depicted in Fig. 9.18, has been the subject of some attention. The presence of a field stop is beneficial in this regard, since one effect of high-pass filtering is to produce spurious image content that extends beyond the extent of a field stop, where true image content is known to be identically zero. Compensation methods address this by placing artificial data into the missing cone to force a reconstructed image to be identically zero outside the bounds of the field stop. A variant of the iterative Gerchberg–Saxton algorithm (Gerchberg and Saxton, 1972) shown in Fig. 9.19 is one numerical approach that can converge to such a solution.

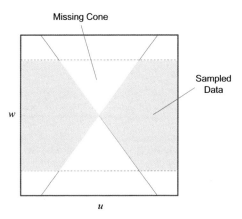

Figure 9.18 Depiction of the missing cone within a vertical slice of a cube Fourier transform.

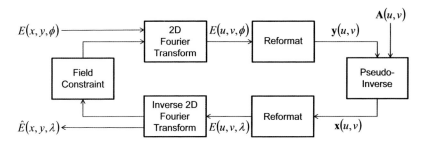

Figure 9.19 Gerchberg–Saxton reconstruction approach to compensate the missing cone.

9.4.2 Multi-order diffraction instrument

An alternate approach to a CTIS instrument captures projections of a hyperspectral data cube without using any moving parts. It accomplishes this by using a set of multi-order phase gratings at different orientations in place of a rotating direct-view prism, such that image irradiance simultaneously represents multiple hyperspectral data cube projections on a single, oversized, high-density FPA, as shown in Figs. 9.20 and 9.21. The primary advantage of this implementation is that it can be used as a snapshot, staring hyperspectral imager to capture rapid spectral events or true hyperspectral video. The primary disadvantages are that it requires a very large-format FPA with a detector count exceeding the product of hyperspectral frame size and number of spectral bands, and that calibration can be quite challenging. Because of the former issue, this technique is limited in practice to either VNIR operation, where extremely large format FPAs are available, or for applications where a limited image size is acceptable. To address the latter issue, the system matrix is often

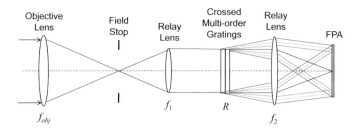

Figure 9.20 Chromotomographic imaging spectrometer using crossed, multi-order gratings.

Figure 9.21 Depiction of image irradiance from the CTIS approach employing multi-order gratings.

empirically estimated. Letting **x** represent the desired hyperspectral data cube as an unrastered vector and **y** be the unrastered data, they correspond through some system matrix **A** and additive noise **n** through the linear relationship

$$\mathbf{y} = \mathbf{A}\,\mathbf{x} + \mathbf{n}. \tag{9.32}$$

By illuminating the instrument with a number of sources with known spectral and spatial characteristics (usually monochromatic point sources) and measuring the corresponding instrument irradiance patterns, it is possible to estimate unknown system matrix **A** and its pseudo-inverse to reconstruct a hyperspectral image cube from any arbitrary data. For further details on CTIS reconstruction, the reader should consult the work of Descour and Dereniak (1995).

9.5 Summary

The field of imaging spectrometry remains a vibrant research area, with new instrument designs being developed for the more traditional dispersive and Fourier transform varieties, as well as other new designs that use innovative approaches for applications for which these traditional methods have limitations. Such applications currently include the need to perform snapshot, staring hyperspectral sensing to capture temporally resolved hyperspectral imagery of dynamic and moving objects. There are also applications for which the size, complexity, and cost of traditional imaging spectrometers are prohibitive; sensing methods that exchange hardware complexity for greater computation can ultimately be beneficial in an age when sophisticated data processing can be performed rapidly and inexpensively with embedded digital electronics. Nearly all of the designs presented in this chapter are proof-of-concept research instruments with a wide range of practical limitations that need to be addressed to take advantage of their unique capabilities. In order to address the needs of these emerging applications, further research and development is required to mature these approaches, as well as to pursue new design concepts not discussed or yet to be invented.

References

Chang, I. C., "Development of an infrared tunable acousto-optic filter," *Proc. SPIE* **131**, 2–10 (1978).

Daly, J. T., Bodkin, W. A., Schneller, W. J., Kerr, R. B., Noto, J., Haren, R., Eismann, M. T., and Karch, B. K., "Tunable narrow-band filter for LWIR hyperspectral imaging," *Proc. SPIE* **3948**, 104–115 (2000) [doi:10.1117/12.382145].

Descour, M. and Dereniak, E., "Computed tomography imaging spectrometer: experimental calibration and reconstruction results," *Appl. Optics* **34**, 4817–4826 (1995).

Elerding, G. T., Thunen, J. G., and Woody, L. M., "Wedge imaging spectrometer: application to drug and pollution law enforcement," *Proc. SPIE* **1479**, 380–392 (1991) [doi:10.1117/12.44546].

Gerchberg, R. W. and Saxton, W. O., "A practical algorithm for the determination of the phase from image and diffraction plane pictures," *Optik* **35**, 237–246 (1972).

Gittins, C. M. and Marinelli, W. J., "AIRIS multispectral imaging chemical sensor," *Proc. SPIE* **3383**, 65–74 (1998) [doi:10.1117/12.317637].

Glenar, D. A., Hillman, J. J., Saif, B., and Bergstralh, J., "Polaris-II: an acousto-optic imaging spectropolarimeter for ground-based astronomy," *Proc. SPIE* **1747**, 92–101 (1992) [doi:10.1117/12.138833].

Jeter, J. W. and Blasius, K. R., "Wedge spectrometer concepts for space IR remote sensing," *Proc. SPIE* **3756**, 211–222 (1999) [doi:10.1117/12. 366375].

Mooney, J. M., Vickers, V. E., An, M., and Brodzick, A. K., "High throughput hyperspectral imaging camera," *J. Opt. Soc. Am. A* **14**, 2951–2961 (1997).

Pedrotti, F. L. and Pedrotti, L. S., *Introduction to Optics*, Prentice-Hall, Englewood Cliffs, NJ (1987).

Reuter, D. C., McCabe, G. H., Dimitrov, R., Graham, S. M., Jennings, D. E., Matsamura, M. M., Rapchun, D. A., and Travis, J. W., "The LEISA/Atmospheric Corrector (LAC) on EO-1," *Proc. IEEE Geosci. Remote Sens.* **1**, 46–48 (2001).

Chapter 10
Imaging Spectrometer Calibration

Raw data that is output by an imaging spectrometer represents a quantity that is ideally related to incident pupil-plane spectral radiance as a function of spatial or angular position and spectral band. Pupil-plane spectral radiance is dependent on the material properties of the scene of interest. However, these data are integrated over an SRF that can be wavelength dependent, perturbed in spatial and spectral scale by optical distortions of the instrument, and also dependent on the effective optical transmission and stray radiance of the optical system as well as the responsivity and dark current of the FPA. Extracting quantitative spectral information about material properties from this raw data requires a method to determine the system response spectrally, spatially, and radiometrically. This method is referred to as spectrometer *calibration*. One theoretical construct for calibration is to precisely measure the characteristics of every system component and determine system response through an extensive model, such as that outlined in Chapter 7. This is generally complex and impractical. An alternate, more practical approach is to directly measure the full system response using sources with known characteristics. The basic methodology of this approach is described in this chapter.

10.1 Spectral Calibration

Spectral calibration refers to the determination of spectral band centers for all samples in a hyperspectral data cube. The case of a dispersive imaging spectrometer is treated first and involves characterizing the center wavelength for each element of a 2D FPA to correct any potential spatial–spectral distortion. To perform a thorough spectral calibration, the spectrometer can be illuminated by a monochromator that produces monochromatic illumination at controllable wavelengths. Theoretically, a monochromator can be stepped across an entire spectral range at a sufficiently fine spectral sampling such that the center wavelength at each detector pixel can be determined as the illuminating

417

wavelength producing a maximum response. Such a thorough spectral calibration is sometimes impractical and generally unnecessary because the spectral–spatial distortion of an imaging spectrometer is often a somewhat well-behaved function.

10.1.1 Spectral mapping estimation

Suppose that a spectrometer slit is uniformly illuminated sequentially or in combination by sources containing a distinct set of N narrowband spectral lines at wavelengths $\{\lambda_1, \lambda_2, \ldots, \lambda_N\}$. These spectral lines should be evident in a single frame of the imaging spectrometer data as a set of N lines nominally oriented along FPA rows. These lines can possess some curvature due to spatial–spectral distortion, as described earlier. Assume that the FPA rows are aligned with the direction of the slit such that, in a perfect system, i corresponds to the spatial coordinate and k corresponds to the spectral coordinate. These contours can be characterized by numerically extracting a number of (row, column) pairs (i, k) that correspond to the contour centers, or points of peak intensity. Assume that M such points are extracted for each spectral line, and let them be denoted as (i_{mn}, k_{mn}), where m and n are corresponding indices for the extracted point and spectral line source, respectively.

It is common to assume that spatial–spectral distortion conforms to some low-order, polynomial surface such as the biquadratic model, given by

$$\lambda(i, k) = ai^2 + bk^2 + cik + di + ek + f, \tag{10.1}$$

where $a, b, c, d, e,$ and f are the parameters that characterize spatial–spectral distortion. The least-squares method described is relevant to higher-order models, although the biquadratic model is generally adequate for well-designed imaging spectrometers. Based on this model, smile is a result of the first term in Eq. (10.1), which produces a spectral deviation across a row. Assume that the FPA consists of I rows (spatial positions) and K columns (spectral bands). According to the quadratic model, smile is given by

$$\frac{\lambda_{quadratic} - \lambda_{linear}}{\delta\lambda} = a\frac{I^2}{4}, \tag{10.2}$$

where $\delta\lambda$ is the spectral sampling. Terms c and d reflect FPA misalignment relative to the spectrometer, while terms $b, e,$ and f model quadratic dispersion.

Based on the series of measurements previously described, estimates for the unknown parameters in Eq. (10.1) are based on an overdetermined system of linear equations described by

$$
\begin{bmatrix}
i_{11}^2 & k_{11}^2 & i_{11}k_{11} & i_{11} & k_{11} & 1 \\
\vdots & \vdots & \vdots & \vdots & \vdots & \vdots \\
i_{M1}^2 & k_{M1}^2 & i_{M1}k_{M1} & i_{M1} & k_{M1} & 1 \\
i_{12}^2 & k_{12}^2 & i_{12}k_{12} & i_{12} & k_{12} & 1 \\
\vdots & \vdots & \vdots & \vdots & \vdots & \vdots \\
i_{M2}^2 & k_{M2}^2 & i_{M2}k_{M2} & i_{M2} & k_{M2} & 1 \\
\vdots & \vdots & \vdots & \vdots & \vdots & \vdots \\
i_{1N}^2 & k_{1N}^2 & i_{1N}k_{1N} & i_{1N} & k_{1N} & 1 \\
\vdots & \vdots & \vdots & \vdots & \vdots & \vdots \\
i_{MN}^2 & k_{MN}^2 & i_{MN}k_{MN} & i_{MN} & k_{MN} & 1
\end{bmatrix}
\begin{bmatrix} a \\ b \\ c \\ d \\ e \\ f \end{bmatrix}
=
\begin{bmatrix} \lambda_1 \\ \vdots \\ \lambda_1 \\ \lambda_2 \\ \vdots \\ \lambda_2 \\ \vdots \\ \lambda_N \\ \vdots \\ \lambda_N \end{bmatrix}.
\tag{10.3}
$$

Denoting this system of equations in matrix-vector form as

$$\mathbf{Ax} = \mathbf{b}, \tag{10.4}$$

the least-squares estimate for the biquadratic model parameters is given by the pseudo-inverse

$$\mathbf{x} = [\mathbf{A}^T\mathbf{A}]^{-1}\mathbf{A}^T\mathbf{b}. \tag{10.5}$$

The spectral band centers are then given as a function of FPA indices (i, k) according to Eq. (10.1).

10.1.2 Spectral calibration sources

A variety of illumination sources are used for spectral calibration measurements. A monochromator, such as the grating instrument depicted in Fig. 10.1, is often used to select specific spectral lines from a broadband illumination source. A monochromator is basically a spectrometer with an additional slit in place of the FPA that passes only a single narrow band of incident light. In the Twyman–Green design illustrated in Fig. 10.1, the grating is mounted on a precision rotation stage to tune the filtered wavelength. The output of the monochromator is then relayed to the spectrometer either via slit-to-slit imaging or by some other means, and the grating is tilted to precisely control the illumination wavelength. To effectively use a monochromator, it must be accurately calibrated. By automating control of the monochromator and mounting an imaging spectrometer on a rotating stage, it is practical to fully characterize the SRF for each spectral band and spatial location.

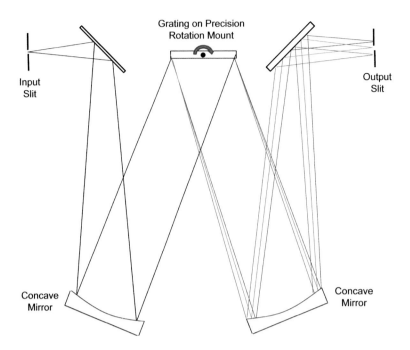

Figure 10.1   Grating monochromator.

An alternate spectral calibration approach is to use gas lamps, which emit radiation at a number of specific spectral lines associated with their atomic and molecular structure. Figure 10.2 illustrates the spectral output for argon (Ar), krypton (Kr), mercury–neon (Hg–Ne), and xenon (Xe) gas discharge lamps that can be used for the VNIR/SWIR spectral region. Finally, lasers can also be used as monochromatic illumination sources. He–Ne (633 nm), Ar (488 and 514 nm), and GaAs (940 nm, various) lasers are commonly used. Although Xe exhibits some emission lines extending into the very short-wavelength end of the MWIR spectral region, sources with fine spectral lines are much more limited in the MWIR and LWIR spectral range. One approach to circumvent this source limitation is to illuminate the system with a blackbody source through a material that possesses fine spectral absorption lines. Transparent polymer films used for packaging have empirically been found to work well over this spectral region.

10.1.3 Spectral-response-function estimation

Two basic methods can be used to estimate the SRF for an imaging grating spectrometer. First, the response function can be determined directly by continuously tuning a sufficiently narrowband monochromator across the spectral range of a spectrometer, while concurrently reading out frames of data for each successive wavelength sample. Measured spectral bandwidth

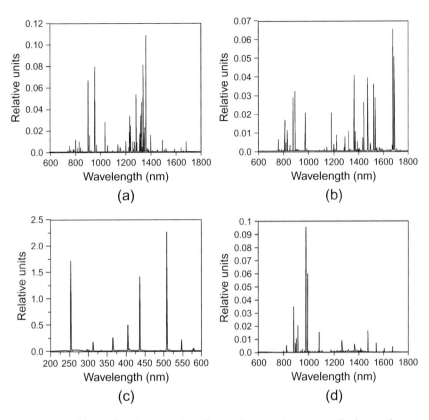

Figure 10.2 Normalized spectral radiance from various gas discharge lamps: (a) argon (Ar), (b) krypton (Kr), (c) mercury–neon (Hg–Ne), and (d) xenon (Xe) (Newport Corporation).

corresponds to the convolution of the spectrometer SRF and that of the monochromator, necessitating that the latter be sufficiently small. An alternate approach is to illuminate the system with a source containing fine spectral lines, and to numerically determine spectral bandwidth based on an assumption of the shape of the SRF that provides the best fit to the measured data. A Gaussian line shape is typically assumed, and nonlinear least-squares estimation algorithms are generally employed to perform such a fit. Both approaches can be applied at a few spatial–spectral locations across the FPA, and spectral resolution can be estimated as a function of FPA indices using the modeling approach described in Section 10.1.1 for a biquadratic model similar to Eq. (10.1).

10.1.4 Spectral calibration example

As an example of the spectral calibration method, consider a VNIR spectrometer instrument assembled for exploratory research in hyperspectral change detection, composed of a Headwall Photonics

Hyperspec VS-25 imaging spectrometer of the Offner grating design, a Dalsa Pantera TF 1M60 1024 × 1024-Si CCD camera with 12-μm detectors, and an Navitar 50-mm lens operated at $f/4$ (Meola, 2006). The VS-25 has a spectral range of 400 to 1000 nm over 6-mm dispersion using a holographic diffraction grating. The quoted spectral resolution is 2 nm using a 12-μm slit, and keystone and smile distortions are less than 0.1% with the system designed at $f/2$. To perform wavelength calibration, the slit was illuminated by a He–Ne laser and Ar, Kr, and Hg–Ne lamps, and various points were extracted from digitally acquired frames to estimate the biquadratic wavelength fit coefficients described in Section 10.1.1. Figure 10.3 shows frames captured for each source. Since the intensity of the emission lines is multiplied by the effective responsivity of the system, one must be very careful when performing this calibration process to properly associate lines in the captured frame data with those of the truth spectra in Fig. 10.2.

The very low spatial–spectral distortion of the VS-25 spectrometer is clearly evident by the lack of any curvature in the spectral lines in Fig. 10.3. In the He–Ne laser case, there appears to be a ghost of the 633-nm line at

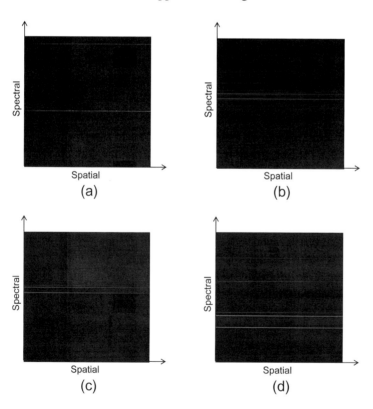

Figure 10.3 Captured VS-25 spectrometer frames for various wavelength calibration sources: (a) He–Ne laser, (b) Ar lamp, (c) Kr lamp, and (d) Hg–Ne lamp.

longer wavelengths; this is actually second-order diffraction and is visible because no order-sorting filter was included on the FPA. Compensation for this is discussed later. These second orders also exist in the lamp data but are not as bright. The results of the biquadratic fit in this case are $a = b = c = 0, d = 0.0016, e = 1.1933$, and $f = 105.7009$, such that mapping becomes

$$\lambda(i, k) = 0.0016\,i + 1.1933\,k + 105.7009. \tag{10.6}$$

The a, b, and c coefficients are all less than 1×10^{-6} and therefore are considered to be effectively zero. This affirms that smile and keystone are at an immeasurably small level. The presence of a nonzero d coefficient indicates a small rotational misalignment of the FPA to the spectral and spatial axes of the spectrometer. In fact, the misalignment is given by

$$\delta\theta = \tan^{-1}(d/e), \tag{10.7}$$

which in this case equals 0.0768 deg. To make subsequent data processing easier, this misalignment can be corrected by numerically derotating collected frames counterclockwise by this measured amount using simple bilinear interpolation. After doing so, the wavelength mapping becomes simply

$$\lambda(i, k) = 1.1933\,k + 106.5198. \tag{10.8}$$

Using these same measurements, the measured spectral line shape can be extracted for each spectral line (i.e., column trace of the data frame) at several spatial positions. Modeling the SRF as a Gaussian, this measurement is assumed to correspond to the convolution of the SRF with the actual line shape of the spectral line, which is known from reference data or a high-resolution, point spectrometer measurement. A nonlinear least-squares algorithm is then used to determine the spectral bandwidth of the Gaussian SRF that produces the best fit between the measured data and the predicted result. This is repeated for all of the measurements, and the measurements are fit to a quadratic model as in Eq. (10.1) to characterize spectral bandwidth as a function of FPA row and column. The SRF is a dimensionless function that peaks at 1.

10.2 Radiometric Calibration

Radiometric calibration is defined here as a process for estimating the pupil-plane spectral radiance corresponding to each measured FPA data value. Before investigating the radiometric calibration of imaging spectrometers, we first consider a conventional imaging system employing a 2D FPA. Such calibration requires sources whose spectral radiance is precisely known and controlled. It is shown that there are limits to how well the nonuniformity of 2D FPAs can be corrected, leaving a residual nonuniformity that acts as spatial noise to broadband imaging systems and spatial–spectral noise for dispersive spectrometers. The calibration methodology is extended for both dispersive and Fourier transform imaging spectrometers.

10.2.1 Nonuniformity correction of panchromatic imaging systems

The relationships in Chapter 6 express the detected photoelectrons for a given FPA element of a conventional imaging system as a linear function of the spatially and spectrally integrated pupil-plane spectral radiance. That is,

$$N(i, j) = g\, L(i, j) + d, \qquad (10.9)$$

where $N(i, j)$ represents the detected photoelectrons for the (i, j) spatial FPA element, $L(i, j)$ represents corresponding spatially and spectrally integrated radiance, g is the effective gain coefficient for the system, and d is the effective dark offset. Typically, $N(i, j)$ represents raw image data in arbitrary units such as digitized data numbers, but the relationship in Eq. (10.9) still applies. In this case, radiometric calibration simply involves determination of the scalar gain and offset coefficients by measuring the detected signal for two known spectral radiance levels. This is referred to as two-point radiometric calibration.

Real FPAs exhibit some level of nonuniformity, which can be modeled as a random variance in gain and offset across the FPA. Nonuniformity can also exist due to the irradiance roll-off function of the imaging system. Compensating for such nonuniformity requires a two-point calibration performed on a detector-by-detector basis. Additionally, FPAs generally exhibit some level of response nonlinearity, which is shown to cause residual spatial nonuniformity even with accurate two-point radiometric calibration (Eismann and Schwartz, 1997).

To mathematically investigate the effectiveness of two-point radiometric calibration in the presence of FPA nonuniformity and nonlinearity, consider a quadratic response model of the form

$$N(i, j) = d(i, j) + g(i, j)\, L(i, j) + c(i, j)\, L(i, j)\, [\, L_{max} - L(i, j)\,] + n(i, j),$$
$$\qquad (10.10)$$

where $N(i, j)$ and $L(i, j)$ have been previously defined, $d(i, j)$ is the dark offset, $g(i, j)$ is the gain response, $c(i, j)$ is the nonlinearity parameter,

$n(i, j)$ is the detector noise realization, and L_{max} is the saturation radiance. The equation is written in the form shown so that there is no deviation from linear response [$c(i, j) = 0$] at radiance levels of zero and L_{max}, as shown in Fig. 10.4.

Two-point calibration ignores nonlinearity and noise components, which are assumed unknown, and determines gain and offset parameters for each detector element based on known (and usually spatially uniform) illumination at two radiance levels. Letting $L_1(i, j)$ and $L_2(i, j)$ represent calibration radiance levels, the corresponding raw image data are given by

$$N_1(i, j) = d(i, j) + g(i, j) L_1(i, j) + c(i, j)L_1(i, j)[L_{max}-L_1(i, j)] + n_1(i, j)$$
$$(10.11)$$

and

$$N_2(i, j) = d(i, j) + g(i, j) L_2(i, j) + c(i, j) L_2(i, j) [L_{max}-L_2(i, j)] + n_2(i, j).$$
$$(10.12)$$

Note that gain and offset patterns are assumed to be fixed in time, while noise is assumed to be uncorrelated between image frames; that is, noise realization differs for the two calibration measurements, while gain and offset patterns are the same.

Applying two-point calibration involves estimating detector element gains and offsets based on Eqs. (10.11) and (10.12), under the assumption that $c(i, j), n_1(i, j)$, and $n_2(i, j)$ are all zero, and then applying these to invert Eq. (10.10) for spectral radiance corresponding to a measured raw image. Under the given assumptions, Eqs. (10.11) and (10.12) are solved to obtain

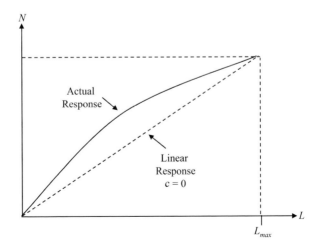

Figure 10.4 Characterization of detector nonlinearity.

estimates of the offset pattern

$$\hat{d}(i, j) = \frac{L_2(i, j)N_1(i, j) - L_1(i, j)N_2(i, j)}{L_2(i, j) - L_1(i, j)}, \tag{10.13}$$

and the gain pattern

$$\hat{g}(i, j) = \frac{N_2(i, j) - N_1(i, j)}{L_2(i, j) - L_1(i, j)}. \tag{10.14}$$

Applying the two-point calibration results in a transformation from the raw image frame to the calibrated spectral radiance frame given by

$$\hat{L}(i, j) = \frac{N(i, j) - \hat{d}(i, j)}{\hat{g}(i, j)}. \tag{10.15}$$

To investigate the effectiveness of two-point radiometric calibration, Eqs. (10.11) and (10.12) are inserted into Eqs. (10.13) and (10.14) to determine estimated offset and gain as a function of true gain and offset. Then these results are inserted into Eq. (10.15) to compute estimated radiance as a function of true radiance. From this, estimates of residual error can be made. Based on Eqs. (10.11) through (10.14), the offset estimate is given by

$$\hat{d}(i, j) = d(i, j) + c(i, j) L_1(i, j) L_2(i, j) + \frac{L_2(i, j) n_1(i, j) - L_1(i, j) n_2(i, j)}{L_2(i, j) - L_1(i, j)}, \tag{10.16}$$

and the gain estimate is given by

$$\hat{g}(i, j) = g(i, j) + c(i, j) [L_{max} - L_2(i, j) - L_1(i, j)] + \frac{n_2(i, j) - n_1(i, j)}{L_2(i, j) - L_1(i, j)}. \tag{10.17}$$

Inserting Eqs. (10.16) and (10.17) into Eq. (10.15) gives

$$\hat{L}(i, j)$$
$$= \frac{g(i, j) L(i, j) + c(i, j) [L(i, j) \{ L_{max} - L(i, j) \} - L_1(i, j) L_2(i, j)] + n(i, j) + \frac{L_2(i,j) n_1(i,j) - L_1(i,j) n_2(i,j)}{L_2(i,j) - L_1(i,j)}}{g(i, j) + c(i, j) [L_{max} - L_2(i, j) - L_1(i, j)] + \frac{n_2(i,j) - n_1(i,j)}{L_2(i,j) - L_1(i,j)}}$$
$$= \frac{L(i, j) + \frac{c(i,j)}{g(i,j)} [L(i, j) \{ L_{max} - L(i, j) \} - L_1(i, j) L_2(i, j)] + \frac{n(i,j)}{g(i,j)} - \frac{L_2(i,j) n_1(i,j) - L_1(i,j) n_2(i,j)}{g(i,j) [L_2(i,j) - L_1(i,j)]}}{1 + \frac{c(i,j)}{g(i,j)} [L_{max} - L_2(i, j) - L_1(i, j)] + \frac{1}{g(i,j)} \frac{n_2(i,j) - n_1(i,j)}{L_2(i,j) - L_1(i,j)}}. \tag{10.18}$$

To further investigate Eq. (10.18), some assumptions are made concerning characteristics of the parameters involved. First, it is assumed that true gain $g(i, j)$ consists of a spatially fixed component g_0 and a spatially random component g_1,

$$g(i, j) = g_0(1 + g_1), \qquad (10.19)$$

with g_1 modeled as a zero-mean, uncorrelated, normally distributed random process with a standard deviation R_{NU}, the relative gain nonuniformity across the FPA, denoted by

$$g_1 \sim N(0, R_{NU}^2). \qquad (10.20)$$

Under the approximation that $g_1 \ll 1$,

$$\frac{n(i, j)}{g(i, j)} \approx \frac{n(i, j)}{g_0} \sim N(0, \text{NESR}^2), \qquad (10.21)$$

$$\frac{n_1(i, j)}{g(i, j)} \approx \frac{n_1(i, j)}{g_0} \sim N(0, \text{NESR}_1^2), \qquad (10.22)$$

and

$$\frac{n_2(i, j)}{g(i, j)} \approx \frac{n_2(i, j)}{g_0} \sim N(0, \text{NESR}_2^2). \qquad (10.23)$$

In some cases, NESR during calibration and scene measurements stays the same, but a more general model is maintained here to accommodate situations where frame averaging is used during calibration to reduce NESR. Also, spatial dependence of the nonlinearity is ignored, such that $c(i, j) = c$, and

$$\frac{c(i, j)}{g(i, j)} = \frac{c}{g_0(1 + g_1)} \approx \frac{c}{g_0}(1 - g_1). \qquad (10.24)$$

Given these assumptions, the third term in the denominator of Eq. (10.18) can be ignored, and an estimate can be written as

$$\hat{L}(i, j) = \frac{L(i, j) + \frac{c}{g_0}(1 - g_1)[L(i, j)\{L_{max} - L(i, j)\} - L_1(i, j)L_2(i, j)] + \frac{n(i,j)}{g_0} - \frac{L_2(i,j)n_1(i,j) - L_1(i,j)n_2(i,j)}{g_0[L_2(i,j) - L_1(i,j)]}}{1 + \frac{c}{g_0}[L_{max} - L_2(i, j) - L_1(i, j)]}.$$

$$(10.25)$$

Assuming that the second term in the denominator of Eq. (10.25) is much less than one,

$$
\hat{L}(i, j) =
\left|
\begin{array}{l}
L(i, j) + \dfrac{c}{g_0}(1 - g_1)\left[L(i, j)\{L_{max} - L(i, j)\} - L_1(i, j)L_2(i, j)\right] \\[2mm]
+ \dfrac{n(i, j)}{g_0} - \dfrac{L_2(i, j)n_1(i, j) - L_1(i, j)n_2(i, j)}{g_0[L_2(i, j) - L_1(i, j)]}
\end{array}
\right|
$$
$$
\cdot \left| 1 - \dfrac{c}{g_0}[L_{max} - L_2(i, j) - L_1(i, j)] \right|.
\tag{10.26}
$$

Keeping only first-order terms of the product,

$$
\hat{L}(i, j) = L(i, j) + \dfrac{c}{g_0}(1 - g_1)\,[L(i, j)\{L_{max} - L(i, j)\} - L_1(i, j)L_2(i, j)
$$
$$
- L(i, j)\{L_{max} - L_1(i, j) - L_2(i, j)\}]
$$
$$
+ \dfrac{n(i, j)}{g_0} - \dfrac{L_2(i, j)n_1(i, j) - L_1(i, j)n_2(i, j)}{L_2(i, j) - L_1(i, j)}.
\tag{10.27}
$$

This can be written in four terms:

$$
\hat{L}(i, j) = L(i, j)
$$
$$
+ \dfrac{c}{g_0}[L(i, j)\{L_1(i, j) + L_2(i, j) - L(i, j)\} - L_1(i, j)L_2(i, j)]
$$
$$
- g_1\dfrac{c}{g_0}[L(i, j)\{L_1(i, j) + L_2(i, j) - L(i, j)\} - L_1(i, j)L_2(i, j)]
$$
$$
+ \dfrac{n(i, j)}{g_0} - \dfrac{L_2(i, j)\,n_1(i, j) - L_1(i, j)\,n_2(i, j)}{L_2(i, j) - L_1(i, j)}.
\tag{10.28}
$$

The first term is the true radiance. The second is the mean calibration error due to FPA nonlinearity. The third is residual spatial nonuniformity, which is seen to be the result of raw FPA nonuniformity and nonlinearity. Finally, the fourth term is random detector noise converted to pupil-plane radiance units, including contributions from calibration source measurements.

Consider the mean calibration error

$$
\Delta L(i, j) = \dfrac{c}{g_0}[L(i, j)\{L_1(i, j) + L_2(i, j) - L(i, j)\} - L_1(i, j)L_2(i, j)].
\tag{10.29}
$$

For the moment, let $L_1(i, j) = 0$, $L_2(i, j) = L_{max}$, and $L(i, j) = L_{max}/2$. Then,

$$
\Delta L = \dfrac{cL_{max}^2}{4g_0}.
\tag{10.30}
$$

This represents the worst-case mean error according to the picture in Fig. 10.4. By defining relative nonlinearity as

$$R_{NL} = \frac{\Delta L}{L_{max}} \qquad (10.31)$$

for this worst-case situation, the nonlinearity coefficient can be solved from Eq. (10.30) to be

$$c = \frac{4g_0 R_{NL}}{L_{max}}. \qquad (10.32)$$

By inserting Eq. (10.32) into Eq. (10.29), the mean error for spatially uniform illumination (removing the spatial dependence of L_1 and L_2) is then given as

$$\Delta L = \frac{4R_{NL}}{L_{max}}[L(L_1 + L_2 - L) - L_1 L_2]. \qquad (10.33)$$

The spatial noise of an imaging system is defined as the random variation across the spatial dimension of a calibrated FPA and is the combination of the last two terms of Eq. (10.28). Letting the illumination terms be spatially uniform, the variance of these terms is

$$\sigma_L^2 = R_{NU}^2 \frac{c^2}{g_0^2}[L(L_1 + L_2 - L) - L_1 L_2]^2 + \text{NESR}^2 + \frac{L_2^2 \text{NESR}_1^2 - L_1^2 \text{NESR}_2^2}{(L_2 - L_1)^2}, \qquad (10.34)$$

where nonuniformity and noise terms are assumed to be statistically independent. Substituting Eq. (10.32) into Eq. (10.34) leads to the final result,

$$\sigma_L^2 = \frac{16R_{NU}^2 R_{NL}^2}{L_{max}^2}[L(L_1 + L_2 - L) - L_1 L_2]^2 \\ + \text{NESR}^2 + \frac{L_2^2 \text{NESR}_1^2 - L_1^2 \text{NESR}_2^2}{(L_2 - L_1)^2}. \qquad (10.35)$$

That is, noise is the combination of three terms: (1) a residual spatial nonuniformity due to the combination of FPA nonuniformity and nonlinearity, (2) the inherent radiance-referred noise floor of the sensor, often denoted as temporal noise, and (3) additive noise introduced by calibration-source measurements. When combined with the other noise terms responsible for temporal noise, we only include the first and third terms in Eq. (10.35) as the modeled spatial noise component that is added to the temporal noise, as described in Chapter 6.

It is interesting to understand the nature of spatial noise as a function of various parameters. Consider an imaging system with an inherent SNR at the maximum illumination level (L_{max}) of 500; that is, NESR = $L_{max}/500$. Assume that the NESR of the calibration measurements is one-fourth of this inherent system NESR, which is typically achieved by coadding frames during collection of calibration-source data. Finally, let L_1 = $L_{max}/4$ and $L_2 = 3L_{max}/4$. Based on these parameters, Fig. 10.5 illustrates the standard deviation of spatial noise relative to L_{max} for 5% relative nonuniformity in gain response as a function of illumination level for several values of relative nonlinearity. Note that spatial noise is minimized at calibration-source illumination levels, as expected, and increases as incident radiance differs from that of the calibration sources.

Because of the inability of two-point calibration to eliminate effects of detector nonuniformity due to detector nonlinearity, the use of multipoint calibration is sometimes considered. Based on the quadratic model used in this analysis, a three-point calibration would provide perfect results from the perspective of eliminating the first term in Eq. (10.35). However, it would also add another noise term as well as some complexity to the calibration process. Also, deviations of detector response from the quadratic (i.e., fourth- and higher-order terms) must also be considered and will cause some residual nonuniformity. More calibration points can be used to mitigate these higher terms, but each comes at the expense of more

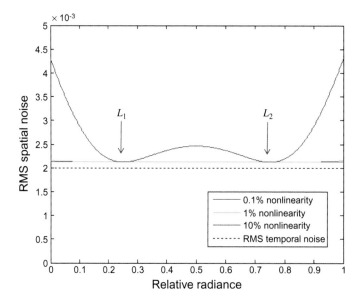

Figure 10.5 Spatial noise as a function of illumination level.

additive noise and greater complexity. Therefore, multipoint calibration is typically only used when nonlinearity is sufficiently large that residual nonunformity based on two-point calibration greatly exceeds the noise terms in Eq. (10.35). Otherwise, the preferred approach to reducing spatial noise is to use detectors that are very linear over the expected illumination range with low-gain nonuniformity.

10.2.2 Radiometric calibration sources

Before applying the previously described radiometric calibration analysis to the case of imaging spectrometers, it can be helpful to understand the types of sources typically used to perform radiometric calibration. When selecting an appropriate source, it is essential to establish whether the goal of calibration for the application of interest is merely to reduce the effects of spatial noise in the imagery, or to produce an image product that maintains absolute radiometric traceability to pupil-plane radiance. The first is referred to as *relative calibration*, while the latter is called *absolute calibration*. Suppose that the relative spatial characteristics of a calibration-source illumination $L_1(i, j)$ and $L_2(i, j)$ are known and perhaps even spatially uniform, but the absolute level in radiance units is unknown. Applying two-point calibration via Eqs. (10.13), (10.14), and (10.15) results in an image $L(i, j)$ that is corrected to the extent possible in terms of residual nonuniformity, but is related to the true pupil-plane radiance by some unknown multiplicative constant. This is the relative calibration case, which requires sources that are either spatially uniform or spatially calibrated but not necessarily radiometrically calibrated. On the other hand, absolute calibration requires sources for which illumination is precisely known in radiometric terms, traceable to some absolute reference. The description of various calibration sources implies the need for absolute calibration, but extension to the simpler case of relative calibration should be clear.

Absolute calibration requires a source that is traceable to a reference standard. In the VNIR/SWIR spectral region, deuterium and quartz–tungsten (QTH) sources are available whose spectral irradiance is calibrated to National Institute of Standards and Technology (NIST) standards. The spectral irradiance characteristics of some source examples are shown in Fig. 10.6. The problem with directly using such a source for calibrating an imaging instrument, however, is that its spatial illumination pattern is relatively nonuniform and directional, meaning that L_1 and L_2 will still be a function of (i, j). Diffuse illumination from such as source can be obtained by viewing scattered radiation from a rough, reflective material such as Spectralon® (Labsphere, Inc.), which has high reflectance and is also a good approximation to a Lambertian scatterer. With the arrangement given in Fig. 10.7, spectral radiance at a measurement point

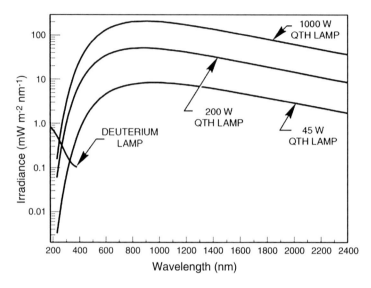

Figure 10.6   Irradiance of various calibrated lamps (Newport Corporation).

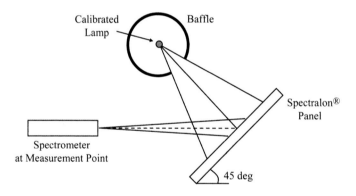

Figure 10.7   Diffuse panel illumination geometry for spectrometer calibration.

is given by

$$L(\lambda) = \frac{\rho_s(\lambda)\, E_s(\lambda)}{\sqrt{2\pi}}, \qquad (10.36)$$

where $E_s(\lambda)$ is the calibrated spectral irradiance of the source at illumination distance d, and $\rho_s(\lambda)$ is the directional hemispherical reflectance of the Spectralon® panel (Zadnik *et al.*, 2004).

   While Eq. (10.36) transfers the calibrated-source spectral irradiance to a diffuse spectral radiance by means of a simple, measurable optical setup, it only holds over a relatively small region of the panel for which illumination distance d matches the calibration conditions. Away from this central point, irradiance falls off according to the square of the source-to-panel distance. Therefore, this arrangement does not produce the spatially uniform spectral radiance desired. When placed near the aperture of an imaging system, uniform irradiance of the image plane effectively results due to the Lambertian nature of the panel. Therefore, such an arrangement can be effective for relative calibration if the FOV of the sensor is narrow. However, this arrangement is less effective for absolute calibration because the spatially integrated radiance over the entrance pupil is not directly measurable and is also limited by the temporal stability of the lamp between measurements.

The use of an *integrating sphere* solves this problem, as it produces illumination that is both spatially uniform and Lambertian. As shown in Fig. 10.8, an integrating sphere consists of a sphere that is coated with a reflective, diffusely scattering material and has a hole for an illuminating source and another for the output. Baffles are also used to ensure that the output radiation is the result of multiple scattering off the walls of the source. In this way, the source provides the desired uniform, diffuse illumination, even for nonuniform or directional source illumination. In these regards, it is close to an ideal calibration source. A problem with the integrating sphere, however, is that the relationship between spectral radiance of the output and spectral radiance of the source is more complex. Nominally, this relationship is given by

$$\tau(\lambda) = \frac{L(\lambda)}{L_s(\lambda)} = \frac{\rho_s(\lambda)A_e/A_s}{1 - \rho_s(\lambda)\left(1 - A_p/A_s\right)}, \tag{10.37}$$

where $L_s(\lambda)$ is the source spectral radiance at the input port of the integrating sphere, $L(\lambda)$ is the source spectral radiance at the output port of the integrating sphere, A_e is the exit area of the integrating sphere, A_s is the surface area of the integrating sphere, A_p is the sum of all port areas, and $\rho_s(\lambda)$ is the directional hemispherical reflectance of the inside coating of the sphere. The transmission of an integrating sphere can be fairly low since the exit area is typically a small fraction of the total sphere surface area. Therefore, large input irradiance is needed to overcome this low transmission, and it is sometimes difficult to achieve an output irradiance that is high enough to cover the dynamic range of the instrument being calibrated. This is the price paid for the uniformity that an integrating sphere provides.

Due to the uncertainty of some of the parameters in Eq. (10.37), integrating spheres are usually calibrated by a method called *transfer calibration* from a NIST traceable source. To understand this basic approach, consider an uncalibrated point spectrometer, with a stable and linear response, that measures the signal associated with the measurement point in the arrangement given in Fig. 10.7 with the calibrated source. This

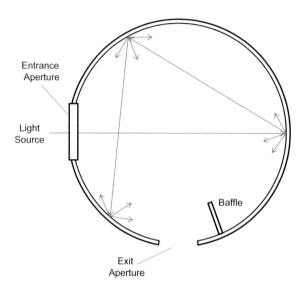

Figure 10.8   Integrating sphere.

signal is then associated with a spectral radiance that is traceable in an absolute sense. This same spectrometer is then used to measure the output of an integrating sphere with a stable illumination source. By virtue of the linearity of the spectrometer response between illumination levels of the two sources, as well as stability between measurements, spectral radiance of the integrating sphere output port can be directly computed by

$$L(\lambda) = \frac{N_s(\lambda)}{N_r(\lambda)} L_r(\lambda), \tag{10.38}$$

where $N_s(\lambda)$ is the dark-subtracted spectrometer measurement for the integrating sphere, $N_r(\lambda)$ is the dark-subtracted spectrometer measurement for the reference geometry in Fig. 10.7, and $L_r(\lambda)$ is the reference spectral radiance computed using Eq. (10.36).

In the MWIR and LWIR spectral regions, blackbody calibration sources are typically used for radiometric calibration as opposed to lamp sources. The most accurate blackbody source is a cavity blackbody, which consists of a metal spherical cavity with an exit port typically coated with a black coating and heated to a known uniform temperature. Spectral radiance of the output is then given by the blackbody spectral radiance function for a known temperature. Because the exit port of a blackbody source is much smaller than the total size of the source, cavity blackbody sources are not directly useful for calibrating systems with relatively large aperture sizes. Several approaches are used to resolve this limitation. First, more-compact cavity sources can be produced by using a conical (as opposed to spherical) structure, such as that designed for the GIFTS sensor described in Chapter 8 and depicted in Fig. 10.9. Second, the source can be placed at an object plane of a collimator and projected into the imaging system. This is effective for a high-f/number collimator with a well-known spectral transmission but results in nonuniform spectral radiance for more-compact projection optics. The final alternative is to use a flat-plate blackbody, which consists of a temperature-controlled surface coated with a very highly emissive paint. Such sources are available with nominally 0.1-K temperature uniformity and stability, with coatings that achieve an emissivity in the range of 0.96 to 0.99. The surfaces of such sources are sometimes corrugated to increase their effective emissivity by the cavity effect.

10.2.3 Dispersive imaging spectrometer calibration

The principles of radiometric calibration are directly extended to the calibration of dispersive imaging spectrometers. A step-by-step methodology is presented so that imaging spectrometer output can be related to absolute pupil-plane spectral radiance as a function of spatial position and wavelength. This methodology is representative of methods reported in the literature; however, variations are used depending on the specific calibration goals, the remote sensing application, the nature of the imaging spectrometer, and the available calibration hardware.

Radiometric calibration is performed in a manner similar to nonuniformity correction of a conventional imaging system, with a few important differences. First, spatial index j in all of the equations is replaced by spectral index k. Calibration sources $L_1(\lambda)$ and $L_2(\lambda)$ are assumed to be uniform in the spatial direction along the slit (index i) but

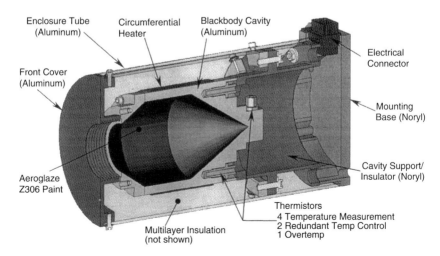

Figure 10.9 Cavity blackbody radiance source designed for the GIFTS sensor (Best *et al.*, 2005).

always exhibit some radiance dependence with wavelength. Because of spatial–spectral distortion, a row (constant k) cannot actually correspond to a constant wavelength. Instead, the center wavelength can be dependent on both FPA coordinates, as determined by the spectral calibration previously shown. Therefore, true spectral radiance for calibration sources as a function of FPA coordinates is determined by the spectral profile of the source, as mapped to FPA spectral mapping estimates.

The preferred method for radiometric calibration is to use a calibrated integrating sphere (in the VNIR/SWIR) and blackbody source (in the MWIR/LWIR) that uniformly flood illuminates the entire entrance aperture. Several calibration frames, on the order of 20 to 50, can be collected and averaged to minimize noise in the calibration source measurement. Accounting for spectral calibration in determining $L_1(i, k)$ and $L_2(i, k)$, the spectrometer is calibrated according to

$$\hat{d}(i, k) = \frac{L_2[\lambda(i, k)] N_1(i, k) - L_1[\lambda(i, k)] N_2(i, k)}{L_2[\lambda(i, k)] - L_1[\lambda(i, k)]}, \qquad (10.39)$$

$$\hat{g}(i, k) = \frac{N_2(i, k) - N_1(i, k)}{L_2[\lambda(i, k)] - L_1[\lambda(i, k)]}, \qquad (10.40)$$

and

$$\hat{L}(i, k) = \frac{N(i, k) - \hat{d}(i, k)}{\hat{g}(i, k)}. \qquad (10.41)$$

Radiometric calibration performance is characterized by the same results as are used in the case of a conventional imaging system, namely Eqs. (10.33) and (10.35), but with the wavelength dependence included in L, L_1, and L_2. Spatial noise corresponds to the along-slit spatial dimension and spectral dimension, while temporal noise corresponds to the frame-to-frame spatial dimension. Residual nonuniformity is correlated from frame to frame.

Again, using a VS-25 imaging spectrometer as an example, the instrument was calibrated using a dark reference ($L_1 = 0$), along with 3062-ft-Lambert reference illumination from the calibrated integrating sphere provided in Fig. 10.10. (Calibration sources are sometimes characterized in photometric units such as foot-Lamberts that must be converted to corresponding radiometric units through photopic response functions). The source spectral radiance is provided across the entire VNIR/SWIR spectral range as well as across the limited spectral range of the imaging spectrometer. When a dark reference source is used, the offset in Eq. (10.39) is simply equal to the corresponding averaged frame, and the gain is simply the ratio of the average frame under the integrating sphere

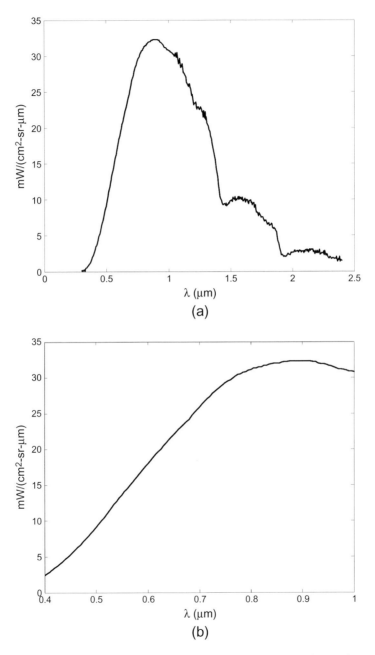

Figure 10.10 Spectral radiance of 3062-ft-Lambert integrating sphere: (a) VNIR/SWIR spectral range and (b) VNIR spectral range.

illumination minus the dark level and sphere spectral radiance. Estimated dark offset and gain profiles are illustrated for this case in Fig. 10.11, while Fig. 10.12 compares a raw spatial–spectral frame, collected for a spatially uniform 3233-ft-Lambert integrating sphere radiance, to the calibrated frame computed using Eq. (10.41) with the estimated offset and gain profiles given in Fig. 10.11.

It is evident from Fig. 10.12 that the calibration not only results in a spectral radiance profile that appears to match the source spectral radiance, but also that the obvious spatial noise in the raw image appears to be significantly reduced. The calibration model predicts that residual spatial uniformity will be eliminated at calibration radiance levels, such that noise measured in the spatial direction should equal that measured in the temporal direction; however, some residual nonuniformity can remain at other radiance levels. To test this theory for one specific case, frames are captured at other integrating-sphere radiance levels and calibrated according to Eq. (10.41). Then, noise variance for measurements at each source radiance is estimated both in spatial (along rows) and temporal (frame to frame) directions. For each spectral band, temporal noise variance is subtracted from spatial noise variance to provide an estimate of residual nonuniformity, which is then averaged over spectral bands, providing an overall estimate to compare to the theory.

Figure 10.13 provides calibrated spectral radiance profiles for several rows corresponding to 3233- and 1255-ft-Lambert radiance levels. Note that the former result, corresponding to calibration radiance, exhibits much less pixel-to-pixel variance in both spatial and spectral directions than the latter result exhibits. It also corresponds better to the calibrated-source radiance denoted by the dotted line and designated by IS, for an integrating sphere. Based on the numerical analysis described earlier, the estimated residual nonuniformity as a function of source illuminance is given in Fig. 10.14. As expected, it is zero at the calibration points and larger in between. The plot does not precisely follow the theoretical result in Fig. 10.5, but exhibits the same nominal behavior.

Because the VS-25 imaging spectrometer was used without an order-sorting filter, another calibration step is required to compensate for the second-order effect. It is preferable from a system performance perspective that order-sorting filters be used, but when they are not, it is possible to compensate the second-order effect by characterizing the relative system response between the first and second orders by illuminating the sensor with filtered radiation. In this case, the specific concern is the second order of 400- to 500-nm radiation overlapping the 800- to 1000-nm region of the FPA. This second-order mixing can be described by extending the standard

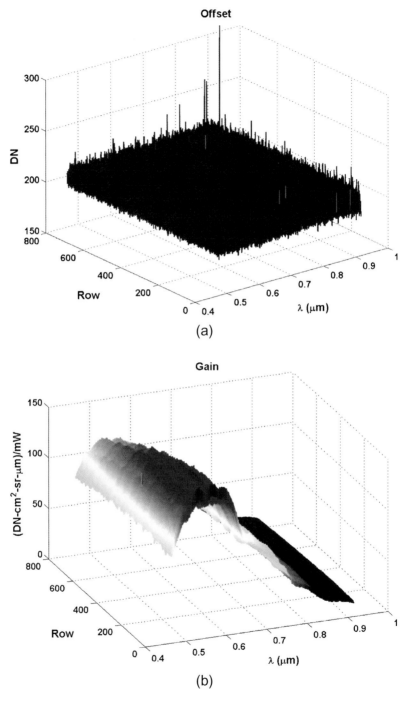

Figure 10.11 Estimated (a) offset and (b) gain profiles for the VS-25 imaging spectrometer using dark and 3233-ft-Lambert integrating sphere reference sources. The offset is given in units of data numbers (DNs), while the gain is expressed in DN per spectral radiance units.

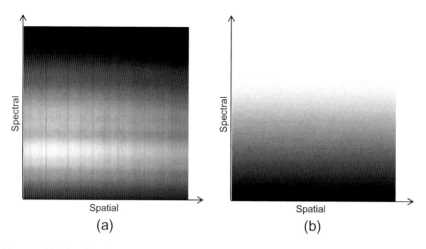

Figure 10.12 Spatial–spectral frames for a 3233-ft-Lambert source (a) before and (b) after radiometric calibration.

linear model to

$$N(i,k) = \begin{cases} g(i,k)\,L(i,k;\lambda) + d(i,k) & \lambda < 800 \text{ nm} \\ g(i,k)\,L(i,k;\lambda) + \alpha(i,k)\,g(i,k)\,L(i,k;\lambda/2) + d(i,k) & \lambda \geq 800 \text{ nm} \end{cases},$$

$$(10.42)$$

where the additional relative gain term $\alpha(i,k)$ accounts for the relative second-order response. After estimating and subtracting the dark offset using a dark reference, the model simplifies to

$$N(i,k) = \begin{cases} g(i,k)\,L(i,k;\lambda) & \lambda < 800 \text{ nm} \\ g(i,k)\,L(i,k;\lambda) + \alpha(i,k)\,g(i,k)\,L(i,k;\lambda/2) & \lambda \geq 800 \text{ nm} \end{cases}.$$

$$(10.43)$$

Now suppose that the spectrometer is illuminated with a source through a filter (similar to that shown in Fig. 10.15) that passes radiation in the 400- to 500-nm band but blocks radiation in the 800- to 1000-nm band. A resulting spatial–spectral frame would appear as in Fig. 10.16, where only the second order exists in the 800- to 1000-nm region of the FPA. The relative gain can then be estimated according to the relationship

$$\alpha(i,k) = \frac{N(i,k)}{N(i_{1/2},k_{1/2})},$$

$$(10.44)$$

where $(i_{1/2}, k_{1/2})$ are FPA coordinates corresponding to one half the center wavelength in (i,k), as dictated by spectral calibration. The second-order-

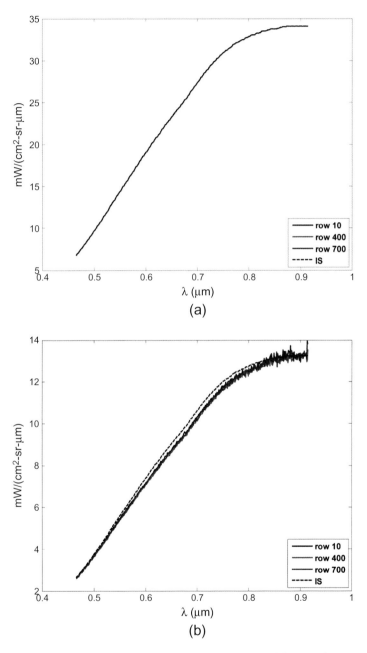

Figure 10.13 Several rows of calibrated spatial–spectral frames for (a) 3233-and (b) 1255-ft-Lambert source illumination compared to the calibrated-source radiance (IS).

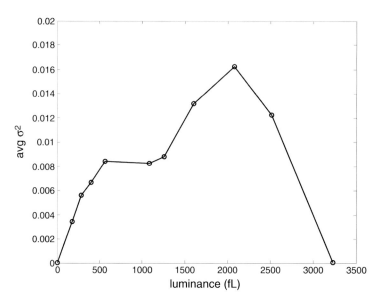

Figure 10.14   Estimated residual nonuniformity variance as a function of source illuminance.

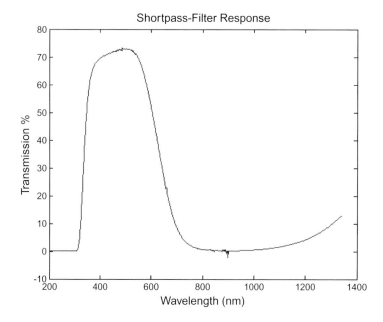

Figure 10.15   Spectral low-pass filter used for second-order compensation.

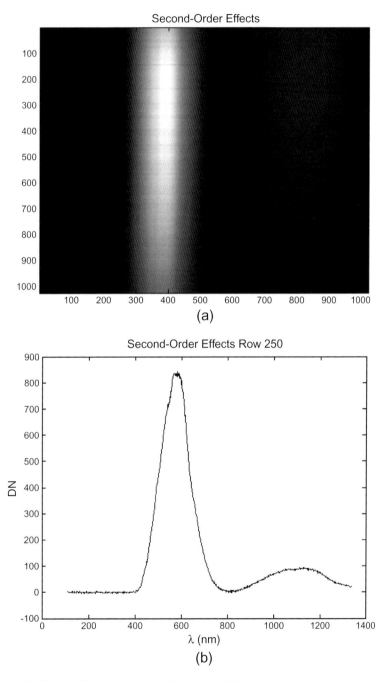

Figure 10.16 (a) Spatial–spectral frame and (b) extracted row showing output from low-pass-filtered radiation. Note that the image in (a) is rotated in this case, so that the spectral dimension is in the horizontal direction (Meola, 2006).

compensated data are then given by

$$N'(i,k) = \begin{cases} N(i,k) & \lambda < 800 \text{ nm} \\ N(i,k) - \alpha(i,k)\,N(i_{1/2},k_{1/2}) & \lambda \geq 800 \text{ nm} \end{cases}. \quad (10.45)$$

These compensated frames are then used for estimating the gain $g(i,k)$ as previously described.

10.2.4 Imaging Fourier transform spectrometer calibration

Because an imaging FTS measures spectral information in a very different manner from a dispersive imaging spectrometer, calibration of such an instrument has some distinct differences. First, spectral calibration is determined through precise knowledge of the OPD that is introduced to create interferograms. This is almost always done by including a monochromatic source such as a He–Ne laser in the system design. The introduced OPD is then precisely measured by determining the phase difference for monochromatic-source intereference fringes. Often, a broadband light source is included to determine the zero-path-difference position according to where the intensity of the white-light source peaks. By accounting for this OPD in a Fourier transform operation, precise spectral calibration is assured. The SRF is directly known by the maximum OPD and the apodization method used.

Radiometric calibration is performed by flood illuminating the front aperture with an integrating sphere or blackbody with known spectral radiance as a function of wavelength. A series of frames are collected to capture an inteferogram for each detector element, and the frame sequences are Fourier transformed to form a hyperspectral data cube corresponding to each calibration source. This data cube is then used to compute gain and offset as a function of all three dimensions (two spatial plus spectral) and applied to a measurement data cube. That is, the calibration process is performed in the hyperspectral image domain as opposed to the frame-to-frame interferogram domain.

For infrared FTIRs that are not cryogenically cooled, it has been empirically determined that phase errors can be introduced by emission from optical components within the interferometer, such as a beamsplitter. When a complex-valued Fourier transform is performed, these phase errors can result in a nonsymmetric interferogram and, therefore, a complex-valued spectrum. To properly calibrate out such phase errors requires that the radiometric calibration be performed in the complex domain (Revercomb *et al.*, 1988). Let $\tilde{N}_1(i,j,k)$ and $\tilde{N}_2(i,j,k)$ correspond to complex hyperspectral data cubes computed by performing a complex-valued Fourier transform for the interferogram data cubes captured for the two calibration sources. Assume that these sources have a spatially uniform spectral radiance $L_1(\lambda)$ and $L_2(\lambda)$, typically blackbody spectral radiance profiles for two different temperatures. The index k corresponds to the spectral axis, and the complex Fourier transform is given by

$$\tilde{N}(i, j, k) = \frac{1}{N} \sum_{n=1}^{N} W(\Delta_n)\, I(i, j, \Delta_n)\, e^{i2\pi\sigma\Delta_n}, \tag{10.46}$$

where $I(i, j, \Delta_n)$ is the raw interferogram data cube. The complex-valued offset and gain are determined by

$$\hat{d}(i, j, k) = \frac{L_2(\lambda_k)\tilde{N}_1(i, j, k) - L_1(\lambda_k)\tilde{N}_2(i, j, k)}{L_2(\lambda_k) - L_1(\lambda_k)}, \tag{10.47}$$

and

$$\hat{g}(i, j, k) = \frac{\tilde{N}_2(i, j, k) - \tilde{N}_1(i, j, k)}{L_2(\lambda_k) - L_1(\lambda_k)}. \tag{10.48}$$

The calibrated hyperspectral data cube is then computed as

$$\hat{L}(i, j, k) = \mathrm{Re}\left\{\frac{\tilde{N}(i, j, k) - \hat{d}(i, j, k)}{\hat{g}(i, j, k)}\right\}. \tag{10.49}$$

The imaginary part of the argument in Eq. (10.46) is theoretically zero. If the magnitude of this component is significantly higher than the noise level, calibration error is indicated.

Calibration of an imaging FTS represents a slightly more formidable challenge relative to a dispersive imaging spectrometer from the perspective that a complete interferogram must be captured. This implies a large sequence of frames as opposed to a single data frame. As in the dispersive case, frame coadding is helpful in reducing the effect of the additive noise in the calibration-source measurement. The upshot is that collecting calibration data for an imaging FTS can be a time-consuming task. The spatial noise predicted in Eq. (10.35) corresponds to the two spatial dimensions of the imaging FTS, while temporal noise corresponds to the spectral dimension.

10.3 Scene-Based Calibration

Because response characteristics of FPAs and sensor electronics are frequently time varying, radiometric calibration is ideally performed onboard the sensor platform. This is often impractical because of the size, weight, power, and expense of calibration-source assemblies. A common compromise is to rely on absolute calibration conducted in a laboratory or on the ground, and to only perform relative calibration onboard on a more frequent time basis. In any of these nonideal situations, it is desirable to augment the limited onboard calibration capabilities with some other reference, such as information from the collected scene content.

10.3.1 Vicarious calibration

First consider the case where onboard calibration is adequate to perform a relative calibration but inadequate for absolute calibration. Theoretically, such calibration could be performed in-flight by viewing objects in the scene for which pupil-plane spectral radiance can be accurately predicted. This is often called vicarious calibration. A good example of the employment of vicarious calibration is based on the NASA AVIRIS that flies on a high-altitude ER-2 aircraft (Green *et al.*, 1990). Vicarious calibration was performed to confirm sensor onboard calibration by flying over a homogenous playa of Rogers Dry Lake, California. The imaged area was extensively instrumented to determine surface reflectance, atmospheric parameters through radiosonde measurements, and solar illumination, to accurately predict pupil-plane spectral radiance using a MODTRAN radiative transfer model. Calibration was confirmed by comparing the measured AVIRIS spectra of the playa with the predicted spectra. Spectral calibration was confirmed based on the relative position of fine atmospheric lines. Spectral bandwidth was estimated using a nonlinear least-squares fit to determine the bandwidth of a Gaussian line shape required to match the width of the predicted atmospheric absorption lines with the measured spectra. Radiometric calibration was confirmed by comparing the absolute measured spectral radiance to the prediction. Site selection for vicarious calibration is important, as stable environmental conditions and a large uniform target are generally needed to assure an accurate model prediction.

10.3.2 Statistical averaging

Another situation where in-scene calibration is helpful is when residual nonuniformity is not effectively reduced by onboard calibration for reasons such as nonlinearity, instability, or an inadequate onboard source. In this case, onboard calibration is maintained in a global, absolute sense for sensor calibration, but scene-based methods are used to compensate residual detector-to-detector nonuniformity. Considering only the dispersive imaging spectrometer case, let $L(i, j, k)$ correspond to a hyperspectral data cube where i corresponds to the along-slit spatial dimension, j corresponds to the frame-to-frame spatial dimension, and k corresponds to the spectral dimension. According to the previous model, residual nonuniformity or spatial noise is expected to be correlated in the frame-to-frame spatial dimension. Therefore, the presence of residual spatial noise typically shows up in the imagery in terms of streaks in this dimension, such as that depicted by the single-band image in Fig. 10.17(a).

Suppose that an image is large enough in the frame-to-frame spatial dimension that image statistics converge to global values, independent of along-slit spatial position. In this case, mean and variance of the ensemble of measurements in the frame-to-frame dimension would be expected to be the same in a statistical sense for every along-slit spatial pixel for a given spectral band. Under these assumptions, residual spatial noise can be compensated for by adjusting the gain and offset of each along-slit pixel so that these statistics are identical. By preserving these statistics in a global sense over each set of along-slit spatial pixels, absolute calibration from onboard or other sources is preserved.

Let I, J, and K be the number of image pixels in i, j, and k directions. The frame-to-frame sample mean for a detector element is defined as

$$\hat{m}(i,k) = \frac{1}{J} \sum_{j=1}^{J} L(i,j,k), \qquad (10.50)$$

and the frame-to-frame sample variance as

$$\hat{\sigma}^2(i,k) = \frac{1}{J} \sum_{j=1}^{J} [L(i,j,k) - m(i,k)]. \qquad (10.51)$$

From these, the global mean

$$\bar{m}(k) = \frac{1}{I} \sum_{i=1}^{I} m(i,k) \qquad (10.52)$$

and global variance

$$\bar{\sigma}^2(k) = \left[\frac{1}{I} \sum_{i=1}^{I} \sigma(i,k) \right]^2 \qquad (10.53)$$

are defined for each spectral band k. In-scene nonuniformity compensation is then performed by

$$\hat{L}(i,j,k) = \frac{L(i,j,k) - [\hat{m}(i,k) - \bar{m}(k)]}{\hat{\sigma}(i,k)/\bar{\sigma}(k)}. \qquad (10.54)$$

Figure 10.17(b) shows the correction of the single-band image in Fig. 10.17(a) based on this approach. For scene-based calibration to be

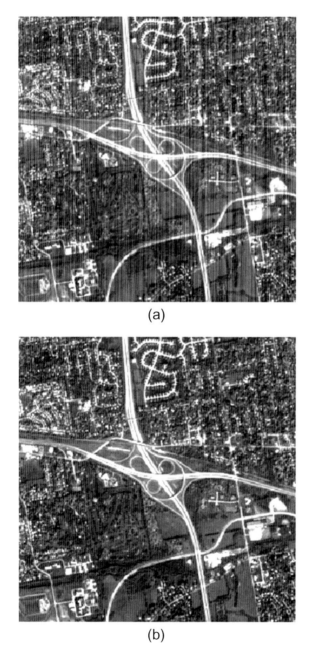

(a)

(b)

Figure 10.17 Example single-band image (a) before and (b) after scene-based calibration.

effective, either long flight lines over spatially heterogeneous terrain or local regions over homogenous terrain such as water bodies are required. If radiance levels or spectral content vary on average across the imaging track, this process can actually do more harm than good.

10.4 Summary

Calibration of raw hyperspectral image data is required to compensate for sensor artifacts such as focal plane array (FPA) nonuniformity and to convert data cubes into meaningful physical units for analysis and information extraction. The first step is spectral calibration, or the estimation of the precise wavelength and perhaps spectral resolution for each pixel in the data cube. The second step is radiometric calibration, which means converting digital data numbers to a physical quantity such as pupil-plane spectral radiance. This can be done in an absolute sense using uniform reference sources of known spectral radiance, such as NIST traceable lamps or blackbody sources. Otherwise, radiometric calibration can be performed in a relative sense, which merely reduces the relative FPA nonuniformity and flattens the field spatially but lacks the global radiometric reference.

Methodology for calibration differs for dispersive and Fourier transform spectrometers. In a dispersive case, spectral calibration based on sources with narrowband features is an essential step. For temporal Fourier transform spectrometers (FTSs), internal monochromatic reference sources driving the interferogram sampling circuitry ensure good spectral calibration.

Radiometric calibration is performed in a similar manner for both, although the temporal FTS case can be more time consuming, since entire calibration data cubes must be collected instead of single calibration frames. Vicarious and statistical in-scene calibration can be used to compensate for sensor instability between laboratory or onboard sensor calibrations if necessary.

A possible model hyperspectral sensor system from the standpoint of spatial, spectral, and radiometric calibration is the AVIRIS sensor described in Chapter 7. Extensive calibration techniques built on the basic principles outlined in this chapter were developed specifically for this instrument and are summarized by Green *et al.* (1998). Readers who are interested in investigating and implementing some of the more subtle details concerning spectrometer calibration can examine the work of Chrien *et al.* (1990, 1995).

References

Best, F. A, Revercomb, H. E., Knuteson, R. O., Tobin, D. C., Ellington, S. D., Werner, M. W., Adler, D. P., Garcia, R. K., Taylor, J. K., Ciganovich,

N. N., Smith, Sr., W. L., Bingham, G. E., Elwell, J. D., and Scottet, D. K., "The Geosynchronous Imaging Fourier Transform Spectrometer (GIFTS) on-board blackbody calibration system," *Proc. SPIE* **5655**, 77–87 (2005) [doi:10.1117/12.579017].

Chrien, T. G., Green, R. O., and Eastwood, M. L., "Accuracy of the spectral and laboratory calibration of the Airborne Visible/Infrared Imaging Spectrometer," *Proc. 2nd Airborne AVIRIS Work*, JPL Publication 90–54, Jet Propulsion Laboratory, Pasadena, CA (1990).

Chrien, T. G., Green, R. O., Chovit, C., and Hajek, P., "New calibration techniques for the Airborne Visible/Infrared Imaging Spectrometer (AVIRIS)," *Sum. Ann. JPL Air. Earth Sci. Work.* **1**, 33–34, JPL Publication 95-1, Jet Propulsion Laboratory, Pasadena, CA (1995).

Eismann, M. T. and Schwartz, C. R., "Focal plane nonlinearity and nonuniformity impacts to target detection with thermal infrared imaging spectrometers," *Proc. SPIE* **3063**, 164–173 (1997) [doi:10.1117/12.276065].

Green, R. O, Conel, J. E., Margolis, J. S., Carrere, V., Bruegge, C. J., Rast, M., and Hoover, G., "In-flight validation and calibration of the spectral and radiometric characteristics of the Airborne Visible/Infrared Imaging Spectrometer," *Proc. SPIE* **1298**, 18–35 (1990) [doi:10.1117/12.21333].

Green, R. O., Eastwood, M. L., Sarture, C. M., Chrien, T. G., Aronsson, M., Chippendale, B. J., Faust, J. A., Pavri, B. E., Chovit, C. J., Solis, M., Olah, M. R., and Williams, O., "Imaging Spectroscopy and the Airborne Visible/Infrared Imaging Spectrometer (AVIRIS)," *Remote Sens. Environ.* **65**, 227–248 (1998).

Labsphere, "Spectralon® optical grade reflectance material," Labsphere, Inc., North Sutton, NH, http://www.labsphere.com (last accessed Oct 2011).

Meola, J., "Analysis of Hyperspectral Change and Target Detection as Affected by Vegetation and Illumination Variations," Master's Thesis, Univ. of Dayton, OH, August 2006.

Newport Corp., "Technical references: Light Sources," Newport Corp., Irvine, CA, http://www.newport.com (last accessed Oct 2011).

Revercomb, H. E., Buijs, H., Howell, H. B., LaPorte, D. D., Smith, W. L., and Sromovsky, L. A., "Radiometric calibration of IR Fourier transform spectrometers: solution to a problem with the High-Resolution Interferometer Sounder," *Appl. Optics* **27**, 3210–3218 (1988).

Zadnik, J., Guerin, D., Moss, R., Orbeta, A., Dixon, R., Simi, C. G., Dunbar, S., and Hill, S., "Calibration procedures and measurements for the COMPASS hyperspectral imager," *Proc. SPIE* **5425**, 182–188 (2004) [doi:10.1117/12.546758].

Chapter 11
Atmospheric Compensation

The general relationships provided in Chapter 5 relate pupil-plane radiance at a given remote viewing orientation to the apparent spectral properties of a material being viewed, along with a number of properties related to the radiative transfer from the material to the sensor. These properties include direct and diffuse illumination, atmospheric transmission, and atmospheric path radiance. The goal of remote sensing is generally to extract information from the apparent spectral properties, such as the reflectance or emissivity spectrum, of the materials being viewed. Therefore, the presence of these unknown illumination and atmospheric parameters is an undesired reality that must be dealt with to achieve this goal. *Atmospheric compensation*, also known as atmospheric correction or normalization, refers to image processing methods intended to invert relationships (such as those from Chapter 5) with limited knowledge of radiative transfer properties to estimate the apparent spectral properties of imaged materials. A by-product of atmospheric compensation is often the estimation of atmospheric spectral properties, which can be useful information in itself. In this case, the terms atmospheric retrieval and atmospheric sounding are often used in the literature.

Atmospheric compensation methods can be loosely separated into two categories: in-scene methods and model-based methods. Model-based methods use a radiative transfer model such as MODTRAN as a basis for atmospheric compensation. While such a model can fairly accurately capture the illumination, transmission, and path radiance characteristics for a specified remote sensing situation, a large number of atmospheric parameters such as gaseous concentration, pressure, and temperature profiles need to be precisely specified for a given situation, and these parameters are typically unknown. On the other hand, possible spectral variation of the atmospheric properties is constrained, making estimation of key unknown parameters a tractable problem in some cases. In-scene methods use some *a priori* knowledge of the spectral nature of materials expected to be present in a scene to guide estimation of atmospheric properties, without regard to a radiative transfer model such as MODTRAN. Because they do not require such sophisticated radiative transfer modeling, in-scene methods are somewhat simpler to understand and less computationally complex; thus, we begin with them first.

11.1 In-Scene Methods

Under the approximations of the diffuse facet model, pupil-plane spectral radiance has the form

$$L_p(\lambda) = \frac{\tau_a(\lambda)}{\pi} [E_s(\theta_s, \phi_s, \lambda) + E_d(\lambda) - \pi B(\lambda, T)] \rho(\lambda)$$

$$+ \tau_a(\lambda) L_{BB}(\lambda, T) + L_a(\lambda), \qquad (11.1)$$

where $\tau_a(\lambda)$ is atmospheric transmission, $E_s(\theta_s, \phi_s, \lambda)$ is direct solar irradiance, $E_d(\lambda)$ is integrated downwelling irradiance, $B(\lambda, T)$ is a blackbody radiance for the material temperature T, $L_a(\lambda)$ is atmospheric path radiance, and $\rho(\lambda)$ is the material apparent reflectance. This approximate form is most applicable for relatively clear atmospheric conditions where the adjacency effects can be ignored. Otherwise, the more complicated form described in Chapter 5, incorporating a factor in the denominator of the model accounting for adjacency effects, provides a more accurate description. Combining all of the unknown distributions, Eq. (11.1) can be expressed in the general linear form

$$L_p(\lambda) = a(\lambda)\rho(\lambda) + b(\lambda). \qquad (11.2)$$

By virtue of the linear relationship, unknown parameters $a(\lambda)$ and $b(\lambda)$ could be estimated by simple linear regression on a band-by-band basis if pupil-plane radiance were measured for at least two materials of known surface reflectance distribution. If such materials were present somewhere in a hyperspectral image, and the unknown parameters were spatially invariant, then surface reflectance could be estimated for all image spectra based on these known materials. That is the essence of in-scene atmospheric compensation.

Variations of in-scene atmospheric compensation methods are the result of differing estimation methods, especially in cases where the "known" materials themselves must be identified in the scene in an unsupervised manner. Further, methods differ in the reflective and emissive spectral regions due to both the nature of radiative transfer and the impact of the material-dependent temperature occurring in the emissive region that can be ignored in the reflective region. As the MWIR region generally includes both solar-reflected and thermal components, it is the hardest to deal with from an atmospheric compensation standpoint.

In the VNIR/SWIR, the relationship between pupil-plane radiance and apparent reflectance is simply

$$L_p(\lambda) = \frac{\tau_a(\lambda)}{\pi} [E_s(\theta_s, \phi_s, \lambda) + E_d(\lambda)] \rho(\lambda) + L_a(\lambda). \qquad (11.3)$$

With regard to the linear relationship in Eq. (11.2),

$$a(\lambda) = \frac{\tau_a(\lambda)}{\pi} [E_s(\theta_s, \phi_s, \lambda) + E_d(\lambda)] \qquad (11.4)$$

and

$$b(\lambda) = L_a(\lambda). \qquad (11.5)$$

In the LWIR, the convention is to use emissivity $\varepsilon(\lambda)$ as the material property to be estimated. In this case, the relationship in Eq. (11.1) can be expressed as

$$L_p(\lambda) = a(\lambda)\varepsilon(\lambda) + b(\lambda), \qquad (11.6)$$

where

$$a(\lambda) = \tau_a(\lambda) [B(\lambda, T) - L_d(\lambda)] \qquad (11.7)$$

and

$$b(\lambda) = \tau_a(\lambda) L_d(\lambda) + L_a(\lambda). \qquad (11.8)$$

Downwelling radiance $L_d(\lambda)$ in Eq. (11.8) is given by

$$L_d(\lambda) = \frac{E_d(\lambda)}{\pi}. \qquad (11.9)$$

The presence of a temperature-dependent blackbody function, which varies from pixel to pixel, is a complication that exists in the LWIR for emissivity estimation that does not occur in the VNIR/SWIR for reflectance estimation. This issue is often addressed by a two-step process of first estimating upwelling radiance from the pupil-plane spectral radiance, and then jointly estimating the temperature and emissivity spectrum from the upwelling radiance. The first is strictly an atmospheric compensation process, while the latter is called temperature–emissivity separation.

In-scene atmospheric compensation methods are generally based on three underlying assumptions that are satisfied to varying degrees in real imagery. The first is the diffuse target assumption that characterizes the BRDF as angle independent and therefore makes reflected radiance in the sensor direction from both directional and diffuse downwelling sources independent of surface orientation. Objects with a significant specular component violate this assumption and can frustrate these methods. The second is the assumption that the viewing range is the same for all points in an image, allowing atmospheric transmission to be a space-invariant quantity. This is often a fair assumption for near-nadir viewing sensors but not for oblique imaging, where there can be significant range variation across a scene. The third is the assumption that direct and diffuse illumination is space invariant, an assumption that can be violated when there is scattered cloud cover or local shadowing due to 3D surface relief. In these situations, space-varying illumination is improperly interpreted as reflectance variation.

11.1.1 Empirical line method

The empirical line method (ELM) is a direct application of the linear relationship in Eq. (11.2) to support reflectance estimation in the VNIR/SWIR spectral regions, based on a number of pixels in a hyperspectral image for which material reflectance spectra are known. Consider first the case of two such image pixels for which reflectance spectra are known to be $\rho_1(\lambda)$ and $\rho_2(\lambda)$ and for which corresponding pupil-plane radiance measurements $L_1(\lambda)$ and $L_2(\lambda)$ are made. The unknown parameters can be estimated from simple slope–intercept relationships for a line as

$$\hat{a}(\lambda) = \frac{L_2(\lambda) - L_1(\lambda)}{\rho_2(\lambda) - \rho_1(\lambda)} \tag{11.10}$$

and

$$\hat{b}(\lambda) = \frac{L_1(\lambda)\rho_2(\lambda) - L_2(\lambda)\rho_1(\lambda)}{\rho_2(\lambda) - \rho_1(\lambda)}. \tag{11.11}$$

Based on these estimates, the apparent reflectance spectrum corresponding to any pupil-plane radiance measurement $L_p(\lambda)$ can be estimated by

$$\hat{\rho}(\lambda) = \frac{L_p(\lambda) - \hat{b}(\lambda)}{\hat{a}(\lambda)}. \tag{11.12}$$

This approach is easily extended to the case of multiple in-stances of known materials by employing a linear least-squares regression analysis. Consider a set of pupil-plane radiance measurements $\{L_1(\lambda), L_2(\lambda), \ldots, L_N(\lambda)\}$ in a hyperspectral image corresponding to a set of known reflectance spectra $\{\rho_1(\lambda), \rho_2(\lambda), \ldots, \rho_N(\lambda)\}$. These measurements are assumed to correspond to a set of linear equations of the form

$$L_i(\lambda) = a(\lambda)\rho_i(\lambda) + b(\lambda), \tag{11.13}$$

where $i = 1, 2, \ldots, N$. This can be placed in standard matrix–vector form

$$\mathbf{Ax} = \mathbf{b}, \tag{11.14}$$

where

$$\mathbf{A} = \begin{bmatrix} \rho_1(\lambda) & 1 \\ \rho_2(\lambda) & 1 \\ \vdots & 1 \\ \rho_N(\lambda) & 1 \end{bmatrix}, \tag{11.15}$$

$$\mathbf{x} = \begin{bmatrix} a(\lambda) \\ b(\lambda) \end{bmatrix}, \tag{11.16}$$

and

$$\mathbf{b} = \begin{bmatrix} L_1(\lambda) \\ L_2(\lambda) \\ \vdots \\ L_N(\lambda) \end{bmatrix}. \tag{11.17}$$

The least-squares estimates can be independently computed on a band-by-band basis for each spectral band using the pseudo-inverse:

$$\mathbf{x} = [\mathbf{A}^T\mathbf{A}]^{-1}\mathbf{A}^T\mathbf{b}. \tag{11.18}$$

Note that the ELM relationships are based simply on the assumptions of sensor response linearity and space invariance of radiative transfer quantities. To use the ELM it is generally unnecessary to calibrate an imaging spectrometer into pupil-plane radiance, as long as sensor response is approximately linear. Therefore, the ELM can be considered a sensor calibration method as well as an atmospheric compensation method. The primary limitation of the ELM, of course, is the need to have a scene with identifiable materials that have known reflectance spectra. In controlled measurement situations, this has been performed by placing large, uniform tarps or panels in the scene with reflectance characteristics that have been precisely measured. The materials must be sufficiently large to ensure that selected image pixels capture only the reference target and not any surrounding background. An alternative application of the ELM is to identify natural materials in the scene, such as uniform areas of a particular land-cover type, for which reflectance spectra are measured by some other source.

The ELM is demonstrated by applying it to VNIR/SWIR hyperspectral data for the scene depicted in Fig. 11.1, composed of several panels in an open grass field near a tree line. Data were collected with the HYDICE imaging spectrometer described in Chapter 8 from an airborne platform over Aberdeen Proving Ground, Maryland, in August 1995 during a data collection campaign known as Forest Radiance I. A series of reflectance panels with increasing broadband reflectance are identified in Fig. 11.1, for which laboratory DHR measurements are provided in Fig. 11.2. The ELM was applied using 4 and 32% reflectance panels as references to derive the gain and offset spectra illustrated in Fig. 11.3. These were then applied to scene spectra extracted from the hyperspectral data at points corresponding to a 16% reflectance panel, green tarp, and grass region, using Eq. (11.12) to derive the apparent spectral reflectance estimates depicted in Fig. 11.4. The gain and offset spectra have the general shape of the product of solar illumination and atmospheric transmission (gain), and path-scattered radiance (offset), as the theory predicts. Scene-derived reflectance estimates agree perfectly for the 4 and 32% reflectance panels by nature of the ELM, but also agree very well for the 16% reflectance panel as well as the grass region. The overall broadband agreement for the green tarp is somewhat poorer, even though the scene-derived reflectance estimate provides an excellent match to the narrowband spectral features. This mismatch could be representative of atmospheric compensation errors but could also be the result of errors in the measurement of the provided laboratory spectral reflectance. In general, the ELM provides good performance when illumination is constant over the scene, and reference panels and objects of interest are in open areas. When illumination varies across the scene due to clouds or local terrain variation, then the basic assumptions of the method break down.

The mismatch between ELM estimates and ground-truth spectra for the green tarp raises some important issues concerning both the model underlying the ELM and the experimental procedures used to collect ground-truth reflectance spectra. As the underlying model assumes perfectly diffuse materials, it is important to understand the mismatch that can occur when materials are not truly diffuse, as is likely the case for the green tarp in the HYDICE image, and the procedures used to deal with the specular component of the material reflectance. Typical laboratory instruments for DHR measurements employ a block in the

Reflectance Panels   Green Tarp

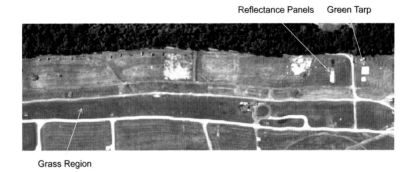

Grass Region

Figure 11.1 Scene used for the VNIR/SWIR hyperspectral data collected by HYDICE imaging spectrometer.

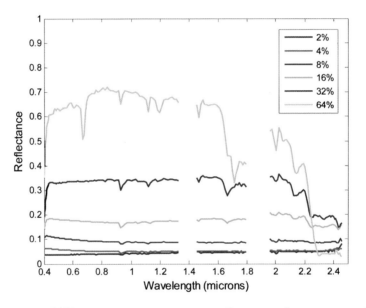

Figure 11.2 DHR measurements corresponding to reflectance panels in VNIR/SWIR hyperspectral data.

IS for the specular component, so that this component is not included in the reflectance measurement. This usually provides the best match to remote sensing measurements, since specular reflection from directional illumination is only captured over a small range of viewing angles. On the other hand, apparent reflectance is much higher if the remote sensor happens to be within the specular lobe, which is atypical. If the object is not a simple flat panel, however, the remote sensor can capture specular reflectance from parts of an object. Thus, a single DHR measurement, whether it is made with or without blocking the specular reflectance component, is fundamentally limited in representing a material spectral

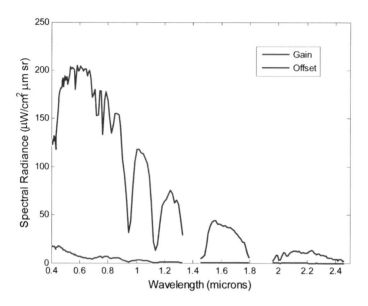

Figure 11.3 ELM gain and offset estimates derived from 4 and 32% reflectance panel data.

signature. While the pedigree of the ground-truth reflectance provided with the HYDICE data for the green tarp is not precisely known, the fact that it appears to be more highly reflective over the entire spectral range relative to the ELM estimate indicates that perhaps the ground-truth measurement included all or part of the specular lobe. This is not uncommon for spectral measurements made with field instruments, which often do not employ an IS and specular block. Fortunately, spectral matching algorithms, such as those presented in Chapter 14, can exhibit some invariance to simple scaling differences between remotely measured and truth spectra, as seen in Fig. 11.4(b); therefore, such mismatches can be tolerable for some applications.

11.1.2 Vegetation normalization

Vegetation normalization is an extension of the ELM to a case where vegetation is used as a reference land cover. It is motivated by the distinctive shape of reflectance spectra of healthy vegetation, along with a preponderance of vegetation in typical imaged scenes. The first step of vegetation normalization is to identify healthy vegetation in the scene using the NDVI,

$$\text{NDVI} = \frac{L_p(860\,\text{nm}) - L_p(660\,\text{nm})}{L_p(860\,\text{nm}) + L_p(660\,\text{nm})}, \tag{11.19}$$

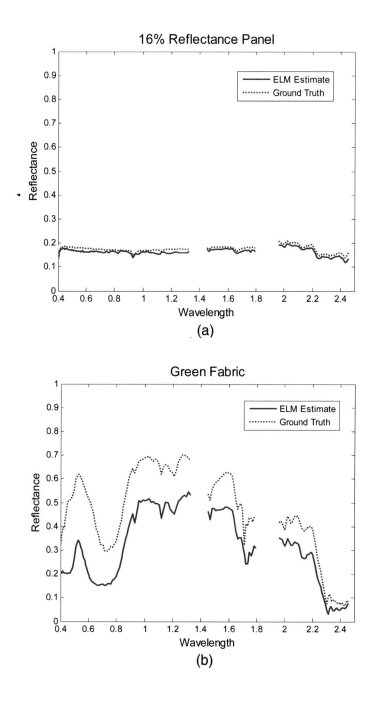

Figure 11.4 Estimated ELM reflectance compared to DHR truth measurements: (a) 16% reflectance panel, (b) green tarp fabric, and (c) grass region.

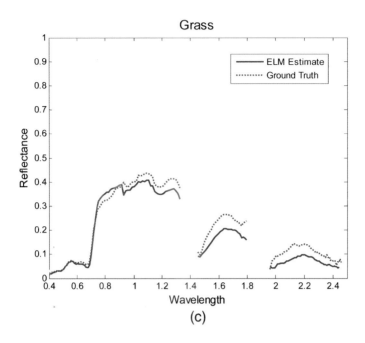

Figure 11.4 (*continued*)

as was discussed in Chapter 4. The NDVI is computed for all image spectra, then spectra with the highest NDVI are considered to correspond to healthy vegetation. From these, a mean spectrum is computed to represent a pupil-plane radiance measurement for vegetation and is used with a corresponding reference vegetation reflectance spectrum to serve as one material for atmospheric compensation with the ELM approach. In practice, solely using the NDVI metric to identify vegetation spectra is not typically very effective because it captures partially illuminated and shadowed vegetation spectra along with fully illuminated vegetation spectra. Due to the space-invariant illumination assumption, the former appears to have lower reflectance and would not properly match the vegetation reference spectrum. To resolve this problem, a brightness criterion can also be employed that filters out only the brightest of the high-NDVI spectra to represent scene vegetation. Average spectral radiance over the full spectral range can serve as this brightness criterion.

Vegetation normalization is also based on the premise that healthy vegetation in a scene is consistent enough in spectral characteristics that it can be represented by a single reflectance spectrum. This, of course, is not a truly accurate assumption. For example, the spectra in Fig. 11.5 for some typical vegetation types illustrate considerable variability in spectral characteristics. However, they are also fairly well correlated, such that

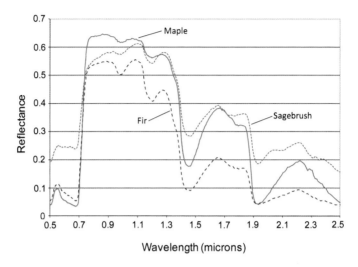

Figure 11.5 Typical reflectance spectra of natural vegetation.

one might expect that some averaging process across high-NDVI pixels in the sensed hyperspectral image, as well as across exemplar vegetation reference spectra, could lead to sufficiently strong correspondence to serve as one known scene material to perform atmospheric compensation using the ELM.

The remaining issue is to determine what to use as the second known material. One approach that is somewhat justified under clear atmospheric conditions is to ignore $b(\lambda)$, which is representative of atmospheric path radiance. Ignoring the Rayleigh scattering contribution of the atmosphere causes reflectance to be overestimated at shorter wavelengths, providing a spectral profile with an increasing trend at the blue end. An alternative approach is to find pixels in the scene that are indicative of a shade point, where measured radiance is expected to be solely due to path radiance. Approaches considered include finding a spectrum in the scene with the lowest mean value over the spectral range, or constructing a shade spectrum composed of the minimum image value in each spectral band. In either case, the spectrum is made to correspond to a reference reflectance that is identically zero. The fact that a shade point is even considered to support vegetation normalization represents a somewhat fundamental flaw in the approach. That is, the presence of shade itself is indicative of an illumination variation, which violates the basic assumption of the formulation. Thus, it should be clear that natural illumination variation causes errors in estimated reflectance spectra. For this reason, one should proceed cautiously when making quantitative interpretations of material spectra derived from remote imagery in this manner. Nevertheless, vegetation normalization has been found to be a useful method for normalizing laboratory reflectance spectra to remotely sensed spectra for signature-based target detection purposes, as discussed further in Chapter 14.

Capabilities and limitations of the vegetation normalization method are portrayed here by applying it to the same VNIR/SWIR hyperspectral data from the HYDICE Forest Radiance collection used to illustrate the ELM. A scatter plot of the NDVI versus brightness for all spectra in the scene, depicted in Fig. 11.6, indicates numerous vegetation spectra with high NDVIs that vary considerably in brightness due to illumination variation. Brightness and NDVI images are displayed in Fig. 11.7, along with a color-coded map specifying the location of the spectra exceeding an NDVI threshold of 0.75 (red), a brightness threshold of 4 mW/cm²µm sr (green), and both (blue). All come almost exclusively from the tree canopy at the top of the image, with variation in brightness of the high-NDVI spectra due to shading that naturally occurs from self-shadowing within the 3D canopy structure.

Using the reference vegetation reflectance spectrum in Fig. 11.8 and estimating the offset from the lowest scene spectral radiance within each spectral band, gain and offset parameters in Fig. 11.9 are derived. These are applied to the data to produce the material reflectance estimates shown in Fig. 11.10. In this case, the results are not quite as accurate as they are for the ELM, where a well-truthed scene reference was available, but agreement is still fairly good. This is especially true for the grass spectra because its reflectance characteristics closely match the tree canopy spectra used as a reference. Again, a significant mismatch occurs

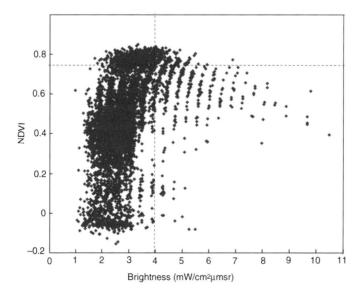

Figure 11.6 Scatter plot of NDVI versus brightness for a VNIR/SWIR hyperspectral dataset example.

NDVI

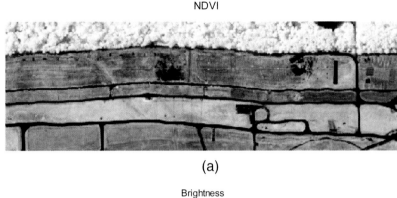

(a)

Brightness

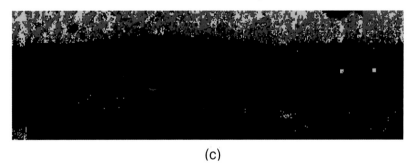

(b)

Red: NDVI > 0.75; Green: L > 4 mW/cm^{2}μmsr; Blue: Both

(c)

Figure 11.7 (a) NDVI and (b) brightness images for the VNIR/SWIR hyperspectral image example, along with (c) a color-coded map identifying spectra with high NDVI (red), high brightness (green) and both high NDVI and high brightness (blue).

between the estimate and ground truth for the green-fabric reflectance spectrum. In this case, the difference is more than just a simple scaling and includes an overall mismatch in the spectral shape. Such errors are more likely to adversely impact spectral matching algorithms than the simple scaling errors seen with the ELM. A fundamental issue that can cause such errors when applying vegetation normalization is the variability

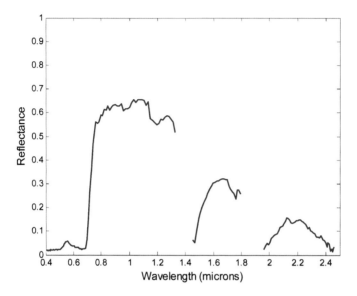

Figure 11.8   Vegetation reference spectrum.

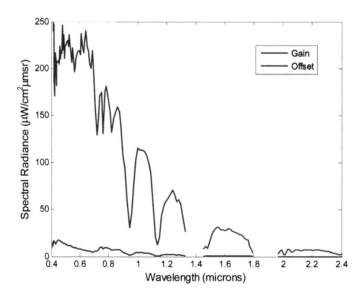

Figure 11.9 Estimated gain and offset spectra based on vegetation normalization.

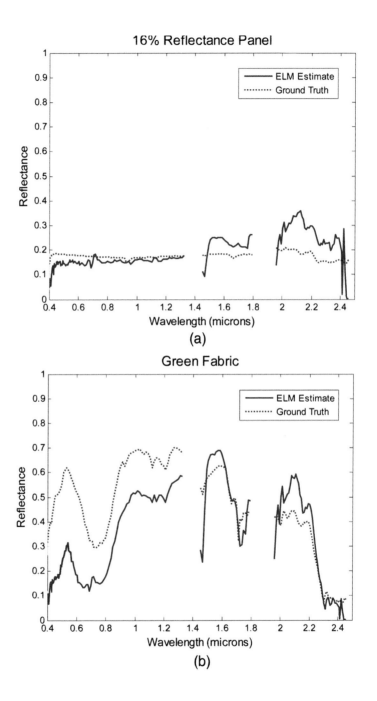

Figure 11.10 Estimated material reflectance spectra using vegetation normalization compared to DHR measurements: (a) 16% reflectance panel, (b) green tarp fabric, and (c) grass region.

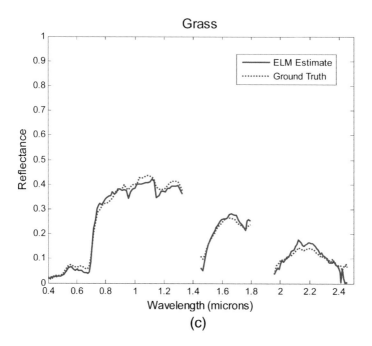

Figure 11.10 (*continued*)

in spectral characteristics that can occur across vegetation types. If a vegetation reference is used that does not match the mean behavior of the set of scene materials selected by the NDVI and brightness thresholding methodology, then all scene-derived spectral reflectance estimates will be in error. A similar effect occurs if there are no deep shadows in the scene for which the scene radiance can be assumed to be effectively zero across the entire spectral range. While these issues can limit the utility of vegetation normalization for precise estimation of reflectance spectra, a subspace target detection algorithm (described in Chapter 14) extends this basic methodology to compensate for these variations. Furthermore, the use of vegetation normalization has proven to be a useful and practical normalization method for real-time target detection using sensors of limited absolute calibration, whereas the model-based approaches described later in this chapter are more difficult to apply.

Other variants of this empirical atmospheric compensation method have been pursued that normalize data with regard to common scene materials other than vegetation. For example, one can normalize to bare soil as opposed to vegetation using a bare-soil index (BI), given by

$$BI = \frac{[L_p(\text{SWIR}) + L_p(R)] - [L_p(\text{NIR}) + L_p(B)]}{[L_p(\text{SWIR}) + L_p(R)] + [L_p(\text{NIR}) + L_p(B)]}, \qquad (11.20)$$

where pupil-plane radiance is integrated over the blue (B), red (R), NIR, and SWIR bands, corresponding nominally to 450 to 500 nm, 600 to 700 nm, 700 to 1000 nm, and 1150 to 2500 nm, respectively. This extends the concept to arid regions where vegetation is sparse but potentially adds more uncertainty in the reference spectra due to the variation in soil spectral characteristics relative to that of vegetation. When atmospheric compensation is performed as part of a target detection or material identification strategy, as is discussed in Chapter 14, this inherent variation can be mitigated using a subspace detection approach; however, this technique is not useful when absolute characterization of the surface apparent reflectance is desired.

Another alternate approach, called quick atmospheric compensation (QUAC), performs normalization based on equalizing the variance of the measured radiance data to reference signatures as opposed to the mean of the measured radiance spectra (Bernstein *et al.*, 2005). This method is based on the empirical observation that the spectral standard deviation of reflectance spectra of a diverse set of materials is a spectrally independent constant. Therefore, any spectral variation in the standard deviation of a set of diverse scene spectra characterizes atmospheric illumination and transmission characteristics. The method proceeds as follows. First, an estimate for offset $b(\lambda)$ is obtained by identifying the shade point in the same manner as was described for vegetation normalization. Next, a set of the most diverse spectra are extracted from the scene, and their standard deviation $\sigma_L(\lambda)$ is computed as a function of wavelength. The gain estimate is then given by

$$\hat{a}(\lambda) = a_0\, \sigma_L(\lambda). \tag{11.21}$$

To determine unknown constant a_0, the gain estimate can be compared to a modeled estimate of the product of direct solar irradiance and atmospheric transmission for a transparent spectral region such as 1.6 to 1.7 μm, where transmission is not as variable. Alternatively, the constant can be estimated by assuming that the peak of the vegetation reflectance is 0.7. In either case, the spectral shape of the gain estimate is dependent on the statistical assumption of diverse material spectra and not dependent on a precise mapping to a particular scene reference material.

11.1.3 Blackbody normalization

For the LWIR spectral region, the issue of atmospheric compensation is first addressed in this section, while the topic of temperature–emissivity separation is deferred to the next section. For an atmospheric compensation problem, the underlying model is simply given by

$$L_p(\lambda) = \tau_a(\lambda)\, L_u(\lambda) + L_a(\lambda), \tag{11.22}$$

where $L_u(\lambda)$ is the upwelling radiance from the material. The in-scene atmospheric compensation scheme outlined in this section is based on finding spectra in a remotely sensed image that most closely represent blackbody radiators, for which upwelling radiance is known except for one parameter, the surface temperature. A regression analysis is then performed on a band-by-band basis from these spectra to estimate path transmission and path radiance. This particular methodology, called in-scene atmospheric compensation (ISAC), was specifically developed by researchers at the Aerospace Corporation for application to LWIR hyperspectral imagery collected by the SEBASS instrument described in Chapter 7 (Johnson, 1998).

To understand this blackbody normalization method, it is helpful to define a quantity called the *spectral apparent temperature*. This is a spectrally dependent quantity that represents the blackbody temperature, for each individual wavelength band, that would produce measured spectral radiance at that specific spectral location. Mathematically, the spectral apparent temperature for a given spectral radiance $L(\lambda)$ is given by inverting the blackbody spectral radiance function according to

$$T_{BB}(\lambda) = \frac{hc}{\lambda k \ln\left(\frac{2hc^2}{\lambda^5 L(\lambda)} + 1\right)}. \tag{11.23}$$

For a true blackbody, the spectral apparent temperature is constant with wavelength and equals the actual surface temperature. For any other material, the spectral apparent temperature has an upper bound given by the true temperature, and a shape that is somewhat correlated with the emissivity spectrum, assuming that the emitted upwelling radiance dominates the reflected component such that the latter can be ignored. If computed from pupil-plane spectral radiance $L_p(\lambda)$, the spectral apparent temperature exhibits strong spectral features due to uncompensated atmospheric path effects. To investigate this behavior with real data, consider the image depicted in Fig. 11.11, which is a false-color composite derived from 8.5-, 9.2-, and 11.2-μm bands of a SEBASS LWIR hyperspectral image, collected from an approximately 5000-ft altitude over the Atmospheric Radiation Measurement (ARM) facility near Lamont, Oklahoma in June 1997 (Young et al., 2002). The pupil-plane radiance spectra from a pond and a road are then extracted from the data. Figure 11.12 compares this spectral radiance to the corresponding spectral apparent temperature. The key to blackbody normalization is to determine

Pond   Panel Array

Road

Figure 11.11 False-color SEBASS LWIR hyperspectral image of the Atmospheric Radiation Measurement (ARM) site based on 8.5-, 9.2-, and 11.2-μm bands.

the atmospheric transmission and path radiance spectral distributions that will flatten the spectral apparent temperature derived from estimated upwelling radiance $L_u(\lambda)$ for such objects, in particular the pond, which should behave effectively as a blackbody radiator.

The first step to applying blackbody normalization is to identify blackbody-like spectra in a scene. This is done by a somewhat heuristic process that begins by determining the spectral band for which atmospheric transmission appears to be the highest, and then selecting scene spectra whose spectral apparent temperature is at a maximum for this reference wavelength. This would indeed occur if the material exhibited a constant emissivity across the spectral range of the sensor. Let a measured hyperspectral image be characterized as a set of pupil-plane radiance spectra $\{L_i(\lambda) : i = 1, 2, \ldots, N\}$. For each measurement, the spectral apparent temperature is computed according to Eq. (11.21), and the maximum is computed according to

$$T_i = \max_{\lambda} T_{BB}[L_i(\lambda)]. \tag{11.24}$$

The wavelength at which this maximum occurs is denoted for each measurement as λ_i, and λ_r represents the most common spectral band for which this occurs. From this set of measurements, a subset of M spectra is selected for which the spectral apparent temperature peaks at λ_r, or $\lambda_i = \lambda_r$. These are considered to be the most-blackbody-like spectra based on the aforementioned premise. Let $j = 1, 2, \ldots, M$ represent indices of these candidate blackbody-like spectra.

Suppose that one were to plot the pupil-plane spectral radiance $L_j(\lambda)$ versus blackbody radiance $L_{BB}(\lambda, T_j)$ for all of the spectra identified as before for a particular wavelength. For spectra that are truly blackbodies, the points would lay on a line with a slope of $\tau_a(\lambda)$ and a y intercept of $L_a(\lambda)$, according to the model in Eq. (11.22), because $L_{BB}(\lambda, T_j)$ would be upwelling radiance $L_u(\lambda)$. That assumes, of course, that atmospheric transmission at λ_r is one; otherwise, T_j would underestimate the true

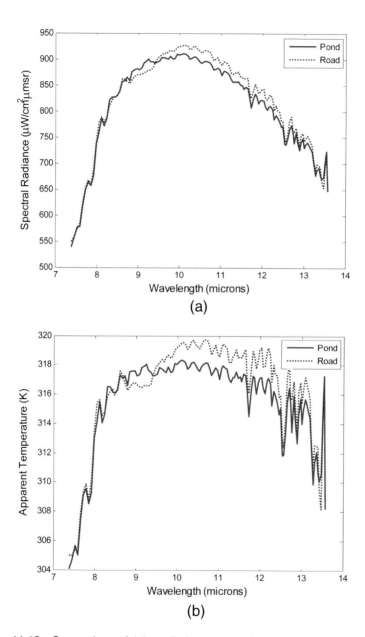

Figure 11.12 Comparison of (a) pupil-plane spectral radiance and (b) spectral apparent temperature data from the SEBASS LWIR hyperspectral image example for pond and road spectra.

surface temperature for all spectra. Materials for which emissivity is less than unity, on the other hand, would fall below this line in some proportion to their emissivity at the particular wavelength. No points could fall above the line because this would imply an emissivity greater than one, which is nonphysical. Figure 11.13 illustrates the distribution for the 8.5-μm band for the SEBASS ARM site data. To estimate path transmission and path radiance for each wavelength, one merely needs to fit a bounding line to the scatter plot and determine its slope and y intercept as indicated. If there is no sensor noise, this can be accomplished by a constrained optimization or a standard least-squares regression, subject to the constraint that no point exceeds the line. However, due to the presence of sensor noise, this could produce erroneous results because it would be sensitive to noisy samples. To achieve better immunity to noise in the estimation process, a methodology using a Kolmogorov–Smirnov (KS) test has been developed and successfully applied. Note that the desired regression line in this case (dotted red line in Fig. 11.13) is different from the standard linear regression line (solid green line).

The KS test is a statistical method for evaluating whether data conform to a particular statistical distribution based on the maximum difference between the sample and model cumulative distribution functions (CDFs). Mathematically, a KS test statistic is defined as

$$D = \max_{u} | P_C(u) - F_C(u) |, \qquad (11.25)$$

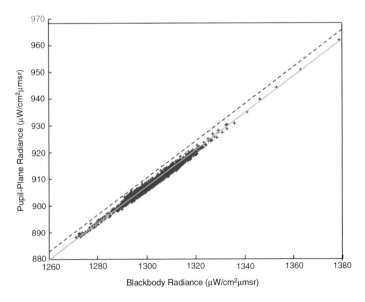

Figure 11.13 Scatter plot used for determining atmospheric transmission and path radiance from the 8.5-μm distribution of candidate blackbody spectra.

where $P_C(u)$ is the sample CDF and $F_C(u)$ is the model CDF (Press *et al.*, 1992). The significance of test statistic D is measured by the function

$$P_{KS}(D, M) = Q_{KS}\left(\sqrt{M}D\right), \tag{11.26}$$

where M is the number of samples from which $P_C(u)$ is estimated, and

$$Q_{KS}(\alpha) = 2\sum_{j=1}^{\infty}(-1)^{j-1}e^{-2j^2\alpha^2}. \tag{11.27}$$

Given a known sensor NESR σ_L for a particular band, noise is assumed to be a zero-mean, normally distributed random variable, such that the CDF at the upper edge of the distribution would be expected to conform to a modeled CDF, given by

$$F_C(u) = erf(u), \tag{11.28}$$

where

$$u = \frac{L - L_m}{\sqrt{2}\sigma_L}, \tag{11.29}$$

L is a spectral radiance sample, and L_m is the mean spectral radiance for the blackbody samples and is the object of estimation. The KS test is applied by finding mean radiance L_m that maximizes the significance $P_{KS}(D, M)$ for each spectral band, and then fitting a line through mean radiance estimates using standard least-squares regression.

Numerical implementation of a KS test is as follows. First, a standard linear regression is performed for each band in the scatter plot depicted in Fig. 11.13, and the data are modified to remove the gross linear trend and offset. The modified scatter plot then appears as depicted in Fig. 11.14, where scatter is due to a combination of sensor noise and emissivity variation of the candidate blackbody spectra. This data cluster is then segregated into a number of bins along the independent axis, typically on the order of 20. Within each bin, an estimate of L_m is made by selecting a data sample within the bin for which the CDF of samples exceeding it provides the best fit to the model CDF, according to the KS significance in Eq. (11.26). The data in each bin are sorted from high to low radiance. P_{KS} is computed starting with the highest two samples ($M = 2$) and proceeding until it falls between some predetermined minimum such as 1×10^{-7}. At some M between the highest and lowest samples, P_{KS} will peak; that sample is taken to represent L_m for the bin. A linear regression is then performed using the L_m estimates from each bin, and

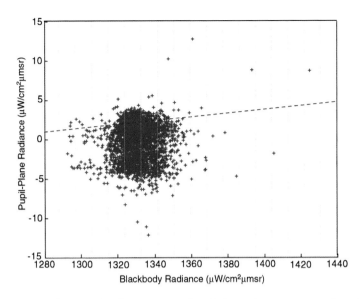

Figure 11.14 Scatter plot from 8.5-μm band data after gross trend and offset removal.

these regression coefficients are then added to the gross coefficients that were initially estimated. The final aggregate slope estimate represents atmospheric transmission for the wavelength band being investigated, and the *y* intercept represents the path radiance estimate. By applying an upper bound linear fit at each wavelength, atmospheric transmission and path radiance spectra can be estimated, and Eq. (11.22) can be inverted to estimate the upwelling radiance spectrum for each image pixel by

$$\hat{L}_u(\lambda) = \frac{L_p(\lambda) - \hat{L}_a(\lambda)}{\hat{\tau}_a(\lambda)}. \qquad (11.30)$$

As an example of ISAC blackbody normalization, the method is applied to SEBASS LWIR hyperspectral imagery collected over the ARM site to derive the atmospheric transmission and path radiance estimates illustrated in Fig. 11.15. While these estimates, denoted as unscaled in the plots, capture the expected spectral characteristics as compared to MODTRAN model predictions, path radiance estimates are negative over a part of the spectral range, making them nonphysical. This can occur because of the fundamental assumption of the method that there are spectra in the scene that are truly blackbodies, and that the spectral apparent temperature at λ_r represents the true surface temperature. If these assumptions are not satisfied at a particular wavelength, path radiance would be underestimated. One way to deal with this problem is to scale the measurements so that atmospheric transmission matches a model-based prediction for a particular wavelength at which a good scene estimate can be derived. The particular wavelength that has proven useful for these purposes is an 11.7-μm water vapor band that is clearly evident in the atmospheric transmission profile.

Let $\tau_{a,u}(\lambda)$ and $L_{a,u}(\lambda)$ represent unscaled atmospheric transmission and path radiance estimates using the methodology described earlier, and assume that the radiative transfer model predicts atmospheric transmission τ_0 at a reference wavelength λ_0. Atmospheric transmission can be made consistent with the model by the simple scaling

$$\hat{\tau}_a(\lambda) = \frac{\tau_0}{\tau_{a,u}(\lambda_0)} \, \tau_{a,u}(\lambda). \tag{11.31}$$

If atmospheric transmission is scaled, however, path radiance must also be scaled to remain consistent with the observed data. Since atmospheric transmission and path radiance were the best fit to the observed radiance variation relative to the spectral apparent temperature, scaling of the transmission means that the estimates will no longer be a best fit over the full range of scene temperatures. One approach is to force path radiance for all wavelengths to be consistent for an assumed scene temperature T_0, and for the reference wavelength to be consistent with an assumed mean atmospheric temperature T_a based on a linear approximation of blackbody radiance versus temperature. The result is (Young *et al.*, 2002)

$$\hat{L}_a(\lambda) = L_{a,u}(\lambda) + \frac{\tau_{a,u}(\lambda)}{\tau_{a,u}(\lambda_0)} \, [\alpha(\lambda, \lambda_0) \, \{ \tau_{a,u}(\lambda_0) - \tau_0 \}$$
$$- \beta(\lambda, \lambda_0) \, \{ L_{a,u}(\lambda_0) - (1 - \tau_0) \, B(\lambda, T_a) \} \,], \tag{11.32}$$

where

$$\alpha(\lambda, \lambda_0) = \left(1 - \frac{\lambda}{\lambda_0} \right) B(\lambda, T_0), \tag{11.33}$$

and

$$\beta(\lambda, \lambda_0) = \frac{\lambda_0}{\lambda} \, \frac{B(\lambda, T_0)}{B(\lambda_0, T_0)}. \tag{11.34}$$

The resulting scaled atmospheric transmission and path radiance are compared to the unscaled estimates in Fig. 11.15 based on an assumed scene temperature of 310 K and atmospheric temperature of 260 K. The 11.7 μm water band is used for scaling, with the reference transmission estimated from MODTRAN. Based on these scaled atmospheric compensation parameters, estimated upwelling spectral radiance for the

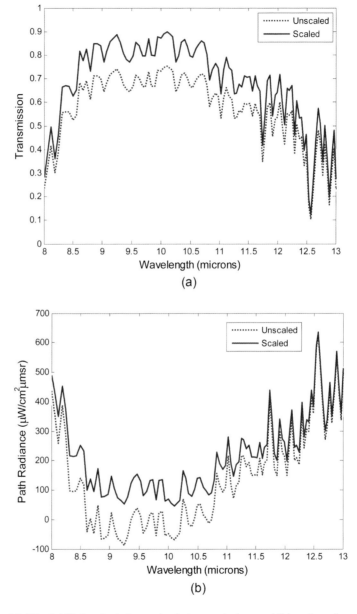

Figure 11.15 (a) Estimated atmospheric transmission and (b) path radiance from SEBASS LWIR hyperspectral data from the ARM site, based on ISAC blackbody normalization.

pond and road measurements is compared to the measured pupil-plane spectral radiance. These are shown in spectral apparent temperature units in Fig. 11.16. The effect of atmospheric compensation is to flatten the spectral apparent temperature for the pond, which ideally would be a straight line at the water temperature due to its blackbody nature. For the road, compensation further enhances the silicate reststrahlen feature in the spectral region around 9 µm.

Some variations in procedures for identifying blackbody-like spectra and for performing a linear fit exist in the employment of blackbody normalization, but this method tends to be a common in-scene atmospheric compensation approach for the LWIR. One such variation is the line-fitting method proposed by Borel (2008) that performs a sequence of linear fits to the scatter plot of blackbody-like spectra and eliminates all data points below the regression line on each iteration (as depicted in Fig. 11.17) to converge toward a fit to the most-blackbody-like spectra at the top of the distribution. The success of blackbody normalization for atmospheric compensation is partly due to the relative preponderance of very high-emissivity materials such as water and natural vegetation in many remotely captured hyperspectral images. Even in relatively arid environments, the method is reported to perform fairly well. Unlike the ELM and vegetation normalization in the VNIR/SWIR, blackbody normalization in the LWIR is not generally effective on spectral data that are not calibrated in an absolute sense because large uncertainties in sensor gain and offset violate assumptions central to blackbody selection and fitting techniques. Nevertheless, the method can be more tolerant of small uncertainties in absolute calibration compared to the model-based approaches described later, making it a practical technique for sensors whose calibration accuracy is limited.

11.1.4 Temperature–emissivity separation

Supposing that upwelling radiance can be accurately estimated in the LWIR using an approach such as that described in the preceding section, the remaining problem is to estimate emissivity from upwelling radiance in the presence of an unknown surface temperature and downwelling radiance. In general, this is very difficult or even impossible to perform because the estimation problem is ill-posed; that is, there are more unknowns than equations for which to estimate the unknowns. To see this, consider a model for the upwelling radiance

$$L_u(\lambda) = \varepsilon(\lambda)\, B(\lambda, T) + [1 - \varepsilon(\lambda)]\, L_d(\lambda). \tag{11.35}$$

If upwelling radiance $L_u(\lambda)$ is estimated from the data using a technique such as blackbody normalization, then surface temperature T and

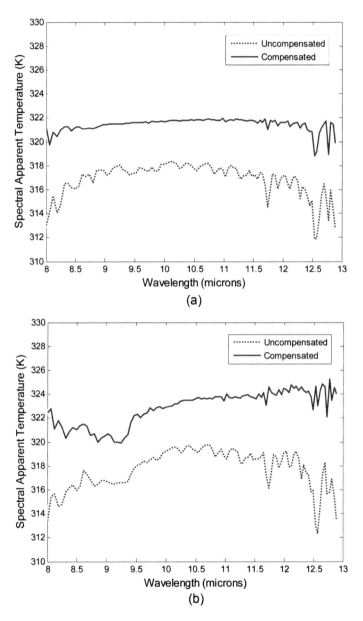

Figure 11.16 Atmospheric compensation results for the ARM (a) site pond spectra and (b) road spectra. The two plots are the spectral apparent temperature before and after application of ISAC with scaling.

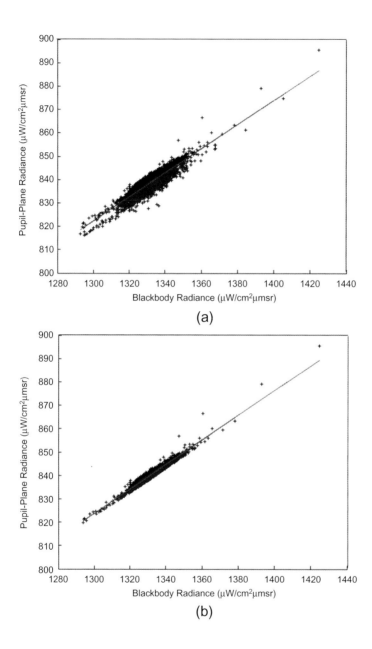

Figure 11.17 Depiction of the sequential line-fitting method for an 8.5-µm band of blackbody-like spectra: (a) first iteration, (b) second iteration, (c) third iteration, and (d) fourth iteration. By eliminating points below the regression line on every iteration, the method converges toward a fit to the most blackbody-like spectra.

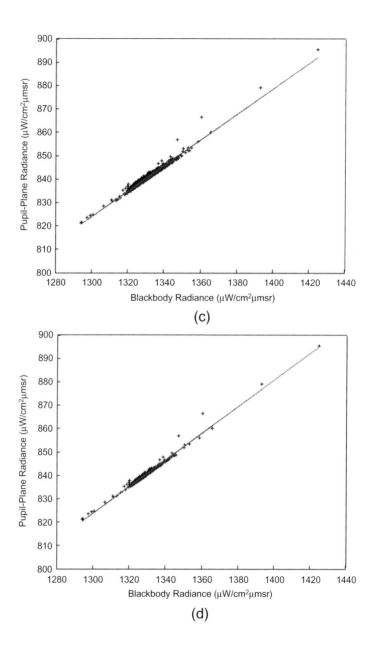

Figure 11.17 (*continued*)

downwelling radiance $L_d(\lambda)$ must be known to derive the emissivity spectrum. If there are K spectral bands, there are $2K + 1$ unknowns to estimate from the K measurements. If the downwelling radiance is space invariant, then it might be possible to estimate it as a common radiance component across the entire scene. Even in that case, there are $K + 1$ unknowns for each scene spectrum and only K measurements. Given some estimate of downwelling radiance and surface temperature, an emissivity spectrum estimate can be made by inverting Eq. (11.35) to

$$\hat{\varepsilon}(\lambda) = \frac{L_u(\lambda) - \hat{L}_d(\lambda)}{B(\lambda, \hat{T}) - \hat{L}_d(\lambda)}. \qquad (11.36)$$

If downwelling radiance is small relative to the emitted radiance because emissivity is high or the surface temperature is significantly larger than the ambient temperature, it can be ignored in both the numerator and denominator. This is not an uncommon practice and is used to make temperature–emissivity separation more tractable. Furthermore, one could simply use the maximum spectral apparent temperature T_{max} associated with upwelling radiance over the spectral range as the temperature estimate, from which the emissivity estimate becomes

$$\hat{\varepsilon}(\lambda) = \frac{L_u(\lambda)}{B(\lambda, T_{max})}. \qquad (11.37)$$

This simple approach leads to an emissivity spectrum that will always have a maximum of unity. That is, there is an unknown scaling in the emissivity estimation due inherently to the ill-posed nature of the estimation problem.

One temperature–emissivity separation method called *alpha emissivity* ignores the reflected component from downwelling radiance and employs Wien's approximation to the blackbody spectral radiance function to estimate emissivity (Kealy and Hook, 1993). That is, it models upwelling radiance as

$$L_u(\lambda) \approx \frac{2hc^2}{\lambda^5} \frac{\varepsilon(\lambda)}{e^{hc/\lambda k_B T}}, \qquad (11.38)$$

where the variable k_B is used for Boltzmann's constant to distinguish it from the spectral index k used later. By taking the logarithm and multiplying both sides by the wavelength λ, Eq. (11.38) reduces to

$$\lambda \ln L_u(\lambda) \approx \lambda \ln \varepsilon(\lambda) + \lambda \ln(2hc^2) - 5\lambda \ln \lambda - \frac{hc}{k_B T}. \qquad (11.39)$$

Subscript k is used to denote the spectral index of upwelling radiance estimate $L_k = L_u(\lambda_k)$ corresponding to a spectrometer center wavelength λ_k, where $k = 1, 2, \ldots, K$. Then Eq. (11.39) can be rewritten as

$$\lambda_k \ln L_k = \lambda_k \ln \varepsilon_k + \lambda_k \ln(2hc^2) - 5\lambda_k \ln \lambda_k - \frac{hc}{k_B T}. \tag{11.40}$$

Taking the spectral average of both sides of Eq. (11.40) leads to

$$\frac{1}{K}\sum_{k=1}^{K}\lambda_k \ln L_k = \frac{1}{K}\sum_{k=1}^{K}\lambda_k \ln \varepsilon_k + \ln(2hc^2)\frac{1}{K}\sum_{k=1}^{K}\lambda_k$$
$$- 5\frac{1}{K}\sum_{k=1}^{K}\lambda_k \ln \lambda_k - \frac{hc}{k_B T}. \tag{11.41}$$

By subtracting Eq. (11.41) from Eq. (11.40) and rearranging terms, the following equation results:

$$\lambda_k \ln \varepsilon_k - \frac{1}{K}\sum_{k=1}^{K}\lambda_k \ln \varepsilon_k = \lambda_k \ln L_k - \frac{1}{K}\sum_{k=1}^{K}\lambda_k \ln L_k + \ln(2hc^2)\frac{1}{K}\sum_{k=1}^{K}\lambda_k$$
$$- \lambda_k \ln(2hc^2) - 5\frac{1}{K}\sum_{k=1}^{K}\lambda_k \ln \lambda_k + 5\lambda_k \ln \lambda_k. \tag{11.42}$$

One defines

$$C_k = \ln(2hc^2)\frac{1}{K}\sum_{k=1}^{K}\lambda_k - \lambda_k \ln(2hc^2) - 5\frac{1}{K}\sum_{k=1}^{K}\lambda_k \ln \lambda_k + 5\lambda_k \ln \lambda_k$$

$$\tag{11.43}$$

as a spectrally varying parameter independent of radiance and emissivity; that is, it is a constant dependent only on spectral band centers and physical constants. By substituting this constant, Eq. (11.42) can be reduced to the form

$$\lambda_k \ln \varepsilon_k - \frac{1}{K}\sum_{k=1}^{K}\lambda_k \ln \varepsilon_k = \lambda_k \ln L_k - \frac{1}{K}\sum_{k=1}^{K}\lambda_k \ln L_k + C_k. \tag{11.44}$$

The expression in Eq. (11.44) relates a normalized function of the material emissivity spectrum to a normalized function of the measured upwelling spectral radiance data. An important characteristic of this relationship is that both sides of the equation are independent of the material surface temperature, as it has cancelled out of the equations. Both sides of the equation are defined as alpha emissivity α_k. That is, the alpha emissivity can be computed directly from the upwelling radiance data as

$$\alpha_k = \lambda_k \ln L_k - \frac{1}{K} \sum_{k=1}^{K} \lambda_k \ln L_k + C_k. \tag{11.45}$$

Surface emissivity can then be estimated from the alpha emissivity according to

$$\hat{\varepsilon}_k = \varepsilon_0 \, e^{\alpha_k / \lambda_k}, \tag{11.46}$$

where

$$\varepsilon_0 = e^{\bar{\alpha}_k / \lambda_k}, \tag{11.47}$$

and

$$\bar{\alpha}_k = \frac{1}{K} \sum_{k=1}^{K} \lambda_k \ln \varepsilon_k \tag{11.48}$$

is a parameter that must be specified to represent the scaling factor ε_0 that the method is not able to estimate.

The alpha emissivity represents a zero-mean quantity that captures the basic spectral shape of the material emissivity and effectively separates this from the temperature dependence of blackbody spectral radiance. From Eq. (11.46), it is possible to compute emissivity from the alpha emissivity, given an estimate of broadband mean emissivity. This makes it a useful quantity for serving as a basis for comparison of upwelling radiance spectra to material emissivity data measured, for example, in a laboratory. As examples, Fig. 11.18 provides emissivity estimates using the alpha emissivity method from ISAC-compensated SEBASS ARM site data based on ε_0 such that peak emissivity is one. The pond estimate is fairly flat but exhibits a general increasing trend with wavelength, while the road spectrum exhibits the expected silicate reststrahlen features. The increasing trend is likely due to errors in the ISAC scaling. The alpha emissivity method reduces the temperature–emissivity separation problem to a model for which there are two unknown parameters: unknown surface temperature T and emissivity scaling factor ε_0. One could consider a variety of other optimization techniques to fit those parameters to the emission-only model for upwelling radiance,

$$L_u(\lambda) = \varepsilon_0 \, e^{\alpha(\lambda)/\lambda} \, B(\lambda, T), \tag{11.49}$$

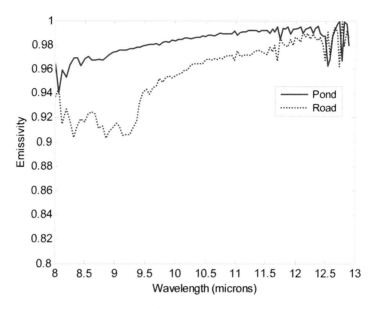

Figure 11.18 Emissivity estimates from compensated SEBASS ARM site data based on the alpha emissivity method.

where T is the unknown surface temperature. There is, however, always some ambiguity between the two because a small change in temperature is indistinguishable from a change in emissivity. Also, errors due to the reflected component are large for low-emissivity materials, or when there is significant downwelling radiance. In these situations, the reflected component needs to be incorporated into the estimation. As the model becomes more complicated, the number of unknowns begins to outweigh the number of measurements, and the optimization problem becomes more ill-conditioned.

To address this ill-conditioned problem, it is necessary to introduce an additional constraint or metric in which to select an optimal surface-temperature estimate among the range of possibilities. As an example of how this can be approached, we consider the method employed by Borel (2008), which has provided good results with LWIR hyperspectral imagery. Examining the observation model,

$$L_p(\lambda) = \tau_a(\lambda)\,\varepsilon(\lambda)B(\lambda, T) + \tau_a(\lambda)\,[1 - \varepsilon(\lambda)]L_d(\lambda) + L_a(\lambda), \quad (11.50)$$

it is apparent that an estimate of downwelling radiance $L_d(\lambda)$ is required in order to have any possibility of performing temperature–emissivity separation in realistic environments for surfaces that are partially reflective. This is addressed by matching in-scene atmospheric parameter estimates to a model. Specifically, atmospheric transmission $\tau_a(\lambda)$ and path radiance $L_a(\lambda)$ are estimated using an in-scene blackbody normalization method similar to ISAC, but then these estimates are compared to a lookup table of MODTRAN predictions for all feasible observation conditions, including imaging geometry and atmospheric modeling parameters, to find the best match based on spectral angle difference criterion (see Chapter 12). Not only are the in-scene atmospheric path transmission and path radiance estimates replaced by these model estimates, but the corresponding model downwelling radiance is used for $L_d(\lambda)$.

At this point, one can estimate the spectral emissivity, given a surface temperature estimate T, as

$$\hat{\varepsilon}(\lambda) = \frac{L_p(\lambda) - \hat{L}_a(\lambda) - \hat{\tau}_a(\lambda)\,\hat{L}_d(\lambda)}{\hat{\tau}_a(\lambda)\,[B(\lambda, T) - \hat{L}_d(\lambda)\,]}. \tag{11.51}$$

As an initial temperature estimate, one can select a temperature at wavelength λ_0 where the atmosphere is highly transmissive (such as 10.1 µm), assume a corresponding surface emissivity $\varepsilon_0 \sim 0.95$, and solve for the corresponding surface temperature consistent with the measurement,

$$T_0 = B^{-1}\left[\lambda_0, \frac{L_p(\lambda_0) - \hat{L}_a(\lambda_0) - \hat{\tau}_a(\lambda_0)\,(1 - \varepsilon_0)\,\hat{L}_d(\lambda_0)}{\hat{\tau}_a(\lambda_0)\,\varepsilon_0}\right], \tag{11.52}$$

where $B^{-1}[\lambda_0, L(\lambda)]$ is an inverse blackbody spectral radiance function of the form given in Eq. (11.23). Inserting Eq. (11.52) into Eq. (11.51), a spectral emissivity consistent with T_0 is provided; however, the estimate is likely to exhibit residual atmospheric features due to the potentially inaccurate initial temperature estimate. The key to the temperature–emissivity separation approach is to search over potential temperatures in the vicinity of T_0 to maximize smoothness of the spectral emissivity estimate, based on the assumption that residual atmospheric features exhibit finer spectral structure than that of surface emissivity. This optimization can be implemented by minimizing the squared error between measured pupil-plane spectral radiance and a model estimate based on a spectrally smoothed rendition of spectral emissivity estimate $\bar{\varepsilon}(\lambda, T)$, expressed as

$$\hat{T} = \min_{T}\left\{\sum_{k=1}^{K}[L_p(\lambda_k) - \hat{L}_p(\lambda_k)]^2\right\}, \tag{11.53}$$

where

$$\hat{L}_p(\lambda) = \tau_a(\lambda)\bar{\varepsilon}(\lambda, T)B(\lambda, T) + \tau_a(\lambda)[1 - \bar{\varepsilon}(\lambda, T)]L_d(\lambda) + L_a(\lambda). \tag{11.54}$$

A simple three-point local spectral averaging filter has been reported to provide acceptable results, but other spectral filters might prove more beneficial, depending on the spectral resolution of the hyperspectral imagery to which the method is applied.

11.2 Model-Based Methods

Model-based atmospheric compensation methods are based on an underlying radiative transfer model, typically MODTRAN, which constrains the space of allowable downwelling radiance, transmission, and path radiance spectra for a given observation. In general terms, such methods attempt to find atmospheric parameters from this feasible set that best correspond to the measured data in some sense. Given the large number of parameters that can be defined to model atmospheric characteristics, this represents a fairly intractable estimation problem. Fortunately, many free variables in an atmospheric model have a similar influence on downwelling and path characteristics. Others do not exhibit significant variability over time, and can be adequately modeled using nominal values. Therefore, the atmospheric retrieval problem can be reduced to an estimation problem over a reasonable set of key underlying parameters, by making judicious choices of these parameters. In this section, model-based methods are first discussed with regard to the VNIR/SWIR spectral region, and are then extended to the LWIR spectral region.

11.2.1 Atmospheric Removal Program

The Atmospheric Removal Program (ATREM) was developed specifically to estimate the apparent surface reflectance from VNIR/SWIR hyperspectral imagery collected with the AVIRIS sensor developed by the NASA Jet Propoulsion Laboratory (JPL) for earth science applications (Green *et al.*, 1993). The approach is based on the MODTRAN radiative transfer code and retrieves atmospheric characteristics by first estimating three key characteristics of the atmosphere: aerosol optical depth, surface pressure elevation, and atmospheric water vapor. These characteristics define the modeling parameters for estimating atmospheric transmission and path radiance that are then coupled with the known solar zenith angle to estimate reflectance spectra from measured pupil-plane radiance data. The underlying model for pupil-plane spectral radiance,

$$L_p(\lambda) = \left[\frac{E_s(\lambda) \cos \theta_s \, \tau_a(\lambda) \rho(\lambda)}{\pi} + L_a(\lambda) \right] * \mathrm{SRF}(\lambda), \qquad (11.55)$$

is similar in form to the diffuse facet model but exhibits a few important differences. First, atmospheric transmission $\tau_a(\lambda)$ is characterized by a two-way transmission from the sun at a zenith angle θ_s to the scene and back up to the sensor, and direct solar irradiance $E_s(\lambda)$ is characterized by the known exo-atmospheric irradiance. Path radiance $L_a(\lambda)$ is due to aerosol scattering and is characterized in the short-wavelength end of the visible spectral range, where the impact is most pronounced. Finally, the modeled pupil-plane spectral radiance is convolved with the known sensor SRF to avoid a spectral resolution mismatch between modeled and measured data in the fitting process.

Aerosol scattering has a significant impact on atmospheric character-istics in the visible spectral range and can exhibit significant variability due to changes in visibility conditions. To estimate aerosol optical depth, pupil-plane radiance data are fit over a 400- to 600-nm spectral range to model radiance data using a nonlinear least-squares-fitting algorithm. The algorithm minimizes the squared error between measured pupil-plane ra-diance spectra and simulated spectra in a MODTRAN-generated lookup table using a simplex search method. Surface reflectance over the spectral range is modeled as

$$\rho(\lambda) = a + b\,\lambda + c\,\rho_v(\lambda), \tag{11.56}$$

where a is the reflectance magnitude parameter, b is the reflectance slope parameter, and $\rho_v(\lambda)$ is a vegetation reflectance distribution scaled by chlorophyll concentration parameter c. Using the model in Eq. (11.50), there are four free parameters over which to optimize the fit between the modeled and measured data: reflectance magnitude, reflectance slope, chlorophyll concentration, and aerosol optical depth that effects both the two-way transmission and path radiance through MODTRAN. As an underlying vegetation model is used for surface reflectance, one would expect better estimation results using vegetated parts of the scene, perhaps based on an NDVI metric.

Figure 11.19 illustrates a radiance fit for the AVIRIS data example, as well as an underlying reflectance fit based on the three-parameter reflectance model. Note that chlorophyll absorption provides the reflectance increase in the green that is characteristic of vegetation. In this particular example, an aerosol optical depth of 0.42 at 500 nm was retrieved. Optical depth is approximated here as the ratio of the transmission decrease $\Delta\tau$ at a specified wavelength to the extrapolated transmission τ_0 expected without a particular absorber or scatterer. Figure 11.20 graphically defines optical depth in the more typical narrowband absorption case as well as this scattering case. Aerosol optical depth values for the 20-km-altitude, down-looking AVIRIS viewing geometry range from 0.27 to 0.53 from mountainous to coastal regions in northern California.

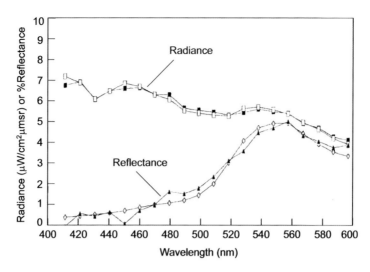

Figure 11.19 Nonlinear least-squares fit for ATREM aerosol optical depth retrieval (Green *et al.*, 1993). The fits of the model correspond to the lines with open data markers.

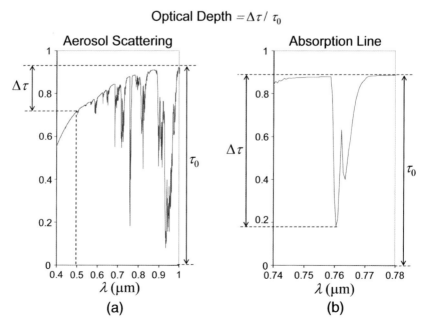

Figure 11.20 Definition of optical depth for (a) aerosol scattering and (b) narrowband absorption.

Surface pressure height is a key parameter for characterizing atmospheric absorption from well-mixed gases and molecular scattering. This parameter is estimated by ATREM from the strength of a 760-nm oxygen absorption line in measured radiance data. This is again performed by a nonlinear least-squares fit of the measured-to-modeled data in the 760-nm region, in this case using pressure elevation as the single free parameter for the MODTRAN transmission and path radiance estimates, and a simpler reflectance model composed solely of reflectance magnitude and reflectance slope parameters. Figure 11.21 shows an example of a fit to AVIRIS data, for which a pressure elevation of 250 m was estimated. Atmospheric water vapor has a considerable influence on path transmission and radiance across the entire VNIR/SWIR spectral region, and its abundance varies greatly both spatially and temporally. To compensate for water vapor absorption, ATREM estimates accumulated water vapor over the entire path, called the water column, for each spatial element of a hyperspectral image. A similar nonlinear least-squares fit is used, but in this case, the estimator is based on the 940-nm water band. The fitting parameters are the water vapor column amount, reflectance magnitude, reflectance slope, and leaf water absorption amount. The underlying surface reflectance is of the same form as in Eq. (11.56) except in this case, vegetation reflectance is replaced by a liquid water reflectance distribution and scaled by the scalar leaf water content parameter. An example of this fit is given in Fig. 11.22.

After estimating these three key atmospheric parameters, estimates for atmospheric transmission $\tau_a(\lambda)$ and path radiance $L_a(\lambda)$ are computed using the MODTRAN forward model. In this case, atmospheric transmission refers to the two-way path from an exo-atmospheric location

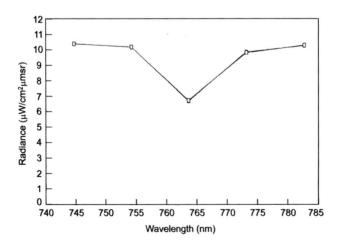

Figure 11.21 Nonlinear least-squares fit for ATREM surface pressure height retrieval (Green *et al.*, 1993). The data and model fit are virtually indistinguishable.

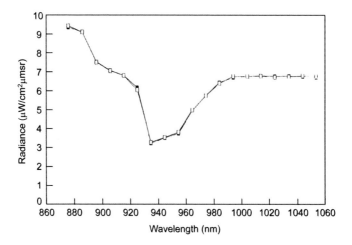

Figure 11.22 Nonlinear least-squares fit for ATREM water vapor column amount retrieval (Green *et al.*, 1993). The data and model fit are virtually indistinguishable.

along the solar illumination direction to the surface and up to the sensor. Downwelling irradiance $E_s(\lambda)$ is defined as the product of the exo-atmospheric solar irradiance and the cosine of the solar zenith angle (to account for the ground projection) and is independent of atmospheric properties. Based on the estimated atmospheric properties, ATREM can be used to normalize a reflectance spectrum to the pupil-plane radiance domain using Eq. (11.55), or to estimate the reflectance spectrum from a pupil-plane radiance measurement by the inverse,

$$\hat{\rho}(\lambda) = \pi \frac{L_p(\lambda) - \hat{L}_a(\lambda)}{E_s(\lambda)\,\cos\theta_s\hat{\tau}_a(\lambda)}. \tag{11.57}$$

Traditionally, it has been applied in the latter fashion to support subsequent data analysis in the reflectance domain.

The description of ATREM in this section is based on the original implementation outlined by Green *et al.* (1993). Its basic approach has been extended to make it more accurate under conditions that do not strictly conform to the model given in Eq. (11.55). Some of these extensions are outlined by Green *et al.* (1996). Further work has been performed by Sanders *et al.* (2001) to generalize the underlying model to include the indirect downwelling radiance absent in Eq. (11.55) to accommodate for atmospheric conditions exhibiting higher aerosol scattering, as well as local environmental radiance from adjacent objects. Instead of detailing these extensions, we will move on to another model-based approach widely used in the hyperspectral remote sensing field that incorporates these factors and is designed also for computational efficiency. The reader should recognize, however, that the specific algorithms described are merely examples of two widely used methods of which there are many variations in practice.

11.2.2 Fast line-of-sight atmospheric analysis of spectral hypercubes

The fast line-of-sight atmospheric analysis of spectral hypercubes (FLAASH) method is a sophisticated, model-based atmospheric-compensation method that attempts to extend the capabilities to more-complicated atmospheric behavior, such as multiple scattering and clouds (Anderson *et al.*, 2002). It is based on a numerical approach that employs large lookup tables of atmospheric profile data over varying conditions to significantly reduce computational time compared to other methods such as ATREM that require numerous forward radiative transfer model runs as part of the nonlinear least-squares fitting. FLAASH is based on a parametric model that differs somewhat from the diffuse facet model, in that it explicitly accounts for local scattering and adjacency effects. As discussed in Chapter 5, it does so by including parameters representative of the average reflectance of the surrounding area and a spherical atmospheric albedo that characterizes the strength of the surface-reflected upwelling radiance that is backscattered back toward the ground. The model has the mathematical form

$$L_p(\lambda) = \frac{A(\lambda)\rho(\lambda)}{1 - S(\lambda)\rho_e(\lambda)} + \frac{B(\lambda)\rho_e(\lambda)}{1 - S(\lambda)\rho_e(\lambda)} + L_s(\lambda), \qquad (11.58)$$

where $\rho(\lambda)$ is the surface reflectance, $\rho_e(\lambda)$ is the average reflectance of the surrounding background, $S(\lambda)$ is the spherical atmospheric albedo, $L_s(\lambda)$ is the radiance backscattered from the atmosphere directly toward the sensor, and $A(\lambda)$ and $B(\lambda)$ are atmospheric parameters related to the model atmosphere. The first term in Eq. (11.58) represents radiance originating from the surface of interest, while the second term represents radiance scattered into the sensor LOS from the surrounding background around the surface along the direct LOS. Parameter $A(\lambda)$ is essentially the product of the total downwelling radiance and atmospheric transmission from the scene to the sensor. The parameter $B(\lambda)$ additionally includes the strength of scattering from the local surround into the sensor LOS. The denominator in the first two terms accounts for increased illumination of the object of interest by the adjacency effect.

The first step in FLAASH is to estimate the water vapor column on a pixel-by-pixel basis. Adjacency effects are ignored in this calculation such that $\rho_e(\lambda)$ is taken to be equal to $\rho(\lambda)$. A regularized lookup table is computed for each wavelength, simulating pupil-plane radiance over a range of water column amounts for a finely spaced set of reflectance values. Also, model parameters $A(\lambda) + B(\lambda)$, $S(\lambda)$, and $L_s(\lambda)$ are computed for each water column amount. From the resulting data, a 2D lookup table is formed, where the water column amount is the dependent variable, and pupil-plane radiance at the center wavelength of an atmospheric water band and the corresponding shoulder wavelength (reference wavelength near the absorption bands) are the independent variables. The basis of the retrieval is the strong correlation between water vapor column amount and the optical depth of the water band, the same feature exploited by ATREM.

The water column is estimated from a measured pupil-plane radiance spectrum by simply searching a gridded, 2D lookup table, where the water band reference radiance and band ratio are the independent variables, and the water vapor column amount is the independent variable. Aerosol retrieval is based on an empirical observation that the reflectance ratio between 0.66 and 2.1-µm spectral bands is typically in the range of 0.4 to 0.5 for natural terrain. Therefore, deviations from this expected ratio averaged over the scene are interpreted as being due to aerosol scattering. Before computing this ratio to retrieve the aerosol, data are spatially convolved with a modified radial exponential PSF to describe the atmospheric scattering of ground-reflected photons. This simulated data cube is referred to as spatially averaged radiance $L_e(\lambda)$. FLAASH also generates a cloud mask by identifying pixels that appear to be cloud pixels based on the brightness, color balance, and low-water-column values computed from the measured radiance spectra. The brightness of the 1.38-µm water band is used to identify cirrus clouds, predicated on the assumption that this water band will be less intense for reflected energy from high-altitude clouds. From atmospheric retrievals, model parameters $A(\lambda)$, $B(\lambda)$, $S(\lambda)$, and $L_s(\lambda)$ are provided for the spatially averaged radiance, which relates to average reflectance according to

$$L_e(\lambda) = \frac{[A(\lambda) + B(\lambda)]\rho_e(\lambda)}{1 - S(\lambda)\rho_e(\lambda)} + L_s(\lambda). \tag{11.59}$$

Average reflectance can be readily computed by solving Eq. (11.59) for $\rho_e(\lambda)$. This provides all of the parameters needed to estimate the reflectance on a pixel-by-pixel basis by algebraically inverting Eq. (11.58) for the reflectance estimate.

As an example of the efficacy of this method, in 1998, the FLAASH atmospheric-compensation model was applied to VNIR/SWIR hyperspectral data collected by the JPL AVIRIS instrument at NASA Stennis Space Center. The data contained a number of ground-truth targets for which surface reflectance measurements were also made (Matthew et al., 2003). Data were collected at a 3-km altitude with an approximately 48-deg solar zenith angle, moderately high humidity (1560 atm-cm water vapor according to a radiosonde measurement), and high visibility. Objects of interest included a black panel, a white panel, grass, and soil. FLAASH was applied at 1-cm^{-1} spectral resolution with an assumed rural haze model and retrieved a visibility of approximately 70 km and average water vapor of 1570 atm-cm. Spectral reflectance estimates for the objects of interest, shown in Fig. 11.23, indicate reasonable agreement with the ground-truth data.

11.2.3 Coupled-subspace model

ATREM and FLAASH atmospheric-compensation methods are effective because spectral characteristics of the atmosphere are distinctive (relative to surface reflectance characteristics), can be characterized by a small number of unknown parameters, and are assumed to be spatially invariant. In these model-based approaches, unknown physical parameters such as aerosol optical depth and water vapor column density are explicitly estimated as part of the compensation process. While it can be a useful by-product of the approaches, this explicit retrieval of atmospheric properties is typically unnecessary. Instead, it is satisfactory to establish accurate estimates of scene irradiance, atmospheric transmission, and atmospheric path radiance. To that end, a coupled-subspace model can be derived that characterizes atmospheric and illumination influences according to basis vectors of the linear diffuse facet model for measured pupil-plane spectral radiance. The atmospheric and illumination properties are then

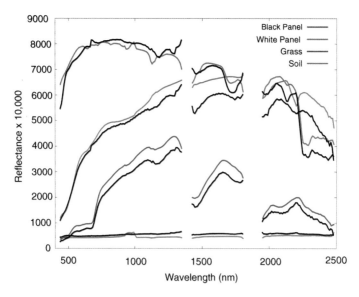

Figure 11.23 Example FLAASH VNIR/SWIR reflectance retrievals from AVIRIS Stennis data compared to surface spectral measurements (Matthew *et al.*, 2003). The black lines are the FLAASH estimates.

approximated by a linear combination of the basis vectors, which are estimated through nonlinear least-squares estimation.

Consider again the form of the diffuse facet model in the reflective spectral region,

$$L_p(\lambda) = \frac{\tau_a(\lambda)}{\pi}[E_s(\theta_s, \varphi_s, \lambda) + E_d(\lambda)]\rho(\lambda) + L_a(\lambda). \qquad (11.60)$$

As for the ELM, this can be written in a linear form

$$L_p(\lambda) = a(\lambda)\rho(\lambda) + b(\lambda), \qquad (11.61)$$

by defining gain and offset parameters,

$$a(\lambda) = \frac{\tau_a(\lambda)}{\pi}[E_s(\theta_s, \varphi_s, \lambda) + E_d(\lambda)] \qquad (11.62)$$

and

$$b(\lambda) = L_a(\lambda). \qquad (11.63)$$

Now suppose that a large set of coupled gain and offset spectra $\{a_i(\lambda), b_i(\lambda); i = 1, 2, \ldots n\}$ are generated using a radiative transfer model such as MODTRAN. This is accomplished by varying the modeling parameters to encompass all possible unknown variations in the imaging geometry, meteorological conditions, and atmospheric and illumination parameters for the particular observation of interest. To determine a coupled basis set for this set of potential realizations, one can form a joint matrix $\mathbf{Z}$, given by

$$\mathbf{Z} = \begin{bmatrix} \mathbf{a}_1 & \mathbf{a}_2 & \cdots & \mathbf{a}_n \\ \mathbf{b}_1 & \mathbf{b}_2 & \cdots & \mathbf{b}_n \end{bmatrix}, \qquad (11.64)$$

where $\mathbf{a}_i$ and $\mathbf{b}_i$ are $1 \times K$ column vector representations of modeled gain and offset spectra,

$$\mathbf{a}_i = \begin{bmatrix} a(\lambda_1) & a(\lambda_2) & \cdots & a(\lambda_K) \end{bmatrix}^T \qquad (11.65)$$

and

$$\mathbf{b}_i = \begin{bmatrix} b(\lambda_1) & b(\lambda_2) & \cdots & b(\lambda_K) \end{bmatrix}^T \qquad (11.66)$$

for K spectral bands. Singular value decomposition is applied to joint matrix $\mathbf{Z}$ to generate optimized basis vectors. These are then decomposed into components corresponding to the gain and offset. More details on performing this dimensionality reduction are provided in Chapter 12. The result is a small set of gain and offset basis spectra $\{A_j(\lambda), B_j(\lambda); j = 1, 2, \ldots J\}$, such that gain and offset parameters for all scene spectra can be approximated by the linear combinations

$$a(\lambda_k) = \sum_{j=1}^{J} \alpha_j A_j(\lambda_k) \qquad (11.67)$$

and

$$b(\lambda_k) = \sum_{j=1}^{J} \alpha_j B_j(\lambda_k). \qquad (11.68)$$

Coefficients α_j are estimated by numerical optimization. Note that in the coupled subspace representation given in Eqs. (11.67) and (11.68), coefficients α_j are the same for gain and offset, are independent of spatial location, and are limited to a smaller number of parameters J than the dimensionality of spectral data K. Estimation of these parameters serves the same essential function as estimating the physical parameters using ATREM or FLAASH. In this case, the dependence on particular atmospheric parameters is implicit. Also, the representation is generalized to any variations for which modeled realizations of gain and offset parameters are generated and is not restricted to specific variations such as water vapor and aerosol content. The spatial-invariance assumption for atmospheric coefficients is a limitation that must be considered when applying this method to imagery with areas of high-terrain relief or other sources of spatial atmospheric variability.

Based on this representation, the diffuse facet model can be approximated as

$$L_p(\lambda_k) = \left[\sum_{j=1}^{J} \alpha_j A_j(\lambda_k)\right] \rho(\lambda_k) + \left[\sum_{j=1}^{J} \alpha_j B_j(\lambda_k)\right]. \qquad (11.69)$$

For any particular scene spectral measurement, there are K measurements and $K + J$ unknown parameters. If there are N spectra in the scene, there will be NK measurements and $NK + J$ unknown parameters, since the basis coefficients are assumed to be space invariant. Despite the space-variant assumption and subspace approximation of atmospheric and illumination characteristics, the estimation problem is ill-posed because of the unknown, space-varying surface reflectance. However, suppose that the surface reflectance can also be approximated by a set of M basis vectors

$\{R_m(\lambda_k); m = 1, 2, \ldots M\}$, which are derived through a similar singular-value decomposition based on a spectral reflectance library of all expected scene materials. In this case, the model can be further approximated as

$$L_p(\lambda_k) = \left[\sum_{j=1}^{J} \alpha_j A_j(\lambda_k) \right] \left[\sum_{m=1}^{M} \beta_m R_m(\lambda_k) \right] + \left[\sum_{j=1}^{J} \alpha_j B_j(\lambda_k) \right], \quad (11.70)$$

where coefficients β_m of the reflectance basis vectors are allowed to vary spatially. The coupled subspace representation in Eq. (11.65) now contains $NM + J$ unknown parameters, which can be estimated from NK measurements using nonlinear least-squares optimization if the surface reflectance variation can be described by a sufficiently small basis set. Another inherent assumption is that basis vectors associated with gain and offset differ enough from those of surface reflectance to distinguish reflectance variation from atmospheric and illumination variation. Published results from HYDICE imagery using Levenberg–Marquardt nonlinear least-squares optimization have indicated reasonably good results using this method (Chandra and Healey, 2008). A derivative of this method is applied to model-based hyperspectral change detection in Chapter 14.

11.2.4 Oblique projection retrieval of the atmosphere

Model-based approaches similar to ATREM and FLAASH have also been pursued in the LWIR spectral region. MODTRAN can again be used as the basis for fitting atmospheric transmission and path radiance to empirical data. In this case, solar irradiance and scattering are unimportant, but atmospheric and environmental temperature profiles become very important. Water vapor content is again a major variable that has a strong influence on atmospheric characteristics; therefore, one typical optimization metric is the fit to narrowband water features in the collected spectra. Also, a spectral smoothness criterion on estimated surface reflectance can provide a useful metric because the apparent spectral reflectance of solid materials generally exhibits much smoother spectral variation than that of the atmospheric gases responsible for the spectral characteristics of path transmission, path radiance, and downwelling radiance.

This section presents a method called oblique projection retrieval of the atmosphere (OPRA) that is based on estimating atmospheric path parameters by spectral projection operators from blackbody-like pixels that are extracted from the image using methods along the lines of those discussed earlier (Villeneuve and Stocker, 2002). The approach is similar to the coupled subspace model in that it performs the estimation of atmospheric parameters using basis vectors established from a set of MODTRAN realizations of atmospheric spectral characteristics. In this case, the method is based on an underlying model, given by

$$L_p(\lambda) = \tau_a(\lambda) \, B(\lambda, T) + L_a(\lambda), \tag{11.71}$$

where a set of pupil-plane radiance spectra are measured for a set of upwelling blackbody spectra of unknown temperature with the same atmospheric path transmission and radiance. Such spectra can be those in the image that are closest to the blackbody-normalization KS regression line discussed in Section 11.1.3. Oblique projection is a signal-processing technique for extracting a lower-dimensional signal from a higher-dimensional measurement space. The signals in this case are spectra, which are described as vectors, and the constrained form is defined as matrices whose columns are basis vectors of the expected variation of the signal and interference terms.

To first understand the underlying signal-processing theory, consider a column vector $\mathbf{x}$ that corresponds to a corrupted observation, including a linear combination of signal term $\mathbf{s}$ and an interference term $\mathbf{b}$. The signal and interference are assumed to reside within subspaces defined by the column spaces of matrices $\mathbf{S}$ and $\mathbf{B}$, respectively. Thus, any vector can be expressed as a linear combination of the form

$$\mathbf{x} = \mathbf{s} + \mathbf{b}$$
$$= \mathbf{S}\alpha + \mathbf{B}\beta, \tag{11.72}$$

where α and β are basis component vectors corresponding to coefficients of the linear combination. In general, $\mathbf{S}$ and $\mathbf{B}$ are not necessarily orthogonal. Linear projection vectors can be used to extract signal and interference components of $\mathbf{x}$. Projection vectors

$$\mathbf{P_S} = \mathbf{S}(\mathbf{S}^T\mathbf{S})^{-1}\mathbf{S}^T \tag{11.73}$$

and

$$\mathbf{P_B} = \mathbf{B}(\mathbf{B}^T\mathbf{B})^{-1}\mathbf{B}^T \tag{11.74}$$

project $\mathbf{x}$ into the signal and interference subspace, respectively. In the case where $\mathbf{S}$ and $\mathbf{B}$ are orthogonal subspaces (i.e., the intersection of the subspaces is the zero vector), the signal projection operator would extract $\mathbf{s}$ from $\mathbf{x}$. The orthogonal subspace projection vectors

$$\mathbf{P_S^\perp} = \mathbf{I} - \mathbf{P_S} = \mathbf{I} - \mathbf{S}(\mathbf{S}^T\mathbf{S})^{-1}\mathbf{S}^T \tag{11.75}$$

and

$$\mathbf{P}_{\mathbf{B}}^{\perp} = \mathbf{I} - \mathbf{P}_{\mathbf{B}} = \mathbf{I} - \mathbf{B}(\mathbf{B}^{T}\mathbf{B})^{-1}\mathbf{B}^{T} \qquad (11.76)$$

perform the complementary operation of extracting the component of $\mathbf{x}$ not residing in the respective subspaces. Suppose that the goal is to provide an estimate of the signal component of $\mathbf{x}$ in the presence of interference for which the subspaces are not orthogonal. The *oblique projection* operator performs this in three steps: (1) it projects the vector into an orthogonal subspace to $\mathbf{B}$ to nullify interference, (2) it projects that vector into the signal subspace $\mathbf{S}$, and (3) it scales the vector to account for the signal part within $\mathbf{B}$ that was nullified in the first step. The operator is given by

$$\mathbf{E}_{\mathbf{SB}} = \mathbf{S}(\mathbf{S}^{T}\mathbf{P}_{\mathbf{B}}^{\perp}\mathbf{S})^{-1}\mathbf{S}\mathbf{P}_{\mathbf{B}}^{\perp}. \qquad (11.77)$$

The theory of oblique projection operators for signal estimation is applied to the atmospheric compensation problem by OPRA in a two-step manner. First, it employs MODTRAN to generate the signal and interference subspace matrices. The underlying physical model in Eq. (11.71) can be written in matrix-vector form as

$$\mathbf{L}_{p} = \boldsymbol{\tau}_{a} \otimes \mathbf{L}_{B} + \mathbf{L}_{a}, \qquad (11.78)$$

where $\otimes$ represents element-by-element multiplication of the $K \times 1$ element spectral column vectors. The first step estimates path radiance by defining the second term in Eq. (11.78) as the signal and the first term as the interference. A subspace matrix $\mathbf{B}_{1}$ is formed by numerically generating an ensemble of pupil-plane radiance data consisting of the product of blackbody radiance functions $\mathbf{L}_{B}$ at varying temperatures and atmospheric transmission profiles $\boldsymbol{\tau}_{a}$ for the range of atmospheric conditions. The data are produced in a Monte Carlo manner by randomly varying the object temperature, water vapor amount, ozone concentration, carbon dioxide concentration, air temperature, and viewing angle across a reasonable range for observation conditions. MODTRAN is used as the underlying model to generate radiance data, and the columns of $\mathbf{B}_{1}$ are formed by a small number of basis vectors (~ 10) that capture the predominant variation of the generated spectra. Dimensionality-reduction methods discussed in Chapter 12 are used for this process. A set of path radiance spectra $\mathbf{L}_{a}$ are generated for the same set of conditions, and these are used to form signal subspace matrix $\mathbf{S}_{1}$ using the same dimensionality-reduction technique. From $\mathbf{S}_{1}$ and $\mathbf{B}_{1}$, the oblique projection operator $(\mathbf{E}_{\mathbf{SB}})_{1}$ can be computed using Eqs. (11.76) and (11.77). Path radiance is then retrieved from the pupil-plane radiance by applying the operator according to

$$\hat{\mathbf{L}}_a = (\mathbf{E}_{\mathbf{SB}})_1 \mathbf{L}_p, \tag{11.79}$$

where $\mathbf{L}_a$ and $\mathbf{L}_p$ are column vectors representing the radiance spectra.

The second step is to extract path transmission, which can be accomplished by recognizing that

$$\ln[L_p(\lambda) - \hat{L}_a(\lambda)] = \ln[\tau_a(\lambda)] + \ln[B(\lambda, T)]. \tag{11.80}$$

In this case, signal matrix $\mathbf{S}_2$ is derived using the same Monte Carlo and dimensionality-reduction approaches as previously described for an ensemble of logarithmic path transmission spectra $\ln \boldsymbol{\tau}_a$. Interference matrix $\mathbf{B}_2$ is derived from an ensemble of logarithmic blackbody radiance spectra $\ln \mathbf{L}_B$. Path transmission spectrum $\boldsymbol{\tau}_a$ is retrieved by applying the oblique projection operator $(\mathbf{E}_{\mathbf{SB}})_2$ for this case, according to

$$\ln \hat{\boldsymbol{\tau}}_a = (\mathbf{E}_{\mathbf{SB}})_2 \ln[\mathbf{L}_p - \hat{\mathbf{L}}_a] = (\mathbf{E}_{\mathbf{SB}})_2 \ln[\mathbf{L}_p - (\mathbf{E}_{\mathbf{SB}})_2\mathbf{L}_p]. \tag{11.81}$$

Once $L_a(\lambda)$ and $\tau_a(\lambda)$ are estimated using Eqs. (11.79) and (11.81), upwelling radiance for any pupil-plane radiance spectrum $L_p(\lambda)$ can be estimated by

$$\hat{L}_u(\lambda) = \frac{L_p(\lambda) - \hat{L}_a(\lambda)}{\hat{\tau}_a(\lambda)}. \tag{11.82}$$

Published results shown in Fig. 11.24 indicate the ability of the method to estimate atmospheric transmission and path radiance of a simulated observed radiance spectrum for a high-altitude sensor and blackbody target of unknown temperature (Villeneuve and Stocker, 2002). The plot in Fig. 11.24(a) shows the simulated sensor measurement and estimated path radiance, while Fig. 11.24(b) compares the estimated atmospheric transmission to the model truth. Additionally, both show the magnitude of the retrieval error in terms of the difference between the simulated and estimated sensor measurements. Most retrieval errors occur in the ozone-absorption region. Application to real hyperspectral imagery was not included in this work.

11.3 Summary

Atmospheric compensation is often an important hyperspectral data processing step when extracting scene spectral characteristics such as apparent surface spectral reflectance, spectral emissivity, or upwelling

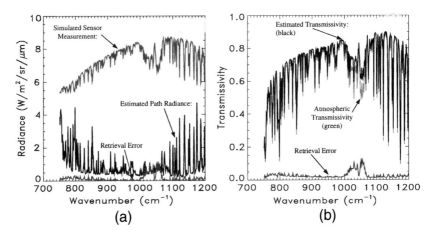

Figure 11.24 Example OPRA retrieval of atmospheric transmission and path radiance of a simulated observed radiance spectrum for a high-altitude sensor and blackbody target of unknown temperature. (a) Simulated sensor measurement and estimated path radiance. (b) Comparison of the estimated atmospheric transmission to the model truth (Villeneuve and Stocker, 2002).

spectral radiance, or when retrieving atmospheric parameters such as path transmission and path radiance. In either case, it provides the link between measured pupil-plane spectral radiance data and the apparent spectral properties of materials in the scene.

In-scene atmospheric compensation methods rely on materials in the scene with known or expected spectral characteristics. Of these, the empirical line method is the simplest but requires specific identifiable materials with known spectral reflectance. This method works well for controlled scenes but has limited applicability to general remote sensing situations. Vegetation normalization has broader applicability due to both the preponderance of vegetation in hyperspectral images and the somewhat consistent nature of vegetation reflectance spectra; however, the absolute reflectance estimation performance is not as effective, largely due to uncertainty in the actual mean vegetation reflectance.

Blackbody normalization is a viable method for use in the LWIR spectral region, again due to the preponderance of water and vegetation with near-blackbody behavior. The methodology for blackbody selection is fairly complicated and heuristic: regression must be performed for the extreme edge of the blackbody data scatter using some statistical test to account for nonblackbody spectra, and path transmission and path radiance spectra must be scaled based on a model for water vapor optical depth. Furthermore, blackbody normalization is only able to estimate upwelling spectral radiance and requires the application of a temperature–emissivity separation algorithm to estimate spectral emissivity. Research continues to refine all of these processes to make them more robust.

In principle, model-based atmospheric compensation methods can provide better estimates of scene spectral properties because they constrain the estimation problem based on realistic atmospheric and illumination characteristics through a radiative transfer model. The disadvantages of these methods include the large amount of computation involved, the necessity of user familiarity with sophisticated radiative transfer models such as MODTRAN, and a greater susceptibility to absolute calibration errors that causes a mismatch between model predictions and actual collected data, even with perfect atmospheric path estimates. This latter issue is the most significant in practice. While model-based methods have been shown to provide excellent results for well-calibrated hyperspectral imagery, in-scene methods are widely used for less-accurately calibrated sensors and for applications where material detection, classification, and identification is not as sensitive to the absolute estimation errors that might result.

References

Anderson, G. P., Felde, G. W., Hoke, M. L., Ratkowski, A. J., Cooley, T., Chetwynd, J. H., Gardner, J. A., Adler-Golden, S. M., Matthew, M. W., Berk, A., Bernsetin, L. S., Acharya, P. K., Miller, D., and Lewis, P., "MODTRAN4-based atmospheric correction algorithm: FLAASH (Fast Line-of-sight Atmospheric Analysis of Spectral Hypercubes)," *Proc. SPIE* **4725**, 65–71 (2002) [doi:10.1117/12.478737].

Bernstein, L. S., Adler-Golden, S. M., Sundberg, R. L., Levine, R. Y., Perkins, T. C., Berk, A., Ratkowski, A. J., Felde, G., and Hoke, M. L., "A new method for atmospheric correction and aerosol optical property retrieval for VIS-SWIR multi- and hyperspectral imaging sensors: QUAC (QUick Atmospheric Correction)," *Proc. Int. Geosci. Remote Sens. Symp.* **5**, 3349–3352 (2005).

Borel, C. C., "Error analysis for a temperature and emissivity retrieval algorithm for hyperspectral imaging data," *Int. J. Remote Sens.* **29**, 5029–5045 (2008).

Chandra, K. and Healey, G., "Using coupled subspace models for reflectance/illumination separation," *IEEE T. Geosci. Remote Sens.* **46**(1), 284–290 (2008).

Green, R. O., Conel, J. E., and Roberts, D. A., "Estimation of aerosol optical depth, pressure elevation, water vapor, and calculation of apparent surface reflectance from radiance measured using the Airborne Visible/Infrared Imaging Spectrometer (AVIRIS) using a radiative transfer code," *Proc. SPIE* **1937**, 2–11 (1993) [doi:10.1117/12.157054].

Green, R. O., Roberts, D. A., and Conel, J. E., "Characterization and compensation of the atmosphere for the inversion of AVIRIS calibrated radiance to apparent surface reflectance," *Proc. AVIRIS Geosci. Work.*, JPL Publication 96–4, Jet Propulsion Laboratory, Pasadena, CA (1996).

Johnson, B. R., *Inscene Atmospheric Compensation: Application to SEBASS Data Collected at the ARM Site*, Part I, Aerospace Report No. ATR-99(8407)-1, The Aerospace Corporation, El Sequndo, CA (1998).

Kealy, P. S. and Hook, S. J., "Separating temperature and emissivity in thermal multispectral scanner data: implications for recovering land surface temperatures," *IEEE T. Geosci. Remote Sens.* **31**, 1155–1164 (1993).

Matthew, M. W., Adler-Golden, S. M., Berk, A., Felde, G., Anderson, G. P., Gorodetsky, D., Paswaters, S., and Shippert, M., "Atmospheric correction of spectral imagery: evaluation of the FLAASH algorithm with AVIRIS data," *Proc. SPIE* **5093**, 474–482 (2003) [doi:10.1117/12.499604].

Press, W. H., Teukolsky, S. A., Vetterling, W. T., and Flannery, B. P., *Numerical Recipes*, Cambridge University Press, Cambridge, MA (1992).

Sanders, L. C., Schott, J. R., and Raqueno, R. V., "A VNIR/SWIR atmospheric correction algorithm for hyperspectral imagery with adjacency effect," *Remote Sens. Environ.*, **78**, 252–263 (2001).

Villeneuve, P. V. and Stocker, A. D., "Oblique projection retrieval of the atmosphere," *Proc. SPIE* **4725**, 95–103 (2002) [doi:10.1117/12.478740].

Young, S. J., Johnson, B. R., and Hackwell, J. A., "An in-scene method for atmospheric compensation of thermal hyperspectral data," *J. Geophys. Res.* **107**(D24), 4774–4793 (2002).

Chapter 12
Spectral Data Models

Hyperspectral imagery contains a wealth of information about the material content of remotely imaged scenes based on the unique spectral characteristics of the objects within them. This information is perturbed by spectral properties of the illumination, atmosphere, and environment, as previously described, and is further corrupted by sensor spectral response and system noise. Extracting information from hyperspectral data for any application requires consideration of all of these factors. While analytical methods can be unique to a particular remote sensing application, they generally share the overall challenge of needing to sort through the huge volume of data that hyperspectral imagery can represent to detect specific materials of interest, classify materials into groupings with similar spectral properties, or estimate specific physical properties based on their spectral characteristics. Whether such an analysis is performed in a fully automated fashion or with significant manual intervention, the ability to perform these primary analytical functions relies on suitable underlying models for the observed data, relative to the expected signal, background, and noise properties. It also requires numerical methods to aid in the extraction of pertinent and useful information from the volume of data. Such models and methods are explored in this chapter. The emphasis is on statistical and geometric models that have been widely employed across hyperspectral remote sensing applications, even if they originated from other fields such as communications, speech processing, and pattern recognition. Since numerical approaches for hyperspectral image classification and target detection are closely coupled with underlying statistical models, this chapter serves as a basis for the two chapters to follow that address these topics.

12.1 Hyperspectral Data Representation

Hyperspectral imagery contains both spatial and spectral information about a scene, and optimal analysis of hyperspectral imagery should exploit both of these attributes. But it is the spectral nature of collected data that is unique to this remote sensing method, relative to conventional passive imaging sensors; therefore, the spectral attribute is generally the primary focus of hyperspectral data analysis, and it is the focus of the treatment given here. With that in mind, data are typically represented as a set of spectral measurements of a scene, $\{\mathbf{x}_i, i = 1, 2, \ldots, N\}$, where i is a spatial index, and each measurement $\mathbf{x}_i$ is a $K \times 1$ element column vector organized as

$$\mathbf{x}_i = \begin{bmatrix} L_i(\lambda_1) & L_i(\lambda_2) & \cdots & L_i(\lambda_K) \end{bmatrix}^T, \qquad (12.1)$$

where $L_i(\lambda_k)$ is a measurement for the spectral band centered at λ_k, and K is the number of spectral bands. It is implied in Eq. (12.1) that $\mathbf{x}_i$ is composed of pupil-plane spectral radiance measurements, which is ideally the case; however, $\mathbf{x}_i$ can also represent some other form of spectral measurement, such as uncalibrated data units or estimated reflectance, emissivity, or upwelling radiance as an output of a particular atmospheric compensation algorithm. An entire hyperspectral image can be placed in a *data matrix*

$$\mathbf{X} = \begin{bmatrix} \mathbf{x}_1 & \mathbf{x}_2 & \cdots & \mathbf{x}_N \end{bmatrix}, \qquad (12.2)$$

where N is the number of collected spectral or spatial pixels in the image. The spatial relationships in the two spatial dimensions of the image are not captured by Eq. (12.2), but it is implied that they are maintained by the rastered ordering of the vectors within the data matrix.

12.1.1 Geometrical representation

Since each spectral measurement $\mathbf{x}_i$ can be represented through Eq. (12.1) as a point in a K-dimensional vector space, it is natural that geometric concepts are used to characterize relationships between spectra. A fundamental property of interest is a measure of the similarity between spectral measurements such as $\mathbf{x}_1$ and $\mathbf{x}_2$, depicted in Fig. 12.1. This can be characterized geometrically as multidimensional distance d_{12}, given by the generalized distance formula

$$d_{12} = |\mathbf{x}_1 - \mathbf{x}_2| = \sqrt{(\mathbf{x}_1 - \mathbf{x}_2)^T (\mathbf{x}_1 - \mathbf{x}_2)}. \qquad (12.3)$$

By this measure, smaller differences infer greater spectral similarity. One problem with this particular measure, however, is that it increases when one measurement is simply scaled relative to the other; that is, $\mathbf{x}_2 = \alpha \mathbf{x}_1$, where α is a scalar. In this case, the underlying spectral shape of both measurements is identical, but the length of one is increased relative to the other by, for example, a change in the strength of illumination. In fact, spectral vector changes in the radial direction often indicate illumination

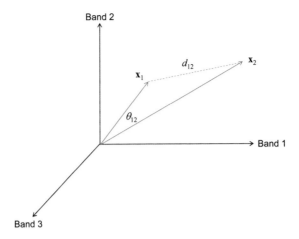

Figure 12.1   Geometrical representation of hyperspectral data and measures of spectral similarity.

changes, while changes in the angular direction are indicative of spectral or color changes. Therefore, it is perhaps more appropriate to use the multidimensional spectral angle θ_{12}, defined by

$$\theta_{12} = \cos^{-1}\left(\frac{\mathbf{x}_1^T \mathbf{x}_2}{\sqrt{(\mathbf{x}_1^T \mathbf{x}_1)(\mathbf{x}_2^T \mathbf{x}_2)}}\right), \tag{12.4}$$

as a measure of spectral similarity rather than using the generalized distance formula.

Another geometric concept concerning hyperspectral data modeling is the spectral mixture that occurs when measurements of a spatial pixel are composed of multiple constituent materials. For example, suppose that $\mathbf{x}_1$ and $\mathbf{x}_2$ represent measurements of two pure materials in a scene, and a third measurement $\mathbf{x}_3$ is made for which the pixel is composed of a fraction, or abundance, α_1 of material 1 and the remainder $\alpha_2 = 1 - \alpha_1$ of material 2. According to a linear mixing model, it is expected that $\mathbf{x}_3$ would fall along the line segment adjoining $\mathbf{x}_1$ and $\mathbf{x}_2$ (as shown in Fig. 12.2) such that $\alpha_1 = d_{13}/d_{12}$ and $\alpha_2 = d_{23}/d_{12}$. Furthermore, if $\mathbf{x}_1$ and $\mathbf{x}_2$ represented the brightest measurements of the two pure materials, then all pure measurements of these materials would fall along the corresponding edges of the triangle depicted in Fig. 12.2, originating at the origin, which would represent a dark point. Any mixture of these two pure materials would fall within this 2D triangular region regardless of how many spectral dimensions existed in the data. If there were additional pure materials, then all mixtures would fall within a simplex formed with the pure materials as vertices. Thus, while the spectral dimensionality of the sensor can be quite

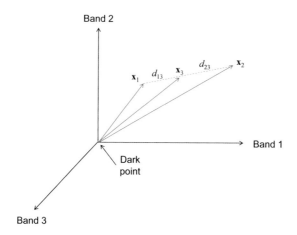

Figure 12.2   Geometrical concept of linear spectral mixing.

large, the dimensionality of the data is limited by the number of distinct materials composing the scene.

Geometric concepts can finally be used to describe the expected transformation of spectral data as a result of changes in illumination and environmental conditions. Under the assumptions of the diffuse facet model, a pupil-plane spectral radiance measurement vector $\mathbf{x}$ is related to the corresponding surface material reflectance vector ρ according to the relationship

$$\mathbf{x} = \mathbf{a} \otimes \rho + \mathbf{b}, \tag{12.5}$$

where $\otimes$ denotes element-by-element vector multiplication. In the reflective spectral region, $\mathbf{a}$ represents the product of the atmospheric transmission and total downwelling radiance, while $\mathbf{b}$ represents path radiance. In the emissive region, both $\mathbf{a}$ and $\mathbf{b}$ are composed of atmospheric transmission, blackbody radiance, and downwelling radiance, while $\mathbf{b}$ also includes path radiance according to the relationships given in Chapter 5. If environmental and illumination conditions change uniformly over the scene, then these characteristics change to $\mathbf{a}'$ and $\mathbf{b}'$ under some other observation condition, and each spectral measurement changes to

$$\mathbf{x}' = \mathbf{a}' \otimes \rho + \mathbf{b}'. \tag{12.6}$$

This transformation occurs similarly for all scene measurements, such that the data are remapped from one region of the multidimensional space to another in some ordered manner, as illustrated in Fig. 12.3. The dimensionality is unchanged, and the linear mixing relationships apply in the same manner for the transformed data as they apply for the original.

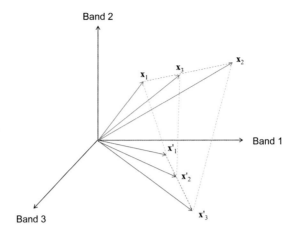

Figure 12.3 Affine transformation of hyperspectral data due to a space-invariant illumination change.

If the transformation is truly linear, then each vector **x** can be mapped to its corresponding vector **x′** with an affine transformation. This is not true if the environmental or illumination changes are space varying, if there is nonlinear mixing due to adjacency effects, if there are pixel-to-pixel surface temperature changes for emissive data, or if for some reason the reflectance vectors change between observations.

12.1.2 Statistical representation

While geometrical concepts are very useful in characterizing spectral similarity, mixing, and change transformation, they fail to model the inherent randomness in hyperspectral data due to sensor noise and the stochastic nature of real scenes. To incorporate this behavior into the model, spectral measurements are considered to be random vector processes characterized by probability density functions (Papoulis, 1984). Typically, a spectral measurement **x** is described by a multidimensional probability density function $p(\mathbf{x}; \theta)$ that is governed by a set of parameters θ. As a simple example, suppose that measurement **x** is composed of a deterministic signal **s** related to the ideal spectrum, along with zero-mean, spatially independent, normally distributed noise **n** with a wavelength-independent variance σ^2. By virtue of the additive model,

$$\mathbf{x} = \mathbf{s} + \mathbf{n}. \tag{12.7}$$

The probability density function for **x** is given by

$$p(\mathbf{x}; \theta) = \frac{1}{(2\pi)^{K/2}} \frac{1}{|\Sigma|^{1/2}} \exp\left\{-\frac{1}{2}[\mathbf{x} - \mu]^T \Sigma^{-1} [\mathbf{x} - \mu]\right\}, \tag{12.8}$$

where the parameters of the distribution $\theta = \{\mathbf{\mu}, \mathbf{\Sigma}\}$ are the mean vector $\mathbf{\mu} = \mathbf{s}$ and covariance matrix $\mathbf{\Sigma} = \sigma^2 \mathbf{I}$, where $\mathbf{I}$ is the identity matrix. This is denoted in shorthand by $\mathbf{x} \sim N(\mathbf{s}, \sigma^2 \mathbf{I})$.

This statistical representation can be extended to characterize not only sensor noise but also the stochastic nature of the scene itself, that is, the fact that there is an inherent randomness to the spectral properties of natural backgrounds and manmade objects. In this case, a simple normal distribution typically does not fully represent the complexity of scene variability, but it still serves as a useful foundation for more complicated statistical models to be discussed in this chapter. Even so, the parameters of the distribution are often not known *a priori* and must be estimated from the collected data from the scene. The concept of maximum likelihood is the standard statistical method employed to perform such estimation. As an example, assume that collected data $\{\mathbf{x}_i, i = 1, 2, \ldots, N\}$ conform to an underlying statistical distribution $p(\mathbf{x}; \theta)$ for which each spectral measurement is statistically independent and identically distributed. The likelihood function for the collected data corresponding to a parameter set θ is then given by

$$L(\mathbf{x}_1, \mathbf{x}_2, \ldots, \mathbf{x}_N; \theta) = \prod_{i=1}^{N} p(\mathbf{x}_i; \theta). \tag{12.9}$$

Maximum-likelihood estimates for unknown parameters in the set $\theta = \{\theta_1, \theta_2, \ldots, \theta_L\}$ are those for which the expression in Eq. (12.9) reaches a maximum, which occurs when

$$\frac{\partial L(\mathbf{x}_1, \mathbf{x}_2, \ldots, \mathbf{x}_N; \theta)}{\partial \theta_l} = 0 \quad l = 1, 2, \ldots, L. \tag{12.10}$$

These are known as the maximum-likelihood equations and can theoretically be solved if the parametric form of probability density function $p(\mathbf{x}; \theta)$ is known or assumed. In many cases, numerical methods are required to perform the maximization.

For many parametric probability density functions, it is more straightforward but equivalent to maximize the log-likelihood function $l(\mathbf{x}_1, \mathbf{x}_2, \ldots, \mathbf{x}_N; \theta)$, which, in the case of the multivariate normal distribution, is given by

$$l(\mathbf{x}_1, \mathbf{x}_2, \ldots, \mathbf{x}_N; \mathbf{\mu}, \mathbf{\Sigma}) = \ln[L(\mathbf{x}_1, \mathbf{x}_2, \ldots, \mathbf{x}_N; \mathbf{\mu}, \mathbf{\Sigma})]$$

$$= -\frac{NK}{2} \ln(2) - \frac{N}{2} \ln|\mathbf{\Sigma}| - \frac{1}{2} \sum_{i=1}^{N} (\mathbf{x}_i - \mathbf{\mu})^T \mathbf{\Sigma}^{-1} (\mathbf{x}_i - \mathbf{\mu}). \tag{12.11}$$

The maximum-likelihood estimates for mean vector $\boldsymbol{\mu}$ and covariance matrix $\boldsymbol{\Sigma}$ can be solved analytically using Eq. (12.10) and are simply the sample statistics

$$\mathbf{m} \equiv \hat{\boldsymbol{\mu}} = \frac{1}{N} \sum_{i=1}^{N} \mathbf{x}_i = \frac{1}{N} \mathbf{X} \mathbf{u} \qquad (12.12)$$

and

$$\mathbf{C} \equiv \hat{\boldsymbol{\Sigma}} = \frac{1}{N} \sum_{i=1}^{N} [\mathbf{x}_i - \mathbf{m}][\mathbf{x}_i - \mathbf{m}]^T = \frac{1}{N} \mathbf{X}\mathbf{X}^T - \mathbf{m}\mathbf{m}^T, \qquad (12.13)$$

where $\mathbf{u}$ is an $N \times 1$ element column vector with unit values. Note that the caret on the statistical parameters designates them as parameter estimates, and that the $\mathbf{m}$ and $\mathbf{C}$ are used to represent sample statistics corresponding to unknown statistical parameters $\boldsymbol{\mu}$ and $\boldsymbol{\Sigma}$. This notation is employed throughout this chapter. A statistic closely related to the sample covariance matrix $\mathbf{C}$ is the correlation matrix:

$$\mathbf{R} = \frac{1}{N} \sum_{i=1}^{N} \mathbf{x}_i \mathbf{x}_i^T = \frac{1}{N} \mathbf{X}\mathbf{X}^T = \mathbf{C} + \mathbf{m}\mathbf{m}^T. \qquad (12.14)$$

While the maximum-likelihood estimate for the mean vector is unbiased, the covariance matrix estimate is biased. In this case, the optimal unbiased estimator is the consistent estimator, given by

$$\hat{\boldsymbol{\Sigma}} = \frac{N}{N-1} \mathbf{C}. \qquad (12.15)$$

For large vales of N, these are essentially the same.

There are complications in statistical representations of hyperspectral data using parametric forms of the probability density function [such as the normal distribution in Eq. (12.8)] due to high data dimensionality. As discussed earlier, the true dimensionality of hyperspectral data is limited by the number of spectrally distinct materials on which the scene is composed. Typically, this is only a fraction of the dimensionality of the data, that is, the number of spectral bands of the instrument. Because of this, the sample covariance matrix computed according to Eq. (12.13) will not be full rank and will theoretically be noninvertible. In the presence of noise, the sample covariance matrix can actually be invertible in practice but will be ill-conditioned, introducing the possibility of large errors in the inverse used to represent the statistical distribution of the data. Because of this dimensionality problem, regularization techniques such as SVD are almost always needed to perform the covariance matrix inversion. Because it appears to be a fundamental property of hyperspectral data, however, this dimensionality issue warrants further investigation, as it seems to indicate that data representation is highly inefficient and overly sensitive to noise.

12.2 Dimensionality Reduction

From the geometric representation in Fig. 12.2, it is apparent from the linear spectral mixing concept that signal content within hyperspectral data is likely restricted to reside within a lower-dimensional subspace of the K-dimensional data space, where subspace dimensionality is dictated by the number of spectrally distinct materials in the scene. Depending on scene complexity, this dimensionality could be small or quite large. In real sensor data, however, spectral measurements are corrupted by noise, which is completely random and not restricted in the same manner. The reduced dimensionality of the information content within hyperspectral imagery can be recognized by the high degree of correlation that typically exists between spectral bands.

For example, Fig. 12.4 illustrates three bands across the full spectral range of a Hyperion satellite image over the intersection of two major highways in Beavercreek, Ohio. These bands capture some of the same spatial features of the scene but also indicate significant spectral differences. The band-to-band correlation is modest, and each band carries a substantial amount of new information not contained within the others. On the other hand, there are other band combinations, such as those depicted in Fig. 12.5, for which correlation is very high and each band carries little additional information relative to the others. To a large degree, the band combinations are redundant. This high degree of correlation, or reduced inherent dimensionality, implies that the data can be represented in a more compact manner. Transforming the data into a reduced-dimensionality representation has multiple benefits. First, it limits the amount of data needed for processing and analysis, an obvious advantage where computer memory, network bandwidth, and computational resources are concerned. Additionally, representing the data according to the primary signal components as opposed to the sensor spectral bands accentuates underlying material content, aiding visualization and analysis. Finally, transforming to a lower-dimensional subspace should provide noise reduction, as this process filters out the noise power in the subspace that is removed.

12.2.1 Principal-component analysis

The principal-component transformation, commonly known as either principal-component analysis (PCA) (Schowengerdt, 1997) or the

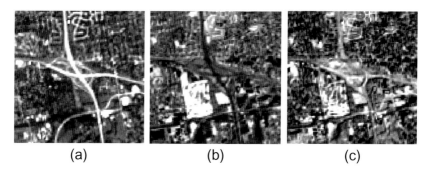

(a)   (b)   (c)

Figure 12.4 Three example bands from a Hyperion VNIR/SWIR hyperspectral image over Beavercreek, Ohio, exhibiting modest spectral correlation: (a) 610 nm, (b) 1040 nm, and (c) 1660 nm.

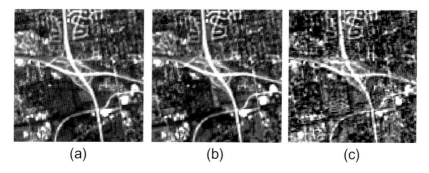

(a)   (b)   (c)

Figure 12.5 Three example bands from a Hyperion VNIR/SWIR hyperspectral image over Beavercreek, Ohio, exhibiting high spectral correlation: (a) 509 nm, (b) 610 nm, and (c) 2184 nm.

Karhunen–Loeve transformation (Karhunen, 1947; Loeve, 1963), addresses the issue of spectral correlation and provides one basis for dealing with data dimensionality. Assume for the moment that the inherent data dimensionality is actually K, such that the covariance matrix is full rank and therefore invertible. Spectral correlation manifests by nonzero, off-diagonal elements of the covariance matrix. Suppose that there was another set of orthogonal coordinate axes in the multidimensional space for which the covariance matrix was actually diagonal. If the data were transformed into this new coordinate system, the spectral correlation between bands would be removed. This is what the PCA transform attempts to perform; it is shown graphically for a simple 2D case in Fig. 12.6, where normally distributed data are presented in the form of a scatter plot. Relative to the sensor spectral bands, the data exhibit a high degree of spectral correlation. However, correlation would be removed in this case if the bands were redefined to correspond to the principal axes of the elliptically shaped scatter distribution, denoted as $\mathbf{v}_1$ and $\mathbf{v}_2$ using dotted vectors.

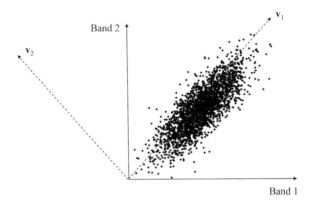

Figure 12.6   2D scatter plot of normally distributed data, illustrating the concept of redefining the bands to correspond to the principal axes in order to remove spectral correlation.

Principal-component transformation is a matter of diagonalizing a sample covariance matrix, a technique that is mathematically performed by determining its eigenvalues and eigenvectors (Strang, 1980). This begins by solving the characteristic equation

$$\det(\mathbf{C} - \sigma^2\mathbf{I}) = 0 \qquad (12.16)$$

for a set of solutions $\{\sigma_j^2, j = 1, 2, \ldots, K\}$, where $\det(\mathbf{A})$ represents the determinant of matrix $\mathbf{A}$, and $\mathbf{I}$ is the $K \times K$ identity matrix. Under the full-rank assumption, K nonzero solutions exist, where each *eigenvalue* σ_j^2 represents the variance of data for a particular eigenvector direction. *Eigenvectors* $\mathbf{v}_j$ correspond to the principal directions for which the spectral correlation is removed and are computed by solving the linear system of equations

$$\mathbf{C}\mathbf{v}_j = \sigma_j^2\mathbf{v}_j \qquad (12.17)$$

for their corresponding eigenvalues. Since eigenvectors can be arbitrarily scaled and still satisfy Eq. (12.17), a unitary basis is chosen, such that

$$\mathbf{v}_j^T\mathbf{v}_j = 1 \qquad (12.18)$$

for all j.

Suppose that diagonal matrix $\mathbf{D}$ is formed by placing the eigenvalues along the diagonal in decreasing order, that is, from highest to lowest variance. The eigenvectors are placed in corresponding order as columns of unitary eigenvector matrix $\mathbf{V}$. It then follows from Eq. (12.17) that

$$\mathbf{CV} = \mathbf{V\,D}. \qquad (12.19)$$

Since the inverse of a unitary matrix is its transpose, it follows that

$$\mathbf{C} = \mathbf{V\,DV}^T, \qquad (12.20)$$

which indicates that the linear transformation represented by eigenvector matrix $\mathbf{V}$ diagonalizes the covariance matrix. Therefore, the principal-component transformation,

$$\mathbf{Z} = \mathbf{V}^T\mathbf{X}, \qquad (12.21)$$

represents a coordinate rotation to principal-component data matrix $\mathbf{Z}$ into an orthogonal basis, such that the new principal-component bands are both uncorrelated and ordered in terms of decreasing variance.

As an example of PCA, consider again the Hyperion Beavercreek image shown in Figs. 12.4 and 12.5. Ranked eigenvalues of the sample covariance matrix are displayed on a logarithmic scale in Fig. 12.7, where it is apparent that variance in the data is predominately captured by a small set of leading principal-component directions. Figure 12.8 provides the first three eigenvectors, while the principal-component band images, corresponding to the Hyperion data, are illustrated in Fig. 12.9. These images capture primary features of the original hyperspectral image with no spectral correlation. Generally, the first principal component corresponds to the broadband intensity variation, while the next few capture the primary global spectral differences across the image. By comparing a three-band false-color composite from these three principal components with an RGB composite from the original 450-, 550-, and 650-nm bands (as illustrated in Fig. 12.10), the ability of the PCA to accentuate scene spectral differences is apparent. Statistically rare spectral features along with sensor noise dominate the low-variance, trailing principal components, three of which are illustrated in Fig. 12.11.

12.2.2 Centering and whitening

Several variations of PCA can be found in the literature and therefore warrant some discussion. The first concerns removal of the sample mean vector $\mathbf{m}$ and scaling of the diagonalized covariance matrix $\mathbf{D}$. Especially when dealing with various target detection algorithms (described in the next chapter), it is sometimes desirable to transform data into an orthogonal coordinate system centered within the data scatter as opposed

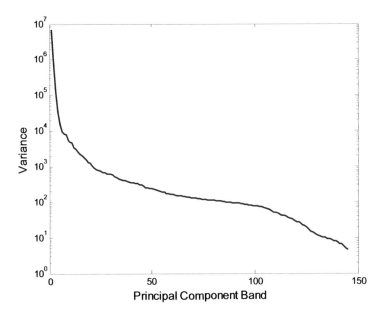

Figure 12.7   Magnitude of the ranked eigenvalues of the Hyperion Beavercreek image.

to the origin defined by the sensor data. This can be performed using the affine transformation,

$$\mathbf{Z} = \mathbf{V}^T(\mathbf{X} - \mathbf{mu}^T) \tag{12.22}$$

in place of the standard principal-component transformation given in Eq. (12.21), or equivalently, by removing the sample mean vector **m** from all original spectral vectors prior to computing and performing the principal-component transformation. This is called centering, demeaning, or mean removal. Another variation is to use the modified transformation to scale the principal-component images such that they all exhibit unit variance:

$$\mathbf{Z} = \mathbf{D}^{-1/2}\mathbf{V}^T(\mathbf{X} - \mathbf{mu}^T). \tag{12.23}$$

This is referred to as whitening the data, and $\mathbf{D}^{-1/2}$ simply refers to a diagonal matrix for which the diagonal elements are all the inverse square root of the corresponding eigenvalues in $\mathbf{D}$. To illustrate the difference between these variants of the principal-component transformation, Fig. 12.12 compares the transformed scatter plot from Fig. 12.6 in the principal component and whitened coordinate system. Data centering is performed in each case.

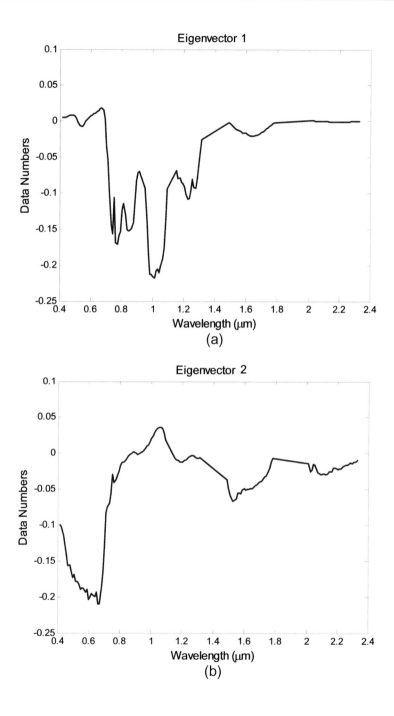

Figure 12.8 Leading eigenvectors for the Hyperion Beavercreek image: (a) principal component (PC) band 1 ($\mathbf{v}_1$), (b) PC band 2 ($\mathbf{v}_2$), and (c) PC band 3 ($\mathbf{v}_3$).

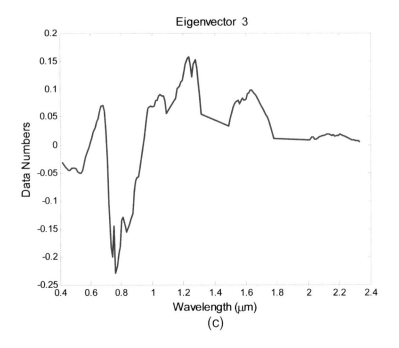

Figure 12.8 (*continued*)

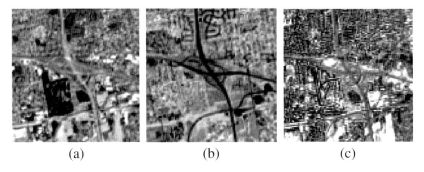

(a)   (b)   (c)

Figure 12.9 Leading principal-component-band images for the Hyperion Beavercreek image: (a) PC band 1, (b) PC band 2, and (c) PC band 3.

12.2.3 Noise-adjusted principal-components analysis

PCA is a standard tool for enhancement of the signal content of hyperspectral imagery. It achieves this enhancement by establishing a ranked ordering of uncorrelated images and capturing the highest-variance image components. However, it does so without any consideration of whether the variance arises from signal or noise. When sensor noise characteristics are known, a variant of PCA sorts the principal-component bands according to the SNR as opposed to total variance to enhance the

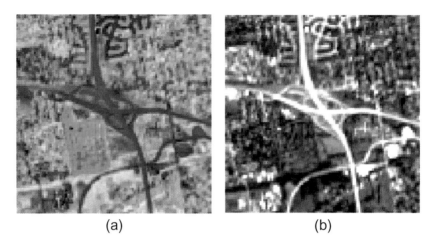

(a)   (b)

Figure 12.10   Comparison of false-color composite image from the leading three principal-component-band images with an RGB composite from the 650-, 550-, and 450-nm spectral bands for the Hyperion Beavercreek image: (a) false color PC composite and (b) RGB spectral composite.

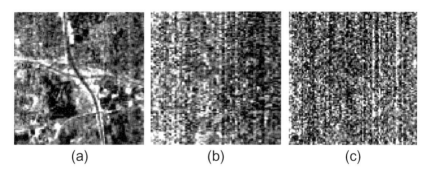

(a)   (b)   (c)

Figure 12.11   Trailing principal-component-band images for the Hyperion Beavercreek image: (a) PC band 8, (b) PC band 10, and (c) PC band 50.

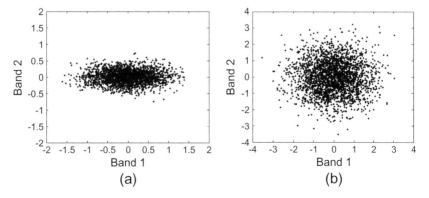

(a)   (b)

Figure 12.12   Comparison of (a) principal component and (b) whitening transforms of the 2D, normally distributed data shown in Fig. 12.6.

signal component relative to noise. This methodology is known in the literature as either the maximum noise fraction (MNF) transformation (Green *et al.*, 1988), or the noise-adjusted principal-component (NAPC) transformation (Lee *et al.*, 1990). The term minimum noise fraction is also used and represented by the same acronym, MNF. In this case, however, the principal components are ordered from high to low SNR, in the reverse order relative to NAPC. This terminology is adopted in this section.

The MNF transform is based on the additive signal model described in Eq. (12.7), where measured data are assumed to be the sum of a signal and a noise component, both of which are considered random vector processes. Assuming independence between these two, the covariance matrix of the data will be the sum

$$\Sigma = \Sigma_s + \Sigma_n, \tag{12.24}$$

where Σ_s is the signal covariance matrix and Σ_n is the noise covariance matrix. MNF seeks to find eigenvectors $\mathbf{v}_j$ that maximize the SNR, defined by

$$\text{SNR} = \frac{\mathbf{v}_j^T \Sigma_s \mathbf{v}_j}{\mathbf{v}_j^T \Sigma_n \mathbf{v}_j} = \frac{\mathbf{v}_j^T \Sigma \mathbf{v}_j}{\mathbf{v}_j^T \Sigma_n \mathbf{v}_j} - 1, \tag{12.25}$$

while simultaneously decorrelating Σ such that

$$\mathbf{v}_i^T \Sigma \mathbf{v}_j = 0 \quad i \neq j. \tag{12.26}$$

This is accomplished in two steps. First, the noise covariance matrix is diagonalized according to

$$\Sigma_n = \mathbf{W} \mathbf{D}_n \mathbf{W}^T, \tag{12.27}$$

where $\mathbf{w}_j$ (the columns of $\mathbf{W}$) are the noise eigenvectors computed by eigen decomposition, as described in Section 12.2.1. The original hyperspectral data matrix $\mathbf{X}$ is then noise-whitened by the transformation

$$\mathbf{Y} = \mathbf{D}_n^{-1/2} \mathbf{W}^T \mathbf{X}, \tag{12.28}$$

then a standard principal-component transformation is performed for the noise-whitened data matrix $\mathbf{Y}$. The resulting principal-component bands are ordered relative to their SNRs, and the diagonal elements of the principal-component covariance matrix equal SNR + 1. Note that Eq. (12.24) represents an additive, signal-independent noise model and ignores the signal dependence of the shot noise that typically dominates high-SNR hyperspectral imagery. This is a common assumption of hyperspectral data models and processing algorithms that is not strictly correct.

If noise is spectrally uncorrelated with equal variance for each spectral band, known as white noise, then $\mathbf{\Sigma}_n = \sigma_n^2 \mathbf{I}$, and the MNF transform is equivalent to the PCA transform but with eigenvalues all scaled by $1/\sigma_n^2$. If it is spectrally uncorrelated but exhibits a spectrally dependent variance, known as colored noise, then the MNF transform simply scales each band in $\mathbf{X}$ by the noise variance prior to performing the PCA transform. This alters the eigenvectors and can result in a difference in ordering of the MNF bands relative to the PCA bands. The difference between MNF and PCA transforms is most pronounced if the sensor noise exhibits some spectral correlation. Unfortunately, whether or not this is the case is not always known from characterization data for a given sensor, making the noise covariance matrix not readily available for the noise whitening in Eq. (12.27). In such cases, it is necessary to estimate the noise covariance matrix from the data itself, which is a challenge because it is typically overwhelmed by the scene variance.

One approach to noise estimation is to estimate noise statistics from sample statistics of the difference between the image and a slightly shifted version of the image, under the assumption that noise is spatially or spectrally uncorrelated over the amount of shift while the signal remains correlated. For panchromatic and multispectral imagery, this is usually implemented by identifying a uniform region of the scene and shifting spatially by one pixel. A problem that occurs is that even a uniform region can exhibit some signal variation over one pixel that is comparable to noise level, thus corrupting the estimate. For hyperspectral imagery, exploiting the spectral correlation of a signal provides an alternate approach (Bioucas-Dias and Nascimento, 2008). Let $\mathbf{Y} = \mathbf{X}^T$ such that $\mathbf{y}_k$ is an $N \times 1$ element column vector containing a k'th spectral band image. Assuming spectral correlation of the signal component, a model

$$\mathbf{y}_k = \mathbf{Y}_{(-k)} \, \boldsymbol{\beta}_k + \mathbf{n}_k \tag{12.29}$$

is assumed for each band, where

$$\mathbf{Y}_{(-k)} = \begin{bmatrix} \mathbf{y}_1 & \mathbf{y}_2 & \cdots & \mathbf{y}_{k-1} & \mathbf{y}_{k+1} & \cdots & \mathbf{y}_K \end{bmatrix} \tag{12.30}$$

is the data matrix with the k'th band removed, $\boldsymbol{\beta}_k$ are the regression coefficients whose least-squares estimates are given by

$$\hat{\boldsymbol{\beta}}_k = [\mathbf{Y}_{(-k)}^T \mathbf{Y}_{(-k)}]^{-1} \mathbf{Y}_{(-k)}^T \, \mathbf{y}_k, \tag{12.31}$$

and $\mathbf{n}_k$ is the realization of noise present in the k'th band. Noise is estimated from these regression coefficients and the data matrix according to

$$\hat{\mathbf{n}}_k = \mathbf{y}_k - \mathbf{Y}_{(-k)}\, \hat{\boldsymbol{\beta}}_k. \qquad (12.32)$$

These noise realizations can then be formed into a data matrix, from which the sample statistics can be computed from Eqs. (12.12)–(12.14) to derive $\mathbf{m}_n$, $\mathbf{C}_n$, and $\mathbf{R}_n$. Sample covariance matrix $\mathbf{C}_n$ is used as the estimate for Σ_n to perform noise whitening for the MNF transform.

While the Bioucas-Dias noise estimation method attempts to compensate nonstationary signal variations, it still works best in regions of uniform spectral content, that is, with spectral samples from a common class. There are other variations of this method, such as that implemented by Roger and Arnold (1996) for modeling noise in AVIRIS imagery. This approach models the signal at each spatial and spectral location as a linear combination of three components: the adjacent spatial sample (in a given direction) for the same spectral band, and the two adjacent spectral samples for the same spatial location.

12.2.4 Independent component analysis

Because PCA and its various extensions are derived solely from second-order statistics of data, there is an implicit assumption that the random vector process underlying hyperspectral imagery is multivariate normal. Principal components are selected to find new data dimensions for which the resulting principal-component band images are decorrelated. However, if the data are not normally distributed, they will exhibit higher-order statistical moments, and the principal-component bands will not necessarily be statistically independent. For such situations, a numerical method called independent component analysis (ICA) seeks to find a different linear decomposition of the spectra on a set of normalized basis vectors, such that the resulting transformed images are not only uncorrelated but also statistically independent. The benefit is that these components have the potential to provide a better representation of fundamental spectral components of the data in situations where this normality assumption is invalid. This section provides a description of the ICA transform and how it can be numerically implemented. For more information, the reader can consult a more detailed reference such as that provided by Hyvärinen and Oja (2000).

Using the same notation as previously used for PCA, ICA represents each spectrum $\mathbf{x}$ by linear decomposition

$$\mathbf{x} = \mathbf{V}\,\mathbf{z}, \qquad (12.33)$$

where the columns of $\mathbf{V}$ are the principal vector directions, and $\mathbf{z}$ is a spectrum in the transformed space where the bands are statistically independent. In PCA, the columns of $\mathbf{V}$ are the eigenvectors of the covariance matrix. In ICA, Eq. (12.33) is written by the inverse

$$\mathbf{z} = \mathbf{W}\,\mathbf{x}, \tag{12.34}$$

where $\mathbf{W}$ is the inverse of $\mathbf{V}$ and is called the *separating matrix*. The goal of ICA is to find a separating matrix that makes bands of the transformed random process $\mathbf{z}$ statistically independent.

Statistical dependence between two random variables z_1 and z_2 is defined by the property

$$p_{12}(z_1, z_2) = p_1(z_1)p_2(z_2), \tag{12.35}$$

where $p_{12}(z_1, z_2)$ is the joint probability density function, and $p_1(z_1)$ and $p_2(z_2)$ are marginal probability density functions. A natural measure of the dependence of K random variables is mutual information I,

$$I(z_1, z_2, \ldots, z_K) = \sum_{k=1}^{K} H(z_k) - H(\mathbf{z}), \tag{12.36}$$

where $H(\mathbf{z})$ is the differential entropy defined by

$$H(\mathbf{z}) = -\int p(\mathbf{z}) \log p(\mathbf{z})\, d\mathbf{z}. \tag{12.37}$$

An important property of mutual information that arises from the linear representation in Eq. (12.34) is

$$I(z_1, z_2, \ldots, z_K) = \sum_{k=1}^{K} H(x_k) - H(\mathbf{x}) - \log | \det \mathbf{W} |. \tag{12.38}$$

If components z_k are chosen such that they are both decorrelated and scaled to unit variance, then it can be further shown that

$$I(z_1, z_2, \ldots, z_K) = C - \sum_{k=1}^{K} J(z_k), \tag{12.39}$$

where C is a constant and $J(\mathbf{z})$ is the negentropy, defined by

$$J(\mathbf{z}) = H(\mathbf{z}_{\mathrm{normal}}) - H(\mathbf{z}), \tag{12.40}$$

where $\mathbf{z}_{normal}$ is a multivariate normal random vector process with the same covariance matrix as $\mathbf{z}$.

It is apparent from Eq. (12.39) that finding a set of orthogonal vector directions $\{\mathbf{w}_k\}$ (forming the rows of $\mathbf{W}$ or columns of $\mathbf{V}$) that maximizes the negentropy sum will result in a transformation that minimizes the statistical dependence of the components. To perform this numerically, it is common to employ an approximation to negentropy, given by

$$J(z) \approx A \left[E\{G(z)\} - E\{G(v)\} \right], \tag{12.41}$$

where A is a constant, $G(z)$ is a nonquadratic function, and v is a normal random variable with zero mean and unit variance. Accuracy of the approximation in Eq. (12.41) is enhanced by a judicious selection of the function $G(z)$. Two choices have been shown to perform well:

$$G_1(z) = \frac{1}{a_1} \log \cosh(a_1 z) \tag{12.42}$$

and

$$G_2(z) = -e^{-z^2/2}, \tag{12.43}$$

where $1 \le a_1 \le 2$.

An algorithm called FastICA finds transformation directions $\{\mathbf{w}_k\}$ by efficiently maximizing the negentropy using a Newton's method solution for the extrema of the expected value $E\{G(\mathbf{w}^T\mathbf{x})\}$. Denoting the derivatives of $G(z)$ by $g(z)$,

$$g_1(z) = \tanh(a_1 z) \tag{12.44}$$

and

$$g_2(z) = z\,e^{-z^2/2}, \tag{12.45}$$

FastICA proceeds as follows. First, hyperspectral data are centered and whitened as previously described. Next, the first direction $\mathbf{w}_1$ is found by the following iteration:

1. Randomly choose an initial weight vector $\mathbf{w}$.
2. Let $\mathbf{w}^+ = E\{\mathbf{x}g(\mathbf{w}^T\mathbf{x})\} - E\{g'(\mathbf{w}^T\mathbf{x})\}\mathbf{w}$, where $g'(z)$ is the first derivative of $g(z)$.
3. Let $\mathbf{w} = \mathbf{w}^+/\|\mathbf{w}^+\|$.
4. Repeats steps 2 and 3 until convergence is achieved.

Note that the expected values in step two are performed by calculating sample statistics of the data; for example,

$$E\{\mathbf{x}\, g(\mathbf{w}^T \mathbf{x})\} = \frac{1}{N} \sum_{i=1}^{N} \mathbf{x}_i g(\mathbf{w}^T \mathbf{x}_i), \qquad (12.46)$$

and $\|\mathbf{w}\|$ represents vector length

$$\|\mathbf{w}\| = \sqrt{\mathbf{w}^T \mathbf{w}}. \qquad (12.47)$$

Convergence is achieved when $(\mathbf{w}^+)^T \mathbf{w}$ is within some minimal difference from unity. The iteration previously outlined is performed successively for each of the K components $\mathbf{w}_k$; however, it is necessary to orthogonalize each new estimate from prior directions using some orthogonalization method. The simplest is the Gram–Schmidt method, where

$$\mathbf{w}_k = \mathbf{w}_k^+ - \sum_{j=1}^{k-1} (\mathbf{w}_k^+)^T \mathbf{w}_j \mathbf{w}_j \qquad (12.48)$$

performs the desired orthogonalization of the new vector from all prior vectors and, by dividing the result by its length, renormalizes to unit length. By forming separating matrix $\mathbf{W}$ from these component directions, ICA is performed by the transformation in Eq. (12.34).

12.2.5 Subspace model

Referring to the Hyperion image eigenvalue distribution in Fig. 12.7, it is apparent that the majority of scene variance is contained within a subspace of much smaller dimensionality than the full number of spectral bands. In this case, trailing principal components would be expected to contain primarily sensor noise. Therefore, it would seem advantageous to describe hyperspectral scene data only with leading principal components. Transformed data matrix $\mathbf{Z}$ from PCA, MNF, ICA, or any other linear transformation onto a new basis set can be written in the form

$$\mathbf{Z} = \begin{bmatrix} \mathbf{L} \\ \mathbf{T} \end{bmatrix}, \qquad (12.49)$$

where $\mathbf{L}$ is an $N \times L$ element matrix representing data in the L-dimensional leading subspace, and $\mathbf{T}$ is an $N \times (K - L)$ element matrix representing data in the $(K - L)$-dimensional trailing subspace. It is assumed in Eq. (12.49) that the principal components are ordered from most to least significant with respect to variance, SNR, or other significance metric. Determination of the leading subspace dimensionality is not always a trivial task but can be approached by selecting the dimensionality that captures a certain percentage of scene variance (i.e., the sum of eigenvalues in leading subspace relative to the total variance) or by selecting the eigenvalue magnitude that is at a particular level relative to a noise estimate as a threshold.

The subspace model is based on the inherent assumption that scene content can be captured within a subspace of the full spectral space, and each spectral vector can be described as a linear combination of a set of vectors that span the scene subspace. Random spectral vector $\mathbf{x}$ is assumed to conform to the model

$$\mathbf{x} = \mathbf{s} + \mathbf{n} = \mathbf{S}\boldsymbol{\alpha} + \mathbf{n}, \tag{12.50}$$

where $\mathbf{S}$ is a matrix of size $K \times L$, whose column vectors span the signal subspace and correspond to the dominant scene spectral components, $\boldsymbol{\alpha}$ is a random vector in the signal subspace containing weights of the linear combination of signal components, and $\mathbf{n}$ is a random vector representing noise content. The first term in Eq. (12.50) is the signal $\mathbf{s} = \mathbf{S}\boldsymbol{\alpha}$. Actual noise generally spans the full dimensionality of the data; however, the components of the noise vector that fall within the signal subspace are typically indistinguishable from the scene content. Therefore, the subspace that is complementary to the scene subspace can be referred to as the noise subspace.

Signal subspace is often determined from data through the PCA, MNF, and ICA methods described in the previous section. After selecting scene dimensionality L by a method to be discussed later, an orthonormal basis for scene and noise subspaces can be formed by decomposing the eigenvector matrix by

$$\mathbf{V} = \begin{bmatrix} \mathbf{S} & \mathbf{S}^{\perp} \end{bmatrix}, \tag{12.51}$$

where $\mathbf{S}$ is an $K \times L$ element matrix whose columns contain the first L-ranked eigenvectors, and $\mathbf{S}^{\perp}$ is a $K \times (K-L)$ element matrix whose columns contain the remaining $(K - L)$-ranked eigenvectors. Since these matrices are both unitary and complementary, it follows that the signal weighting vector and noise can be estimated by

$$\hat{\boldsymbol{\alpha}} = \mathbf{S}^{T}\mathbf{x} \tag{12.52}$$

and

$$\hat{\mathbf{n}} = (\mathbf{S}^{\perp})^{T}\mathbf{x}. \tag{12.53}$$

In some cases, signal subspace is determined by another method and is not in the form of an orthonormal basis. For example, consider a situation where a set of spectral vectors is manually selected from the image, or is given based on some *a priori* information, and it is desired to use them to represent the signal subspace. This can be done by placing them as columns in the matrix $\mathbf{S}$. While the column space of $\mathbf{S}$ spans the scene subspace, it is not necessarily invertible. The least-squares estimates for the scene and noise subspace vectors, however, can be obtained through the pseudo-inverse

$$\hat{\boldsymbol{\alpha}} = (\mathbf{S}^T\mathbf{S})^{-1}\mathbf{S}^T\mathbf{x}, \tag{12.54}$$

and noise can be estimated from the complementary subspace

$$\hat{\mathbf{n}} = [\,\mathbf{I} - \mathbf{S}(\mathbf{S}^T\mathbf{S})^{-1}\mathbf{S}^T\,]\,\mathbf{x}. \tag{12.55}$$

When $\mathbf{S}$ is unitary, Eq. (12.54) easily reduces to Eq. (12.52). If signal dimensionality L is less than data dimensionality K, then signal weighting vector $\boldsymbol{\alpha}$ provides a compact representation of $\mathbf{x}$ with reduced noise.

Another way to use the subspace model is based on the fact that this model decomposes spectral vector $\mathbf{x}$ into the sum of two components, one projected into the signal subspace and the other projected into the noise subspace, or

$$\mathbf{x} = \hat{\mathbf{s}} + \hat{\mathbf{n}} = \mathbf{P}_{\mathbf{S}}\mathbf{x} + \mathbf{P}_{\mathbf{S}}^{\perp}\mathbf{x}. \tag{12.56}$$

For a given matrix $\mathbf{S}$ whose column space spans the signal subspace, the projection operators that directly follow from Eqs. (12.54) and (12.55) are given by

$$\mathbf{P}_{\mathbf{S}} = \mathbf{S}(\mathbf{S}^T\mathbf{S})^{-1}\mathbf{S}^T \tag{12.57}$$

and

$$\mathbf{P}_{\mathbf{S}}^{\perp} = \mathbf{I} - \mathbf{P}_{\mathbf{S}} = \mathbf{I} - \mathbf{S}(\mathbf{S}^T\mathbf{S})^{-1}\mathbf{S}^T. \tag{12.58}$$

When the column space of $\mathbf{S}$ forms an orthonormal basis, these simplify to

$$\mathbf{P}_{\mathbf{S}} = \mathbf{S}\mathbf{S}^T \tag{12.59}$$

and

$$\mathbf{P}_{\mathbf{S}}^{\perp} = \mathbf{I} - \mathbf{S}\mathbf{S}^T. \tag{12.60}$$

12.2.6 Dimensionality estimation

The subspace model provides a compact representation of scene content in a hyperspectral image by projecting data into a subspace of reduced dimensionality. It retains dominant signal components and reduces noise by projecting out the trailing subspace that is assumed to contain only noise. The approach is not without limitations, however, since small amounts of important spectral information can fall within the noise subspace because the variance corresponding to this information fails to significantly impact the scene statistics on which the basis vectors are estimated. Increasing leading subspace dimensionality L decreases this likelihood but forfeits the compactness of the representation and amount of noise reduction. So the question that naturally arises is whether there is some inherent dimensionality for a particular hyperspectral image, and if so, how it can be determined. This is a question that has received some attention in the hyperspectral data processing field and remains somewhat vexing. Nevertheless, some useful methods have been adopted for estimating this inherent dimensionality and are briefly discussed here.

According to the subspace model in Eq. (12.50), each spectral measurement is the sum of a signal vector and noise vector that are assumed to be uncorrelated with each other. Therefore, the covariance matrix for the data can be expressed as the sum

$$\Sigma_{\mathbf{x}} = \Sigma_{\mathbf{s}} + \Sigma_{\mathbf{n}}. \tag{12.61}$$

Suppose that the noise is an independent, identically and normally distributed, white random process, such that

$$\Sigma_{\mathbf{n}} = \sigma_n^2 \mathbf{I}, \tag{12.62}$$

and the data covariance is diagonalized by eigenvector matrix $\mathbf{V}$. In this case, the covariance matrix of transformed data $\mathbf{z}$ exhibits a diagonal covariance matrix

$$\mathbf{D}_{\mathbf{z}} = \mathbf{D}_{\mathbf{s}} + \sigma_n^2 \mathbf{I}, \tag{12.63}$$

and the eigenvalues are

$$(\sigma_{\mathbf{z}}^2)_k = (\sigma_{\mathbf{s}}^2)_k + \sigma_n^2 \mathbf{I}. \tag{12.64}$$

If the signal has an inherent dimensionality L, then the eigenvalues from $L + 1$ to K should be constant and equal to the sensor noise variance, and the inherent dimensionality can be determined by identifying the principal component where the eigenvalue distribution flattens to a constant value.

Unfortunately, the eigenvalue distribution for real data, such as the Hyperion eigenvalue distribution depicted in Fig. 12.7, does not display this behavior, and identifying the noise level is not straightforward. The reason is two-fold. First, noise variance can change with the spectral band (i.e., colored noise) or even exhibit some band-to-band correlation, such that the simple white noise model is not valid. Second, diagonalization of the covariance matrix is performed based on sample statistics instead of on the true covariance matrix. It has been shown that the resulting eigenvalue distribution derived from the sample statistics of white noise is not constant but has a functional form resembling the trailing eigenvalue distribution for the Hyperion data and known as a Silverstein distribution (Silverstein and Bai, 1995). It is apparent that identifying the inherent dimensionality with a noise level that varies with the principal-component band is an ambiguous task.

There are two simple ways to select L. The first selects the principal-component band for which the eigenvalue equals the expected sensor noise variance, disregarding the issues discussed in the preceding paragraph. Another selects the principal-component band for which the leading subspace captures an appropriately high fraction of the total image variance. A useful metric in this regard is the fraction of total variance captured with the leading L dimensions, given by

$$F = \frac{\sum_{k=1}^{L} (\sigma_{\mathbf{z}}^2)_k}{\sum_{k=1}^{K} (\sigma_{\mathbf{z}}^2)_k}. \tag{12.65}$$

This metric is shown in Fig. 12.13 for the Hyperion eigenvalue distribution. It is clear that a high percentage of total variance is captured with just a few principal components. What is not as clear is what total variance fraction is high enough for a given application.

Another proposed approach for determining L is a method called *virtual dimensionality* (Chang and Du, 2004) that examines the eigenvalues of both sample covariance matrix $\mathbf{C}$ and sample correlation matrix $\mathbf{R}$, the difference between which is simply the presence of the data mean in the correlation matrix. By assuming that the signal component of the data has a nonzero mean while the noise is zero mean, one would expect to see an appreciable difference between the eigenvalues of $\mathbf{R}$ relative to $\mathbf{C}$ in the signal subspace, but no such difference in the noise subspace. Therefore, the difference in eigenvalues can be used as a metric for dimensionality estimation. Again, the problem is defining the arbitrary threshold in some meaningful way. A formalism based on a Neyman–Pearson detection criterion has been developed for determining the threshold with respect

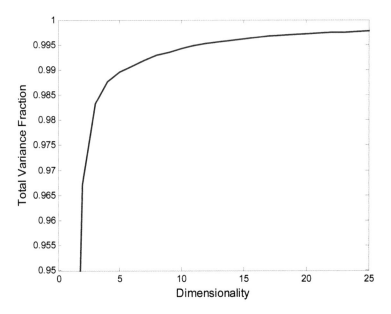

Figure 12.13 Total variance fraction as a function of inherent dimensionality L for the Hyperion image.

to a user-defined false-alarm rate, but the relevance of this arbitrary parameter is not exactly clear. The fundamental problem with all of the dimensionality approaches described thus far is thresholding a monotonic function at some arbitrary level, as opposed to maximizing or minimizing some metric, the results of which would be much less ambiguous.

With this in mind, a technique known as hyperspectral signal identification by minimum error (Bioucas-Dias and Nascimento, 2008) addresses the arbitrary thresholding issue by casting the dimensionality estimation problem as a minimization of the least-squared error between the true signal and subspace representation. Given the subspace model in Eq. (12.50), the mean squared error between the true signal $\mathbf{s}$ and a subspace estimate $\hat{\mathbf{s}}_L$ based on dimensionality L is given by

$$
\begin{aligned}
\text{MSE} &= E\left[(\mathbf{s} - \hat{\mathbf{s}}_L)^T(\mathbf{s} - \hat{\mathbf{s}}_L)\right] \\
&= E\left[(\mathbf{s} - \mathbf{s}_L - \mathbf{P_s}\,\mathbf{n})^T(\mathbf{s} - \mathbf{s}_L - \mathbf{P_s}\,\mathbf{n})\right] \\
&= E\left[(\mathbf{P_s^\perp}\,\mathbf{s} - \mathbf{P_s}\,\mathbf{n})^T(\mathbf{P_s^\perp}\,\mathbf{s} - \mathbf{P_s}\,\mathbf{n})\right] \\
&= \text{trace}(\mathbf{P_s^\perp}\,\mathbf{R_s}) + \text{trace}(\mathbf{P_s}\,\hat{\mathbf{R}}_\mathbf{n}) \\
&= \text{trace}(\mathbf{P_s^\perp}\,\hat{\mathbf{R}}_\mathbf{x}) + 2\,\text{trace}(\mathbf{P_s}\,\hat{\mathbf{R}}_\mathbf{n}) + \text{constant}, \qquad (12.66)
\end{aligned}
$$

where the projection matrices in Eq. (12.66) are based on dimensionality L, $\hat{\mathbf{R}}_\mathbf{x}$ is the estimated data correlation matrix (or sample correlation matrix), and $\hat{\mathbf{R}}_\mathbf{n}$ is the noise correlation matrix estimated by a method similar to that described in Section 12.2.3. Inherent dimensionality is estimated by finding L, so that the mean square error is minimized, given by

$$\hat{L} = \arg \min_{L} \{ \text{trace}(\mathbf{P}_s^{\perp} \hat{\mathbf{R}}_x) + 2 \, \text{trace}(\mathbf{P}_s \, \hat{\mathbf{R}}_n) \}. \qquad (12.67)$$

As L is increased from zero, the first term in Eq. (12.67) decreases due to improved signal representation, while the second term increases due to the higher noise captured by the larger leading subspace. The optimal representation is defined at the point where these competing influences balance, thus providing an unambiguous dimensionality estimate. However, the method relies on a valid estimate of the sample correlation matrix for noise, and this estimate can be difficult to obtain in practice. While theoretically intriguing, the practical utility of the dimensionality estimation approach relative to the other approaches described in this chapter has not been established.

12.3 Linear Mixing Model

One of the limitations of the subspace model is that the component vectors, or the column vectors of $\mathbf{S}$, estimated from the data through PCA are, by definition, uncorrelated. It is therefore unexpected that they would represent actual constituent scene materials because real material spectra generally exhibit some level of spectral correlation between material types. While grounded in a fairly solid statistical basis, at least so far as the multivariate normal assumption is concerned, the subspace model is not grounded in a physical basis.

A linear mixing model (LMM) is widely used to analyze hyperspectral data and has become an integral part of a variety of hyperspectral classification and detection techniques (Keshava and Mustard, 2002). The physical basis of the linear mixing model is that hyperspectral image measurements often capture multiple material types in an individual pixel, and that measured spectra can be described as a linear combination of the spectra of the pure materials from which the pixel is composed. The weights of the linear combination correspond to relative abundances of the various pure materials. This assumption of linear mixing is physically well founded in situations where, for example, sensor response is linear, illumination across a scene is uniform, and there is no scattering. Nonlinear mixing can occur when, for example, there is multiple scattering of light between elements in the scene, such as through the adjacency effects discussed in Chapter 5.

Despite the potential for nonlinear mixing in real imagery, the linear mixing model has been found to be a fair representation of hyperspectral data in many situations. An example 2D scatter plot of simulated data conforming to an LMM is depicted in Fig. 12.14 to describe the basic method. These data simulate noiseless mixtures of three pure materials, where relative mixing is uniformly distributed. They form a triangular region, where the vertices of the triangle represent the spectra (two bands in this case) of pure materials, which are called *endmembers*. The edges of the triangle represent mixtures of two endmembers, and interior points represent mixtures of more than two endmembers. The *abundances* of any particular sample are the relative distances of the sample data point to the respective endmembers. Theoretically, no data point can reside outside the triangular region, because this would imply abundance values that violate the physical model. The scatter plot example shown in Fig. 12.14 can be conceptually extended to the multiple dimensions of hyperspectral imagery by recognizing that data composed of M endmembers will reside within a subspace of dimension $M - 1$, at most, and will be contained within a simplex with M vertices. The relative abundances of a sample point within this simplex are again given by the relative distances to these M endmembers.

According to the LMM, each spectral measurement $\mathbf{x}_i$ can be described as a linear mixture of a set of endmember spectra $\{\boldsymbol{\varepsilon}_m, m = 1, 2, \ldots, M\}$, with abundances $a_{i,m}$ and the addition of sensor noise described by the

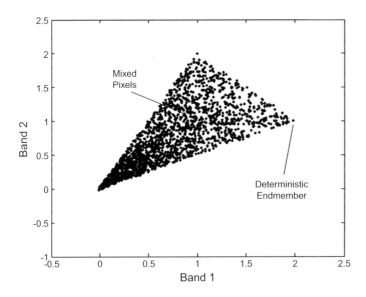

Figure 12.14 Scatter plot of dual-band data that strictly conforms to a linear mixing model.

random vector process **n**, or

$$\mathbf{x}_i = \sum_{m=1}^{M} a_{i,m}\, \boldsymbol{\varepsilon}_m + \mathbf{n}. \tag{12.68}$$

The noise process **n** is a stationary process according to the fundamental model description and is generally assumed to be zero mean. Because of the random nature of **n**, each measured spectrum $\mathbf{x}_i$ should also be considered a realization of a random vector process. It is also appropriate to consider the abundances as a random component; however, an inherent assumption of the LMM is that the endmember spectra $\boldsymbol{\varepsilon}_m$ are deterministic. Equation (12.68) can be extended in data matrix form,

$$\mathbf{X} = \mathbf{EA} + \mathbf{N}, \tag{12.69}$$

by arranging the image spectra as columns in the $K \times N$ data matrix **X** and the endmember spectra as columns in the $K \times M$ endmember matrix **E**, defining the $M \times N$ abundance matrix **A** as $(a_{i,m})^T$, and representing the noise process by **N**. This notation is used to simplify the mathematical description of the estimators for unknown endmember spectra and abundances.

12.3.1 Endmember determination

The fundamental problem of establishing an LLM, called spectral unmixing, is to determine the endmember and abundance matrices in Eq. (12.69), given a hyperspectral image. While there are methods that attempt to estimate these unknown model components simultaneously, this is usually performed by first determining endmember spectra and then estimating abundances on a pixel-by-pixel basis. Often, endmember determination is performed in a manual fashion by selecting areas in an image either known or suspected to contain pure materials, or by selecting spectra that appear to form vertices of a bounding simplex of a data cloud using multidimensional data visualization methods. This section focuses exclusively on fully automated endmember determination methods.

A general strategy for endmember determination is to select spectra in a hyperspectral image that appear to be the purest, based on some metric relating to their extreme locations in a multidimensional data cloud. One such metric is called the Pixel Purity Index^{TM} (PPI^{TM}). The PPI is computed by randomly generating lines in multidimensional data space, projecting data along these lines, and identifying spectra that consistently fall at extreme points on the lines. This is repeated for numerous (e.g., several hundred) random projections, and the number of times each data spectrum is identified to be an extremum is counted. Spectra with the highest count are declared to be pure pixels and are candidate endmembers. Some automated or manual process is then needed to eliminate spectrally similar, pure pixels that represent the same endmembers. The basic process can be intuitively understood by considering the projection of the 2D simplex onto three lines presented in Fig. 12.15.

Another conceptually simple approach to endmember determination is shown in Fig. 12.16 and involves selecting M spectra from columns of $\mathbf{X}$ that most closely represent vertices of the scatter plot. For noiseless data, one should be able to exactly determine pure spectra if they exist in the image. The presence of sensor noise will cause an error in the determined endmembers on the order of the noise variance. If there are no pure spectra in the image for a particular endmember, this approach will select vertices that do not quite capture the full extent of the true LMM simplex. This is referred to as an *inscribed simplex*. From a mathematical perspective, endmember determination in this manner is a matter of maximizing the volume of a simplex formed by any subset of M spectra selected from the full dataset. The volume of a simplex formed by a set of endmembers $\{\boldsymbol{\varepsilon}_m, m = 1, 2, \ldots, M\}$ in a spectral space of dimension $M - 1$ is given by

$$V(\mathbf{E}) = \frac{1}{(M-1)!} \begin{vmatrix} 1 & \cdots & 1 \\ \boldsymbol{\varepsilon}_1 & \cdots & \boldsymbol{\varepsilon}_M \end{vmatrix}. \tag{12.70}$$

According to a procedure known as N-FINDR (Winter, 2007), all possible image vector combinations are recursively searched to find the

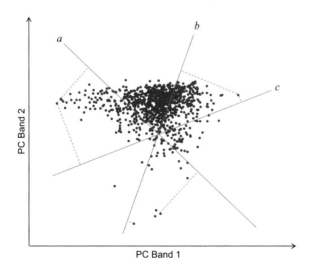

Figure 12.15 Graphical description of PPI by identification of extreme projections on lines $a, b,$ and c.

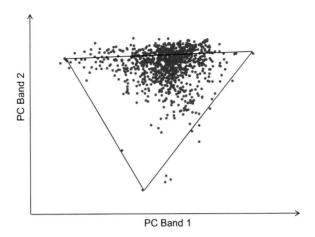

PC Band 1

Figure 12.16 Graphical description of simplex maximization, where data spectra are selected to represent endmembers so that they maximize the volume of the inscribed simplex.

subset of spectra from the image that maximize the simplex volume given in Eq. (12.70). To use this approach, however, the dimensionality of the data must be reduced from the K-dimensional hyperspectral data space to an $M - 1$ dimensional subspace, within which the simplex volume can be maximized. PC analysis is one potential method for that. While the procedure described is sufficient for ultimately arriving at the image M spectra that best represent the endmembers according to the simplex volume metric, it can be significantly improved in terms of computational complexity by using a recursive equation for updating the determinant in Eq. (12.70).

An alternate approach to selecting image spectra that maximize the inscribed simplex volume is to find a *superscribed* simplex with minimum volume and arbitrary vertices, within which all data resides, as illustrated in Fig. 12.17. This approach to endmember determination is often called *shrinkwrapping* and has the advantage that it does not require realizations of pure endmember spectra to exist in the image. While conceptually simple, implementation of this approach is more complicated than simplex volume maximization (Craig, 1994). One particular strategy uses gradient descent optimization of an objective function that includes not only the simplex volume but also a penalty term related to how far the simplex sides are separated from the data (Fuhrmann, 1999). Mathematically, the objective function to be minimized has the form

$$H(\mathbf{E}, \mathbf{A}) = V(\mathbf{E}) + \alpha\, F(\mathbf{E}, \mathbf{A}), \tag{12.71}$$

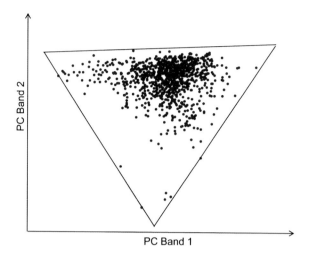

Figure 12.17 Graphical description of shrinkwrapping, where a simplex of minimum volume superscribes the data.

where $V(\mathbf{E})$ is defined in Eq. (12.70), α is a tuning parameter, and $F(\mathbf{E}, \mathbf{A})$ is the penalty function given by

$$F(\mathbf{E}, \mathbf{A}) = \sum_{m=1}^{M} \sum_{i=1}^{N} \frac{1}{\hat{a}_{i,m}}. \qquad (12.72)$$

Given a candidate set of endmembers forming $\mathbf{E}$ and data matrix $\mathbf{X}$, an unconstrained abundance matrix $\mathbf{A}$ is estimated by

$$\hat{\mathbf{A}} = \begin{bmatrix} \boldsymbol{\varepsilon}_1 & \cdots & \boldsymbol{\varepsilon}_M \\ 1 & \cdots & 1 \end{bmatrix}^{-1} \begin{bmatrix} \mathbf{x}_1 & \cdots & \mathbf{x}_N \\ 1 & \cdots & 1 \end{bmatrix}. \qquad (12.73)$$

Note that penalty term $F(\mathbf{E}, \mathbf{A})$ decreases as the simplex sides are forced away from the data cloud (because there are fewer near-zero abundances), while simplex volume $V(\mathbf{E})$ increases. A small value for tuning parameter α favors a large simplex with data clustered near the center, while a large value favors a tightly fitting simplex around the data. Since the abundances estimated by Eq. (12.73) are not constrained to be less than one, an excessively small tuning parameter can result in the simplex becoming inscribed. The algorithm is initialized with a large simplex that easily captures all of the data within it and proceeds by iteratively estimating a new endmember matrix using the gradient algorithm for iteration p:

$$\mathbf{E}^{(p+1)} = \mathbf{E}^{(p)} - \mu_1 \nabla V[\mathbf{E}^{(p)}] - \mu_2(p) \nabla F[\mathbf{E}^{(p)}, \hat{\mathbf{A}}^{(p)}], \qquad (12.74)$$

where ∇V is the simplex volume gradient, ∇F is the penalty function gradient, μ_1 is a fixed constant, and μ_2 is a constant that is successively decreased toward zero such that the final simplex fits tightly around the data. The simplex volume and penalty function gradients can be shown as

$$\nabla V(\mathbf{E}) = V(\mathbf{E}) \left[\frac{\mathbf{b}_1}{|\mathbf{b}_1|^2} \quad \frac{\mathbf{b}_2}{|\mathbf{b}_2|^2} \quad \cdots \quad \frac{\mathbf{b}_M}{|\mathbf{b}_M|^2} \right] \tag{12.75}$$

and

$$\nabla F(\mathbf{E}, \hat{\mathbf{A}}) = \left(\left[\begin{matrix} \boldsymbol{\varepsilon}_1 & \cdots & \boldsymbol{\varepsilon}_M \\ 1 & \cdots & 1 \end{matrix} \right]^{-1} \right)^T \mathbf{D}\,\hat{\mathbf{A}}^T, \tag{12.76}$$

where $\mathbf{b}_i$ is a normal vector from the simplex side opposing the vertex at $\boldsymbol{\varepsilon}_i$, and $\mathbf{D}$ is a matrix of squared abundance inverses:

$$\mathbf{D} = (1/a_{i,m}^2). \tag{12.77}$$

12.3.2 Abundance estimation

Given the determination of endmember matrix $\mathbf{E}$ by one of the methods already described, abundance matrix $\mathbf{A}$ in Eq. (12.69) is generally solved using least-squares methods on a pixel-by-pixel basis. Defining $\mathbf{a}_i$ as the M-element abundance vector associated with image spectrum $\mathbf{x}_i$, the least-squares estimate is given by

$$\hat{\mathbf{a}}_i = \min_{\mathbf{a}_i} \|\mathbf{x}_i - \mathbf{E}\mathbf{a}_i\|_2^2. \tag{12.78}$$

Estimated abundance vectors are usually formed into images corresponding to the original hyperspectral image and are termed abundance images for each respective endmember. According to the physical model underlying the LMM, abundances are constrained by positivity

$$a_{i,m} > 0 \quad \forall i \tag{12.79}$$

and full additivity

$$\mathbf{u}^T \mathbf{a}_i = 1 \quad \forall i, \tag{12.80}$$

where $\mathbf{u}$ is an M-dimensional column vector whose entries are all unity. Different least-squares solution methods are used based on the extent to which these physical constraints are applied.

The unconstrained least-squares solution ignores positivity and full additivity constraints and can be expressed in closed form as

$$\hat{\mathbf{a}}_i = (\mathbf{E}^T \mathbf{E})^{-1} \mathbf{E}^T \mathbf{x}_i. \qquad (12.81)$$

This solution is of limited utility because it fails to provide a reasonable solution in practical cases where there is non-negligible system noise and error in endmember spectra estimates. A closed-form solution can also be derived in the case where the full additivity constraint is employed, the positivity constraint is ignored, and white noise is assumed. Using the method of Lagrange variables, the consistent solution of Eqs. (12.78) and (12.80) under these assumptions is given as

$$\hat{\mathbf{a}}_i = (\mathbf{E}^T \mathbf{E})^{-1} \mathbf{E}^T \mathbf{x}_i - \frac{\mathbf{u}^T (\mathbf{E}^T \mathbf{E})^{-1} \mathbf{E}^T \mathbf{x}_i - 1}{\mathbf{u}^T (\mathbf{E}^T \mathbf{E})^{-1} \mathbf{u}} (\mathbf{E}^T \mathbf{E})^{-1} \mathbf{u}. \qquad (12.82)$$

An alternate approach uses nonlinear least squares to solve for the case where the positivity constraint is employed but full additivity is not. Strictly, however, a quadratic programming approach must be used to find an optimum solution with respect to both constraints. This can be done using the active set method or some other constrained optimization method. For further details, the reader should consult a reference on constrained optimization algorithms, such as Fletcher (1987).

As an example of the application of spectral unmixing, an LMM with four endmembers was developed for the Hyperion image using simplex maximization to determine endmembers and quadratic programming to determine fully constrained abundance maps. The resulting abundance maps shown in Fig. 12.18 appear to capture the primary material features of the image, except for perhaps the fourth endmember. The selected endmembers are denoted by plus marks in the scatter plots shown in Fig. 12.19. These data appear to conform well to an LMM, although two of the endmembers (3 and 4) appear to correspond to outliers in the data. This problem is commonly encountered when selecting endmembers from image data by simplex maximization and is the motivation for the more sophisticated endmember determination methods discussed. The problem can also be addressed by coupling a method such as PPI[TM] or N-FINDR with an outlier rejection algorithm. For example, one could remove selected endmembers from the data, run the simplex maximization algorithm consecutively on the reduced set of data, and retain only those endmembers that show up within some spectral proximity in all solution sets. This would tend to remove the outliers and leave endmembers at the edges of the scatter plot, but still close to several other data samples. For

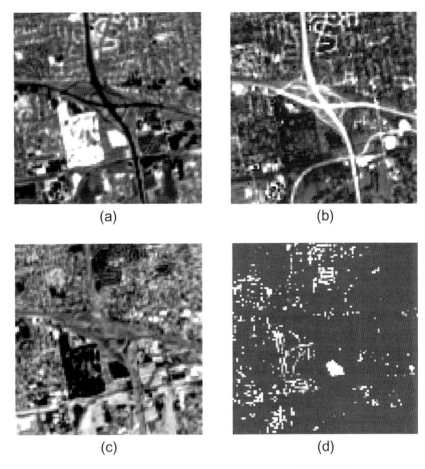

(a)   (b)

(c)   (d)

Figure 12.18 Abundance maps for the Hyperion VNIR/SWIR hyperspectral image, based on a fully constrained LMM composed of four endmembers selected by simplex maximization: (a) endmember 1, (b) endmember 2 (c) endmember 3, and (d) endmember 4.

example, endmembers 3 and 4 in Fig. 12.19 would likely be rejected by this approach.

12.3.3 Limitations of the linear mixing model

The linear mixing model is widely used because of the strong tie between mathematical foundations of the model and physical processes of mixing that result in much of the variance seen in hyperspectral imagery. However, situations exist where basic assumptions of the model are violated and it fails to accurately represent the nature of hyperspectral imagery. The potential for nonlinear mixing is one situation that has already been mentioned. Another is the assumption that mixtures of a small number of deterministic spectra can be used to represent all non-noise variance

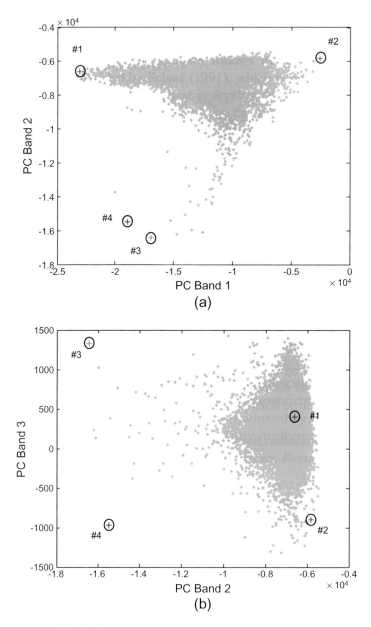

Figure 12.19 Principal-component scatter plots of Hyperion data indicating the selected LMM endmember spectra: (a) PC Bands 1 and 2 and (b) PC Bands 2 and 3.

in hyperspectral imagery. Common observations would support the case that there is natural variation in almost all materials; therefore, one would expect a certain degree of variability to exist for any materials selected as endmembers for a particular hyperspectral image. This is first driven by the fact that few materials in a scene will actually be pure, and endmembers usually only represent classes of materials that are spectrally very similar. For example, one endmember can represent grass, but even a single grass field is usually composed of a variety of species and variations in health, moisture, fertilization, underlying soil quality, and so forth that result in a texture of visual and spectral variation. Manmade materials such as concrete roads exhibit analogous variations due to, for example, construction, surface characteristics, contaminants, and weathering.

Given the great diversity of materials that compose a typical hyperspectral image, variation in endmember characteristics is generally even larger than that suggested earlier because there is a practical limit the number of endmembers that can be used to represent a scene. Therefore, endmembers usually represent even broader classes of spectrally similar materials. When intraclass variance (variance within endmember classes) is very small relative to interclass variance (variance between endmembers), the deterministic endmember assumption on which the linear mixing model is based might remain valid. However, when intraclass variance becomes appreciable relative to interclass variance, the deterministic assumption becomes invalid, and the endmembers should be treated as random vectors.

Another source of variance in endmember spectra is illumination. In reflective hyperspectral imagery, this is generally treated by capturing a dark endmember and assuming that illumination variation is equivalent to mixing between the dark endmember and each respective fully illuminated endmember. In thermal imagery, the situation is more complicated because upwelling radiance is driven by both material temperature and downwelling radiance. Temperature variations cause a nonlinear change in upwelling radiance spectra, and the concept of finding a dark endmember (analogous to the reflective case) does not carry over into the thermal domain because of the large radiance offset that exists. Also, downwelling illumination has a much stronger local component (i.e., emission from nearby objects) and is not adequately modeled as a simple scaling of endmember spectra. Statistical modeling of endmember spectra is one way to deal with variations of this nature that are not well represented in a linear mixing model.

12.4 Extensions of the Multivariate Normal Model

Compared to the geometric representation provided by the LMM, the multivariate normal model addresses the fundamental random nature of hyperspectral data. Unfortunately, it is too simple to effectively represent the complexity of real hyperspectral data, greatly limiting its applicability. This is readily apparent by examining the scatter plots of the Hyperion image data for various combinations of principal-component bands illustrated in Fig. 12.20. For the leading principal-component bands shown in Fig. 12.20(a) and (b), the data is highly structured and non-normal because of the nonstationary nature of the image structure. On the other hand, the trailing principal-component bands shown in Fig. 12.20(c) and (d) appear to be somewhat normally distributed because they are dominated by noise. To effectively represent statistics of the leading principal-component subspace, it is necessary to extend the multivariate normal model to illustrate the nonstationary nature of the scene.

12.4.1 Local normal model

A common extension of the multivariate normal model that accounts for this nonstationarity is the spatially adaptive or local normal model, for which second-order statistics are assumed to be only locally homogenous. For example, consider the situation depicted in Fig. 12.21, where there is interest in describing the nature of the local background surrounding a particular location, corresponding to spatial index i. Under the assumption of local homogeneity, local statistics can be described by

$$p(\mathbf{x} \mid i) \sim N(\boldsymbol{\mu}_i, \boldsymbol{\Sigma}_i), \tag{12.83}$$

where spatially varying mean vector $\boldsymbol{\mu}_i$ and covariance matrix $\boldsymbol{\Sigma}_i$ can be estimated from sample statistics $\mathbf{m}_i$ and $\mathbf{C}_i$ of the spectra contained within the local background window that surrounds the point of interest, using Eqs. (12.12) and (12.13). The size of the window represents a balance between the homogeneity of the background statistics and the sample size needed to confidently estimate the covariance matrix. A rule of thumb for the number of samples required to estimate a $K \times K$ covariance matrix is K^2. Since more samples are required to produce a statistically confident estimate of the covariance matrix than of the mean vector, one could use a smaller window for the mean vector estimate and a larger window for the covariance matrix estimate to better balance these competing demands.

In practice, computation of the spatially varying statistics can be a formidable numerical task because they must be computed for each pixel in the image. There also can be significant redundancy because of overlap in the local region for neighboring pixel locations. To exploit this redundancy, the parameters can be updated recursively to reduce computational complexity. If ω_i represents a set of spectra composing the local region surrounding the i'th pixel, and ω_j represents a set of spectra composing the local region surrounding the j'th pixel, then mean vector

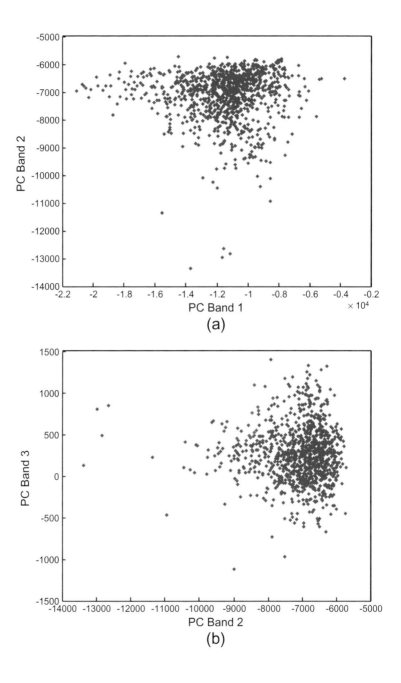

Figure 12.20 Scatter plots of Hyperion VNIR/SWIR hyperspectral data for various PC band combinations: (a) PC bands 1 and 2, (b) PC bands 2 and 3, (c) PC bands 10 and 11, and (d) PC bands 30 and 31.

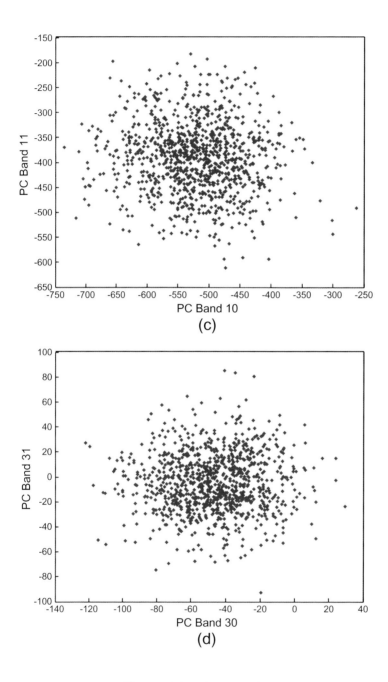

Figure 12.20   (*continued*)

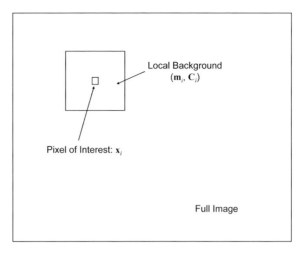

Figure 12.21 Depiction of the spatially adaptive normal model.

and covariance matrix estimates can be updated by the recursion

$$\hat{\boldsymbol{\mu}}_j = \hat{\boldsymbol{\mu}}_i + \frac{1}{N}\left[\sum_{\mathbf{x}_l \in \omega_{new}} \mathbf{x}_l - \sum_{\mathbf{x}_l \in \omega_{old}} \mathbf{x}_l\right], \tag{12.84}$$

$$\mathbf{R}_j = \mathbf{R}_i + \frac{1}{N-1}\left[\sum_{\mathbf{x}_l \in \omega_{new}} \mathbf{x}_l \mathbf{x}_l^T - \sum_{\mathbf{x}_l \in \omega_{old}} \mathbf{x}_l \mathbf{x}_l^T\right], \tag{12.85}$$

and

$$\hat{\boldsymbol{\Sigma}}_j = \mathbf{R}_j - \hat{\boldsymbol{\mu}}_j \hat{\boldsymbol{\mu}}_j^T, \tag{12.86}$$

where N is the number of spectra contained in the local window, $\omega_{new} = \omega_j - \omega_i \cap \omega_j$, and $\omega_{new} = \omega_i - \omega_i \cap \omega_j$. If the computation is structured such that there is maximum overlap between successive statistical estimates, then the summations in Eqs. (12.84) and (12.85) are minimized. For real-time processing systems with typical pushbroom or whiskbroom dispersive imaging spectrometers, the approach depicted in Fig. 12.21 presents a causality problem; that is, the full local background region is not collected until sometime after the point of interest. In such situations, it is common to perform block processing to update the statistics in a causal and adaptive manner. A data format from such a sensor is depicted in Fig. 12.22. In this case, the recursive relationships in Eqs. (12.84) and (12.85) could be used to update the statistics for each block, but that would require spectra from prior blocks to be stored and retrieved from memory. Another approach that can be employed in this case is to use successive parameter estimation by linearly combining sample statistics of the new

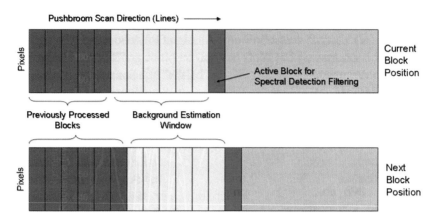

Figure 12.22 Implementation of a spatially adaptive normal model based on block processing (Eismann *et al.*, 2009).

block of data with local estimates from the previous block, according to the recursion

$$\hat{\boldsymbol{\mu}}_n = \frac{\left[\frac{1}{N}\sum_{i=1}^{N}\mathbf{x}_i\right] + \frac{N_0}{N}\hat{\boldsymbol{\mu}}_{n-1}}{1 + \frac{N_0}{N}} \tag{12.87}$$

and

$$\hat{\boldsymbol{\Sigma}}_n = \frac{\left[\frac{1}{N-1}\sum_{i=1}^{N}(\mathbf{x}_i - \hat{\boldsymbol{\mu}}_n)\right] + \frac{N_0}{N}\hat{\boldsymbol{\Sigma}}_{n-1}}{1 + \frac{N_0}{N}}, \tag{12.88}$$

where N is the number of spectra in a block and n is the block index. Arbitrary parameter N_0 weights the contribution to statistics of the new block relative to preceding blocks and can be thought of as the effective number of spectra in the fading statistical memory into which new data samples are merged. A low value of N_0, on the order of N, provides more spatial adaptivity, while a relatively high value of N_0 provides more statistical confidence at the expense of adaptivity.

12.4.2 Normal mixture model

The normal mixture model is an extension of the multivariate normal model that addresses nonstationarity by modeling the multiconstituent nature of scene information through image segmentation. It is based on the assumption that each spectrum $\mathbf{x}_i$ in an image is a member of one of Q classes defined by a class index q, where $q = 1, 2, \ldots, Q$. Each spectrum has a prior probability $P(q)$ of belonging to each respective class, and the class-conditional probability density function $p(\mathbf{x}|q)$ is normally distributed and completely described by mean vector $\boldsymbol{\mu}_{\mathbf{x}|q}$ and covariance matrix $\boldsymbol{\Sigma}_{\mathbf{x}|q}$. The probability density function $p(\mathbf{x})$ then becomes multimodal, given by

$$p(\mathbf{x}) = \sum_{q=1}^{Q} P(q)\, p(\mathbf{x} \mid q), \qquad (12.89)$$

where

$$p(\mathbf{x} \mid q) \sim N(\boldsymbol{\mu}_{\mathbf{x}|q}, \boldsymbol{\Sigma}_{\mathbf{x}|q}). \qquad (12.90)$$

The statistical representation in Eqs. (12.89) and (12.90) is referred to as normal mixture distribution. The term *mixture* in this context is used in a different manner than its use with respect to the LMM, and it is very important to understand the distinction. In the LMM, mixing represents the modeling of spectra as a combination of various endmembers. In the normal mixture model, however, each spectrum is ultimately assigned to a single class, and mixing of this sort (i.e., weighted combinations of classes) is not modeled. The term *mixture*, in this case, merely refers to the representation of the total probability density function $p(\mathbf{x})$ as a linear combination of the class conditional distributions, weighted by the prior probabilities.

Prior probability values $P(q)$ are subject to the constraints of positivity,

$$P(q) > 0 \quad \forall q, \qquad (12.91)$$

and full additivity,

$$\sum_{q=1}^{Q} P(q) = 1, \qquad (12.92)$$

in a manner similar to the abundances in the linear mixing model. This common characteristic is somewhat coincidental, however, since these parameters represent very different quantities in the respective models. Figure 12.23 illustrates a dual-band scatter plot of data that conforms well to a normal mixture model. The ovals are intended to represent the one-standard-deviation contours of the class-conditional distributions for each class.

A normal mixture model is fully represented by a set of parameters $\{P(q), \boldsymbol{\mu}_{\mathbf{x}|q}, \boldsymbol{\Sigma}_{\mathbf{x}|q}, q = 1, 2, \ldots, Q\}$. When these parameters are known,

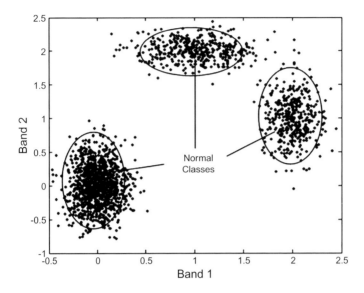

Figure 12.23 Scatter plot of dual-band data that conforms well to a normal mixture model.

or estimates of these parameters exist, each image spectrum $\mathbf{x}_i$ can be assigned to a class using a distance metric from respective classes. The most common alternatives are the Euclidean distance,

$$d_q = \| \mathbf{x}_i - \boldsymbol{\mu}_{\mathbf{x}|q} \| = \sqrt{(\mathbf{x}_i - \boldsymbol{\mu}_{\mathbf{x}|q})^T (\mathbf{x}_i - \boldsymbol{\mu}_{\mathbf{x}|q})}, \qquad (12.93)$$

and the Mahalanobis distance,

$$d_q = (\mathbf{x}_i - \boldsymbol{\mu}_{\mathbf{x}|q})^T \boldsymbol{\Sigma}_{\mathbf{x}|q}^{-1} (\mathbf{x}_i - \boldsymbol{\mu}_{\mathbf{x}|q}). \qquad (12.94)$$

The assignment process is *classification*, with these two alternatives referred to as linear and quadratic classification, respectively. Normal mixture modeling includes not only the classification of the image spectra, but also the concurrent estimation of the model parameters, which can be determined either through the use of training data or by finding the modes of data automatically. The latter process is typically called *clustering*; a variety of methods for performing clustering are described in the literature. A common iterative methodology used for clustering follows:

1. Initialize parameters of the Q classes in some manner.
2. Classify all image spectra into Q classes using a distance measure such as Eq. (12.93) or Eq. (12.94).
3. Estimate the parameters of Q classes based on sample statistics of the spectra assigned to each respective class.

4. Repeat steps (2) and (3) until class assignments converge.

The Linde–Buzo–Gray (LBG) clustering approach (Linde *et al.*, 1980) employs this basic iteration using the Euclidean distance or another similar distortion metric.

A problem that arises with this simple clustering approach is that it tends to bias against overlapping class-conditional distributions. This occurs because of the hard decision rules in classification. That is, even though a spectrum can have comparable probability of being part of more than one class, it is always assigned to the class that minimizes the respective distance measure. Stochastic expectation maximization (SEM) is a quadratic clustering algorithm that addresses the bias against overlapping class-conditional probability density functions by employing a Monte Carlo class assignment (Masson and Pieczynski, 1993). Functionally, SEM is composed of the following steps:

1. Initialize the parameters of Q classes using uniform priors and global sample statistics (i.e., the same parameters for all classes).
2. Estimate the posterior probability for combinations of the N image spectrum and Q classes based on current model parameters.
3. Assign N image spectra among Q classes by a Monte Carlo method based on posterior probability estimates.
4. Estimate the parameters of Q classes based on sample statistics of the spectra assigned to each respective class.
5. Repeat steps (2) to (4) until the algorithm converges.
6. Repeat step (2) and classify the image spectra to the classes exhibiting the maximum posterior probability.

Further details on LBG (also known as K-means) and SEM clustering, as well as additional clustering approaches, are provided in Chapter 13 on hyperspectral image classification.

Basic characteristics of normal mixture models derived from linear and quadratic clustering can be illustrated by applying LBG and SEM clustering to the Hyperion hyperspectral image. In each case, normal mixture model parameters are iteratively estimated, assuming that the image is composed of four distinct spectral classes. The corresponding classification maps are shown in Fig. 12.24, where map colors correspond to the four classes into which data are classified. Scatter plots of data in the leading three principal-component bands are also shown in Figs. 12.25 and 12.26, where the samples are color coded with respect to the assigned class. The difference in clustering methods is very clear, as the linear approach results in classes that are primarily separated along the first principal component, since it exhibits the highest variance, while the SEM result produces overlapping classes that capture much of the scatter through differences in the second-order statistics. In this case, the

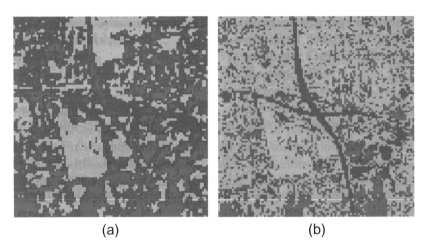

(a)   (b)

Figure 12.24 Normal mixture model classification maps for the Hyperion VNIR/SWIR hyperspectral image derived from (a) linear (LBG) and (b) quadratic (SEM) clustering.

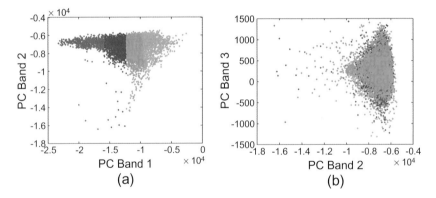

(a)   (b)

Figure 12.25 Dual-band scatter plots of the Hyperion image for leading PC band combinations that are color coded to illustrate class assignments for the normal mixture model based on LBG clustering: (a) PC bands 1 and 2 and (b) PC bands 2 and 3.

SEM approach results in a classification map that appears more closely correlated to actual material differences in the scene. For this example, however, there is substantial spectral mixing, and the classes do not separate like the idealized scatter plot shown in Fig. 12.23. Therefore, while the normal mixture model provides necessary degrees of freedom to provide a better fit to the non-normal probability density function, the classes are not distinct and should not be expected to correspond to pure spectral constituents in the scene. These issues are further discussed in Chapter 13.

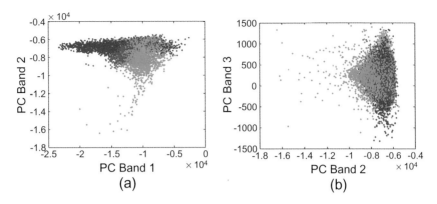

Figure 12.26 Dual-band scatter plots of the Hyperion image for leading PC band combinations that are color coded to illustrate class assignments for the normal mixture model based on SEM clustering: (a) PC bands 1 and 2 and (b) PC bands 2 and 3.

12.4.3 Generalized elliptically contoured distributions

Normal distribution is a special case of elliptically contoured distributions that are unimodal and exhibit elliptical contours of equal probability. Given mean vector $\boldsymbol{\mu}$ and covariance matrix $\boldsymbol{\Sigma}$ of the data, normal distribution can be expressed as

$$\mathbf{x} \sim N_K(\boldsymbol{\mu}, \boldsymbol{\Sigma}) \rightarrow p(\mathbf{x}) = \frac{1}{(2\pi)^{K/2}} \frac{1}{|\boldsymbol{\Sigma}|^{1/2}} e^{-d(\mathbf{x})/2}, \qquad (12.95)$$

where $d(\mathbf{x})$ is the Mahalanobis distance,

$$d(\mathbf{x}) = (\mathbf{x} - \boldsymbol{\mu})^T \boldsymbol{\Sigma}^{-1} (\mathbf{x} - \boldsymbol{\mu}). \qquad (12.96)$$

Based on the normal distribution for $\mathbf{x}$, the Mahalanobis distance would be chi-squared distributed,

$$d \sim \chi_K^2 \rightarrow p(d) = \frac{1}{2^{K/2} \, \Gamma(K/2)} d^{K/2-1} e^{-d/2}, \qquad (12.97)$$

where $\Gamma(n)$ is a gamma function; $\Gamma(n) = (n-1)!$ in the integer case. Observations of empirical data, however, typically indicate that statistical distributions of the Mahalanobis distance do not fall as rapidly as the normal distribution in Eq. (12.95) predicts. Instead, they exhibit longer tails due to outlier spectra in natural scenes; these tails can become even more pronounced in culturally cluttered areas.

To account for this heavy tail behavior, the hyperspectral data can be modeled using a K-dimensional t-distribution $t_K(\nu, \boldsymbol{\mu}, \mathbf{R})$ of the form

$$\mathbf{x} \sim t_K(\nu, \boldsymbol{\mu}, \mathbf{C}) \rightarrow p(\mathbf{x}) = \frac{\Gamma\left(\frac{K+\nu}{2}\right)}{(\pi\nu)^{K/2}\Gamma\left(\frac{\nu}{2}\right)}$$

$$\times \frac{1}{|\mathbf{C}|^{1/2}} \left[1 + \frac{1}{\nu}(\mathbf{x} - \boldsymbol{\mu})^T \mathbf{C}^{-1} (\mathbf{x} - \boldsymbol{\mu})\right]^{-\frac{K+\nu}{2}}. \tag{12.98}$$

The number of degrees of freedom ν controls how rapidly the distribution falls off. As ν approaches infinity, the t-distribution asymptotically approaches multivariate normal distribution. Decreasing ν increases the tail probabilities, leading to a Cauchy distribution for $\nu = 1$. The mean vector is given by $\boldsymbol{\mu}$, and the parameter $\mathbf{C}$ is related to the covariance matrix according to

$$\boldsymbol{\Sigma} = \frac{\nu}{\nu - 2}\mathbf{C} \quad \nu \geq 3. \tag{12.99}$$

We retain notation from the reference literature concerning parameter $\mathbf{C}$ but warn the reader not to confuse it with the sample covariance matrix described earlier in this chapter. Distribution of the Mahalanobis-like distance

$$\delta(\mathbf{x}) = \frac{1}{\nu}(\mathbf{x} - \boldsymbol{\mu})^T \mathbf{C}^{-1} (\mathbf{x} - \boldsymbol{\mu}), \tag{12.100}$$

is an F-distribution,

$$\delta \sim F_{K,\nu} \rightarrow p(\delta) = \frac{\Gamma\left(\frac{K+\nu}{2}\right)}{\Gamma(K/2)\Gamma(\nu/2)} \frac{K^{K/2}\nu^{\nu/2}}{} \frac{\delta^{K/2-1}}{(\nu + K\delta)^{(K+\nu)/2}}, \tag{12.101}$$

when $\mathbf{x}$ is t-distributed.

Figure 12.27 illustrates an empirical distribution of Fort A.P. Hill AVIRIS imagery, relative to various F-distribution models (Manolakis et al., 2006). It is readily apparent that the data do not conform well to the normal model due to the heavy tails of empirical data. To model this heavy-tail behavior, a fairly low degree of freedom ($\nu \sim 9$) must be used in the model fit; however, this causes the body of the model distribution (i.e., smaller Mahalanobis distance) to deviate significantly from empirical distribution. To address this dilemma, a mixture density of the form

$$\mathbf{x} \sim w\, t_K(\nu_1, \boldsymbol{\mu}, \mathbf{C}) + (1 - w)\, t_K(\nu_2, \boldsymbol{\mu}, \mathbf{C}) \tag{12.102}$$

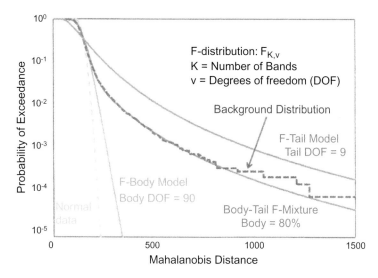

Figure 12.27 Modeling the distribution of VNIR/SWIR hyperspectral data of Fort A.P. Hill from the AVIRIS sensor as a mixture of F-distributions [reprinted with permission from Manolakis *et al.* (2006) © 2006 IEEE].

can be used to accurately model both regions of the distribution, where the weight $0 < w < 1$ balances the body and tail contributions, v_1 is optimized to fit the body of the distribution, and v_2 is optimized to fit the tail. For the data illustrated, $w = 0.8, v_1 = 90$, and $v_2 = 9$. All parameters are estimated globally in this example. To extend the utility to more-complex, nonstationary backgrounds, the *t*-distribution model can be applied as either a locally adaptive model, such as a local normal model, or as a more complex mixture distribution. The locally adaptive approach can be problematic due to computational overhead associated with adaptively estimating additional parameters of the *t*-distribution. A more extensive treatment of mixture *t*-distributions for hyperspectral data modeling is given by Marden and Manolakis (2003). The use of such models can provide more-realistic estimates of the false-alarm-rate performance of target detection algorithms described in Chapter 14.

12.5 Stochastic Mixture Model

The stochastic mixing model (SMM) is a linear mixing model that treats endmembers as random vectors as opposed to deterministic spectra (Stocker and Schaum, 1997). While potential exists for statistically representing endmembers through a variety of probability density functions, only the use of normal distributions is detailed in this section. As in a normal mixture model, the challenge in stochastic mixture modeling lies more in the estimation of model parameters than in the representation

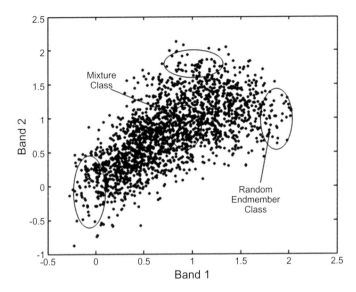

Figure 12.28 Scatter plot of dual-band data that conforms well to a stochastic mixing model.

of data. The SMM is similar to the LMM in that it attempts to decompose spectral data in terms of a linear combination of endmember spectra. However, in the case of an SMM, data are represented by an underlying random vector $\mathbf{x}$ with endmembers $\boldsymbol{\varepsilon}_m$ that are $K \times 1$ normally distributed random vectors, parameterized by their mean vector $\mathbf{m}_m$ and covariance matrix $\mathbf{C}_m$. Variance in the hyperspectral image is then interpreted by the model as a combination of both endmember variance and subpixel mixing; this contrasts with LMM for which all variance is modeled as subpixel mixing and noise. Figure 12.28 illustrates dual-band data that conforms well to an SMM. The ovals indicate contours of endmember class distributions that linearly mix to produce the remaining data scatter.

Hyperspectral data are treated as a spatially independent set of realizations of underlying random vector $\mathbf{x}$, which is related to random endmembers according to the linear mixing relationship

$$\mathbf{x} = \sum_{m=1}^{M} a_m \, \boldsymbol{\varepsilon}_m + \mathbf{n}, \tag{12.103}$$

where a_m are random mixing coefficients constrained by positivity and full additivity as in Eqs. (12.79) and (12.80), and $\mathbf{n}$ represents the sensor noise as in Eq. (12.68). Because of the random nature of endmembers, sensor noise can be easily incorporated as a common source of variance to all endmembers. This allows the removal of noise process $\mathbf{n}$ in Eq. (12.103) without losing any generality in the model. While the SMM provides the benefit of modeling inherent endmember variance (relative to the LMM), this benefit comes at the cost of turning a fairly standard least-squares estimation problem into a much more complicated estimation problem. This arises due to the doubly stochastic nature of Eq. (12.103); that is, the fact that $\mathbf{x}$ is the sum of products of two random variables a_m and $\boldsymbol{\varepsilon}_m$.

Two fundamentally different approaches have been developed for solving the problem of estimating random parameters underlying the SMM from a hyperspectral image (Eismann and Stein, 2007). In the first, the problem is constrained by requiring mixing coefficients to be quantized to a discrete set of mixing levels. By performing this quantization, the estimation problem can be turned into a quadratic clustering problem similar to that described in Section 12.4.2. The difference in the SMM case is that the classes are interdependent due to linear mixing relationships. A variant of SEM is employed to self-consistently estimate endmember statistics and abundance images. The abundance quantization that is fundamental to this method provides an inherent limitation in model fidelity, and the achievement of finer quantization results in significant computational complexity problems. This solution methodology is termed the *discrete stochastic mixing model*.

The second approach employs no such abundance quantization. A continuous version of the SMM is termed the *normal compositional model* (NCM). In this terminology, *normal* refers to the underlying class distribution, and *compositional* refers to the property of the model that represents each datum as a convex combination of underlying classes. The term compositional is in contrast to mixture, as in normal mixture model, in which each datum emanates from one class that is characterized by a normal probability distribution. The NCM is identified as a hierarchical model, and the Monte Carlo expectation maximization algorithm is used to estimate the parameters. The Monte Carlo step is performed by using Monte Carlo Markov chains to sample from the posterior distribution (i.e., the distribution of abundance values, given the observation vector and current estimate of the class parameters). Only the discrete SMM is described further.

12.5.1 Discrete stochastic mixture model

In the discrete SMM, a finite number of mixture classes Q are defined as linear combinations of endmember random vectors,

$$\mathbf{x} \mid q = \sum_{m=1}^{M} a_m(q)\, \boldsymbol{\varepsilon}_m, \qquad (12.104)$$

where $a_m(q)$ are fractional abundances associated with the mixture class, and $q = 1, 2, \ldots, Q$ is the mixture-class index. M of the Q mixture classes will be pure endmember classes, while the remaining $(Q - M)$ classes will be mixtures of at least two endmembers. Fractional abundances corresponding to each mixture class are assumed to conform to the physical constraints of positivity,

$$a_m(q) \geq 0 \quad \forall m, q, \tag{12.105}$$

and full additivity,

$$\sum_{m=1}^{M} a_m(q) = 1 \quad \forall q. \tag{12.106}$$

Also, abundances are quantized into L levels (actually $L + 1$ levels when zero abundance is included), such that

$$a_m(q) \in \left\{ 0, \frac{1}{L}, \frac{2}{L}, \ldots, 1 \right\} \quad \forall q. \tag{12.107}$$

This is referred to as the quantization constraint. For a given number of endmembers M and quantization levels L, there is a finite number of combinations of abundances $\{a_1(q), a_2(q), \ldots, a_M(q)\}$ that simultaneously satisfy the constraints in Eqs. (12.105)–(12.107). This defines Q discrete mixture classes.

Due to the linear relationship in Eq. (12.104), the mean vector and covariance matrix for the q'th mixture class are functionally dependent on the corresponding endmember statistics by

$$\boldsymbol{\mu}_{\mathbf{x}|q} = \sum_{m=1}^{M} a_m(q) \, \boldsymbol{\mu}_m \tag{12.108}$$

and

$$\boldsymbol{\Sigma}_{\mathbf{x}|q} = \sum_{m=1}^{M} a_m^2(q) \, \boldsymbol{\Sigma}_m. \tag{12.109}$$

The probability density function is described by the normal mixture model given in Eqs. (12.89) and (12.90). The mixing model parameters that must be estimated from measured data are the prior probability values for all mixture classes and the mean vectors and covariance matrices of endmember classes. In this case, however, the relationship between mixture class statistics is constrained through Eqs. (12.108) and (12.109). Therefore, the methodology used for parameter estimation must be modified to account for this fact. It is important to recognize the difference between prior probability values $P(q)$ and mixture-class fractional abundances $a_m(q)$ used in this model. Prior probability characterizes the expectation that a randomly selected image pixel will be classified as part of the q'th mixture class, while fractional abundances define the estimated proportions of endmembers of which the pixel is composed, if it is contained in that mixture class. All of the mixture classes are represented with prior probability values, not just endmember classes.

12.5.2 Estimation algorithm

A modified SEM approach is used to self-consistently estimate model parameters $\boldsymbol{\mu}_{\mathbf{x}|q}, \boldsymbol{\Sigma}_{\mathbf{x}|q}$, and $P(q)$, and abundance vectors $a_m(q)$, for the measured hyperspectral scene data. Estimation in a lower-dimensional subspace of hyperspectral data, using one of the dimensionality reduction methods discussed in Section 12.2, is preferred and perhaps even necessary for reasons of reducing computational complexity and improving robustness against system noise. The primary steps of this stochastic unmixing algorithm are detailed next.

Through Eqs. (12.108) and (12.109), mean vectors and covariance matrices of all mixture classes are completely dependent on the corresponding statistical parameters of endmember classes, as well as the set of feasible combinations of mixture class abundances. Therefore, to initialize the SMM, it is only necessary to initialize these parameters. For endmember mean vector initialization, a methodology based on the simplex maximization algorithm can be used. That is, M image spectra are selected from the image data that maximize the simplex volume metric, and these are used to define the initial endmember class mean vectors. Use of the global covariance matrix of data for all initial endmember classes has been found to be an adequate approach for covariance initialization. Initialization also involves specifying the set of feasible abundance vectors that are to represent mixed classes, where an abundance vector is defined as

$$\mathbf{a}(q) = [a_1(q), a_2(q), \ldots, a_M(q)]^T. \tag{12.110}$$

Note that a unique abundance vector is associated with each mixture class index q, and the set of all abundance vectors over Q mixture classes encompasses all feasible abundance vectors, subject to the constraints given by Eqs. (12.97)–(12.99). Therefore, these abundance vectors need to be defined and associated with mixture class indices only once. As the iterative SEM algorithm progresses, these fixed abundance vectors will maintain the relationship between endmember and mixture class statistics given by Eqs. (12.108) and (12.109). The final estimated abundances for a given image spectrum are determined directly by the associated abundance vector for the mixture class into which that spectrum is ultimately classified.

A structured, recursive algorithm has been used to determine the complete, feasible set of abundance vectors. It begins by determining the complete, unique set of M-dimensional vectors containing ordered (nonincreasing) integers between zero and the number of levels L whose elements add up to L. A matrix is then constructed where rows consist of all possible permutations of each of the ordered vectors. The rows are sorted, and redundant rows are removed by comparing adjacent rows. After dividing the entire matrix by L, the remaining rows contain the complete, feasible set of abundance vectors for the given quantization level and number of endmembers.

The resulting number of total mixture classes Q is equivalent to the unique combinations of placing L items in M bins, or

$$Q = \binom{L + M - 1}{L} = \frac{(L + M - 1)!}{(L)!(M - 1)!}. \qquad (12.111)$$

An extremely large number of mixture classes can occur when the number of quantization levels and/or number of endmembers increases; this can make the stochastic mixing model impractical in these instances. Even when the number of endmembers in a scene is large, however, for real hyperspectral imagery, it is unexpected that any particular pixel will contain a large subset of the endmembers. Rather, it is more reasonable to expect that they will be composed of only a few constituents, although specific constituents will vary from pixel to pixel. This physical expectation motivates the idea of employing a scarcity constraint on the fraction vectors to allow only a subset (say, M_{max}) of the total set of endmembers to be used to represent any pixel spectrum. This can be performed by adding this additional constraint in the first step of determining a feasible abundance vector set. The equivalent combinatorics problem then becomes placing L items in M endmember classes, where only M_{max} classes can contain any items. The resulting number of mixture classes in this case is

$$Q = \sum_{j=1}^{\min(M_{max}, L)} \binom{M}{j}\binom{L - 1}{j - 1}. \qquad (12.112)$$

Figure 12.29 illustrates the way in which limiting the mixture classes in this way reduces the total number of mixture classes in the model for a six-endmember case example. As the number of endmembers increases, reduction is even more substantial.

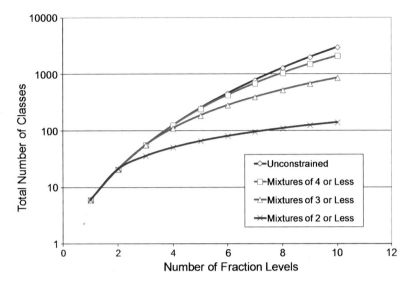

Figure 12.29 Impact of a mixture constraint on the number of SMM mixture classes (Eismann and Hardie, 2005).

After initialization, the model is updated by iterating the following three SEM steps:

1. Estimate the posterior probability, given by

$$\hat{P}^{(n)}(q \mid \mathbf{x}_i) = \frac{\hat{P}^{(n)}(q)\hat{p}^{(n)}(\mathbf{x}_i \mid q)}{\sum\limits_{q=1}^{Q} \hat{P}^{(n)}(q)\hat{p}^{(n)}(\mathbf{x}_i \mid q)}, \qquad (12.113)$$

on iteration n for all combinations of the N image spectra $\mathbf{x}_i$ and Q mixture classes based on the current model parameters; that is, mean vectors and covariance matrices of all mixture classes that define conditional probability density functions. To avoid a bias against sparsely populated mixture classes, prior probabilities are uniformly reset to $1/Q$ for all classes.

2. Assign N image spectra among the Q classes by a Monte Carlo method based on posterior probability estimates. For each spectrum $\mathbf{z}_i$, a random number R is generated with a probability distribution of

$$P_R(R) = \begin{cases} \hat{P}^{(n-1)}(q \mid \mathbf{x}_i) & q = 1, 2, \ldots, Q \\ 0 & \text{otherwise.} \end{cases} \qquad (12.114)$$

The class index estimate is then given by

$$\hat{q}^{(n)}(\mathbf{x}_i) = R. \qquad (12.115)$$

This is independently repeated for each image spectrum. Because of the Monte Carlo nature of this class assignment, the same spectrum can be assigned to different classes on successive iterations, even when the model parameters do not change; this is what allows the resulting classes to overlap.

3. Estimate second-order statistics of the Q classes based on sample statistics $\boldsymbol{\mu}_m$ and $\boldsymbol{\Sigma}_m$ of the spectra assigned to M endmember classes, and then update the remaining mixture-class statistics using Eqs. (12.108) and (12.109).

SEM iterations are terminated based on convergence of a log-likelihood or separability metric (see Chapter 13). Final pixel class assignments are made by making a final posterior probability estimate according to Eq. (12.113), and then assigning each pixel to the class $\hat{q}_i$ that maximizes posterior probability. Estimated prior probability values for all mixture classes,

$$\hat{P}^{(n)}(q) = N_q/N, \qquad (12.116)$$

are used in this step, where N_q is the mixture-class population instead of the uniformly distributed values used during parameter estimation. The estimated abundance vector for each sample $\mathbf{x}_i$ is then given by abundances $a_m(\hat{q}_i)$ corresponding to this mixture class. These abundance vectors can be formatted into abundance images corresponding to each endmember and are quantized at the initially selected level.

12.5.3 Examples of results

As an example of the efficacy of this method, an SMM with four endmembers was developed for the Hyperion VNIR/SWIR hyperspectral image of Beavercreek, Ohio. The resulting abundance maps are shown in Fig. 12.30, with the selected pure classes denoted by colored marks in the scatter plots shown in Fig. 12.31. Three pure classes correspond well to LMM endmembers, and the large scatter of the "blue" class is captured by pure class covariance. This remedies the outlier problem exhibited by the LMM and illustrates the underlying formulation of the SMM.

12.6 Summary

When processing hyperspectral data, it is common to represent spectra as vectors in a multidimensional data space. In doing so, spectra can be compared using geometrical concepts such as distances and angles, and changes in illumination and other observation conditions can be characterized by linear, affine, and nonlinear transformations. By combining this geometrical representation with a physical model of

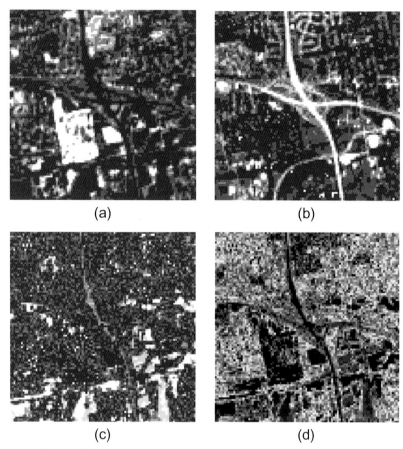

Figure 12.30 Abundance maps for the Hyperion VNIR/SWIR hyperspectral image based on the SMM: (a) endmember 1, (b) endmember 2, (c) endmember 3, and (d) endmember 4.

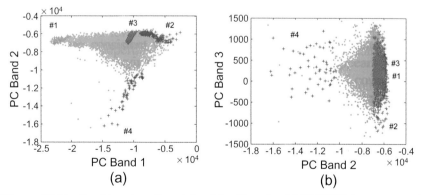

Figure 12.31 Dual-band scatter plots of the Hyperion VNIR/SWIR hyperspectral image in leading PC bands, indicating pure SMM endmember classes: (a) PC bands 1 and 2 and (b) PC bands 2 and 3.

linear spectral mixing, the linear mixing model characterizes variance in hyperspectral imagery as random mixtures of pure endmember spectra. These endmember spectra, in an unsupervised processing setting, can be estimated as vertices of a multidimensional simplex, and fractional abundances for each scene spectrum can be estimated through constrained optimization.

An alternate data-modeling approach is to employ statistical representations such as the multivariate normal model, local normal model, normal mixture model, and extensions based on generalized elliptically contoured distributions. These methods can better represent real imagery in situations where the strict assumptions of the linear mixing model are violated, especially in cases that concern the existence of pure, deterministic endmembers and use only linear mixing. In the process, however, statistical methods tend to become divorced from the physical modeling relationships underlying the scene spectral information. The stochastic mixing model is one example of an approach that preserves this physical mixture model in a statistical context.

12.7 Further Reading

Numerous journal papers expand on the primary spectral models outlined in this chapter, several of which are cited in the text. A topical treatment of hyperspectral data models and processing techniques is provided by Chang (2007), which provides extensive references to this larger body of work.

References

Bioucas-Dias, J. M. and Nascimento, J. M. P., "Hyperspectral subspace identification," *IEEET. Geosci. Remote Sens.* **46**, 2435–2445 (2008).

Chang, C.-I (Ed.), *Hyperspectral Data Exploitation: Theory and Applications*, Wiley, Hoboken, NJ (2007).

Chang, C.-I and Du, Q., "Estimation of number of spectrally distinct signal sources in hyperspectral imagery," *IEEET. Geosci. Remote Sens.* **42**, 608–619 (2004).

Craig, M. D., "Minimum-volume transforms for remotely sensed data," *IEEE T. Geosci. Remote Sens.* **32**, 542–552 (1994).

Eismann, M. T., Stocker, A. D., and Nasrabadi, N. M., "Automated hyperspectral cueing for civilian search and rescue," *Proc. IEEE* **97**, 1031–1055 (2009).

Eismann, M. T., and Hardie, R. C., "Improved initialization and convergence of a stochastic spectral unmixing algorithm," *Appl. Optics* **43**, 6596–6608 (2005).

Eismann, M. T. and Stein, D. W. J., "Stochastic Mixture Modeling," Chapter 5 in Chang, C.-I (Ed.), *Hyperspectral Data Exploitation: Theory and Applications*, Wiley, Hoboken, NJ (2007).

Fletcher, R., *Practical Methods of Optimization*, John Wiley & Sons, Chichester, UK (1987).

Fuhrmann, D. R., "A simplex shrink-wrap algorithm," *Proc. SPIE* **3718**, 501–511 (1999) [doi:10.1117/12.359990].

Green, A. W., Berman, B., Switzer, P., and Craig, M. D., "A transformation for ordering multispectral data in terms of image quality with implications for mean removal," *IEEET. Geosci. Remote Sens.* **26**(1), 65–74 (1988).

Hyvärinen, A. and Oja, E., "Independent component analysis: algorithms and applications," *Neural Networks* **13**, 411–430 (2000).

Karhunen, K., "Über lineare Methoden in der Wahrscheinlichkeitsrechnung," *Ann. Acad. Sci. Fenn. A1: Math.-Phys.* **37** (1947).

Keshava, N. and Mustard, J. F., "Spectral unmixing," *IEEE Signal Proc. Mag.*, 44–57 (2002).

Lee, J. B., Woodyatt, A. S., and Bermon, B., "Enhancement of high spectral resolution remote-sensing data by a noise-adjusted principal components transform," *IEEET. Geosci. Remote Sens.* **28**, 295–304 (1990).

Linde, Y., Buzo, A., and Gray, R. M., "An algorithm for vector quantization," *IEEE T. Commun. Theory* **28**(1), 84–95 (1980).

Loeve, M., *Probability Theory*, Van Nostrand, New York (1963).

Manolakis, D., Rosacci, M., Cipar, J., Lockwood, R., Cooley, T., and Jacobsen, J., "Statistical characterization of natural hyperspectral backgrounds," *Proc. IEEE Geosci. Remote Sens. Symp.*, 1624–1627 (2006).

Marden, D. and Manolakis, D., "Modeling hyperspectral imaging data," *Proc. SPIE* **5093**, 253–262 (2003) [doi:10.1117/12.485933].

Masson P. and Pieczynski, W., "SEM algorithm and unsupervised statistical segmentation of satellite images," *IEEE T. Geosci. Remote Sens.* **31**, 618–633 (1993).

Papoulis, A., *Probability, Random Variables and Stochastic Processes*, McGraw-Hill, New York (1984).

Roger, R. E. and Arnold, J. F., "Reliably estimating the noise in AVIRIS hyperspectral images," *Int. J. Remote Sens.* **17**, 1951–1962 (1996).

Schowengerdt, R. A., *Remote Sensing: Models and Methods for Image Processing*, Elsevier, Burlington, MA (1997).

Silverstein, J. W. and Bai, Z. D., "On the empirical distribution of eigenvalues for a class of large dimensional random matrices," *J. Multivariate Anal.* **54**, 175–192 (1995).

Stocker, A. D. and Schaum, A. P., "Application of stochastic mixing models to hyperspectral detection problems," *Proc. SPIE* **3071**, 47–60 (1997) [doi:10.1117/12.280584].

Strang, G., *Linear Algebra and its Applications*, Second Edition, Academic Press, Orlando, FL (1980).

Winter, M. E., "Maximum Volume Transform for Endmember Spectra Determination," Chapter 7 in Chang, C.-I (Ed.), *Hyperspectral Data Exploitation: Theory and Applications*, Wiley, Hoboken, NJ (2007).

Chapter 13
Hyperspectral Image Classification

One objective of hyperspectral data processing is to classify collected imagery into distinct material constituents relevant to particular applications, and produce classification maps that indicate where the constituents are present. Such information products can include land-cover maps for environmental remote sensing, surface mineral maps for geological applications and precious mineral exploration, vegetation species for agricultural or other earth science studies, or manmade materials for urban mapping. The underlying models and standard numerical methods to perform such image classifications have already been described to some extent in Chapter 12. This chapter formalizes the general classification strategy and underlying mathematical theory, provides further detail on the implementation of specific classification algorithms common to the field, and shows examples of the results of some of the methods. Hyperspectral image classification research is an intense field of study, and a wide variety of new approaches have been developed to optimize performance for specific applications that exploit both spatial and spectral image content. This chapter focuses on the most widely used and reported spectral classification approaches and includes a very limited discussion of the incorporation of spatial features in a classification algorithm.

13.1 Classification Theory

A general strategy for spectral image classification is illustrated in Fig. 13.1. Input data are assumed to be uncalibrated sensor output, calibrated pupil-plane spectral radiance, or estimated reflectance or emissivity derived by an atmospheric compensation algorithm. For some classification algorithms, the level of calibration or atmospheric compensation can make a difference, especially when the approach involves a human operator identifying constituents in the scene to extract training data or label classes. For many algorithms, however, the level of calibration or atmospheric compensation is not as important because the methods are generally invariant to these simple linear or affine transformations of the data.

The basic steps of a material classification algorithm are feature extraction, classification, and labeling. Feature extraction involves transforming data into a domain that efficiently captures salient information that supports separation of the data into distinct classes of interest. Classification is the actual process where each spectrum in a hyperspectral image is assigned to a different material class by some similarity metric. Finally, labeling is the process of associating each class with a physical identification pertinent to the particular application, such as land-cover type, surface mineral, or manmade material. Each of these steps can be accomplished either completely autonomously, in which case it is called unsupervised, or with some level of manual supervision to identify salient features, extract training spectra from the scene on which to base the material classes, or provide class labels. For supervised classifiers, classification rules are designed to carve up the multidimensional space of data or feature vectors, such that the training data are optimally separated by some criterion. For unsupervised classifiers, no such training data exist, and classification rules are derived to separate data into clusters, where spectral variation within a cluster is small, and spectral variation between clusters is large. For this reason, unsupervised classification is often called clustering.

If we choose to model data as a random vector, the optimum classification strategy in the sense of minimizing Baye's risk with uniform costs is Bayesian hypothesis testing, which assigns a spectrum $\mathbf{x}$ to class ω_q that maximizes *a posteriori* probability $P(q \mid \mathbf{x})$. Given Q classes (or hypotheses), the probability density function of $\mathbf{x}$ is given by

$$p(\mathbf{x}) = \sum_{q=1}^{Q} P(q)\, p(\mathbf{x} \mid q), \qquad (13.1)$$

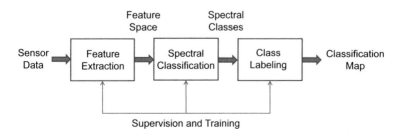

Figure 13.1 General strategy for spectral image classification.

where $P(q)$ is a prior probability, $p(\mathbf{x} \mid q)$ is a class-conditional probability density function, and q is a class index. In this case, the *a posteriori* probability for a particular class is related to the mixture model parameters by the Bayes theorem,

$$P(q \mid \mathbf{x}) = \frac{P(q)p(\mathbf{x} \mid q)}{p(\mathbf{x})} = \frac{P(q)p(\mathbf{x} \mid q)}{\sum\limits_{q=1}^{Q} P(q)p(\mathbf{x} \mid q)}. \tag{13.2}$$

If these underlying class distributions are known, the Bayesian classification rule simply becomes

$$\mathbf{x}_i \in \omega_q \quad \text{if } P(q)p(\mathbf{x}_i \mid q) > P(r)p(\mathbf{x}_i \mid r) \quad \forall\, r \neq q, \tag{13.3}$$

where $\mathbf{x}_i$ is the spectrum to be assigned, and ω_q represents the subset of image spectra assigned to the q'th class. The classification rule in Eq. (13.3) can be restated in terms of the log-likelihood function as

$$\mathbf{x}_i \in \omega_q \quad \text{if } \log P(q) + \log p(\mathbf{x}_i \mid q) > \log P(r) + \log p(\mathbf{x}_i \mid r) \quad \forall\, r \neq q. \tag{13.4}$$

In many situations, no *a priori* information exists concerning prior probabilities. Consequently, uniform priors are assumed, and the classification rule simplifies to

$$\mathbf{x}_i \in \omega_q \quad \text{if } \log p(\mathbf{x}_i \mid q) > \log p(\mathbf{x}_i \mid r) \quad \forall\, r \neq q. \tag{13.5}$$

It is atypical for most practical applications, however, to know the underlying conditional density functions of the classes, making it difficult to use the Bayesian classification rule. Instead, either some form of the density functions must be assumed with the parameters of the distribution estimated from the data, or some other classification methodology must be employed.

If mixture classes are assumed to be multivariate normal, such that $p(\mathbf{x} \mid q) \sim N(\boldsymbol{\mu}_q, \boldsymbol{\Sigma}_q)$, then it is convenient to define a discriminant function

$$d_q(\mathbf{x}) = -2 \log p(\mathbf{x} \mid q) + \frac{K \log 2\pi}{2} = (\mathbf{x} - \boldsymbol{\mu}_q)^T \boldsymbol{\Sigma}_q^{-1} (\mathbf{x} - \boldsymbol{\mu}_q) + \log |\boldsymbol{\Sigma}_q|, \tag{13.6}$$

such that the Bayesian classification rule becomes

$$\mathbf{x}_i \in \omega_q \quad \text{if } d_q(\mathbf{x}_i) > d_r(\mathbf{x}_i) \quad \forall\, r \neq q. \tag{13.7}$$

Considering two classes ω_q and ω_r, we can substitute Eq. (13.6) into Eq. (13.7) and simplify the result to determine decision surface $d_{q,r}(\mathbf{x}) = d_q(\mathbf{x}) - d_r(\mathbf{x})$ between the two classes,

$$
\begin{aligned}
d_{q,r}(\mathbf{x}) = &(\mathbf{x} - \boldsymbol{\mu}_q)^T \boldsymbol{\Sigma}_q^{-1} (\mathbf{x} - \boldsymbol{\mu}_q) \\
&- (\mathbf{x} - \boldsymbol{\mu}_r)^T \boldsymbol{\Sigma}_r^{-1} (\mathbf{x} - \boldsymbol{\mu}_r) + \log|\boldsymbol{\Sigma}_q| - \log|\boldsymbol{\Sigma}_r|.
\end{aligned}
\tag{13.8}
$$

In general, Eq. (13.8) represents a quadratic decision surface, and the resulting classification rule is known as the quadratic classifier. For the two-class problem, it simply becomes

$$
\mathbf{x}_i \in \omega_q \quad \text{if } d_{q,r}(\mathbf{x}_i) < 0
\tag{13.9}
$$

and

$$
\mathbf{x}_i \in \omega_r \quad \text{if } d_{q,r}(\mathbf{x}_i) > 0.
\tag{13.10}
$$

For the multiclass problem, such decision surfaces (shown in Fig. 13.2) are drawn between all classes based on their mean vectors and covariance matrices, and each image spectrum is assigned to a class for which the expressions in Eqs. (13.9) and (13.10) hold true for every class with which it is paired. In the illustrated case, a test spectrum would be assigned to material class ω_q. Note that there will be an additional bias in each decision surface equal to the difference in the *log* of prior probabilities if they are not all assumed equal.

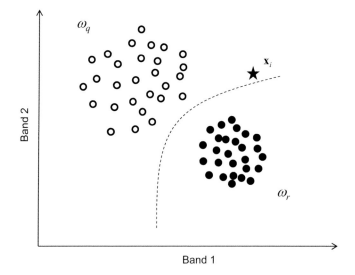

Figure 13.2   Illustration of a quadratic classifier.

A special case occurs when the covariance matrices are all equal: $\Sigma_q = \Sigma_r = \Sigma$. In this case, decision surfaces can be simplified to

$$d_{q,r}(\mathbf{x}) = -2(\boldsymbol{\mu}_q - \boldsymbol{\mu}_r)^T \Sigma^{-1} \mathbf{x} + \boldsymbol{\mu}_q^T \Sigma^{-1} \boldsymbol{\mu}_q - \boldsymbol{\mu}_r^T \Sigma^{-1} \boldsymbol{\mu}_r, \quad (13.11)$$

which are recognized as planar surfaces. This special case is known as a linear classifier and is illustrated in Fig. 13.3. If the covariance matrices are both equal and white, $\Sigma = \sigma^2 \mathbf{I}$, then the discriminant functions simply become

$$d_q(\mathbf{x}) = (\mathbf{x} - \boldsymbol{\mu}_q)^T (\mathbf{x} - \boldsymbol{\mu}_q) = \| \mathbf{x} - \boldsymbol{\mu}_q \|^2, \quad (13.12)$$

and the resulting classification rule becomes the linear distance classifier. Under the assumption of equal class covariance matrices, it should be noted that whitening the data allows a simple linear distance classifier to be used, even when the actual class covariance matrices are nonwhite. In the illustration, the test spectrum that was classified into ω_q by the quadratic classifier is now classified into ω_r.

To apply classification rules based on multivariate normal classes, it is necessary to have estimates of class parameters: mean vectors in the case of a linear distance classifier, and mean vectors and covariance matrices for a quadratic classifier. In supervised classification, these estimates can be based on training data manually extracted from the scene. In an unsupervised case, however, these parameters must be estimated from the data itself, without knowledge of the correct class assignments. This

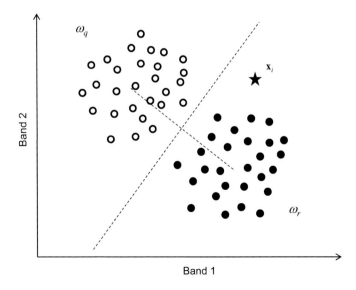

Figure 13.3 Illustration of a linear classifier.

estimation problem represents the fundamental challenge of classifier design and is responsible for the variety of clustering methods described later in this chapter and throughout the literature. There is an alternate classification strategy, however, that is not based on an assumed underlying form of class-conditional density functions and does not require the estimation of density parameters. Instead, nonparametric classifiers derive typically nonlinear decision surfaces based simply on local relationships of image spectra in multidimensional space. Two such methods, the k nearest neighbor (kNN) method and the kernel support vector machine (SVM), are described later in this chapter.

A final theoretical consideration of spectral classification is the method for characterizing the efficacy of a classification algorithm. If there is ground truth that specifies correct class assignments of the image spectra, then one could produce a confusion matrix of the form depicted in Fig. 13.4 that compares the assigned class to the true class in terms of the number of spectra assigned to each. In a perfect classifier, all of the off-diagonal elements of the confusion matrix would be zero. In a real classifier, some proportion of spectra will be misclassified into off-diagonal elements of the confusion matrix, and only a fraction of all spectra correctly classified into diagonal elements. The ultimate efficacy of the classifier is therefore characterized by the probability of correct classification P_{cc} and probability of misclassification P_{mc}, estimated by applying the classifier to imagery for which ground truth is available. The probability of correct classification is the ratio of the trace of the confusion matrix to the total number of samples, while the probability of misclassification is the ratio of the sum of off-diagonal elements to the total. If true distributions of the classes are known, then it is possible to estimate P_{cc} and P_{mc} from a decision theory; otherwise, they can only be estimated through empirical testing. Normal conditional distributions

		\multicolumn{4}{c}{**Assigned Class**}
		1	2	3	4
Truth Class	1	23	1	0	1
	2	2	27	1	0
	3	0	0	38	2
	4	1	0	0	19

Figure 13.4 Example of a confusion matrix for which $P_{cc} = 0.92$.

are often assumed to gauge the expected classification performance as a function of second-order class statistics, which are known as class parameters.

While a confusion matrix and classification probabilities can help characterize classifier performance in a test setting, these metrics provide little value in determining whether mixture parameters conform well to the data from which they are estimated, especially when no such ground truth exists. In this regard, useful metrics are the likelihood function

$$L(\mathbf{X} \mid \hat{\theta}) = \prod_{i=1}^{N} \sum_{q=1}^{Q} P(q) p(\mathbf{x}_i \mid q, \hat{\theta}_q), \qquad (13.13)$$

which is a measure of probability that image data matrix $\mathbf{X}$ could have been randomly drawn from the mixture model defined by class parameter estimates $\hat{\theta}_q$, and the log-likelihood function

$$l(\mathbf{X} \mid \hat{\theta}) = \log L(\mathbf{X} \mid \hat{\theta}) = \sum_{i=1}^{N} \log \left[\sum_{q=1}^{Q} P(q) p(\mathbf{x}_i \mid q, \hat{\theta}_q) \right]. \quad (13.14)$$

Parameter set θ is the union of all class parameters θ_q for all classes. A higher likelihood or log-likelihood function indicates a better model fit. Both of these representations assume spatial independence of hyperspectral data required to write the likelihood function of an entire image as a product of likelihood functions of individual spectra.

13.2 Feature Extraction

The objective of feature extraction is to reduce image data to an efficient representation of salient features, called feature vectors, that capture essential information in the data to support better classification. Such features can be determined based on some physical knowledge of important characteristics pertinent to the particular spectral classification problem, such as the location of absorption features and band edges, or are driven by statistics of the collected data. From a statistical perspective, dimensionality reduction techniques such as PCA, maximum noise fraction (MNF), and ICA perform a feature extraction function by transforming data to a coordinate system where importance of the bands is quantified by corresponding eigenvalues, characterizing either image variance or SNR. By maintaining only the leading subspace, signal content is more efficiently captured, which is one objective of feature extraction. Unfortunately, this approach does nothing with regard to favoring features that maximize separation between spectral classes, another important aspect of feature extraction. Without any *a priori* information about material classes, however, there is little basis to address this class separability aspect, and these dimensionality reduction methods often serve as a feature extraction approach.

The LMM represents a physical construct for feature extraction, in that it identifies pure materials in the scene in either a supervised or unsupervised manner that would be expected to form the basis of spectral image classification. Feature vectors in this case are endmember spectra, and abundance images represent the transformation of data into reduced-dimensionality feature space. It should be readily apparent that the set of abundance images should accentuate differences between endmember material classes, to the extent that the LMM is appropriate for hyperspectral data. The use of linear spectral unmixing as a basis for spectral image classification has been applied to several remote sensing problems, including vegetation species mapping (Dennison and Roberts, 2003), ecology (Ustin *et al.*, 2004), and fire danger assessment (Roberts *et al.*, 2003).

13.2.1 Statistical separability

Suppose that *a priori* class information does exist for a particular hyperspectral image. In this case, a general measure of separability between two classes ω_q and ω_r defined by their conditional density functions is given by (Fukunaga, 1990)

$$J_{q,r} = \int \left[p(\mathbf{x} \mid q) - p(\mathbf{x} \mid r) \right] d\mathbf{x}. \qquad (13.15)$$

If density functions were known, this expression could serve as a theoretical basis for optimizing data features. Such extensive information about material classes, however, is generally not available, and even if it were, optimization would likely be numerically intractable. On the other hand, suppose again that multivariate normal class density functions are assumed and some *a priori* information concerning class mean vectors $\boldsymbol{\mu}_q$, covariance matrices $\boldsymbol{\Sigma}_q$, and prior probabilities $P(q)$ for the Q expected classes were available. Such information could be available from class statistics of training data in a supervised classification setting. In this case, we can define a within-class scatter matrix

$$\mathbf{S}_w = \sum_{q=1}^{Q} P(q)\, \boldsymbol{\Sigma}_q, \qquad (13.16)$$

a between-class scatter matrix,

$$\mathbf{S}_b = \sum_{q=1}^{Q} P(q) (\boldsymbol{\mu}_q - \boldsymbol{\mu}_0) (\boldsymbol{\mu}_q - \boldsymbol{\mu}_0)^T, \qquad (13.17)$$

and a mixture scatter matrix,

$$\mathbf{S}_m = \mathbf{S}_w + \mathbf{S}_b, \qquad (13.18)$$

where $\boldsymbol{\mu}_0$ is the overall mean vector,

$$\boldsymbol{\mu}_0 = \sum_{q=1}^{Q} P(q) \boldsymbol{\mu}_q. \qquad (13.19)$$

A useful measure of the separability of classes is separability metric J, defined by

$$J = \text{trace}(\mathbf{S}_m^{-1} \mathbf{S}_b). \qquad (13.20)$$

Given training data for several classes, the scatter matrices and separability metric can be estimated from training class sample statistics.

Based on the normal mixture model assumption implied by the preceding separability analysis, optimum linear features from the perspective of maximizing separability metric J are the leading $Q - 1$ eigenvectors of the eigenvalue problem

$$(\mathbf{S}_w^{-1} \mathbf{S}_b) \phi_q = \lambda_q \phi_q \{ \phi_1, \phi_2, \ldots, \phi_{Q-1} : \lambda_1 > \lambda_2 > \cdots > \lambda_{Q-1} > \lambda_j \ j \geq Q \}. \qquad (13.21)$$

Such eigenvectors are known as Fischer's linear discriminants. If these $L - 1$ eigenvectors are placed as columns into eigenvector matrix $\boldsymbol{\Phi}$, then data matrix $\mathbf{X}$ can be transformed into feature space matrix $\mathbf{Z}$ by the linear transform

$$\mathbf{Z} = \boldsymbol{\Phi}^T \mathbf{X}, \qquad (13.22)$$

where each column of $\mathbf{Z}$ is the feature vector of the corresponding spectral vector in $\mathbf{X}$. Performing feature extraction in this manner reduces the dimensionality of data while optimizing class separability, in so far as the normal mixture model assumptions are supported by the data.

13.2.2 Spectral derivatives

From a more physical perspective, feature extraction can involve identification of specific spectral reflectance, emission, or transmission features unique to material classes to be separated, such as optical depth measures at particular wavelengths, ratios between spectral bands near characteristic band edges, or normalized differential indices such as the NDVI. An example is the Tetracorder algorithm developed by the United States Geological Survey (USGS) to perform geological and vegetation mapping (Clark *et al.*, 2003). This algorithm is perhaps somewhat heuristic in that it keys off of specific, diagnostic spectral feature locations known to correspond to materials of interest; that is, the identifiable narrowband absorption features apparent in the empirical reflectance plots displayed in Chapter 4. The feature extraction process in this case is the manual identification of feature locations associated with each particular material of interest. While perhaps not elegant from a statistical pattern recognition perspective, this manual process was determined to be extremely effective in terms of enhancing classification accuracy.

The image classification strategy employed in Tetracorder basically computes the correlation coefficient between the test spectrum and a number of reference spectra representing material classes of interest, then classifies the spectrum to one with the highest correlation coefficient. This simple approach, however, was found to provide poor classification accuracy because the coarse spectral shape of the measured spectra varied considerably within a material class and influenced the correlation coefficient as much or more than fine diagnostic spectral features. Therefore, the spectral matching algorithm was extended to only compute the correlation coefficient for the spectral location of diagnostic features for each class, and to normalize both the test and reference spectra by removing the continuum (i.e., subtracting the continuum slope by a linear fit through spectral samples, at wavelengths slightly above and below the diagnostic feature location) and linearly normalizing the spectra to each other. Where a particular material class of interest exhibits a single diagnostic feature, the correlation coefficient in this spectral location is used as the classification metric. If the material class exhibits multiple diagnostic features, a linear combination of the individual correlation coefficients is used, where the weights are relative fractional areas (depth by spectral width) of the reference continuum-removed spectral features. Example results from this spectral classification approach are illustrated in Chapter 1.

Since diagnostic material spectral features often involve sharp emission peaks, absorption features, and band edges, they can be enhanced and located by taking spectral derivatives of the image spectra. Numerically, the n'th spectral derivative of $\mathbf{x}$ at the k'th spectral band can be represented as

$$\left(\frac{\partial^n \mathbf{x}}{\partial \lambda^n}\right)_k = \frac{\partial}{\partial \lambda}\left(\frac{\partial^{n-1}\mathbf{x}}{\partial \lambda^{n-1}}\right)_k$$

$$\approx \begin{cases} \dfrac{x_{k+n/2} - x_{k-1+n/2} \cdots + x_{k-n/2}}{\Delta\lambda^n} & n = \text{even} \\[2ex] \dfrac{x_{k+n/2-1/2} - x_{k-1+n/2-1/2} \cdots + x_{k-n/2-1/2}}{\Delta\lambda^n} & n = \text{odd} \end{cases}, \quad (13.23)$$

where $\Delta\lambda$ is the spectral sampling period of the data. This equation assumes uniform spectral sampling of data; for nonuniform sampling, $\Delta\lambda$ is k dependent. Figures 13.5 and 13.6 illustrate both laboratory reflectance and pupil-plane radiance spectra for vegetation and a green tarp, along with first and second derivatives computed by Eq. (13.23). For laboratory reflectance spectra, the derivatives accentuate fine material spectral features and suppress the low-order differences in reflectance magnitude, a positive feature from the standpoint of feature extraction. For pupil-plane radiance spectra, however, the derivative analysis also accentuates atmospheric features and increases noise, both of which are negative attributes. In any case, a PCA can be performed on derivative spectral images to isolate dominant features. The dominant PCA bands from the derivative image can be added to leading-order PCA bands from the original hyperspectral image as an input to the classifier. It should be noted that the derivative images carry no additional information relative to the hyperspectral image; instead, they merely accentuate fine spectral structure to make it more prominent in the global statistics driving the PCA process.

13.3 Linear Classification Algorithms

As described in Section 13.1, a linear classifier segments the K-dimensional spectral or feature space into Q regions by a set of hyperplanes, as depicted in Fig. 13.3. We can arbitrarily describe decision surfaces between classes ω_q and ω_r by the plane

$$d_{q,r}(\mathbf{x}) = \mathbf{a}_{q,r}^T \mathbf{x} + b_{q,r}, \quad (13.24)$$

where parameters $\mathbf{a}_{q,r}$ and $b_{q,r}$ are related to class statistics according to Eq. (13.11). If there is no *a priori* information concerning class statistics, then locations of the decision surfaces must be optimized with regard to the data. In a supervised situation, that means locating the planes between the training classes to maximize their separation. In an unsupervised situation, it means locating the planes that separate actual scene data into separated clusters. Use of the nearest mean classification rule in Eq. (13.12) is a

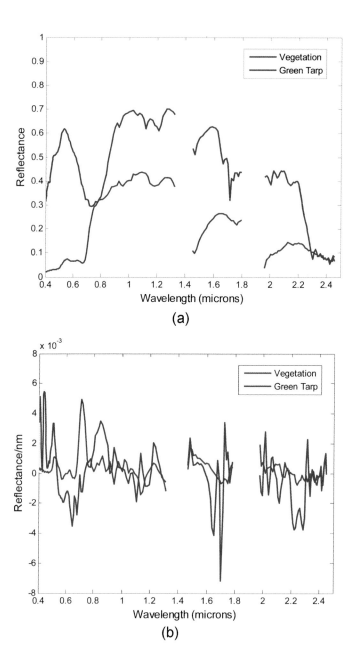

Figure 13.5 Spectral derivatives of laboratory spectra for vegetation and a green tarp: (a) reflectance spectra, (b) first derivatives, and (c) second derivatives.

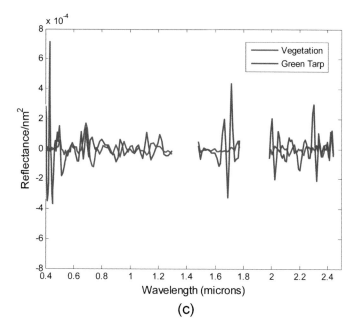

(c)

Figure 13.5   (*continued*)

common linear classification strategy, but it requires some manner of also estimating class mean vectors and perhaps even the number of classes in an unsupervised setting.

Even when classes are normally distributed, a linear classifier will only be optimal when covariance matrices of the classes are identical. However, it is possible to optimize locations of the classification surfaces in the case of unequal class covariance matrices by defining an appropriate optimization criterion. One such criterion is the Fischer criterion, which measures the difference of class means normalized by the average variance. An optimum linear classifier based on this criterion can be derived from the general Bayesian classification rule in Eq. (13.3), assuming a common global covariance matrix for all classes (Fukunaga, 1990). The discriminant function for class ω_q is given by

$$d_q(\mathbf{x}) = \boldsymbol{\mu}_q^T \boldsymbol{\Sigma}^{-1} \mathbf{x} - \frac{1}{2} \boldsymbol{\mu}_q^T \boldsymbol{\Sigma}^{-1} \boldsymbol{\mu}_q + \log P(q), \qquad (13.25)$$

where

$$\boldsymbol{\Sigma} = \sum_{q=1}^{Q} P^2(q) \, \boldsymbol{\Sigma}_q. \qquad (13.26)$$

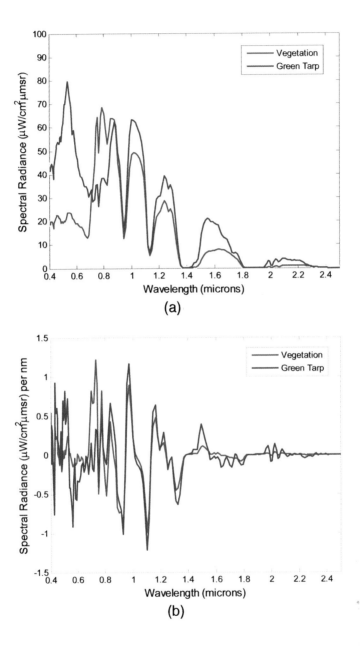

Figure 13.6 Spectral derivatives of pupil-plane radiance spectra for vegetation and a green tarp: (a) radiance spectra, (b) first derivatives, and (c) second derivatives.

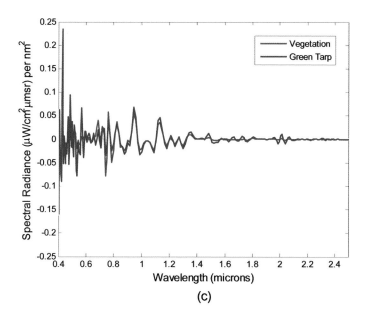

Figure 13.6 *(continued)*

The multiclass classification rule becomes one of selecting the class that maximizes the linear discriminant for the test spectrum:

$$\mathbf{x}_i \in \omega_q \quad \text{if } d_q(\mathbf{x}_i) > d_r(\mathbf{x}_i) \quad \forall \, r \neq q. \tag{13.27}$$

This form of optimum linear classifier, also known as the Fischer linear classifier, includes knowledge of not only mean vectors and covariance matrices for all classes, but also prior probability values. When the latter information is not available, prior probability values are all assumed to be equal. Incorporation of the prior probability reduces the likelihood of classifying test spectra into less populated or rarer classes.

To illustrate the characteristics of many of the spectral classification algorithms described in this chapter, these characteristics are applied to the VNIR hyperspectral image illustrated in Fig. 13.7 that was collected on 25 August 2005 by the Hyperspec imaging spectrometer described in Chapter 10. Specifications of the hyperspectral instrument and collection conditions are summarized in Table 13.1. This image provides a good case for characterizing classification algorithm performance because of its relative simplicity. It is basically composed of six material classes: the foreground grass field, background tree line, and four distinctly different manmade panels. These classes are identified in the scatter plots for the first three principal components illustrated in Fig. 13.8. All of the algorithms to be described are applied to the leading ten-dimensional

Table 13.1 Hyperspec instrument specifications and image collection conditions.

Characteristic	Specification
Spectral range	0.46 to 0.9 μm
Spectral sampling	3.6 nm
Spectral resolution (nominal)	8 nm
Number of spectral bands	52
Image format	400×512
Angular sampling	0.24 mrad
Spatial sampling (nominal)	6 cm
Sensor-to-scene range (nominal)	250 m
Sensor elevation	25.9 m
Depression angle	6 deg
Date	25 August 2005
Time	11:00 a.m.

Figure 13.7 Hyperspec VNIR panel image used for characterizing image classification performance.

principal-component subspace of data. The trailing subspace is dominated by noise.

The results of applying the optimum linear classifier in Eq. (13.27) to the leading ten-dimensional principal-component subspace of the Hyperspec image are conveyed by the color-coded class map and leading principal-component scatter plots illustrated in Figs. 13.9 and 13.10. The results indicate almost perfect image classification with the optimum linear classifier based on known class statistics; that is, the sample statistics of data (including prior probability estimates) based on the truth classification

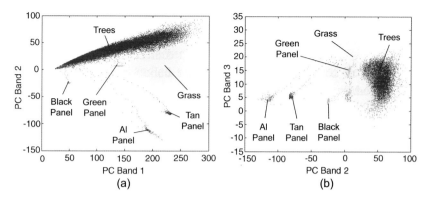

Figure 13.8 Identification of spectral classes in scatter plots of the leading three principal components of the Hyperspec VNIR panel image data: (a) PC bands 1 and 2 and (b) PC bands 2 and 3.

Figure 13.9 Color-coded class map for Hyperspec VNIR panel image based on the optimum linear classifier algorithm with six classes.

map. Differences are due to border pixels (for which the ground truth is ambiguous) and a small number of misclassified spectra in the grass and tree regions. Since class statistics are generally not known *a priori*, these test-on-train results are better than normally expected in a practical setting, especially in an unsupervised setting where there is no *a priori* class information. However, the results do support the adequacy of linear classification surfaces to separate the six scene classes in this simple case.

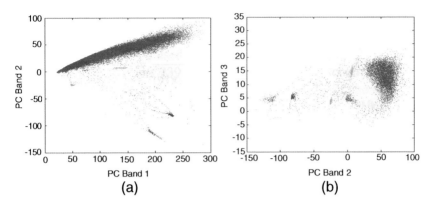

Figure 13.10 Color-coded scatter plot of leading three principal components for Hyperspec VNIR panel image based on the optimum linear classifier algorithm with six classes: (a) PC bands 1 and 2 and (b) PC bands 2 and 3.

13.3.1 k-means algorithm

The nearest mean classification rule is the foundation of a specific linear classification algorithm, known alternatively as the k-means algorithm (Tso and Mather, 2009), Linde–Buzo–Gray (LBG) clustering (Linde *et al.*, 1980), or vector quantization, and can be shown to minimize the mean-squared error (MSE) defined by

$$\text{MSE} = \frac{1}{N} \sum_{q=1}^{Q} \sum_{\mathbf{x}_i \in \omega_q} (\mathbf{x}_i - \boldsymbol{\mu}_q)^T (\mathbf{x}_i - \boldsymbol{\mu}_q). \qquad (13.28)$$

Note that this error criterion is completely independent of any other class statistics or prior probabilities and would therefore be expected to perform best with whitened class statistics. The k-means algorithm iteratively estimates class mean vectors based on class sample statistics, and makes class assignments based on the nearest mean classification rule in an interleaved fashion. The basic steps are as follows:

1. Define the number of classes Q to represent the data.
2. Set the initial class mean vectors $\{\boldsymbol{\mu}_q: q = 1, 2, \ldots, Q\}$ from either *a priori* data or by randomly selecting Q image spectra or feature vectors.
3. For each image spectrum or feature vector $\mathbf{x}_i$, compute the Euclidean distance d_q in Eq. (13.12) for each class mean vector, and assign the sample to class ω_q with minimal distance.
4. After all samples have been assigned, update the mean vector for each class to the sample mean of all spectra assigned to the class.
5. Determine the amount of change in terms of either the number of spectra that switched classes or the average shift in class mean vectors.

6. Repeat steps 3, 4, and 5 until convergence based on the change is appreciably small.

Although the k-means algorithm is designed to operate unsupervised, one can include training spectra in the calculation of the class sample statistics and force them to remain assigned to known classes. Training spectra would also be used to initialize classes in the supervised case.

To illustrate the features of this approach, the k-means algorithm is applied to the leading ten-dimensional principal-component subspace of the Hyperspec VNIR data using six classes. The results are provided in terms of color-coded class maps, along with corresponding color-coded scatter plots of the first three principal components in Figs. 13.11 and 13.12. Convergence characteristics for the six-class case are also shown in terms of separability metric J from Eq. (13.20) and changes in class assignments in Fig. 13.13. The performance of the k-means algorithm with these data is somewhat poor relative to the truth. This is due primarily to the high variance difference between principal components and the highly colored nature of class covariance matrices. The classifier primarily segments data into regions along the first two principal components because of the high variance in this subspace. This has the effect of segmenting the image based on broadband irradiance differences, the primary driver of the first principal component, more than spectral differences. The resulting classification decision surfaces show little to no resemblance to those of the optimum linear classifier in Fig. 13.10.

Figure 13.11 Color-coded class map for Hyperspec VNIR panel image based on the k-means algorithm with six classes.

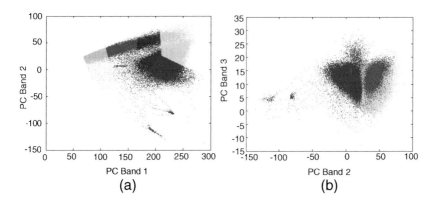

Figure 13.12 Color-coded scatter plot of leading three principal components for Hyperspec VNIR panel image based on the k-means algorithm with six classes: (a) PC bands 1 and 2 and (b) PC bands 2 and 3.

Convergence of the mean-based approach is very rapid, with negligible change in separability after only three iterations and few class changes after seven iterations.

As the nearest mean classification rule is based on an assumption of whitened classes, performance can be improved by transforming data to whitened space prior to applying the k-means classifier. In the general case, this is done by whitening transform

$$\mathbf{Z} = \mathbf{\Lambda}^{-1/2}\mathbf{\Phi}^T\mathbf{X}, \tag{13.29}$$

where $\mathbf{\Phi}$ and $\mathbf{\Lambda}$ are the eigenvector and diagonalized eigenvalue matrix corresponding to the mixture scatter matrix $\mathbf{S}_m$. If there is no *a priori* class information from which to estimate the mixture scatter matrix, then one can at least whiten the principal-component data using eigenvectors and eigenvalues of the sample covariance matrix before applying the k-means algorithm. The results of whitening for the Hyperspec panel image, prior to classifying the data with six k-means classes, are provided in Figs. 13.14 and 13.15. Compared to the six-class results without whitening in Figs. 13.11 and 13.12, whitening accentuates the spectral differences between classes over broadband irradiance differences. In this case, however, the method is still unable to separate the actual scene classes. Because the classes are not actually white, mean-based clustering still separates the dominant natural background classes into subclasses along the directions of highest variance and mixes the more sparsely populated panel spectra among these dominant classes.

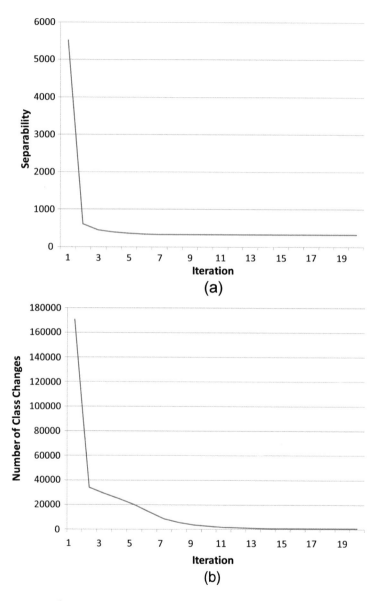

Figure 13.13 *k*-means algorithm convergence for six classes based on the MSE metric, the log-likelihood metric, and number of class changes: (a) separability metric and (b) number of class changes.

Figure 13.14 Color-coded class map for Hyperspec VNIR panel image based on the k-means algorithm with six classes and prewhitening.

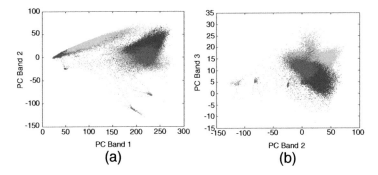

Figure 13.15 Color-coded scatter plot of leading three principal components for Hyperspec VNIR panel image based on the k-means algorithm with six classes and prewhitening: (a) PC bands 1 and 2 and (b) PC bands 2 and 3.

13.3.2 Iterative self-organizing data analysis technique

An important limitation of the k-means algorithm, as well as many other classifiers, is the need to arbitrarily define a number of classes. In a practical setting, the number of scene classes may not be known, and a poor choice of the number of classes can adversely impact classification performance. To address this limitation, variations of the linear classifier have been developed that automatically split clusters that grow to be too large in spectral extent and merge clusters that are close together. One such variation is the iterative self-organizing data analysis technique (ISODATA), which is common in the field of remote sensing data processing (Jensen, 1996). To support cluster splitting and merging, there are several parameters that the user must define for ISODATA to proceed. These include:

Q_0 = the desired number of classes.

I = the maximum number of iterations allowed.

P = the maximum number of class pairs that can be merged during a single iteration.

N_{min} = the minimum allowable number of samples in a class.

σ_{min} = the threshold class standard deviation for splitting.

d_{min} = the minimum class separation for merging.

α = the splitting parameter.

Based on these user definitions, the algorithm proceeds as follows:

1. Set the initial class mean vectors $\{\boldsymbol{\mu}_q: q = 1, 2, \ldots, Q\}$ from either *a priori* data or by randomly selecting Q_0 image spectra or feature vectors. Set $Q = Q_0$.
2. Assign all spectra to one of the Q classes by the nearest mean classification rule. Let N_q be the number of samples assigned to class ω_q.
3. Check all classes for population. Discard class ω_q if $N_q < N_{min}$, and decrement Q by one. The samples in the class will be reassigned to another class on the next iteration.
4. Update the class mean vectors to the sample mean of N_q spectra assigned to each class.
5. Compute the average distance $\bar{d}_q$ of the spectra in each class ω_q to the class mean vector according to

$$\bar{d}_q = \frac{1}{N_q} \sum_{\mathbf{x}_i \in \omega_q} \| \mathbf{x}_i - \boldsymbol{\mu}_q \|. \qquad (13.30)$$

6. Compute the overall average distance of the samples from the class mean vectors according to

$$\bar{d} = \frac{1}{N} \sum_{q=1}^{Q} N_q \bar{d}_q. \qquad (13.31)$$

7. If $Q \leq Q_0/2$, split all classes for which $\bar{d}_q > d_{min}$, $N_q > 2N_{min}$, and $\max_k \sigma_{q,k} > \sigma_{max}$, where $\boldsymbol{\sigma}_q$ represents the band-by-band standard deviation of the class, defined by

$$\boldsymbol{\sigma}_q = [\sigma_{q,1} \quad \sigma_{q,2} \quad \cdots \quad \sigma_{q,K}]^T, \tag{13.32}$$

where

$$\sigma_{q,k} = \sqrt{\frac{1}{N_q} \sum_{x_i \in \omega_q} (x_{i,k} - \mu_{i,k})^2}. \tag{13.33}$$

A class ω_q is split into classes ω_r and ω_p by setting $\boldsymbol{\mu}_r = \boldsymbol{\mu}_q - \alpha\boldsymbol{\sigma}_q$ and $\boldsymbol{\mu}_p = \boldsymbol{\mu}_q + \alpha\boldsymbol{\sigma}_q$, and incrementing Q by one.

8. If $Q > Q_0$, merge up to P class pairs with the smallest pairwise distance

$$\bar{d}_{r,p} = \| \boldsymbol{\mu}_r - \boldsymbol{\mu}_p \|. \tag{13.34}$$

Two classes ω_r and ω_p are merged into a class ω_q by decrementing Q by one and defining

$$\boldsymbol{\mu}_q = \frac{1}{N_r + N_p}(L_r \boldsymbol{\mu}_r + L_p \boldsymbol{\mu}_p). \tag{13.35}$$

9. Repeat steps 2 through 8 until either there are no further changes in class assignments or the maximum number of iterations I is reached.

13.3.3 Improved split-and-merge clustering

Despite the widespread use of ISODATA for multispectral and hyperspectral remote sensing data analysis, it can be difficult to implement because the user must set a number of somewhat arbitrary parameters to define splitting, merging, and stopping criteria. The optimal settings for these parameters can actually vary as a function of the characteristics of the data, such that a standard set of default settings does not exist. To address these limitations, an improved split-and-merge clustering (ISMC) algorithm was developed that both limits the number of user-defined parameters and normalizes the parameters to the statistics of the data, so that default settings can have widespread applicability (Simpson *et al.*, 2000). Two parameters, ε_{split} and ε_{merge}, are set to define split and merge thresholds, where default values are $\varepsilon_{split} = 0.075$ and $\varepsilon_{merge} = 0.005$. Absolute split and merge thresholds τ_{split} and τ_{merge} are defined as $\tau_{split} = \alpha\varepsilon_{split}$ and $\tau_{merge} = \alpha\varepsilon_{merge}$, where α is an image-specific scaling factor defined by the band-averaged, maximal squared difference of the data,

$$\alpha = \sum_{k=1}^{K} \left(\max_i x_{i,k} - \min_i x_{i,k} \right)^2. \tag{13.36}$$

The third and final user-specified parameter is the stopping parameter τ_{stop}, which can be set to default value $\tau_{stop} = 0.05$.

Based on these three parameters, the ISMC algorithm proceeds as follows:

1. Initialize with a single class $Q = 1$ with a mean vector $\boldsymbol{\mu}_1$ given by the sample mean of all data.
2. Split classes successively until they are all below the splitting threshold. To do so for class ω_q, a few parameters are used for characterizing the extent of the class. First, matrix $\mathbf{B}_q$ is defined whose columns $\mathbf{b}_j$ contain all 2^K bounding vectors of the class, where each bounding vector defines a vertex in a K-dimensional polygon that encloses all data within it. The $K \times 2^K$ element matrix $\mathbf{B}_q$ has the form

$$\mathbf{B}_q = \begin{bmatrix} \min_{x_i \in \omega_q} x_{i,1} & \min_{x_i \in \omega_q} x_{i,1} & \min_{x_i \in \omega_q} x_{i,1} & \cdots & \max_{x_i \in \omega_q} x_{i,1} \\ \min_{x_i \in \omega_q} x_{i,2} & \min_{x_i \in \omega_q} x_{i,2} & \min_{x_i \in \omega_q} x_{i,2} & \cdots & \max_{x_i \in \omega_q} x_{i,2} \\ \vdots & \vdots & \vdots & \ddots & \vdots \\ \min_{x_i \in \omega_q} x_{i,K-1} & \min_{x_i \in \omega_q} x_{i,K-1} & \max_{x_i \in \omega_q} x_{i,K-1} & \cdots & \max_{x_i \in \omega_q} x_{i,K-1} \\ \min_{x_i \in \omega_q} x_{i,K} & \max_{x_i \in \omega_q} x_{i,K} & \max_{x_i \in \omega_q} x_{i,K} & \cdots & \max_{x_i \in \omega_q} x_{i,K} \end{bmatrix}. \tag{13.37}$$

Next, the two spectra within the class with minimal and maximal Euclidean distances from each bounding vector $\mathbf{b}_j$ are identified and included in the peripheral subset π_q of the class, which will ultimately have 2×2^K elements. An exhaustive search of all pairs of spectra within this subset is then made to find the two decision spectra $\mathbf{x}_r$ and $\mathbf{x}_p$ with maximal separation; that is,

$$\mathbf{x}_r, \mathbf{x}_p \in \pi_q : \| \mathbf{x}_r - \mathbf{x}_p \| > \| \mathbf{x}_i - \mathbf{x}_j \| \quad \forall \mathbf{x}_i, \mathbf{x}_j \in \pi_q \ i, j \neq r, p. \tag{13.38}$$

The class is split if $\| \mathbf{x}_r - \mathbf{x}_p \| > \tau_s$, in which case the remaining class spectra are assigned to the new class containing the nearest decision spectrum according to the Euclidean distance. For each class split, increment Q by one.

3. Remove any null classes.

4. Merge classes successively until they are all above the merging threshold. Compute the pair-wise distance between all pairs of class mean vectors, and merge classes r an p if

$$d_{r,p} = \| \boldsymbol{\mu}_r - \boldsymbol{\mu}_p \| < \tau_m. \tag{13.39}$$

For each class merger, decrement Q by one.

5. Assign all spectra to one of the Q classes by the nearest mean classification rule. Define N_q as the number of samples assigned to class ω_q.

6. Update the class mean vectors to the sample mean of N_q spectra assigned to each class.

7. Estimate the between-class scatter matrix,

$$\hat{\mathbf{S}}_b = \frac{1}{N} \sum_{q=1}^{Q} N_q \, (\boldsymbol{\mu}_q - \boldsymbol{\mu}_0) \, (\boldsymbol{\mu}_q - \boldsymbol{\mu}_0)^T, \tag{13.40}$$

and compute the separability metric $J^{(n)}$ for the n'th iteration as

$$J^{(n)} = \mathrm{trace}(\hat{\mathbf{S}}_b). \tag{13.41}$$

This separability is an alternative to the separability metric in Eq. (13.20).

8. Repeat steps 2 through 7 until the separability metric converges according to

$$\frac{J^{(n)} - J^{(n-1)}}{J^{(n)}} < \tau_s. \tag{13.42}$$

Results of image classification of the Hyperspec VNIR panel image with the ISMC algorithm are shown in Figs. 13.16 and 13.17 for the case of whitened data. The algorithm converged to nine classes, partly because it placed small, anomalous natural background areas into separate classes. In this case, the added process of splitting and merging was able to separate three of the panels into distinct classes. The green panel, which is most spectrally similar to the natural background, was merged with a single dominant vegetation class that encompassed most of the grass and tree spectra. Parts of the background that are separately classified from this dominant class are somewhat spectrally distinct. Therefore, while the results do not necessarily match the truth from the standpoint of known object and land-cover material classes, it does a fairly good job of classifying spectrally similar materials together and keeping small classes distinct from dominant classes.

Figure 13.16 Color-coded class map for Hyperspec VNIR panel image based on the ISMC algorithm with prewhitening.

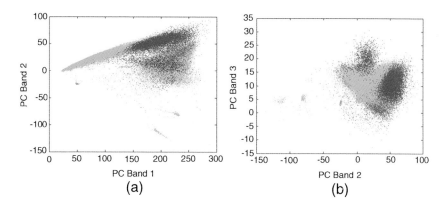

Figure 13.17 Color-coded scatter plot of leading three principal components for Hyperspec VNIR panel image based on the ISMC algorithm with prewhitening: (a) PC bands 1 and 2 and (b) PC bands 2 and 3.

13.3.4 Linear support vector machine

Consider the conceptual classification problem depicted by the 2D scatter plot illustrated in Fig. 13.18, where classes do not exhibit normal behavior. Specifically in this case, they include an asymmetric distribution of outlier spectra that do not fit well, even with a general covariance matrix. Performing classification based simply on the mean vectors results in a decision surface (indicated by the solid line), which misclassifies a number of samples. In this example, there is indeed a linear decision surface

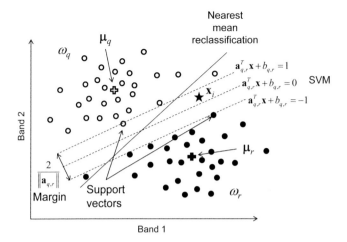

Figure 13.18 Illustration of the linear SVM concept with separable, 2D training data.

that can perfectly separate data into two classes (indicated by the central dotted line), but it differs significantly from that corresponding to the mean classification rule. It is apparent that this optimal decision surface is dictated more by samples on the periphery of the two classes than those exhibiting more mean behavior, which is why the mean classification rule is not effective. In an unsupervised classification setting, there is no additional information to distinguish the classes to which the peripheral samples belong; therefore, it is impossible to determine this optimal decision surface. In a supervised setting, however, training samples can exist that represent not only the mean behavior but also such outliers. By identifying the outliers in each class that are closest to the neighboring classes, decision surfaces can be derived to optimally separate these most-adjacent training samples, a strategy that should outperform the mean classification rule as long as the training samples provide a good representation of the actual imagery. These adjacent samples are called support vectors, and the classification algorithm based on this strategy is called a support vector machine (SVM). This has become a common method for classification of hyperspectral remote sensing data (Melgani and Bruzzone, 2004).

Consider supervised binary classification as classes ω_q and ω_r for which there are M training samples, each designated as $\mathbf{x}_j$. Each training sample is tagged with a class assignment, denoted by $y_j = -1$ for ω_r or $y_j = +1$ for ω_q, where $r = 1$ and $q = 2$. The general linear decision surface is given by

$$d_{q,r}(\mathbf{x}) = \mathbf{a}_{q,r}^T \mathbf{x} + b_{q,r}, \tag{13.43}$$

where $d_{q,r}(\mathbf{x}) < 0$ corresponds to ω_r, and $d_{q,r}(\mathbf{x}) > 0$ corresponds to ω_p. Therefore, an optimum linear decision surface would choose $\mathbf{a}_{q,r}$ and $b_{q,r}$, such that

$$(\mathbf{a}_{q,r}^T \mathbf{x} + b_{q,r}) y_j > 0 \tag{13.44}$$

for all training samples. The SVM approach finds the optimal linear decision surface, or hyperplane, that maximizes the distance between the closest training samples to it (as mentioned, these samples are the support vectors). By defining the support vectors to satisfy the condition

$$\min_j (\mathbf{a}_{q,r}^T \mathbf{x} + b_{q,r}) y_j \geq 1, \tag{13.45}$$

the distance between the support vectors and the hyperplane will equal $1/\|\mathbf{a}_{q,r}\|$. This distance is called the *margin*. The objective of an SVM is to maximize the margin, subject to the support vector condition in Eq. (13.45). This can be expressed as the following convex quadratic programming problem with a linear constraint:

$$\begin{cases} \text{minimize} & \dfrac{1}{2} \|\mathbf{a}_{q,r}^T\|^2 \\ \text{subject to} & (\mathbf{a}_{q,r}^T \mathbf{x} + b_{q,r}) y_j \geq 1 \quad \forall j. \end{cases} \tag{13.46}$$

Using a Lagrangian formulation, this can be rewritten into the dual problem

$$\begin{cases} \text{minimize} & \displaystyle\sum_{j=1}^{M} \alpha_j - \frac{1}{2} \sum_{i=1}^{M} \sum_{j=1}^{M} \alpha_i \alpha_j (\mathbf{x}_i^T \mathbf{x}_j) y_i y_j \\ \text{subject to} & \displaystyle\sum_{j=1}^{M} \alpha_j y_j = 0 \quad \text{and} \quad \alpha_j \geq 0 \quad \forall j, \end{cases} \tag{13.47}$$

where Lagrange multipliers α_i effectively weight each training vector according to its importance in determining the decision surface. Vectors with nonzero weights are support vectors that are assumed to compose training subset S. Quadratic programming methods are used to determine all of the weights in Eq. (13.47) to minimize the objective function subject to the constraints. The optimal decision surface is then defined as

$$d_{q,r}(\mathbf{x}) = \sum_{\mathbf{x}_j \in S} \alpha_j (\mathbf{x}_j^T \mathbf{x}) y_j + b_{q,r}, \tag{13.48}$$

where the bias $b_{q,r}$ is a free parameter optimally selected as part of the quadratic programming.

If the training data are not completely separable with a linear decision surface, as illustrated in Fig. 13.19, then there will be no solution to the constrained minimization problems given in Eq. (13.46) or (13.47). Special consideration must be given for samples that are misclassified. These are called nonmargin support vectors to differentiate them from support vectors that are correctly classified (called margin support vectors). Each nonmargin support vector is associated with a slack variable ξ_j relating to its distance from the decision surface $\xi_j/\|\mathbf{a}_{q,r}\|$. With this modification, the minimization problem becomes

$$
\begin{cases}
\text{minimize} & \dfrac{1}{2}\|\mathbf{a}_{q,r}^T\|^2 + C\displaystyle\sum_{j=1}^{M}\xi_j \\[2mm]
\text{subject to} & y_j(\mathbf{a}_{q,r}^T\mathbf{x} + b_{q,r}) \geq 1 - \xi_j \quad \text{and} \quad \xi_j \geq 0 \quad \forall j.
\end{cases}
\tag{13.49}
$$

Since the SVM is inherently a binary classier, further extension is needed to handle the multiclass problems typical in remote sensing. One multiclass strategy is to determine decision surface $d_{q,r}(\mathbf{x})$ between all pairs of classes, as described earlier, and to maximize a score function that counts the number of favorable and unfavorable votes that a particular test vector $\mathbf{x}$ obtains when classified between class q and its $L-1$ competing

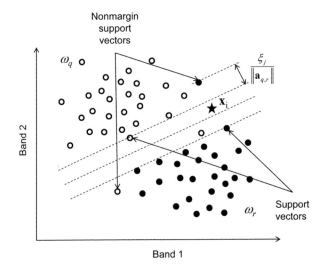

Figure 13.19 Illustration of the linear SVM concept with nonseparable, 2D training data. Margin and nonmargin support vectors are identified by the bold samples.

classes. The score function $S_q(\mathbf{x})$ is given by

$$S_q(\mathbf{x}) = \sum_{\substack{r=1 \\ r \neq q}}^{L} \text{sgn}[d_{q,r}(\mathbf{x})], \tag{13.50}$$

and the one-against-one, winner-takes-all classification rule becomes

$$\mathbf{x}_i \in \omega_q \quad \text{if } S_q(\mathbf{x}_i) < S_r(\mathbf{x}_i) \quad \forall r \neq q. \tag{13.51}$$

There are other multiclass strategies. The one-against-all strategy, which is a variation of the one-against-one strategy, merges all competing classes into one. Other strategies include hierarchical tree-based approaches, which structure the binary classification steps into organized trees for improved efficiency. When using this approach, it is necessary to optimize the tree structure for the classification problem as well as the linear decision surfaces. Further information is given by Melgani and Bruzzone (2004).

13.4 Quadratic Classification Algorithms

As described in Section 13.1, a quadratic classifier segments K-dimensional spectral or feature space into Q regions by a set of quadratic decision surfaces (as depicted in Fig. 13.2) that take into account the differing covariance matrices of the classes. In this case, the general decision surface between classes ω_q and ω_r is given by

$$d_{q,r}(\mathbf{x}) = (\mathbf{x} - \boldsymbol{\mu}_q)^T \boldsymbol{\Sigma}_q^{-1} (\mathbf{x} - \boldsymbol{\mu}_q) - (\mathbf{x} - \boldsymbol{\mu}_r)^T \boldsymbol{\Sigma}_r^{-1} (\mathbf{x} - \boldsymbol{\mu}_r)$$
$$+ \log \frac{|\boldsymbol{\Sigma}_q|}{|\boldsymbol{\Sigma}_r|} - 2 \log \frac{P(q)}{P(r)}, \tag{13.52}$$

where the prior probabilities are included in the equation. While these decision surfaces provide more degrees of freedom to separate classes with differing underlying distributions, quadratic classification represents a more difficult estimation problem because the covariance matrices of the classes need to be confidently estimated. To do so from class sample statistics (the normal approach taken) requires sufficiently large class populations and represents a greater computational challenge, making quadratic classification more difficult to perform in real-time processing settings.

An optimum quadratic classifier has a similar form to that of an optimum linear classifier but includes covariance matrix $\boldsymbol{\Sigma}_q$ for each class.

The discriminant function for each class is modified to

$$d_q(\mathbf{x}) = -\frac{1}{2}(\mathbf{x} - \boldsymbol{\mu}_q)^T \boldsymbol{\Sigma}_q^{-1}(\mathbf{x} - \boldsymbol{\mu}_q) - \frac{1}{2}\log|\boldsymbol{\Sigma}_q| + \log P(q), \quad (13.53)$$

and the classification rule is the same:

$$\mathbf{x}_i \in \omega_q \quad \text{if } d_q(\mathbf{x}_i) > d_r(\mathbf{x}_i) \quad \forall\, r \neq q. \tag{13.54}$$

The test-on-train results for the Hyperspec panel data using an optimum quadratic classifier are essentially indistinguishable from the optimum linear classifier results given in Figs. 13.9 and 13.10 because the optimum linear decision surfaces are already effective at properly classifying the data relative to the sample statistics.

13.4.1 Simple quadratic clustering

The most straightforward implementation of unsupervised quadratic classification is to iteratively apply the quadratic classification rule based on Eq. (13.49), along with parameter estimation, to compute mixture model parameters for each class based on sample statistics. This methodology, often denoted as quadratic clustering, proceeds as follows:

1. Define the number of classes Q to represent the data.
2. Randomly assign all spectra to the Q classes or start with some other initial classification of the data.
3. Estimate mean vector $\boldsymbol{\mu}_q$ and covariance matrix $\boldsymbol{\Sigma}_q$ of each class ω_q based on the sample statistics of the spectra assigned to it. Estimate the class prior probability as

$$\hat{P}(q) = \frac{N_q}{N}, \tag{13.55}$$

   where N_q is the number of spectra assigned to the q'th class, and N is the total number of spectra.
4. Assign each image spectrum $\mathbf{x}_i$ based on the quadratic classification rule

$$\mathbf{x}_i \in \omega_q \quad \text{if } d_q(\mathbf{x}_i) > d_r(\mathbf{x}_i) \quad \forall\, r \neq q, \tag{13.56}$$

   where

$$d_q(\mathbf{x}) = (\mathbf{x} - \hat{\boldsymbol{\mu}}_q)^T \hat{\boldsymbol{\Sigma}}_q^{-1}(\mathbf{x} - \hat{\boldsymbol{\mu}}_q) + \log|\hat{\boldsymbol{\Sigma}}_q| - 2\log\hat{P}(q). \tag{13.57}$$

5. Repeat steps 3 and 4 until no spectra change from one class to another.

Figures 13.20 and 13.21 present the results of applying the quadratic algorithm to the leading ten-dimensional principal-component subspace of the Hyperspec VNIR panel data based on six classes in terms of color-coded class maps, along with corresponding color-coded scatter plots of the first three principal components. These results can be directly compared to the previously illustrated k-means and ISMC algorithm results. In this case, all four panels are separated into a single class, and the natural background classes are separated into subclasses due to their non-normal distributions. This is arguably an improvement over the k-means approach for these data, but not as effective as ISMC in separating the panels into distinct classes.

13.4.2 Maximum-likelihood clustering

A quadratic clustering algorithm essentially classifies spectra by assigning them to the class that minimizes their class-conditional Mahalanobis distance. Normally distributed classes correspond to a Bayesian classifier, where the classification boundary is at a point where conditional probability density functions are equal, under the assumption that the prior probabilities and ratio of covariance matrix determinants are equal. As opposed to making this decision based on the Mahalanobis distance, one could equivalently make this decision based on posterior probability, as outlined in Eqs. (13.2) and (13.3).

Figure 13.20 Color-coded class map for Hyperspec VNIR panel image based on the simple quadratic clustering algorithm with six classes.

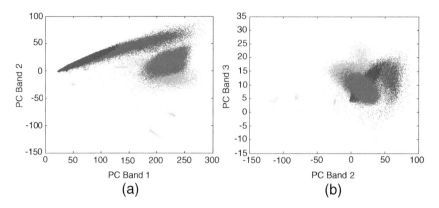

Figure 13.21 Color-coded scatter plot of leading three principal components for Hyperspec VNIR panel image based on the simple quadratic clustering algorithm with six classes: (a) PC bands 1 and 2 and (b) PC bands 2 and 3.

One problem of the quadratic clustering approach, as depicted in Fig. 13.22, is that it makes a hard classification between classes and then uses sample statistics based on these class assignments to estimate class statistics. If the classes exhibit some overlap, then it is apparent from the figure that true class statistics will not be properly estimated from the assigned samples due to the truncation of distributions that occurs. That is, removal of the tails of the distributions will alter both mean vectors and covariance matrices in the general K-dimensional case. Therefore, while this truncation might seem a reasonable step to take with regard to final class assignments, it does not seem appropriate with regard to the parameter estimation that is iteratively performed.

An alternate quadratic classification strategy that avoids this truncation problem is the maximum likelihood (ML) or expectation maximization (EM) algorithm, which uses posterior probability as a basis for parameter estimation instead of class sample statistics and does not perform the spectral classification step until the end. Given estimates of prior probability, mean vector, and covariance matrices for the underlying normal mixture model, the posterior probability of spectrum $\mathbf{x}_i$ belonging to class q can be computed for an iteration n using the formula

$$\hat{P}^{(n)}(q \mid \mathbf{x}_i) = \frac{\hat{P}^{(n)}(q)\hat{p}^{(n)}(\mathbf{x}_i \mid q)}{\sum\limits_{q=1}^{Q} \hat{P}^{(n)}(q)\hat{p}^{(n)}(\mathbf{x}_i \mid q)}, \tag{13.58}$$

where $p(\mathbf{x}_i|q) \sim N(\boldsymbol{\mu}_q, \boldsymbol{\Sigma}_q)$. Prior probability, mean vector, and covariance matrix for each class are updated for the next iteration based on posterior

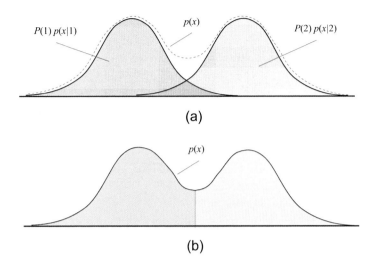

Figure 13.22 Truncation of overlapping, normally distributed classes performed by quadratic decision surface: (a) true overlapping class-conditional distributions and (b) estimated class-conditional distributions.

probability values for each spectrum according to

$$\hat{P}^{(n+1)}(q) = \frac{1}{N} \sum_{i=1}^{N} \hat{P}^{(n)}(q \mid \mathbf{x}_i), \tag{13.59}$$

$$\hat{\boldsymbol{\mu}}_q^{(n+1)} = \frac{1}{N} \sum_{i=1}^{N} \hat{P}^{(n)}(q \mid \mathbf{x}_i) \, \mathbf{x}_i, \tag{13.60}$$

and

$$\hat{\boldsymbol{\Sigma}}_q^{(n+1)} = \frac{1}{N} \sum_{i=1}^{N} \hat{P}^{(n)}(q \mid \mathbf{x}_i) \, (\mathbf{x}_i - \hat{\boldsymbol{\mu}}_q^{(n+1)}) \, (\mathbf{x}_i - \hat{\boldsymbol{\mu}}_q^{(n+1)})^T. \tag{13.61}$$

This ML clustering algorithm proceeds as follows:

1. Define the number of classes Q to represent the data.
2. Randomly assign all spectra to the Q classes or start with some other initial classification of the data.
3. Estimate mean vector $\boldsymbol{\mu}_q$ and covariance matrix $\boldsymbol{\Sigma}_q$ of each class ω_q, based on the sample statistics of spectra assigned to it. Estimate class prior probability for each class using Eq. (13.55).
4. Estimate posterior probability for each image spectrum based on current prior probability, mean vector, and covariance matrix estimates for all classes using Eq. (13.58).

5. Update prior probability, mean vector, and covariance matrix estimates for all classes using Eqs. (13.59)–(13.61).
6. Repeat steps 4 and 5 until convergence occurs in terms of minimal change in either a separability metric or posterior probability values.

After convergence, final class assignments can be made in two ways. First, the Bayesian classification rule can be employed, where each spectrum is assigned to the class that maximizes posterior probability:

$$\mathbf{x}_i \in \omega_q \quad \text{if } P(q \mid \mathbf{x}_i) > P(r \mid \mathbf{x}_i) \quad \forall\, r \neq q. \tag{13.62}$$

This enforces a hard classification that truncates the resulting class distributions. However, by leaving this hard classification until the final step and not allowing it to influence the iteratively estimated class statistics, the effect should not be as severe as that which occurs with the simple quadratic clustering approach. Another approach is to randomly assign spectra to classes based on their posterior probability to allow the resulting class distributions to overlap. In this case, a spectrum with nonzero posterior probability for multiple classes could end up being assigned to any of the classes. This can be implemented by drawing a uniformly distributed random number R between zero and one for each spectrum $\mathbf{x}_i$, and assigning the spectrum based on the random draw according to

$$\mathbf{x}_i \in \omega_q \quad \text{if } \sum_{r=1}^{q-1} P(r \mid \mathbf{x}_i) < R \leq P(q \mid \mathbf{x}_i) + \sum_{r=1}^{q-1} P(r \mid \mathbf{x}_i). \tag{13.63}$$

Results from the application of the ML algorithm to the leading ten-dimensional principal-component subspace of the Hyperspec VNIR panel data based on six classes are provided in terms of color-coded class maps, along with corresponding color-coded scatter plots of the leading three principal components, in Figs. 13.23 and 13.24. Results are similar to those of the quadratic clustering method, although the green panel is now mixed into a tree class. Again, background classes are separated into overlapping subclasses to represent their non-normal behavior.

13.4.3 Stochastic expectation maximization

The stochastic expectation maximization (SEM) algorithm (Masson and Pieczynski, 1993) is an extension of the ML clustering algorithm because it uses posterior probabilities to drive class assignments; however, class statistics are updated based on class sample statistics instead of posterior probability values. This is implemented by performing random assignments based on posterior probability [as designated by Eq. (13.63)]

Figure 13.23 Color-coded class map for Hyperspec VNIR panel image based on the ML clustering algorithm with six classes.

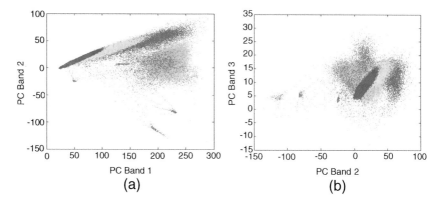

Figure 13.24 Color-coded scatter plot of leading three principal components for Hyperspec VNIR panel image based on the ML clustering algorithm with six classes: (a) PC bands 1 and 2 and (b) PC bands 2 and 3.

as part of each iteration. The idea is that the randomness introduced during parameter estimation processes should allow the algorithm to avoid being trapped in solutions too dependent on algorithm initialization. This algorithm has been widely implemented for hyperspectral data processing, particularly as part of target detection algorithms based on the normal mixture model for describing background clutter statistics. These target detection algorithms are described further in Chapter 14.

The SEM algorithm consists of the following steps:

1. Define the number of classes Q to represent the data.
2. Randomly assign all spectra to the Q classes or start with some other initial classification of the data.
3. Estimate mean vector $\boldsymbol{\mu}_q$ and covariance matrix $\boldsymbol{\Sigma}_q$ of each class ω_q based on the sample statistics of spectra assigned to it. Estimate class prior probability using Eq. (13.55).
4. Using Eq. (13.58), estimate posterior probability for each image spectrum based on current prior probability, mean vector, and covariance matrix estimates for all classes.
5. Randomly assign all spectra among the classes using the Monte Carlo method described in Eq. (13.63).
6. Update prior probability, mean vector, and covariance matrix estimates for all classes for the next iteration using

$$\hat{P}^{(n+1)}(q) = \frac{N_q^{(n)}}{N}, \tag{13.64}$$

$$\hat{\boldsymbol{\mu}}_q^{(n+1)} = \frac{1}{N_q^{(n)}} \sum_{\mathbf{x}_i \in \omega_q^{(n)}} \mathbf{x}_i, \tag{13.65}$$

and

$$\hat{\boldsymbol{\Sigma}}_q^{(n+1)} = \frac{1}{N_q^{(n)} - 1} \sum_{\mathbf{x}_i \in \omega_q^{(n)}} [\mathbf{x}_i - \hat{\boldsymbol{\mu}}_q^{(n+1)}][\mathbf{x}_i - \hat{\boldsymbol{\mu}}_q^{(n+1)}]^T. \tag{13.66}$$

7. Repeat steps 4, 5, and 6 until convergence occurs in terms of minimal change in either a separability metric or posterior probability values.
8. Perform a final class assignment by repeating steps 4 and 5.

Results from the application of the SEM algorithm to the leading ten-dimensional principal-component subspace of the Hyperspec VNIR panel data based on six classes are provided in terms of color-coded class maps, along with corresponding color-coded scatter plots of the leading three principal components, in Figs. 13.25 and 13.26. While the resulting classes are different, the results exhibit the same overall features as those of the other two quadratic methods, with degrees of freedom of the multiple classes applied more toward modeling subtle modes of the dominant natural background classes, while also in this case mixing rarer panel classes together with some of the more anomalous background spectra. Overall, this classification method appears to be better at modeling non-normal backgrounds than separating spectrally distinct material classes. Convergence characteristics in terms of separability metric J and changes in class assignments are shown in Fig. 13.27. Convergence of quadratic methods is typically slower than that of mean-based methods, and the

Figure 13.25 Color-coded class map for Hyperspec VNIR panel image based on the SEM clustering algorithm with six classes.

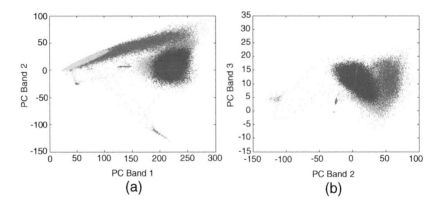

Figure 13.26 Color-coded scatter plot of leading three principal components for Hyperspec VNIR panel image based on the SEM clustering algorithm with six classes: (a) PC bands 1 and 2 and (b) PC bands 2 and 3.

required estimation and inversion of class covariance matrices makes them much more computationally intensive.

Although the split-and-merge concepts employed in the ISODATA and ISMC variants of the linear classification algorithms could theoretically be applied in the quadratic clustering case, such variations are not widely reported in the literature. Instead, SEM is generally implemented with a user-defined number of classes, which is limiting if the number of classes is incorrectly specified. As an example, Figs. 13.28 and 13.29 illustrate

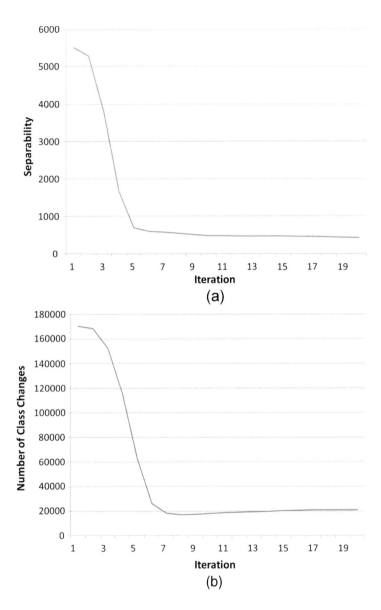

Figure 13.27 SEM algorithm convergence for six classes based on (a) the separability metric and (b) the number of class changes.

Figure 13.28 Color-coded class map for Hyperspec VNIR panel image based on the SEM clustering algorithm with four classes.

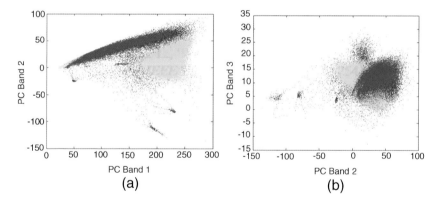

Figure 13.29 Color-coded scatter plot of leading three principal components for Hyperspec VNIR panel image based on the SEM clustering algorithm with four classes: (a) PC bands 1 and 2 and (b) PC bands 2 and 3.

SEM classification results for the Hyperspec VNIR panel data with only four classes (too few), and Figs. 13.30 and 13.31 illustrate what occurs with eight classes (too many). The four-class results nicely separate the grass and tree classes, but the panels are again combined with outliers in the vegetation classes. The eight-class results continue the trend of the six-class results in terms of using additional degrees of freedom to better model non-normal behavior of dominant natural background classes than to separate out rarer manmade material classes.

Figure 13.30 Color-coded class map for Hyperspec VNIR panel image based on the SEM clustering algorithm with eight classes.

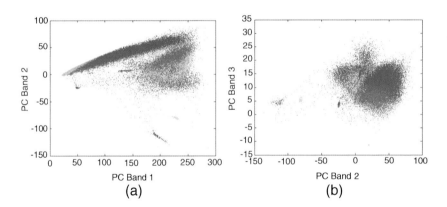

Figure 13.31 Color-coded scatter plot of leading three principal components for Hyperspec VNIR panel image based on the SEM clustering algorithm with eight classes: (a) PC bands 1 and 2 and (b) PC bands 2 and 3.

The SEM results described in this section illustrate the algorithm's ability to model statistical behavior of high-occurrence spectra, but also its tendency to incorporate low-occurrence spectra into more dominant spectral classes. This latter tendency can be disadvantageous for an image classifier, but advantageous for use in modeling background clutter statistics in a target detection algorithm. In such a case, it is desired to accurately model the probability density function of more-dominant clutter spectra, and less accurately model rare target spectra, so that the target spectra can be easily separated from the clutter. The incorporation of SEM in hyperspectral target detection is further discussed in Chapter 14.

13.5 Nonlinear Classification Algorithms

In some situations, the in-class variability of hyperspectral data is such that classes cannot be separated by linear or even quadratic decision surfaces. A simple illustration of such a case is given in Fig. 13.32, where a nonlinear boundary separates two data clusters. Optimization in this case is not simply that of estimating a small number of parameters defining the decision surface, but involves defining the optimal functional form of the decision surface itself. Since this can take on an endless variety of forms, the strategy for approaching this type of optimization is not clear. One potential approach is to employ Bayesian hypothesis testing with more-complex class-conditional probability density functions, which incorporate higher-order class statistics. Again there is some uncertainty regarding which form of distribution function to assume. Additionally, there is the problem of confidently estimating higher-order statistics from scene data, or in a supervised situation, training data that are often limited in extent.

The two nonlinear classification methods described in this section take different approaches to the problem. The first approach represents class-conditional distributions by nonparametric density functions that conform to data scatter. The classification rules correlate to determining minima in the density functions, possibly resulting in arbitrary decision surfaces. Such an approach can be implemented in either a supervised

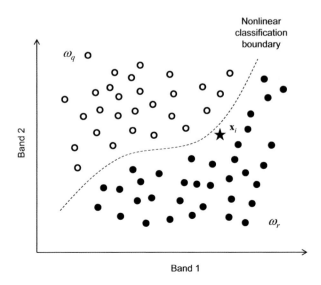

Figure 13.32   Illustration of a nonlinear classification problem.

or unsupervised setting. The second approach is to transform data into a higher-dimensional space and perform linear classification in this transformed space. Linear decision surfaces in higher-dimensional space map to nonlinear surfaces in lower-dimensional space. The particular method described in this section is based on a linear SVM with a kernel transform to the higher-dimensional space. As such, it relies on training data and can only operate in a supervised mode.

13.5.1 Nonparametric classification

The nonparametric classification approach is based on the concept that an underlying density function for multidimensional data can be estimated locally by data samples without needing to resort to some parametric form. That is, regions in multidimensional space where there are many data samples imply a high density, while sparse regions correspond to low density. Two basic ideas for transforming a discrete set of data samples into a continuous probability density function estimate are illustrated in one dimension in Fig. 13.33. The first concept is to convolve discrete samples with some kernel function $\kappa(\mathbf{x})$, whose size is optimally selected based on sparsity of the data. The density estimate is relatively smooth in regions of moderate and high sample density and falls to minima or even zero in regions of low density. This is known as the Parzen density estimate. The second idea is to identify the nearest neighbors from a Euclidean distance perspective from each data sample, and then determine the size of local region $v(\mathbf{x})$ from the sample to its k'th nearest neighbor. The density at the sample location is inversely proportional to the local region size. This estimation concept is known as the k nearest neighbor (kNN) density estimate. Classifiers based on both density estimation methods are outlined next. While each classification rule can be derived from nonparametric density estimates, the classification can be implemented without explicitly solving for multidimensional, class-conditional probability density functions, which would otherwise be computationally prohibitive.

First consider the Parzen estimate, expressed mathematically as

$$\hat{p}(\mathbf{x}) = \frac{1}{N} \sum_{i=1}^{N} \kappa(\mathbf{x} - \mathbf{x}_i), \qquad (13.67)$$

where kernel function $\kappa(\mathbf{x})$ is normalized, such that

$$\int \kappa(\mathbf{x}) \, d\mathbf{x} = 1. \qquad (13.68)$$

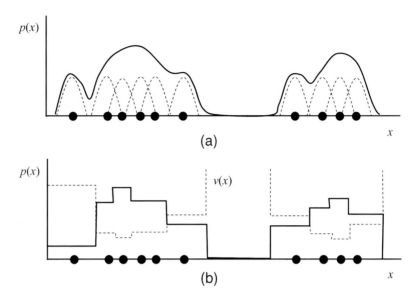

Figure 13.33 Depiction of nonparametric density estimation in one dimension: (a) Parzen density estimate and (b) kNN density estimate.

The shape and extent of the kernel function can be arbitrarily chosen; however, it is necessary to match them to the sparsity of data to obtain an estimate that is relatively smooth in regions of high and moderate density but approaches zero in regions of sparsity. A useful kernel function from this perspective is

$$\kappa(\mathbf{x}) = \frac{m\,\Gamma\left(\frac{K}{2}\right)\Gamma^{K/2}\left(\frac{K+2}{m}\right)}{(K\pi)^{K/2}\Gamma^{K/2+1}\left(\frac{K}{2m}\right)}\frac{1}{r^K|\Sigma|^{1/2}}\exp\left[\frac{\Gamma\left(\frac{K+2}{2m}\right)}{K\,\Gamma\left(\frac{K}{2m}\right)}\mathbf{x}^T(r^2\Sigma)^{-1}\mathbf{x}\right]^m,$$

(13.69)

where m is a shape parameter, r is a size parameter, $\Gamma(n)$ is the gamma function, K is data dimensionality, and Σ is the covariance matrix of the data. The shape parameter m determines the rate of fall off of the kernel function. Specifically, a Gaussian kernel function corresponds to $m = 1$, while the kernel function approaches a uniform hyper-elliptical kernel as m becomes large. This is illustrated in one dimension in Fig. 13.34. While r controls the size of the kernel, it is shaped in the multidimensional case by covariance matrix Σ. That is, it has larger extent in dimensions of higher variance.

To apply the kernel function in Eq. (13.69) for an arbitrary shape parameter m, it is first necessary to optimally select size parameter r for the data. This can be done by minimizing the MSE between the density function estimate and an assumed form for the probability density

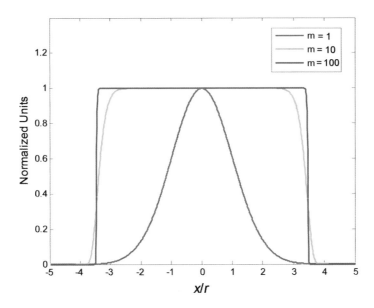

Figure 13.34 Normalized 1D Parzen kernel function for various shape parameters.

function, such as a normal density. While it is understood that the actual data do not conform to this parametric form (if they did conform, this would obviate the need for nonparametric estimation in the first place), this at least provides some basis for selecting r to appropriately conform to data scatter. It is shown by Fukunaga (1990) that the optimum size parameter based on a uniform kernel and normal density function is

$$r = \left[\frac{2^{K+2} \Gamma \left(\frac{K+2}{2} \right)}{(K+2)^{K/2+1}} \right]^{\frac{1}{K+4}} N^{-\frac{1}{K+4}}. \tag{13.70}$$

Under a normal assumption, the optimum size parameter depends only on data dimensionality K and number N of data samples.

A nonparametric classifier can be derived based on the Bayesian classification rule in Eq. (13.3). When the Parzen density estimate with a uniform kernel function (large m) is employed, it can be shown that

$$\hat{P}(q)\hat{p}(\mathbf{x}_i \mid q) = \frac{k(\mathbf{x}_i \mid q)}{N v}, \tag{13.71}$$

where $k(\mathbf{x}_i \mid q)$ is the number of samples within class ω_q in a fixed local volume v around spectrum $\mathbf{x}_i$ with a radius r, the optimum size parameter.

In this case, the Bayesian classification rule corresponds to

$$\mathbf{x}_i \in \omega_q \quad \text{if } k(\mathbf{x}_i \mid q) > k(\mathbf{x}_i \mid r) \quad \forall\, r \neq q. \tag{13.72}$$

To employ a volume with the same extent in all dimensions, it is necessary to whiten the data before applying the classification rule.

In a supervised classifier, Eq. (13.72) can be implemented by determining the number of training samples within the local volume, assuming that there are enough training samples to support this approach. In an unsupervised setting, an iterative valley-seeking algorithm is used, where class assignments are iteratively adjusted until decision surfaces fall toward the minima of the Parzen density estimates. The procedure is as follows:

1. Whiten the data based on the sample covariance matrix.
2. Define the number of classes Q.
3. Determine optimum local volume size r for the number of data samples and data dimensionality, according to Eq. (13.70).
4. Randomly assign all spectra to the Q classes or start with some other initial classification of the data.
5. For each spectrum $\mathbf{x}_i$ and class ω_q, count the number of samples $k(\mathbf{x}_i \mid q)$ with a Euclidean distance from $\mathbf{x}_i$ less than r.
6. Assign each spectrum $\mathbf{x}_i$ to the class ω_q for which $k(\mathbf{x}_i \mid q)$ is maximum.
7. Repeat steps 5 and 6 until there are no further changes in class assignments.

Results from applying the Parzen nonparametric classification algorithm to the leading 3D principal-component subspace of the Hyperspec VNIR panel data based on six classes are provided in terms of color-coded class maps, along with corresponding color-coded scatter plots of the leading three principal components, in Figs. 13.35 and 13.36. The use of a lower-dimensional subspace for this example is due to the high computational complexity of the algorithm. The results are very interesting because three primary classes arise: one containing three of the panels, one containing the grass foreground with small patches of trees, and the third containing most of the trees. The other three classes become trapped on outlier spectra, and the green panel is assigned to the grass class.

Now consider the kNN density estimation approach, for which the density estimate can be expressed mathematically as

$$\hat{p}(\mathbf{x}) = \frac{k-1}{N\,v(\mathbf{x})}, \tag{13.73}$$

Figure 13.35 Color-coded class map for Hyperspec VNIR panel image based on the Parzen nonparametric classification algorithm with six classes.

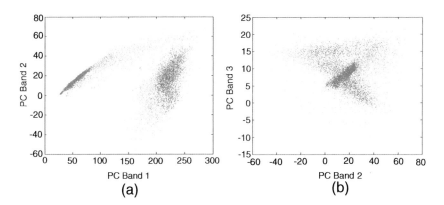

Figure 13.36 Color-coded scatter plot of leading three principal components for Hyperspec VNIR panel image based on the Parzen nonparametric classification algorithm with six classes: (a) PC bands 1 and 2 and (b) PC bands 2 and 3.

where $v(\mathbf{x})$ is the local region volume required to capture the k'th nearest neighbor for a data sample at $\mathbf{x}$. As in the Parzen case, the local region must be shaped consistently with the data scatter. If the data are whitened, then the region can simply be spherical, for which the volume simplifies to

$$v(\mathbf{x}) = c \, \| \mathbf{x} - \mathbf{x}_{kNN}(\mathbf{x}) \|^K, \qquad (13.74)$$

where c is a constant and $\mathbf{x}_{kNN}(\mathbf{x})$ is the k'th nearest-neighbor data sample. The parameter k should be optimally chosen; this choice can be made in a manner analogous to the Parzen case by minimizing the MSE between the kNN density estimate and an assumed parametric form of the probability density function. Based on an assumed normal distribution, the optimum k is shown by Fukunaga (1990) to be

$$ k = \left[\frac{(K+2)^2(K-2)^{K/2+2}}{\Gamma^{4/K}\left(\frac{K+2}{2}\right) K^{K/2+1}(K^2 - 6K + 16)} \right]^{\frac{K}{K+4}} N^{\frac{4}{K+4}}. \qquad (13.75) $$

The Bayesian classification rule can be implemented based on the kNN density estimate, recognizing that

$$ \hat{P}(q)\hat{p}(\mathbf{x}_i \mid q) = \frac{N_q}{N}\frac{k-1}{N_q\,\hat{v}(\mathbf{x}_i)} = \frac{k-1}{N\,c\,\|\mathbf{x}_i - \mathbf{x}_{kNN}(\mathbf{x}_i \mid r)\|^K}, \qquad (13.76) $$

where $\mathbf{x}_{kNN}(\mathbf{x}_i \mid q)$ is the k'th nearest neighbor in class ω_q. This means that the Bayesian classification rule simply corresponds to minimization of the distance to the k'th nearest neighbor in each class:

$$ \mathbf{x}_i \in \omega_q \quad \text{if } \|\mathbf{x}_i - \mathbf{x}_{kNN}(\mathbf{x}_i \mid q)\| < \|\mathbf{x}_i - \mathbf{x}_{kNN}(\mathbf{x}_i \mid r)\| \quad \forall\, r \neq q. $$
$$ (13.77) $$

In a supervised setting, nearest neighbors may be selected from within the training set. For unsupervised classification, the following iterative valley-seeking algorithm can be employed:

1. Whiten the data based on the sample covariance matrix.
2. Define the number of classes Q.
3. Determine the optimum k for the number of data samples and data dimensionality according to Eq. (13.75).
4. Randomly assign all spectra to the Q classes or start with some other initial classification of the data.
5. For each spectrum $\mathbf{x}_i$ and class ω_q, determine the Euclidean distance from $\mathbf{x}_i$ to the k'th nearest neighbor in the class.
6. Assign each spectrum $\mathbf{x}_i$ to the class ω_q for which the kNN distance is minimized.
7. Repeat steps 5 and 6 until there are no further changes in class assignments.

Results from applying the kNN nonparametric classification algorithm to the leading 3D principal-component subspace of the Hyperspec VNIR panel data based on six classes are provided in terms of color-coded class maps, along with corresponding color-coded scatter plots of the leading three principal components, in Figs. 13.37 and 13.38. Again, a lower dimensionality was used for this example due to the computational complexity of the algorithm. Outcomes in this case do not appear to provide reasonable results. The likely reason for this is that the algorithm is quite sensitive to the initial classification. In this case, a random initial classification is used, and the result appears to maintain much of this randomness due to local clustering that can occur by its nonparametric nature. From that standpoint, the method might be more beneficial as a postprocessing stage for a parametric classifier; this would allow general perturbations from linear or quadratic classifications, while maintaining some semblance of parametric segmentation of the multidimensional data space.

13.5.2 Kernel support vector machine

An alternative to defining nonlinear decision surfaces in K-dimensional data space using nonparametric methods is to transform data to a feature space with higher dimensionality, so that classes can be separated by linear decision surfaces. At first glance, this simply replaces the problem of defining arbitrary nonlinear decision surfaces with the problem of defining a nonlinear mapping to the feature space. However, kernel-based methods (Schölkopf and Smola, 2002) address this issue by using a so-called kernel trick that allows a linear algorithm to be represented in terms of a kernel

Figure 13.37 Color-coded class map for Hyperspec VNIR panel image based on the kNN nonparametric classification algorithm with six classes.

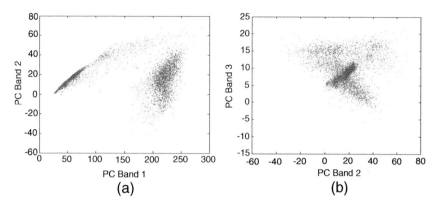

Figure 13.38 Color-coded scatter plot of leading three principal components for Hyperspec VNIR panel image based on the kNN nonparametric classification algorithm with six classes: (a) PC bands 1 and 2 and (b) PC bands 2 and 3.

function without any knowledge of actual nonlinear mapping. Consider a nonlinear mapping $\Phi(\mathbf{x})$ from a K-dimensional spectral space to a higher-dimensionality feature space. The basic idea of kernel-based methods is to define mapping functions so that the dot product between two nonlinearly mapped spectra $\mathbf{x}_i$ and $\mathbf{x}_j$ can be expressed by a kernel function $K(\mathbf{x}_i, \mathbf{x}_j)$; that is,

$$\Phi(\mathbf{x}_i) \cdot \Phi(\mathbf{x}_j) = K(\mathbf{x}_i, \mathbf{x}_j) \quad \forall i, j. \tag{13.78}$$

This is known as Mercer's condition; a number of kernel functions exhibit this property, including polynomial functions of any order,

$$K(\mathbf{x}_i, \mathbf{x}_j) = [\,(\mathbf{x}_i \cdot \mathbf{x}_j) + c\,]^d, \tag{13.79}$$

and the Gaussian radial basis function,

$$K(\mathbf{x}_i, \mathbf{x}_j) = e^{-\|\mathbf{x}_i - \mathbf{x}_j\|^2 / \sigma^2}, \tag{13.80}$$

where c, d, and σ are constants.

A key attribute of kernel tricks is the ability to express the linear classification algorithm as a function of dot products. In doing so, application in nonlinear transform space can be expressed by replacing dot products with the kernel function due to Mercer's condition, avoiding the need to explicitly define or compute nonlinear mappings. Since the Lagrangian form of the SVM in Eq. (13.44) is based on minimizing a function of dot products, it can be readily expressed in kernel form

(Melgani and Bruzzone, 2004) as

$$
\begin{cases}
\text{minimize} \quad \sum_{j=1}^{M} \alpha_j - \frac{1}{2} \sum_{i=1}^{M} \sum_{j=1}^{M} \alpha_i \alpha_j y_i y_j K(\mathbf{x}_i, \mathbf{x}_j) \\
\text{subject to} \quad \sum_{j=1}^{M} \alpha_j y_j = 0 \quad \text{and} \quad 0 \leq \alpha_j \leq c \quad \forall j.
\end{cases}
\tag{13.81}
$$

Kernel SVM implementation involves optimally selecting the kernel parameters (c, d, and σ) to either maximize the margin or minimize some measure of error (Vapnik, 1998). Hyperspectral image classification results that use kernel SVM, as well as other kernel-based classification methods, are provided by Camps-Valls and Bruzzone (2005).

13.6 Summary

Standard mathematical methods of hypothesis testing and statistical pattern recognition can be applied to hyperspectral image data to classify imagery into spectrally similar material classes. When *a priori* knowledge of spectral classes exists in the form of class statistics or training data, classification accuracy can be quite good if the material classes are indeed distinct. The example used in this chapter indicates excellent performance using even a simple multiclass linear classifier, and performance can be enhanced using linear and nonlinear SVMs given representative training data. In an unsupervised setting, however, the prognosis for accurately identifying and separating distinct material classes is not as good, primarily due to the inherent ambiguity that exists in the way in which non-normal data distribution can be statistically represented. While the approaches described can be effective in separating dominant background classes, they all appear to struggle in distinctly identifying rarer material classes, especially when they are not very well separated from more dominant classes. In general, manual supervision and *a priori* training data are necessary to support this level of classifier fidelity.

While the examples of results in this chapter provide some insights into the strengths, limitations, and performance of the image classification methods described, one should exercise caution when generalizing the basic observations, as classifier performance can be very dependent on characteristics of the data. When applied to real hyperspectral remote sensing problems, the current trend in the literature is to use supervised linear or nonlinear SVM methods, coupled with spatial or other contextual information to enhance classifier performance. Unsupervised methods are more readily applicable to modeling the non-normal statistical behavior of natural or even manmade background data as a component of target detection algorithms used to find small targets in hyperspectral imagery, which is the subject of the next (and final) chapter.

13.7 Further Reading

Most of the algorithms described in the chapter are general statistical pattern recognition methods tailored to hyperspectral data processing applications. A more detailed treatment is provided by Fukunaga (1990), including theoretical performance aspects not presented here. Tso and Mather (2009) expands on these and other methods, with specific applications for remote sensing.

References

Camps-Valls, G. and Bruzzone, L., "Kernel-based methods for hyperspectral image application," *IEEE T. Geosci. Remote Sens.* **43**, 1351–1362 (2005).

Clark, R. N., Swayze, G. A., Livo, K. E., Kokaly, R. F., Sutley, S. J., Dalton, J. B., McDougal, R. R., and Gent, C. A., "Imaging spectroscopy: Earth and planetary remote sensing with the USGS Tetracorder and expert systems," *J. Geophys. Res.* **108**, 5131 (2003).

Dennison, P. E. and Roberts, D. A., "The effects of vegetation phenology on endmember selection and species mapping in southern California chaparral," *Remote Sens. Environ.* **87**, 295–309 (2003).

Fukunaga, K., *Introduction to Statistical Pattern Recognition*, Academic Press, San Diego (1990).

Linde, Y., Buzo, A., and Gray, R. M., "An algorithm for vector quantization," *IEEE T. Commun. Theory* **28**(1), 84–95 (1980).

Jensen, J. R., *Introductory Image Processing: A Remote Sensing Perspective*, Prentice Hall, Upper Saddle River, NJ (1996).

Masson P. and Pieczynski, W., "SEM algorithm and unsupervised statistical segmentation of satellite images," *IEEE T. Geosci. Remote Sens.* **31**, 618–633 (1993).

Melgani, F. and Bruzzone, L., "Classification of hyperspectral remote sensing images with support vector machines," *IEEE T. Geosci. Remote Sens.* **42**, 1778–1790 (2004).

Roberts, D. A., Dennison, P. E., Gardner, M., Hetzel, Y., Ustin, S. L., and Lee, C. T., "Evaluation of the potential of Hyperion for fire danger assessment by comparison to the Airborne Visible/Infrared Imaging Spectrometer," *IEEE T. Geosci. Remote Sens.* **41**, 1297–1310 (2003).

Schölkopf, B. and Smola, A. J., *Learning with Kernels*, MIT Press, Cambridge, MA (2002).

Simpson, J. J., McIntire, T. J., and Sienko, M., "An improved hybrid clustering algorithm for natural scenes," *IEEE T. Geosci. Remote Sens.* **38**, 1016–1032 (2000).

Tso, B. and Mather, P. M., *Classification Methods for Remotely Sensed Data*, Second Edition, CRC Press, Boca Raton, FL (2009).

Ustin, S. L., Roberts, D. A., Gamon, J. A., Asner, G. P., and Green, R. O., "Using imaging spectroscopy to study ecosystem processes and properties," *Bioscience* **54**, 523–534 (2004).

Vapnik, V. N., *Statistical Learning Theory*, Wiley, New York (1998).

Chapter 14
Hyperspectral Target Detection

The objective of hyperspectral target detection is to find objects of interest within a hyperspectral image by using the particular spectral features associated with an object's surface-material content. In some cases, the objects of interest make up a significant enough portion of the image to be treated as a material class, and the classification methods described in the previous chapter can be well suited for identifying them. In other cases, the objects of interest are more scarcely populated in the scene, and the problem of detection requires a different method of treatment. Typically, this involves hypothesis testing, where one of two decisions is made for each image spectrum: (1) the spectrum corresponds to the object of interest, called a target, or (2) it corresponds to something other than the target, referred to as background clutter.

This chapter describes many of the basic methods applied to the problems associated with hyperspectral target detection. As this problem has very strong parallels in fields such as communications, radar signal processing, and speech recognition, methods are generally drawn from the broader field of statistical signal processing and then adapted to the nature of hyperspectral imagery. The chapter begins with a brief review of target detection theory and then introduces various methods within a taxonomy centered on the use of a generalized likelihood ratio test for target detection.

14.1 Target Detection Theory

The target detection problem is known in statistical signal processing references as a hypothesis test between a *null hypothesis* H_0 that asserts that the spectrum under test is associated with background clutter, and an *alternative hypothesis* H_1 that asserts that the spectrum under test is a target (Scharf, 1991). Under the null hypothesis, spectrum $\mathbf{x}$ is an element of acceptance region A of a multidimensional data space, while under the alternative hypothesis, it is an element of rejection region R, meaning that the null hypothesis is rejected. One can identify an *indicator function* $\phi(\mathbf{x})$ that defines the detector according to

$$\phi(\mathbf{x}) = \begin{cases} 1 & \mathbf{x} \in R \\ 0 & \mathbf{x} \in A \end{cases}. \qquad (14.1)$$

Of course, the detector may not be perfect, and there are two interdependent metrics that define the efficacy of a detector. The first, known sometimes as detector power, is the probability that an actual target will be detected. This probability of detection is defined by

$$\begin{aligned} P_D &= E\{\phi(\mathbf{x})|H_1\} \\ &= P\{[\phi(\mathbf{x}) = 1]|H_1\}, \end{aligned} \qquad (14.2)$$

where the expected value E and total probability P are characterized across all target spectra associated with alternative hypothesis H_1. The second metric is the probability that a background clutter spectrum will erroneously be declared a target. This probability of false alarm is defined as

$$\begin{aligned} P_{FA} &= E\{\phi(\mathbf{x})|H_0\} \\ &= P\{[\phi(\mathbf{x}) = 1]|H_0\}. \end{aligned} \qquad (14.3)$$

Most detection algorithms have an arbitrary threshold parameter that can be varied to trade off between detection probability and the probability of false alarm. In that case, there is a one-to-one correspondence between the detection and false-alarm probabilities, a function known as the receiver operating characteristic (ROC) curve. The ROC curve is the essential performance metric of a target detection algorithm.

14.1.1 Likelihood ratio test

According to Bayesian hypothesis testing for a two-class detection problem in which there is no prior probability information, a sufficient test statistic for target detection is the likelihood ratio

$$l(\mathbf{x}) = \frac{p(\mathbf{x}|H_1)}{p(\mathbf{x}|H_0)}, \qquad (14.4)$$

where $p(\mathbf{x}|H_1)$ is the probability density function (or likelihood function) for the random hyperspectral image vector $\mathbf{x}$ under alternative hypothesis H_1, and $p(\mathbf{x}|H_0)$ is the probability density function for the random hyperspectral image vector $\mathbf{x}$ under null hypothesis H_0. The

Neyman–Pearson lemma states that the most powerful test for target detection is the likelihood ratio test (LRT), given by the indicator function

$$\phi(\mathbf{x}) = \begin{cases} 1 & l(\mathbf{x}) > \xi \\ 0 & l(\mathbf{x}) \le \xi \end{cases}, \tag{14.5}$$

where ξ is a scalar threshold that controls the false-alarm probability. The term "most powerful" refers to the fact that this strategy provides the highest P_D for a given P_{FA}. In that respect, the LRT is considered to be the optimum detection strategy in the absence of decision costs and prior probability information about the hypotheses.

Under the LRT, probability density functions are assumed to be known, and variation of (P_D, P_{FA}) with ξ provides the detector ROC curve. Typically, this is set to achieve a tolerable false-alarm probability for a given application. Detection and false-alarm probabilities are determined by integrating distributions of the likelihood ratio test statistic for null and alternative hypotheses above threshold:

$$P_D = P\{ [l(\mathbf{x}) > \xi] | H_1 \}$$
$$= \int_{\xi}^{\infty} p(l \mid H_1) \, dl \tag{14.6}$$

and

$$P_{FA} = P\{ [l(\mathbf{x}) > \xi] | H_1 \}$$
$$= \int_{\xi}^{\infty} p(l \mid H_0) \, dl. \tag{14.7}$$

An interpretation of P_D and P_{FA} is illustrated in Fig. 14.1 for two simple, normal conditional probability density functions.

Since any monotonic function of $l(\mathbf{x})$ carries the same information, it is common to transform the likelihood function into some other form to simplify the mathematics for given probability density functions. The most common variety is the log-likelihood ratio

$$r(\mathbf{x}) = \log l(\mathbf{x})$$
$$= \log p(\mathbf{x} \mid H_1) - \log p(\mathbf{x} \mid H_0), \tag{14.8}$$

where $r(\mathbf{x})$ is called a discriminant function or *detection statistic*. Transforms other than a simple logarithm can also be beneficial for certain situations, but the log-likelihood ratio is particularly useful when conditional probability density functions are assumed to be normal. The

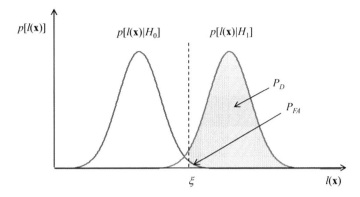

Figure 14.1 Interpretation of probabilities of detection and false alarm.

indicator function then becomes

$$\phi(\mathbf{x}) = \begin{cases} 1 & r(\mathbf{x}) > \eta \\ 0 & r(\mathbf{x}) \le \eta \end{cases}, \tag{14.9}$$

where $\eta = \log \xi$. The P_D and P_{FA} relationships are identical to Eqs. (14.6) and (14.7), with l replaced by r and ξ replaced by η.

14.1.2 Multivariate normal model

As an example, consider the classical problem of detecting the presence of a known vector signal $\mathbf{s}$ in multivariate, normally distributed, additive noise $\mathbf{n}$, with a known mean vector $\boldsymbol{\mu}$ and covariance matrix $\boldsymbol{\Sigma}$. The hypothesis testing problem for this example can be written as

$$\begin{aligned} H_0 &: \mathbf{x} = \mathbf{n} \\ H_1 &: \mathbf{x} = \mathbf{s} + \mathbf{n}, \end{aligned} \tag{14.10}$$

where $\mathbf{n} \sim N(\boldsymbol{\mu}, \boldsymbol{\Sigma})$. Therefore, $\mathbf{x} \mid H_0 \sim N(\boldsymbol{\mu}, \boldsymbol{\Sigma}), \mathbf{x} \mid H_1 \sim N(\boldsymbol{\mu} + \mathbf{s}, \boldsymbol{\Sigma})$, and the likelihood ratio becomes

$$\begin{aligned} l(\mathbf{x}) &= \exp\left\{ -\frac{1}{2}(\mathbf{x} - \mathbf{s} - \boldsymbol{\mu})^T \boldsymbol{\Sigma}^{-1}(\mathbf{x} - \mathbf{s} - \boldsymbol{\mu}) + \frac{1}{2}(\mathbf{x} - \boldsymbol{\mu})^T \boldsymbol{\Sigma}^{-1}(\mathbf{x} - \boldsymbol{\mu}) \right\} \\ &= \exp\left\{ \mathbf{s}^T \boldsymbol{\Sigma}^{-1} \mathbf{x} - \frac{1}{2}\mathbf{s}^T \boldsymbol{\Sigma}^{-1} \mathbf{s} - \mathbf{s}^T \boldsymbol{\Sigma}^{-1} \boldsymbol{\mu} \right\}. \end{aligned} \tag{14.11}$$

In this case, it is advantageous to use the log-likelihood ratio,

$$r(\mathbf{x}) = \mathbf{s}^T \boldsymbol{\Sigma}^{-1} \mathbf{x} - \frac{1}{2}\mathbf{s}^T \boldsymbol{\Sigma}^{-1} \mathbf{s} - \mathbf{s}^T \boldsymbol{\Sigma}^{-1} \boldsymbol{\mu}, \tag{14.12}$$

as the detection statistic, which is known as a linear matched filter.

Since $r(\mathbf{x})$ is a linear transform of a normally distributed random vector $\mathbf{x}$, it is also normally distributed with second-order statistics under the two hypotheses, given by

$$\mu_r \mid H_0 = E[\, r(\mathbf{x} \mid H_0)\,] = -\frac{1}{2}\mathbf{s}^T \boldsymbol{\Sigma}^{-1}\mathbf{s}, \tag{14.13}$$

$$\mu_r \mid H_1 = E[\, r(\mathbf{x} \mid H_1)\,] = \frac{1}{2}\mathbf{s}^T \boldsymbol{\Sigma}^{-1}\mathbf{s}, \tag{14.14}$$

$$\sigma_r^2 \mid H_0 = E[\,\{\, r(\mathbf{x} \mid H_0) - \mu_r \mid H_0 \,\}^2\,] = \mathbf{s}^T \boldsymbol{\Sigma}^{-1}\mathbf{s}, \tag{14.15}$$

and

$$\sigma_r^2 \mid H_1 = E[\,\{\, r(\mathbf{x} \mid H_1) - \mu_r \mid H_1 \,\}^2\,] = \mathbf{s}^T \boldsymbol{\Sigma}^{-1}\mathbf{s}. \tag{14.16}$$

If the SNR is defined as

$$\text{SNR} = \sqrt{\mathbf{s}^T \boldsymbol{\Sigma}^{-1}\mathbf{s}}, \tag{14.17}$$

then $r \mid H_0 \sim N(-\text{SNR}^2/2, \text{SNR}^2)$ and $r \mid H_1 \sim N(\text{SNR}^2/2, \text{SNR}^2)$. Therefore,

$$
\begin{aligned}
P_D &= \int_\eta^\infty p(r \mid H_1)\, dr \\
&= \int_\eta^\infty \frac{1}{\sqrt{2\pi}\,\text{SNR}^2} \exp\left\{ -\frac{1}{2\,\text{SNR}^2}\left(r - \frac{\text{SNR}^2}{2} \right)^2 \right\} dr \\
&= \int_{\frac{1}{\text{SNR}}\left(\eta - \frac{\text{SNR}^2}{2}\right)}^\infty \frac{1}{\sqrt{2\pi}} e^{-z^2/2}\, dz \\
&= \frac{1}{2}\operatorname{erfc}\left(\frac{1}{2\sqrt{2}}\frac{2\eta - \text{SNR}^2}{\text{SNR}} \right),
\end{aligned}
\tag{14.18}
$$

and

$$
\begin{aligned}
P_{FA} &= \int_\eta^\infty p(r \mid H_0)\, dr \\
&= \int_\eta^\infty \frac{1}{\sqrt{2\pi}\,\text{SNR}^2} \exp\left\{ -\frac{1}{2\,\text{SNR}^2}\left(r + \frac{\text{SNR}^2}{2} \right)^2 \right\} dr \\
&= \int_{\frac{1}{\text{SNR}}\left(\eta + \frac{\text{SNR}^2}{2}\right)}^\infty \frac{1}{\sqrt{2\pi}} e^{-z^2/2}\, dz
\end{aligned}
$$

$$= \frac{1}{2}\mathrm{erfc}\left(\frac{1}{2\sqrt{2}}\frac{2\eta + \mathrm{SNR}^2}{\mathrm{SNR}}\right), \tag{14.19}$$

where $\mathrm{erfc}(x)$ is the complementary error function.

The ROC curves for this classical linear matched filter example are shown in Fig. 14.2 with the SNR as a free parameter. As expected, detection performance is improved with higher SNR as the P_D increases for the same P_{FA} (or the P_{FA} decreases for the same P_D). In hyperspectral target detection problems, it is typically necessary to operate with a P_{FA} on the order of 10^{-5}, so it is more useful to display the ROC curve with a logarithmic independent (P_{FA}) axis, as illustrated in Fig. 14.3. It is also common to characterize the false-alarm rate (FAR) in units for false alarms per unit area (e.g., FA/km^2), which nominally equals P_{FA}/GSD^2 for simple pixel-level target detection. In a hyperspectral target detection case, it is often background clutter that limits detection, and the operative parameter is the signal-to-clutter ratio (SCR) rather than the SNR; however, this issue is deferred until later in the chapter.

14.1.3 Generalized likelihood ratio test

While the LRT is a useful theoretical construct for target detection, it is of limited practical value because, even when the underlying parametric form of probability density functions of data can be assumed, the parameters of the distribution are not known *a priori*. In such instances, a common detection strategy is to replace unknown parameters with maximum-likelihood estimates (MLEs) derived from the observed image data. This

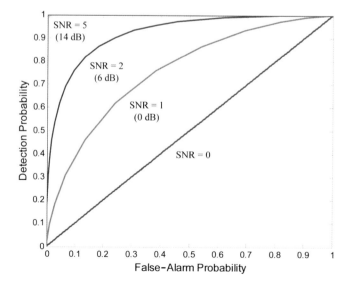

Figure 14.2   ROC curves as a function of SNR for a classical linear matched filter.

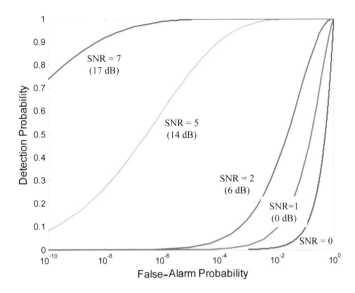

Figure 14.3 ROC curves as a function of SNR for a classical linear matched filter with a logarithmic probability of false-alarm axis.

is known as the generalized likelihood ratio test (GLRT), expressed as

$$l(\mathbf{x}) = \frac{\max_{\theta_1} p(\mathbf{x} \mid H_1)}{\max_{\theta_0} p(\mathbf{x} \mid H_0)}, \qquad (14.20)$$

where θ_0 and θ_1 represent sets of unknown parameters under the null and alternative hypotheses. Like the LRT, the GLRT can be expressed by monotonic transformations of the likelihood ratio to simplify mathematics, with no effect on the detector other than altering the relationship between detector threshold levels and the FAR. The GLRT does not exhibit the same optimality properties as the LRT but is nevertheless a reasonable and commonly used strategy for target detection.

The benefit of the GLRT is that it extends the LRT methodology to more sophisticated target, background, and noise models, where the MLE is used to determine model parameters. The remainder of this chapter explores a variety of such detectors, where model assumptions vary based on different levels of *a priori* information as well as different expected hyperspectral data characteristics. As was observed in Chapter 12, all of the various spectral data models exhibit strengths and limitations, and their efficacy is often data dependent. Thus, one would also expect similar data dependence in the performance of detection algorithms based on these different underlying data models.

Detection algorithms described in this chapter are organized into five basic categories: anomaly detection, signature-matched detection, false-alarm mitigation, subspace-matched detection, and change detection. Anomaly detection, signature-matched detection, and subspace-matched detection are distinguished by the level of *a priori* target information. Anomaly detection refers to the situation where there is no *a priori* target information; signature-matched detection applies to the situation where a target is defined by a single reference spectrum; and subspace-matched detection deals with the case where a target is represented by a set of basis vectors that account for potential signal variation. False-alarm mitigation refers to a set of methods specifically tailored toward improving the selectivity of signature-matched methods that often produce false alarms due to anomalous scene spectra. Finally, change detection uses reference imagery to detect targets as image differences between observations at different times due to, for example, the insertion, deletion, or movement of targets in the scene.

14.2 Anomaly Detection

Anomaly detection refers to a condition where there is no *a priori* target information on which to make a detection decision. In this case, the decision is based solely on nonconformity to a background model, since no other information exists. Different anomaly detectors can be derived from the GLRT by employing different background and noise models. Throughout this chapter, let $\mathbf{s}$ represent the signal (or target), $\mathbf{b}$ represent background (or clutter), and $\mathbf{n}$ represent noise. When represented by subspace models, $\mathbf{S}$ and $\mathbf{B}$ are used to represent the corresponding signal and background subspace matrices, and $\boldsymbol{\alpha}$ and $\boldsymbol{\beta}$ are used to represent the corresponding basis coefficient vectors. This section follows the taxonomy given in Table 14.1 that summarizes the assumed signal, background, and noise models, along with known (or assumed) and unknown (or estimated) model parameters for each anomaly detection algorithm to be described.

14.2.1 Mahalanobis distance detector

Consider the detection problem

$$
\begin{aligned}
H_0 &: \mathbf{x} = \mathbf{b} \\
H_1 &: \mathbf{x} = \mathbf{s},
\end{aligned}
\tag{14.21}
$$

where background clutter $\mathbf{b}$ is assumed to be normally distributed with a mean $\boldsymbol{\mu}$ and covariance matrix $\boldsymbol{\Sigma}$, and nothing is known about signal $\mathbf{s}$. Noise is assumed to be part of the background clutter. Under the null

Table 14.1 Anomaly detection algorithm taxonomy.

Algorithm	Models			Parameters	
	Signal	Background	Noise	Known	Unknown
Mahalanobis distance	None	Normal	Part of background		$\mathbf{s}, \boldsymbol{\mu}_b, \boldsymbol{\Sigma}_b$
Reed–Xiaoli (RX) detector	None	Local normal	Part of background		$\mathbf{s}, \boldsymbol{\mu}_{local}, \boldsymbol{\Sigma}_{local}$
Subspace RX detector	None	Subspace	Local normal	L	$\mathbf{s}, \mathbf{B}, \boldsymbol{\mu}_{local}, \boldsymbol{\Sigma}_{local}$
Complementary subspace detector	Subspace	Subspace	Normal	L, M	$\mathbf{S}, \mathbf{B}, \boldsymbol{\alpha}, \boldsymbol{\beta}, \boldsymbol{\Sigma}_n$
Gaussian mixture model detector	None	Normal mixture	Part of background	Q	$\mathbf{s}, \boldsymbol{\mu}_q, \boldsymbol{\Sigma}_q, P(q)$
Gaussian mixture RX detector	None	Normal mixture	Part of background	Q	$\mathbf{s}, \boldsymbol{\mu}_q, \boldsymbol{\Sigma}_q, P(q)$
Cluster-based anomaly detector	None	Normal mixture	Part of background	Q	$\mathbf{s}, \boldsymbol{\mu}_q, \boldsymbol{\Sigma}_q, P(q)$
Fuzzy cluster-based anomaly detector	None	Normal mixture	Part of background	Q	$\mathbf{s}, \boldsymbol{\mu}_q, \boldsymbol{\Sigma}_q, w(q)$

hypothesis,

$$p(\mathbf{x} \mid H_0) = \frac{1}{(2\pi)^{K/2}} \frac{1}{|\hat{\boldsymbol{\Sigma}}|^{1/2}} \exp\left\{-\frac{1}{2}(\mathbf{x} - \hat{\boldsymbol{\mu}})^T \hat{\boldsymbol{\Sigma}}^{-1} (\mathbf{x} - \hat{\boldsymbol{\mu}})\right\}, \quad (14.22)$$

where the caret over the background statistics indicates that the MLEs are inserted in place of the unknown parameters. In this case, the MLEs are simply the sample statistics of the entire image. Under the alternative hypothesis, the MLE for the expected value of signal $E[\mathbf{s}]$ is the measurement $\mathbf{x}$, such that

$$p(\mathbf{x} \mid H_1) = \text{constant.} \quad (14.23)$$

Substituting Eqs. (14.22) and (14.23) into a logarithmic form of the GLRT leads to the detection statistic

$$r_{MD}(\mathbf{x}) = (\mathbf{x} - \hat{\boldsymbol{\mu}})^T \hat{\boldsymbol{\Sigma}}^{-1} (\mathbf{x} - \hat{\boldsymbol{\mu}}), \quad (14.24)$$

which is known alternately as the Mahalanobis distance (MD) or the multivariate energy detector.

The MD is the distance between vector $\mathbf{x}$ and $\hat{\boldsymbol{\mu}}$ relative to the standard deviation of the background distribution, in the direction from $\hat{\boldsymbol{\mu}}$ to $\mathbf{x}$. This is depicted in Fig. 14.4 for a 2D case, where the ellipse represents a contour of constant MD. With this in mind, the threshold associated with the GLRT corresponds to such an elliptical surface, and measurement vectors outside the surface are declared targets. Changing the threshold

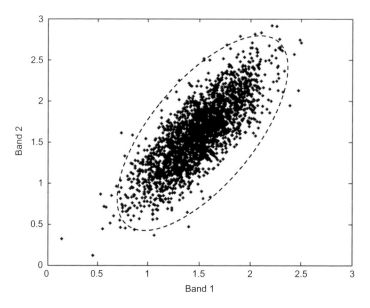

Figure 14.4 Plot of the Mahalanobis distance (MD) detector perfomance for a 2D case.

value corresponds to changing the size (but not shape) of the decision surface. This concept directly extends to multiple dimensions, even though it is not readily visualized. Through covariance normalization, the MD measures the difference from the background distribution relative to its scatter in a particular direction.

If a diagonalized form of covariance matrix

$$\hat{\Sigma} = \hat{V}\hat{D}\hat{V}^T \tag{14.25}$$

is substituted in Eq. (14.24), the MD can be written as

$$r_{MD}(\mathbf{x}) = [\hat{D}^{-1/2}\hat{V}^T(\mathbf{x} - \hat{\mu})]^T [\hat{D}^{-1/2}\hat{V}^T(\mathbf{x} - \hat{\mu})]. \tag{14.26}$$

By recognizing the whitened form of the data vector as

$$\mathbf{z} = \hat{D}^{-1/2}\hat{V}^T(\mathbf{x} - \hat{\mu}), \tag{14.27}$$

it is seen that the MD reduces in the whitened space as

$$r_{MD}(\mathbf{z}) = \|\mathbf{z}\|^2; \tag{14.28}$$

that is, decision surfaces become simple spheres in the whitened space, as illustrated in Fig. 14.5. The MD detector, therefore, is simply a measure of the whitened power of the data.

The performance of the MD detector under a normal background assumption can be analytically evaluated, since detection statistic $r_{MD}(\mathbf{x})$ will be chi-squared distributed under such assumptions if the MLEs are accurate. Specifically, the detection statistic distribution under the null hypothesis is a central chi-squared distribution,

$$r_{MD}(\mathbf{x}) \mid H_0 \sim \chi_K^2(r), \qquad (14.29)$$

which exhibits a probability density function described by

$$p(r; K) = \frac{1}{\Gamma(K/2)\, 2^{K/2}}\, r^{K/2-1}\, e^{-r/2}. \qquad (14.30)$$

Under the alternative hypothesis, the detection statistic distribution exhibits a noncentral chi-squared distribution,

$$r_{MD}(\mathbf{x}) \mid H_1 \sim \chi_K^2(r; \lambda^2), \qquad (14.31)$$

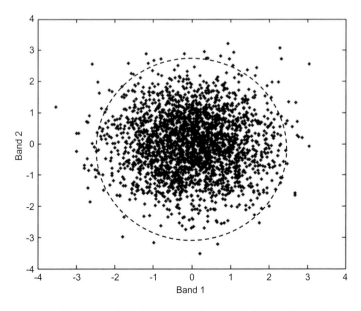

Figure 14.5 Plot of the MD detector performance for a whitened 2D case.

which has the probability density function

$$p(r; K, \lambda^2) = \sum_{n=0}^{\infty} e^{-\lambda^2/2} \frac{(\lambda^2/2)^n}{n!} \frac{1}{\Gamma(K/2 + n) \, 2^{K/2+n-1}} r^{K/2+n-1} e^{-r/2}$$

$$= \frac{\lambda e^{-\lambda^2/2}}{2 \, (\lambda^2 r)^{K/4}} r^{(K-1)/2} e^{-r/2} I_{K/2-1}\left(\sqrt{\lambda^2 r}\right), \tag{14.32}$$

where $I_\nu(x)$ is a modified Bessel function of the first kind (Lebedev, 1972). Noncentrality parameter λ^2 in Eqs. (14.31) and (14.32) is given by the squared SCR:

$$\lambda^2 = \text{SCR}^2 = (\mathbf{s} - \hat{\boldsymbol{\mu}})^T \hat{\boldsymbol{\Sigma}}^{-1} (\mathbf{s} - \hat{\boldsymbol{\mu}}). \tag{14.33}$$

Theoretical ROC performance for the MD detector is derived by integrating the tails of the chi-squared distributions in Eqs. (14.31) and (14.32),

$$P_D = \int_\eta^\infty p(r \mid H_1) \, dr \tag{14.34}$$

and

$$P_{FA} = \int_\eta^\infty p(r \mid H_0) \, dr, \tag{14.35}$$

as a function of threshold level η for different values of the SCR. The results are provided in Fig. 14.6.

When applied to real hyperspectral image data, detection statistic distributions for the MD detector are not chi-squared because the background clutter, as seen in Chapter 12, is not well described by a multivariate normal distribution. It is also difficult to unambiguously state what the false-alarm performance of an anomaly detector is when considering real data because, in some respects, anything that exceeds the threshold is by definition an anomaly. Setting these semantics aside, the practical application of anomaly detection is usually aimed at finding specific objects of interest and not just spectrally anomalous scene pixels. Therefore, anomalies that are not objects of interest are defined as false alarms.

As an example, consider the HYDICE Forest Radiance I image depicted as a false-color image of the leading three principal components in Fig. 14.7. This is the same image used to illustrate the in-scene VNIR/SWIR atmospheric compensation algorithms in Chapter 11. The image contains a number of vehicles, target panels, and calibration panels,

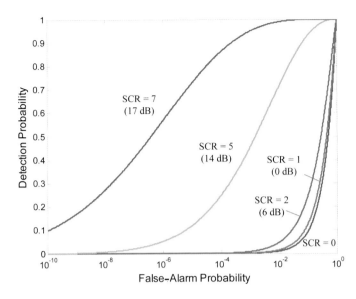

Figure 14.6 Theoretical ROC curves as a function of SCR for the MD detector.

Figure 14.7 Color-composite image of the leading three principal components for the HYDICE Forest Radiance VNIR/SWIR hyperspectral image used for illustrating various detection algorithms.

including a large green tarp, arrayed in a grass field near a forested area and surrounded by a road network. One might consider the various manmade objects meant to be detected targets as anomalous to the natural background regions composed of the trees, grass, and roads.

A detection statistic image for the MD detector applied to this image is provided in Fig. 14.8. Recall that a high (bright) value of the detection statistic is indicative of a target, while a low (dark) value is indicative of background clutter. With this in mind, the detector identifies most of the panel array objects and calibration target panels, along with a fair number of the vehicles, as anomalous. It also detects a few other objects that appear to be manmade, but it detects the roads as anomalous, as well as some image pixels that apparently are corrupted by sensor artifacts such as noisy

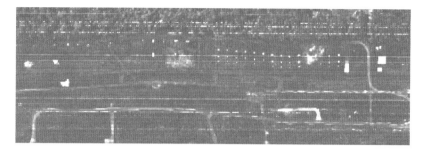

Figure 14.8 Detection statistic image of a HYDICE Forest Radiance image using a Mahalanobis distance (MD) detector.

detectors. The corresponding detection statistic distribution provided in Fig. 14.9 roughly resembles theoretical chi-squared distribution but has a much heavier tail distribution and is displaced in terms of the peak of the distribution.

14.2.2 Reed–Xiaoli detector

A major limitation of the MD detector is that it assumes that background clutter is spatially stationary, meaning that the statistics are space invariant. This assumption is almost never true, as spatially stationary statistics would imply an unstructured scene. The term *structured* clutter refers to a spatial distribution that is spatially correlated and whose statistics are nonstationary. The Reed–Xiaoli (RX) algorithm is a widely used

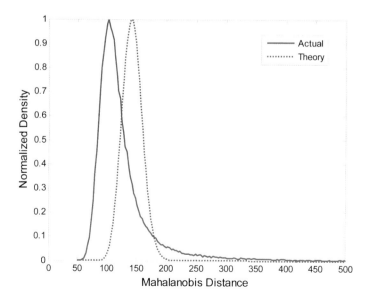

Figure 14.9 Detection statistic distribution of the Hyperspec tarp bundles image using an MD detector.

anomaly detector for multidimensional data that addresses this limitation by adapting mean-vector and covariance-matrix estimates to the local statistics for each spectrum under test (Reed and Yu, 1990). The basic formulation includes a spatial filtering component typically not employed when applied to hyperspectral image data; otherwise, it is basically a spatially adaptive version of the MD detection statistic that replaces the normal background model with a local normal model. The adaptive component is based on the use of sample statistics of a local background region surrounding the pixel under test for the mean vector and covariance matrix in Eq. (14.24). That is,

$$r_{RX}(\mathbf{x}) = (\mathbf{x} - \hat{\boldsymbol{\mu}}_{local})^T \hat{\boldsymbol{\Sigma}}_{local}^{-1} (\mathbf{x} - \hat{\boldsymbol{\mu}}_{local}), \qquad (14.36)$$

where

$$\hat{\boldsymbol{\mu}}_{local} = \frac{1}{N_{local}} \sum_{i \in \Omega_{local}} \mathbf{x}_i, \qquad (14.37)$$

and

$$\hat{\boldsymbol{\Sigma}}_{local} = \frac{1}{N_{local} - 1} \sum_{i \in \Omega_{local}} (\mathbf{x}_i - \hat{\boldsymbol{\mu}}_{local}) (\mathbf{x}_i - \hat{\boldsymbol{\mu}}_{local})^T; \qquad (14.38)$$

local background region Ω_{local} is depicted in Fig. 14.10, and N_{local} is the number of pixels contained in Ω_{local}.

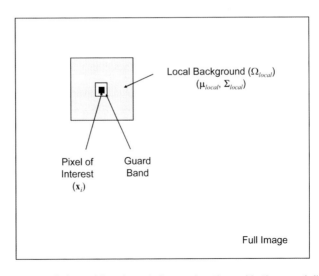

Figure 14.10 Depiction of local statistics estimation with the spatially adaptive RX algorithm.

Succesfully implementing the RX detector involves a trade-off in the size of the local background region. Selecting a region that is too large reduces the adaptive nature of the method with regard to inhomogenous background clutter, and selecting a region that is too small can result in poor performance due to covariance estimation errors. A general rule of thumb for a decent covariance is to make $N_{local} > K^2$. Therefore, reducing data dimensionality is one method to address this trade-off. Covariance matrix regularization using singular-value decomposition or diagonal matrix loading can be necessary for dealing with matrix inversion problems when the local region is insufficiently large.

Another way to approach such matrix inversion issues is to use a quasi-local covariance matrix as proposed by Caefer *et al.* (2008). In this approach, the global covariance matrix is first diagonalized, and data are transformed to principal-component spectra $\mathbf{z}$ according to the PCA method described in Chapter 12 using the eigenvector matrix $\mathbf{V}$. The global covariance matrix in this domain is diagonal matrix $\mathbf{D}$. The quasi-local covariance method uses the global covariance matrix structure but only alters the variances for each band locally. That is, global diagonalized covariance matrix $\mathbf{D}$ is replaced by local but diagonal covariance matrix $\mathbf{D}_{local}$, estimated in the principal-component domain. The detection statistic is given as

$$r_{RX}(\mathbf{x}) = (\mathbf{x} - \hat{\boldsymbol{\mu}}_{local})^T \hat{\mathbf{V}} \, \hat{\mathbf{D}}_{local}^{-1} \, \hat{\mathbf{V}}^T (\mathbf{x} - \hat{\boldsymbol{\mu}}_{local}). \qquad (14.39)$$

If the detection statistic is actually computed with the principal-component spectra, then Eq. (14.30) simplifies to

$$r_{RX}(\mathbf{z}) = (\mathbf{z} - \hat{\boldsymbol{\mu}}_{local})^T \hat{\mathbf{D}}_{local}^{-1} (\mathbf{z} - \hat{\boldsymbol{\mu}}_{local}), \qquad (14.40)$$

where both local statistics are computed in the principal component domain. The eigenvector matrix $\mathbf{V}$ is spatially stationary, maintaining the global form of the covariance matrix while allowing variances to adapt locally.

Theoretically, the RX detector performance is the same as that of the MD detector for normal background clutter with spatially stationary statistics. When these statistics are not stationary, however, local adaptation achieved by the RX detector normalizes the threshold level to the background clutter level, making the FAR more spatially uniform. This constant FAR (CFAR) feature of a spatially adaptive detector such as the RX detector is not shared by detectors that employ global statistics, such as the MD detector. Within the limit of local normality, the distribution of an RX-detection statistic should approach the chi-squared distributions predicted for an MD detector.

An RX-detection statistic image based on the HYDICE Forest Radiance image for a background window size of 49×49 is provided in Fig. 14.11 to illustrate the effect of local adaptivity. In this case, most of the panel array targets and vehicles are still detected as anomalous along with the smaller calibration targets. The detection statistic is weaker for the larger calibration panels and manmade objects and is heavily suppressed relative to the global MD detector for the roads; the anomalous detections from sensor artifacts are nonexistent. It is clear that this provides a superior detector for this particular image. This assertion is further supported by the detection statistic distribution (illustrated in Fig. 14.12), which provides a much better match to the theoretical distribution. Even in this case, however, the actual distribution exhibits a heavier tail distribution, which is a result of the non-normal nature of hyperspectral imagery.

14.2.3 Subspace Reed–Xiaoli detector

Suppose that the detection problem underlying the RX detector is now modified to incorporate a structured background model, characterized by subspace matrix **B**. Specifically, it is cast as

$$H_0 : \mathbf{x} = \mathbf{B}\boldsymbol{\beta} + \mathbf{n}$$
$$H_1 : \mathbf{x} = \mathbf{s}, \tag{14.41}$$

where $\boldsymbol{\beta}$ is the basis coefficient vector for a given background clutter spectrum, and **n** represents a combination of the sensor noise and unstructured background clutter. We assume that $\mathbf{n} \sim N(\boldsymbol{\mu}_{local}, \boldsymbol{\Sigma}_{local})$, where the local statistics are spatially varying and can be estimated in a similar manner as described for the RX algorithm. Again, there is no *a priori* signal information, thus the MLE for **s** is the test spectrum, and the density function under the alternative hypothesis is constant.

Suppose that data are transformed into the principal-component domain based on global sample statistics. In this domain, the null and alternative

Figure 14.11 Detection statistic image of the HYDICE Forest Radiance image using an RX detector.

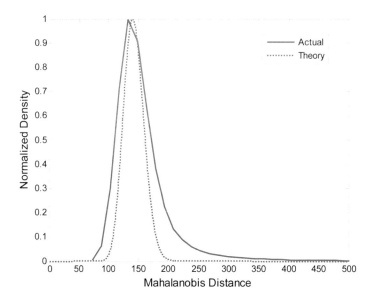

Figure 14.12 Detection statistic distribution of the HYDICE Forest Radiance image using an RX detector.

hypotheses are identical to Eq. (14.41), with spectral vector $\mathbf{x}$ replaced by corresponding principal component vector $\mathbf{z}$. We choose to estimate $\mathbf{B}$ using the first L eigenvectors of the principal-component transformation matrix $\mathbf{V}$, where L is arbitrarily chosen to capture background structure while hopefully not capturing target spectral content. In this case, the structured component in Eq. (14.41) can be eliminated by projecting data into the trailing subspace. This can be written mathematically by the orthogonal subspace projection operator

$$\mathbf{P}_{\mathbf{B}}^{\perp} = \mathbf{I} - \mathbf{B}\mathbf{B}^{T}. \qquad (14.42)$$

The operator can be written in this simplified form because of the orthonormality of estimated background eigenvectors.

After performing a principal-component transformation and orthogonal projection, the detection hypotheses reduce to

$$\begin{aligned} H_0 &: \mathbf{P}_{\mathbf{B}}^{\perp}\mathbf{z} = \mathbf{n} \\ H_1 &: \mathbf{P}_{\mathbf{B}}^{\perp}\mathbf{z} = \mathbf{s}, \end{aligned} \qquad (14.43)$$

where we further assume that $\mathbf{s}$ and $\mathbf{n}$ are orthogonal to subspace $\mathbf{B}$ by definition. This is inconsequential, since $\mathbf{s}$ is unknown and $\mathbf{n}$ is arbitrarily defined. Under these assumptions, the RX detector becomes the subspace

RX (SSRX) detector,

$$r_{SSRX}(\mathbf{z}) = (\mathbf{P}_{\mathbf{B}}^{\perp}\mathbf{z} - \mathbf{P}_{\mathbf{B}}^{\perp}\hat{\mathbf{u}}_{local})^{T}\,\hat{\mathbf{D}}_{local}^{-1}\,(\mathbf{P}_{\mathbf{B}}^{\perp}\mathbf{z} - \mathbf{P}_{\mathbf{B}}^{\perp}\hat{\mathbf{u}}_{local}), \qquad (14.44)$$

where local statistics are estimated in the orthogonal subspace of background clutter in the same manner as the RX detector local statistics. Orthogonal subspace projection is implemented by simply eliminating the leading L principal-component bands from the data and computing the RX detection statistic in the trailing subspace. The selection of L is a somewhat arbitrary trade-off between eliminating structured background clutter as well as components of the signal.

An SSRX-detection statistic image for structured background dimensionality $L = 8$ is illustrated for the HYDICE Forest Radiance image with a 49×49 pixel window size in Fig. 14.13. The result is very similar to the RX detection statistic image for this example, presumably because the background structure in the leading subspace is already suppressed in the RX detector, due to large elements of the inverse covariance matrix corresponding to these bands. In some cases, the SSRX detector can provide better background suppression relative to the RX detector if the target content falls outside the leading subspace.

14.2.4 Complementary subspace detector

If subspace models for both the target and background are assumed (Schaum, 2001), the detection problem can be cast as

$$\begin{aligned} H_0 &: \mathbf{x} = \mathbf{B}\,\boldsymbol{\beta} + \mathbf{n} \\ H_1 &: \mathbf{x} = \mathbf{S}\,\boldsymbol{\alpha} + \mathbf{n}, \end{aligned} \qquad (14.45)$$

where $\mathbf{S}$ and $\mathbf{B}$ are matrices containing columns that span target and background subspaces, respectively, and $\mathbf{n}$ is assumed to be normally distributed with zero-mean and covariance matrix $\boldsymbol{\Sigma}_{\mathbf{n}}$. This is typically

Figure 14.13 Detection statistic image of HYDICE Forest Radiance image using an SSRX detector.

considered to represent additive sensor noise, but often the sample covariance matrix estimated from the data is used for Σ_n. In that case, n is considered to represent the residual background clutter outside of B in addition to the sensor noise. The vectors α and β are unknown parameters of the model containing the basis coefficients for a given test spectrum.

The corresponding probability density functions for the model in Eq. (14.45) are

$$p(x \mid H_0, \beta) = \frac{1}{(2\pi)^{K/2}} \frac{1}{|\Sigma_n|^{1/2}} \exp\left\{-\frac{1}{2}(x - B\beta)^T \Sigma_n^{-1}(x - B\beta)\right\},$$
(14.46)

and

$$p(x \mid H_1, \alpha) = \frac{1}{(2\pi)^{K/2}} \frac{1}{|\Sigma_n|^{1/2}} \exp\left\{-\frac{1}{2}(x - S\alpha)^T \Sigma_n^{-1}(x - S\alpha)\right\}.$$
(14.47)

By symbolically writing noise whitening as the operation $\Sigma_n^{-1/2}$, the problem of estimating β can be considered in terms of solving for the least-squares solution of

$$\Sigma_n^{-1/2} B\beta = \Sigma_n^{-1/2}x - \Sigma_n^{-1/2}n$$
(14.48)

in the whitened space, where additive noise is spectrally uncorrelated. The resulting MLE is the least-squares estimate

$$\hat{\beta} = (B^T \Sigma_n^{-1} B)^{-1} B^T \Sigma_n^{-1} x.$$
(14.49)

Similarly for the target hypothesis,

$$\hat{\alpha} = (S^T \Sigma_n^{-1} S)^{-1} S^T \Sigma_n^{-1} x.$$
(14.50)

The estimates in Eqs. (14.49) and (14.50) maximize the respective density functions in Eqs. (14.46) and (14.47). Therefore, a GLRT detection statistic can be formed as the logarithm of the ratio of Eqs. (14.46) and (14.47) using these estimates. This can be shown to be the joint subspace detector (JSD):

$$r_{JSD}(x) = x^T \Sigma_n^{-1} S (S^T \Sigma_n^{-1} S)^{-1} S^T \Sigma_n^{-1} x - x^T \Sigma_n^{-1} B (B^T \Sigma_n^{-1} B)^{-1} B^T \Sigma_n^{-1} x.$$
(14.51)

To better understand Eq. (14.51), suppose that the operations are performed on noise-whitened data $\mathbf{z}$ as opposed to original data $\mathbf{x}$. Also consider a situation where background and target subspace matrices $\mathbf{S}$ and $\mathbf{B}$ are orthonormal. In this case, Eq. (14.51) reduces to

$$r_{JSD}(\mathbf{z}) = \mathbf{z}^T \mathbf{S} \, \mathbf{S}^T \mathbf{z} - \mathbf{z}^T \mathbf{B} \, \mathbf{B}^T \mathbf{z}$$
$$= \| \mathbf{S}^T \mathbf{z} \|^2 - \| \mathbf{B}^T \mathbf{z} \|^2. \qquad (14.52)$$

The detection statistic is merely the difference between the whitened power of the measurement vector projected into the target subspace and that projected into the background subspace.

In the case of an anomaly detector, no *a priori* knowledge exists from which to form the target subspace. Therefore, it must be estimated from the data. One methodology, known as the complementary subspace detector (CSD), estimates background subspace as the leading principal-component subspace and considers its orthogonal complement to be the target subspace. Mathematically, eigenvector matrix $\mathbf{V}$ from image sample statistics is simply decomposed as

$$\mathbf{V} = \begin{bmatrix} \mathbf{B} & \mathbf{S} \end{bmatrix} \qquad (14.53)$$

for some assumed background dimensionality L. The CSD algorithm biases against spectral content in the leading subspace, as it is consistent with the background model, unlike the SSRX algorithm, which merely projects it out. One problem that can arise with this approach is that target information falls within the leading subspace. In this case, it can be advantageous to use entire eigenvector matrix $\mathbf{V}$ to represent $\mathbf{S}$, while using only the leading subspace for $\mathbf{B}$. This approach can be extended to incorporate *a priori* information about the target into subspace matrix $\mathbf{S}$ to develop a matched subspace detector. This extension is further discussed in Section 14.5.

Another limitation of this basic implementation of CSD is that it is not always robust to noise, which corrupts the trailing subspace assumed to be part of the target subspace. This can be balanced by modifying the decomposition to

$$\mathbf{V} = \begin{bmatrix} \mathbf{B} & \mathbf{S} & \mathbf{N} \end{bmatrix} \qquad (14.54)$$

for an assumed background dimensionality L and signal dimensionality M. That is, target subspace is determined only by either a middle set of eigenvectors ($\mathbf{S}$), or alternatively, to compensate for the loss of target information in the leading subspace, the leading $L + M$ eigenvectors. A limitation of this method is that determination of background and signal dimensionality is somewhat arbitrary.

A complementary-subspace-detection statistic image for the HYDICE Forest Radiance image is shown in Fig. 14.14. In this example, **B** is estimated from the leading ten eigenvectors, and **S** is estimated from the leading 40 eigenvectors. Since there are a large number of targets in this case resulting in target information in the leading subspace, **S** and **B** are chosen to be nonorthogonal. If the leading subspace were excluded from **S** to achieve orthogonality, detection performance would be worse in this case. In this example, the CSD appears superior to the MD detector but inferior to RX and SSRX detectors. On the other hand, the CSD requires less computation than both the spatially adaptive RX and SSRX detectors, which require computation of local statistics.

14.2.5 Normal mixture model detectors

By using a normal mixture model for background clutter to represent its non-normal behavior, the anomaly detection model becomes the same as Eq. (14.12), except that now the probability density function under the null hypothesis for the composite background plus noise is the normal mixture

$$p(\mathbf{x} \mid H_0) = \sum_{q=1}^{Q} P(q)p(\mathbf{x} \mid q, H_0), \tag{14.55}$$

where

$$p(\mathbf{x} \mid q, H_0) = \frac{1}{(2\pi)^{K/2}} \frac{1}{|\Sigma_q|^{1/2}} \exp\left\{-\frac{1}{2}(\mathbf{x} - \boldsymbol{\mu}_q)^T \Sigma^{-1} (\mathbf{x} - \boldsymbol{\mu}_q)\right\}. \tag{14.56}$$

As the density function under the alternative hypothesis is again constant due to the lack of *a priori* target information, the GLRT detection

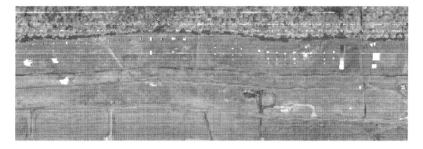

Figure 14.14 Detection statistic image of the HYDICE Forest Radiance image using a complementary subspace detector.

statistic is simply equal to the inverse of Eq. (14.55), or the logarithm thereof, where MLEs are inserted for, in general, Q and $\{P(q), \boldsymbol{\mu}_q, \Sigma_q, q = 1, 2, \ldots, Q\}$. Different varieties of detectors derived from this basic formulation employ varying assumptions and techniques for parameter estimation. Some are briefly described here.

All of the methods described assume that the number of normal modes or classes Q used to represent background clutter are either known or simply assumed. The Gaussian mixture model (GMM) detector (Stein *et al.*, 2002) estimates remaining parameters from an entire image using the stochastic expectation maximization (SEM) algorithm described in Chapter 13. An important assumption is that the proportion of target spectra is sufficiently small, such that they do not influence the background model parameter estimates in Eq. (14.55) or form their own class. Employing the log-likelihood GLRT, the GMM detection statistic is given by

$$r_{GMM}(\mathbf{x}) = -\log\left[\sum_{q=1}^{Q} P(q)\frac{1}{(2\pi)^{K/2}}\frac{1}{|\hat{\Sigma}_q|^{1/2}}\right.$$

$$\left. \times \exp\left\{-\frac{1}{2}(\mathbf{x} - \hat{\boldsymbol{\mu}}_q)^T \hat{\Sigma}_q^{-1}(\mathbf{x} - \hat{\boldsymbol{\mu}}_q)\right\}\right]. \qquad (14.57)$$

This represents the statistical distance of the test spectrum from the composite normal mixture model. The detection statistic varies smoothly between modes of mixture density, as illustrated for a 2D, two-class example in Fig. 14.15.

Application of the GMM detection algorithm to the HYDICE Forest Radiance image is illustrated in Figs. 14.16 and 14.17. In this case, the image is segmented into four classes using the SEM algorithm, resulting in the color-coded classification map depicted in Fig. 14.16. The second-order class samples statistics and prior probability estimates are then used to form the detection statistic in Eq. (14.48), with the results depicted in Fig. 14.17. Performance is fairly poor in this case, as there are large regions of background and sensor artifacts (specifically the dark-blue class indicated in Fig. 14.16) that are just as anomalous as the targets. This results from a low prior probability for this class, producing a high detection statistic through Eq. (14.48). That is, the GMM detector treats sparsely populated background classes as anomalous and detects them as such.

An alternate approach is to estimate class index q for each spectrum, and instead of using the composite statistical difference, perform the detection based on the class-conditional MD. This is known as the Gaussian mixture Reed–Xiaoli (GMRX) detector (Stein *et al.*, 2002). Again, SEM can be

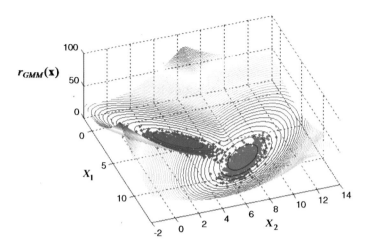

Figure 14.15   GMM detection statistic behavior for 2D, two-class example (Hytla *et al.*, 2010).

Figure 14.16   Color-coded classification map for the HYDICE Forest Radiance image based on 20 iterations of the SEM algorithm with four classes.

Figure 14.17   Detection statistic image of the HYDICE Forest Radiance image using a GMM detector with four classes.

used to derive parameters of the mixture model from Eqs. (14.55) and (14.56), but then the class index is estimated according to

$$\hat{q}(\mathbf{x}) = r \quad \text{if } \hat{P}(r \mid \mathbf{x}) > \hat{P}(s \mid \mathbf{x}) \; \forall \, r \neq s, \tag{14.58}$$

where the *a posteriori* probability estimate for class index q is given by

$$\hat{P}(q \mid \mathbf{x}) = \frac{\hat{P}(q)\hat{p}(\mathbf{x} \mid q)}{\sum\limits_{r=1}^{Q} \hat{P}(r)\hat{p}(\mathbf{x} \mid r)}. \tag{14.59}$$

The GMRX detector is then given by the class-conditional MD

$$r_{GMRX}(\mathbf{x}) = (\mathbf{x} - \hat{\boldsymbol{\mu}}_{\hat{q}(\mathbf{x})})^T \hat{\boldsymbol{\Sigma}}_{\hat{q}(\mathbf{x})}^{-1} (\mathbf{x} - \hat{\boldsymbol{\mu}}_{\hat{q}(\mathbf{x})}). \tag{14.60}$$

Unlike the GMM algorithm, the GMRX detection statistic can exhibit discontinuities at class-conditional boundaries. These do not occur in the illustration in Fig. 14.18 because the classes are normally distributed. They do occur for non-normal classes.

A detection statistic image example based on application of the GMRX detection algorithm to the HYDICE Forest Radiance image with the same four SEM classes is illustrated in Fig. 14.19. Performance is much improved relative to the GMM result, primarily because of elimination of prior probability in the detection statistic. There is still greater background clutter leakage from the road network and sensor artifacts relative to

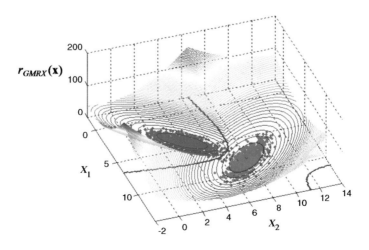

Figure 14.18 GMRX detection statistic behavior for a 2D, two-class example (Hytla *et al.*, 2010).

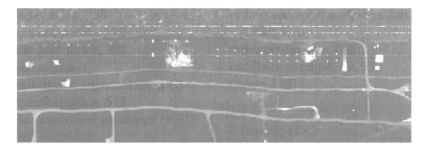

Figure 14.19 Detection statistic image of the HYDICE Forest Radiance image using a GMRX detector with four classes.

the RX, SSRX, and CSD detectors. This most likely represents non-normal class behavior and could be limited by using more background classes. Adding background classses, however, increases the risk that class statistics might be corrupted by target information, at which point the targets are no longer anomalous to the background model.

A similar approach to GMRX is the cluster-based anomaly detector (CBAD) described by Carlotto (2005). The primary differences between the two are that CBAD uses k-means classification to estimate class index q for each test spectrum and then estimates $\boldsymbol{\mu}_q$ and $\boldsymbol{\Sigma}_q$ from the sample statistics of each class. The detection statistic is mathematically identical to the GMRX detection statistic in Eq. (14.60), but it exhibits a more discontinuous nature (as shown in Fig. 14.20) because the k-means classification takes no consideration of class covariance matrices.

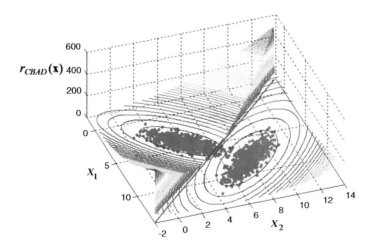

Figure 14.20 CBAD detection statistic behavior for a 2D, two-class example (Hytla *et al.*, 2010).

Application of the CBAD detection algorithm to the HYDICE Forest Radiance image is illustrated in Figs. 14.21 and 14.22. In this case, the image is segmented into four classes using the k-means algorithm, resulting in the color-coded classification map depicted in Fig. 14.21. Second-order class samples statistics are then used to form the detection statistic in Eq. (14.60), with the results depicted in Fig. 14.22. Use of k-means classification leads to better suppression of the road network and a detection performance that appears comparable to—and perhaps better than—the CSD algorithm. While it appears to remain inferior to RX and SSRX detectors for this sample image, the combination of k-means classification and class-conditional detection is still more computationally efficient relative to these spatially adaptive detectors.

To address class boundary discontinuities without resorting to the numerical and computational complexity of the GMM detection statistic and SEM classification, a modification of the CBAD called a fuzzy cluster-based anomaly detector (FCBAD) was proposed by Hytla *et al.* (2010) that uses a membership weighting function as opposed to hard class

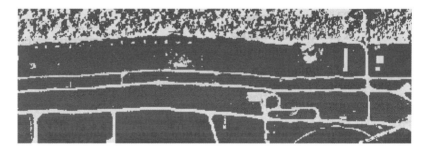

Figure 14.21 Color-coded classification map for the HYDICE Forest Radiance image based on 20 iterations of the k-means algorithm with four classes.

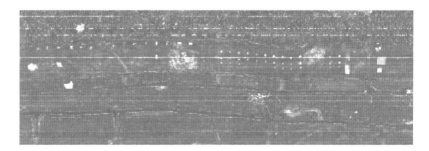

Figure 14.22 Detection statistic image of the HYDICE Forest Radiance image using a CBAD detector with four classes.

assignments. Membership function $w_q(\mathbf{x})$ is expressed as

$$w_q(\mathbf{x}) = \frac{1}{[(\mathbf{x} - \hat{\boldsymbol{\mu}}_q)^T \hat{\boldsymbol{\Sigma}}^{-1} (\mathbf{x} - \hat{\boldsymbol{\mu}}_q)]^p + 1}, \qquad (14.61)$$

where p is a tuning parameter. Membership function ranges from zero (indicating that the spectrum is distant from a particular class) to one (indicating it is central to it). Although the k-means classifier on which it is based does not consider class covariance matrices when forming class assignments, membership function is influenced by them. A higher tuning parameter strengthens its influence, although algorithm sensitivity to p is reported as fairly weak, and a value of $p = 2$ is recommended. The FCBAD detection statistic is given by

$$r_{FCBAD}(\mathbf{x}) = \sum_{q=1}^{Q} w_q(\mathbf{x})(\mathbf{x} - \hat{\boldsymbol{\mu}}_q)^T \hat{\boldsymbol{\Sigma}}_q^{-1} (\mathbf{x} - \hat{\boldsymbol{\mu}}_q), \qquad (14.62)$$

which is a weighted class-conditional MD detector. FCBAD detection statistic behavior corresponding to the 2D, two-class example is illustrated in Fig. 14.23.

Results of applying the FCBAD detector to the HYDICE Forest Radiance image is illustrated for three different tuning parameters in Fig. 14.24. The result for $p = 0$ appears very similar to the GMRX result; detection performance is poor in this example with $p = 1$ and $p = 2$. Other reported results in the literature have shown that this algorithm can

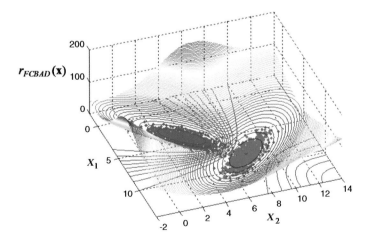

Figure 14.23 FCBAD detection statistic behavior for a 2D, two-class example (Hytla *et al.*, 2010).

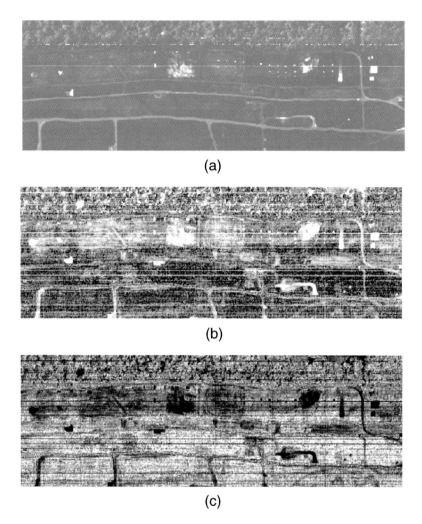

(a)

(b)

(c)

Figure 14.24 Detection statistic images of the HYDICE Forest Radiance image using an FCBAD detector for three different tuning parameters: (a) $p = 0$, (b) $p = 1$, and (c) $p = 2$.

provide better performance for cases where the targets are larger (Hytla *et al.*, 2010).

As can be seen from this section, FARs associated with hyperspectral anomaly detection algorithms tend to be fairly high. This is intuitively understandable since there are many spectrally anomalous objects in nature; therefore, anomaly detectors are limited as algorithms for finding specific objects of interest. However, such algorithms are useful as preprocessing stages for signature-matched target detection algorithms (described later in this chapter) and as tools for spectral analysts to nominate potential targets. These targets can be screened by association with other remote sensing information, such as photo-interpretation, and then used as in-scene extracted signatures for target detection algorithms. In this way, materials identified in a scene can be automatically detected in broader hyperspectral imagery with the help of an anomaly detection algorithm. It is seen in the next section that such in-scene-derived target signatures can work better than laboratory or field reflectance spectra in signature-matched detection because they circumvent errors associated with atmospheric normalization. This semi-automated target nomination approach is a common use of anomaly detection in practice.

14.3 Signature-Matched Detection

The performance of anomaly detection algorithms is fundamentally limited by the fact that no *a priori* information exists for the target signature being sought. Some improvement is possible through more-sophisticated background modeling, but FARs are often too high for practical application. With signature-matched detection there is a known reference signature for a target. This detection model usually takes the form of a reflectance spectrum measured in the laboratory or field, or perhaps a radiance spectrum measured from a different image, implying an ability to perform atmospheric and environmental normalization of a reference signature that corresponds to collected data. In this section, it is assumed that a reference spectrum for the target of interest is available and has been appropriately normalized to the data.

As in the case of anomaly detectors, a family of signature-based detectors can be developed based on the GLRT and other test statistics using different underlying background clutter models and model assumptions coupled with various parameter estimation methods. In all cases, we assume the availability of a single, deterministic reference spectrum **s** that represents the target signature normalized to hyperspectral data units. Variance or uncertainty in the reference signature is not accommodated in this section but is addressed in following sections. Table 14.2 summarizes the model assumptions for the methods described.

14.3.1 Spectral angle mapper

Consider detection problem

$$
\begin{aligned}
H_0 &: \mathbf{x} = \mathbf{b} \\
H_1 &: \mathbf{x} = \alpha\,\mathbf{s} + \mathbf{b},
\end{aligned}
\tag{14.63}
$$

where **b** represents a combination of background clutter and noise, and α is an unknown parameter that accounts for uncertainty in the strength of known signal **s**, introduced under the alternative hypothesis. Variation in α

Table 14.2 Signature-matched detection algorithm taxonomy.

Algorithm	Models			Parameters	
	Signal	Background	Noise	Known	Unknown
Spectral angle mapper	Deterministic	White	Part of background	s	α, σ
Spectral matched filter	Deterministic	Normal	Part of background	s	α, μ, Σ
Constrained energy minimization	Deterministic	Normal	Part of background	s	R
Adaptive coherence/cosine estimator	Deterministic	Normal	Part of background	s	$\alpha, \beta, \mu, \Sigma$
Class-conditional spectral matched filter	Deterministic	Normal Mixture	Part of background	s, Q	μ_q, Σ_q, q
Pairwise adaptive linear matched filter	Deterministic	Normal Mixture	Part of background	s, Q	μ_q, Σ_q, q
Orthogonal subspace projection	Deterministic	Subspace	White	s	$\alpha, \beta, \mathbf{B}, \sigma_n$

can be due to illumination variation or subpixel mixing between the target and background clutter. The random background and noise **b** is assumed to be zero mean, white, and normally distributed with an unknown variance σ^2, such that $\mathbf{b} \sim N(0, \sigma^2 \mathbf{I})$. The unknown parameters that must be estimated are α and σ^2. The MLE for α is the normalized projection

$$\hat{\alpha} = \frac{\mathbf{s}^T \mathbf{x}}{\mathbf{s}^T \mathbf{s}}. \tag{14.64}$$

Using the logarithmic form of the GLRT, the test statistic under the model assumptions is

$$r(\mathbf{x}) = \frac{\mathbf{x}^T \mathbf{x}}{\sigma^2} - \frac{(\mathbf{x} - \hat{\alpha} \, \mathbf{s})^T (\mathbf{x} - \hat{\alpha} \, \mathbf{s})}{\sigma^2}$$
$$= \frac{2\hat{\alpha} \, \mathbf{s}^T \mathbf{x} - \hat{\alpha}^2 \, \mathbf{s}^T \mathbf{s}}{\sigma^2}. \tag{14.65}$$

Inserting Eq. (14.64) into Eq. (14.65) leads to

$$r(\mathbf{x}) = \frac{1}{\sigma^2} \frac{(\mathbf{s}^T \mathbf{x})^2}{(\mathbf{s}^T \mathbf{s})}. \tag{14.66}$$

To achieve CFAR characteristics, one must estimate clutter/noise variance σ^2, which is assumed to vary spatially across the scene. A simple approach is to adaptively estimate variance by the square of spectrum magnitude

$\mathbf{x}^T\mathbf{x}$. Under this assumption, the result is

$$r(\mathbf{x}) = \frac{(\mathbf{s}^T\mathbf{x})^2}{(\mathbf{s}^T\mathbf{s})(\mathbf{x}^T\mathbf{x})},\qquad(14.67)$$

which is the square of the normalized projection of the data spectrum onto the reference spectrum. Since both the square root and inverse cosine functions are monotonic, an equivalent detection statistic is

$$r_{SAM}(\mathbf{x}) = -\cos^{-1}\left(\frac{\mathbf{s}^T\mathbf{x}}{\sqrt{(\mathbf{s}^T\mathbf{s})(\mathbf{x}^T\mathbf{x})}}\right).\qquad(14.68)$$

This is a measure of the angle between the data and reference spectral vectors, as depicted in Fig. 14.25, and is known as the spectral angle mapper (SAM). The negative sign is used because the inverse cosine function is monotonically decreasing, making small angles more target-like than large angles.

As an example of the application of signature-matched detectors, the HYDICE Forest Radiance scene is used again, but in this case, an attempt is made to detect the green tarp in the image, for which there is a reference spectrum that can be normalized to the radiance domain. The ELM and vegetation-normalization atmospheric-compensation methods described in Chapter 11 are used to normalize the laboratory reflectance reference signature. Reference signatures transformed to pupil-plane

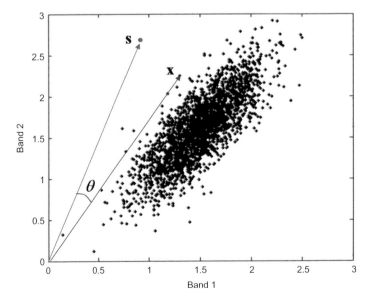

Figure 14.25 Plot of the SAM detector performance.

spectral radiance data are compared to the sample mean vector extracted from the target area in the scene in Fig. 14.26. We see that both the ELM and vegetation normalization produce a brighter reference signature relative to the actual in-scene signature, even though the estimated spectra appear to capture finer spectral features. Detection statistic images based on Eq. (14.68) resulting from the direct application of these normalized reference signatures, as well as the signature estimated from the hyperspectral data, are illustrated in Fig. 14.27, with the corresponding ROC curves compared in Fig. 14.28. Even with the in-scene signature, the SAM detector cannot detect more than 15% of the target pixels with a low FAR, although it does exhibit good detection performance for more than 80% of the target pixels. The reference signature mismatch from atmospheric compensation (or perhaps even signature ground truth) errors causes the SAM detection results based on the ELM normalized signature to be a bit worse, and the result based on the vegetation-normalized signature to be poor.

14.3.2 Spectral matched filter

While it is usually an improvement over anomaly detectors, SAM is limited in utility due to zero-mean, white background clutter assumptions. However, this limitation can be addressed by prewhitening the data. We can also theoretically extend the algorithm by removing the zero-mean, white background assumption from the detection model in Eq. (14.63) and

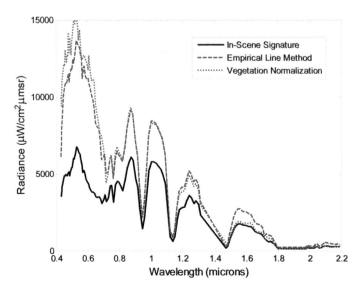

Figure 14.26 Three different reference signatures, normalized to pupil-plane spectral radiance for the green tarp in the HYDICE Forest Radiance image, used to characterize various signature-matched target detection algorithms.

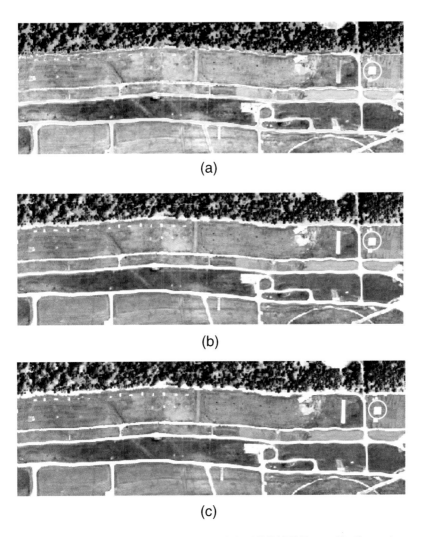

(a)

(b)

(c)

Figure 14.27 Detection statistic images of the HYDICE Forest Radiance image using an SAM detector with three different reference signatures: (a) in-scene mean spectrum, (b) empirical line method, and (c) vegetation normalization. The location of the target is circled in each image.

allowing a background distribution to be a general normal distribution, $\mathbf{b} \sim N(\boldsymbol{\mu}, \boldsymbol{\Sigma})$, where $\boldsymbol{\mu}$ and $\boldsymbol{\Sigma}$ are unknown parameters. We use sample statistics of the image as MLEs for both hypotheses, such that

$$p(\mathbf{x} \mid H_0) = \frac{1}{(2\pi)^{K/2}} \frac{1}{|\hat{\boldsymbol{\Sigma}}|^{1/2}} \exp\left\{-\frac{1}{2}(\mathbf{x} - \hat{\boldsymbol{\mu}})^T \hat{\boldsymbol{\Sigma}}^{-1} (\mathbf{x} - \hat{\boldsymbol{\mu}})\right\}, \quad (14.69)$$

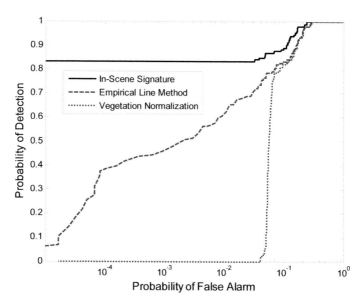

Figure 14.28 ROC curves from applying the SAM detector to the HYDICE Forest Radiance image using the three different reference signatures.

and

$$p(\mathbf{x} \mid H_1) = \frac{1}{(2\pi)^{K/2}} \frac{1}{|\hat{\boldsymbol{\Sigma}}|^{1/2}} \exp\left\{-\frac{1}{2}(\mathbf{x} - \hat{\alpha}\,\mathbf{s} - \hat{\boldsymbol{\mu}})^T \hat{\boldsymbol{\Sigma}}^{-1}\,(\mathbf{x} - \hat{\alpha}\,\mathbf{s} - \hat{\boldsymbol{\mu}})\right\}.$$
$$(14.70)$$

The MLE for α is determined by maximizing Eq. (14.61),

$$\frac{\partial}{\partial \alpha} p(\mathbf{x} \mid H_1) = 0, \qquad (14.71)$$

or minimizing the argument,

$$\frac{\partial}{\partial \alpha}\left[-\frac{1}{2}(\mathbf{x} - \hat{\alpha}\,\mathbf{s} - \hat{\boldsymbol{\mu}})^T \boldsymbol{\Sigma}^{-1}\,(\mathbf{x} - \hat{\alpha}\,\mathbf{s} - \hat{\boldsymbol{\mu}})\right] = 0. \qquad (14.72)$$

The solution is shown to be

$$\hat{\alpha} = \frac{\mathbf{s}^T \hat{\boldsymbol{\Sigma}}^{-1}(\mathbf{x} - \hat{\boldsymbol{\mu}})}{\mathbf{s}^T \hat{\boldsymbol{\Sigma}}^{-1}\mathbf{s}}, \qquad (14.73)$$

which reduces to Eq. (14.64) when $\boldsymbol{\Sigma} = \mathbf{I}$ and $\boldsymbol{\mu} = 0$.

The logarithmic form of the GLRT detection statistic in this case is identical to the linear matched filter in Eq. (14.12), except for inclusion of the parameter estimate for α,

$$r(\mathbf{x}) = \hat{\alpha}\, \mathbf{s}^T \hat{\Sigma}^{-1}\mathbf{x} - \frac{1}{2}\hat{\alpha}^2\, \mathbf{s}^T \hat{\Sigma}^{-1}\mathbf{s} - \mathbf{s}^T \hat{\Sigma}^{-1}\hat{\mu}. \qquad (14.74)$$

Inserting Eq. (14.73) into Eq. (14.74) leads to

$$r(\mathbf{x}) = -\frac{1}{2}\frac{[\mathbf{s}^T \hat{\Sigma}^{-1}(\mathbf{x} - \hat{\mu})]^2 \mathbf{s}^T \Sigma^{-1}\mathbf{s}}{\mathbf{s}^T \hat{\Sigma}^{-1}\mathbf{s}} + \frac{[\mathbf{s}^T \hat{\Sigma}^{-1}(\mathbf{x} - \hat{\mu})]^2}{\mathbf{s}^T \hat{\Sigma}^{-1}\mathbf{s}}, \qquad (14.75)$$

where the first and second terms from Eq. (14.74) are combined. Ignoring the factor of $\frac{1}{2}$ and simplifying, we arrive at the standard form for a spectral matched filter (SMF):

$$r_{SMF}(\mathbf{x}) = \frac{[\mathbf{s}^T \hat{\Sigma}^{-1}(\mathbf{x} - \hat{\mu})]^2}{\mathbf{s}^T \hat{\Sigma}^{-1}\mathbf{s}}. \qquad (14.76)$$

There are two problems with the SMF that must be considered when applying it to hyperspectral imagery. The first is that it is based on the additive model in Eq. (14.63), commonly used in communications theory; this is problematic for hyperspectral data because the expected value of $\mathbf{x}$ under the alternative hypothesis with $\alpha = 1$ is $\mathbf{s} + \mu$ when it should be $\mathbf{s}$. The problem is the presence of the background term under this hypothesis. Removing it would imply no uncertainty under the alternative hypothesis and would lead to the simple Euclidean distance (ED) detector

$$r_{ED}(\mathbf{x}) = \|\mathbf{x} - \mathbf{s}\|. \qquad (14.77)$$

This is typically a poor spectral detector unless the reference spectrum is highly accurate and unique. An alternate approach to this additive detection modeling problem is to adjust the reference spectrum to $\mathbf{s} - \mu$ to compensate for the background mean, at least in terms of expected value.

Another problem is that SMF formulation applies no constraint to parameter α as part of the MLE. Since $\alpha < 0$ would represent a spectrum in the opposite direction of $\mathbf{s}$ relative to background mean vector μ, the two-sided SMF in Eq. (14.76) would detect spectra less similar to the reference spectrum than even the background mean. This is physically not sensible for this application of the detector. A solution is to use the positive square root of Eq. (14.76), which becomes a one-sided detection statistic and alleviates this problem, since test spectra in this opposing direction would result in a negative test statistic (Eismann *et al.*, 2009). By making these two modifications to the standard form of the SMF, we arrive at a modified detection statistic,

$$r_{SMF}(\mathbf{x}) = \frac{(\mathbf{s} - \hat{\boldsymbol{\mu}})^T \hat{\boldsymbol{\Sigma}}^{-1}(\mathbf{x} - \hat{\boldsymbol{\mu}})}{\sqrt{(\mathbf{s} - \hat{\boldsymbol{\mu}})^T \hat{\boldsymbol{\Sigma}}^{-1}(\mathbf{s} - \hat{\boldsymbol{\mu}})}}, \tag{14.78}$$

which can be applied based on either global background sample statistics, or in a spatially adaptive manner using local sample statistics, as in the RX algorithm. In a global case, the denominator in Eq. (14.78) can be ignored, and the SMF can be simplified to the form

$$r_{SMF}(\mathbf{x}) = (\mathbf{s} - \hat{\boldsymbol{\mu}})^T \hat{\boldsymbol{\Sigma}}^{-1}(\mathbf{x} - \hat{\boldsymbol{\mu}}). \tag{14.79}$$

To understand the SMF, consider whitened data

$$\mathbf{z} = \hat{\boldsymbol{\Sigma}}^{-1/2}(\mathbf{x} - \hat{\boldsymbol{\mu}}) \tag{14.80}$$

and corresponding whitened reference spectrum

$$\mathbf{s}_w = \hat{\boldsymbol{\Sigma}}^{-1/2}(\mathbf{s} - \hat{\boldsymbol{\mu}}). \tag{14.81}$$

By substituting Eqs. (14.80) and (14.81) into Eq. (14.79), the detection statistic in the whitened space is shown to be

$$r_{SMF}(\mathbf{z}) = \mathbf{s}_w^T \mathbf{z}; \tag{14.82}$$

that is, the SMF is simply a spectral projection of data along the direction of the reference spectrum in whitened space, as illustrated in Fig. 14.29. In nonwhitened space, the SMF represents a balance between a linear projection in the direction of the reference spectrum, represented by the parameter estimate for α, and a linear projection in the direction of minimal background variance. Normalization in Eq. (14.78) is such that $r_{SMF}(\mathbf{x})$ equals the SNR, as defined in Eq. (14.17) when $\mathbf{x} = \mathbf{s}$.

Theoretical SMF performance follows the analysis provided in Section 14.1.2, with a slight modification to account for mean subtraction. Specifically, detection statistics in Eq. (14.78) or (14.79) under both hypotheses are normally distributed, given normal background clutter statistics, due to their linear nature. When the actual signal matches the reference, $\alpha = 1$ and the variance is given by the squared SCR,

$$SCR^2 = (\mathbf{s} - \boldsymbol{\mu})^T \boldsymbol{\Sigma}^{-1}(\mathbf{s} - \boldsymbol{\mu}). \tag{14.83}$$

The mean should be zero under the null hypothesis and equal to SCR^2 under the alternative hypothesis; the theoretical ROC curves follow

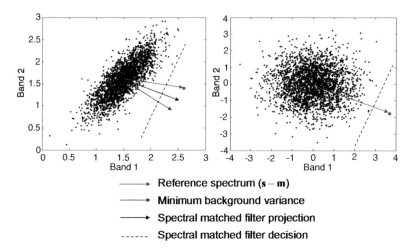

Figure 14.29 Plots of the SMF detector performance in (left) spectral space and (right) whitened data space.

those illustrated in Fig. 14.2, with the SNR replaced by the SCR in Eq. (14.83). When α is not zero, then the SCR scales linearly with α, and detection performance changes accordingly. Real background clutter, however, is typically not normally distributed, so the statistics and detector performance can vary significantly from theory. Example SMF detection statistic images are shown in Fig. 14.30 for the HYDICE Forest Radiance detection example described in the previous section using the three different reference signatures. When using the in-scene reference signature, detection performance appears excellent, with the few false alarms due mainly to some of the various panels in the scene. There is again an appreciable degradation using ELM and vegetation-normalized signatures due to the mismatch with in-scene data. This confirms the importance of good atmospheric compensation when using signature-matched target detectors. In this case, false alarms begin to be driven by sensor noise as the noisier principal components are boosted by the inverse covariance matrix.

The corresponding ROC curves are compared in Fig. 14.31. Considering the nominal SCR of 25 for the green-tarp target in the imagery, the ROC performance using the in-scene signature (which is excellent) is still inferior to model predictions because of the manmade false-alarm sources that do not conform to the theoretical normal distribution. The ROC curves for the laboratory reference signatures normalized by ELM and vegetation normalization are reasonably good down to a FAR of about 10^{-3}, at which point the spectral projection of the target falls into the noise.

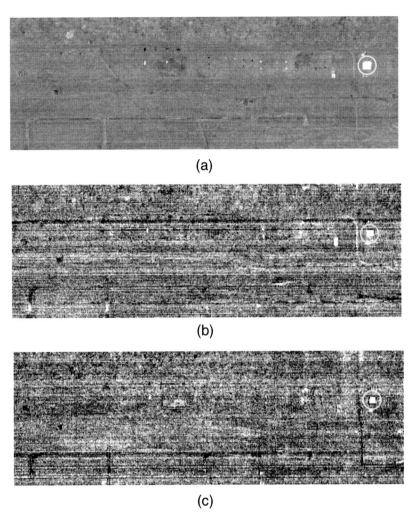

(a)

(b)

(c)

Figure 14.30 Detection statistic images of the HYDICE Forest Radiance image using an SMF detector with three different reference signatures: (a) in-scene mean spectrum, (b) empirical line method, and (c) vegetation normalization. The location of the target is circled in each image.

In the derivation of the SMF, an implicit assumption is made that the sample covariance matrix of the image provides an MLE for both the null and alternative hypotheses. It was pointed out by Kelly (1986) in an application to detect radar pulses, however, that this assumption is invalid, and therefore the SMF is not strictly a GLRT detector. Kelly provides a more theoretically correct GLRT derivation, where the additive signal model in Eq. (14.54) is again assumed, but, in this case, background $\mathbf{b}$ is explicitly assumed to be zero mean. It is also assumed that there is a training set of N spectra $\{\mathbf{y}_i\}$ that correspond to the null hypothesis, along

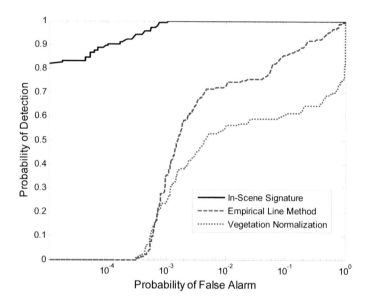

Figure 14.31 ROC curves from applying an SMF detector to the HYDICE Forest Radiance image using three different reference signatures.

with the spectrum under test **x** that can jointly be used to derive MLEs for the unknown scalar α, as well as unknown covariance matrices Σ_0 and Σ_1, corresponding to the null and alternative hypotheses.

If we define $\hat{\Sigma}$ as the estimated covariance matrix of the training data (simply the sample covariance matrix), then Σ_0 and Σ_1 are estimated by maximizing the joint probability density functions of the training and test data, which are assumed jointly normal under both hypotheses due to the assumed normality of **b**. The resulting estimates are shown by Kelly (1986) to be

$$\hat{\Sigma}_0 = \frac{1}{N+1}[\,\mathbf{x}\mathbf{x}^T + N\,\hat{\Sigma}\,] \tag{14.84}$$

and

$$\hat{\Sigma}_1 = \frac{1}{N+1}[\,(\mathbf{x} - \alpha\,\mathbf{s})\,(\mathbf{x} - \alpha\,\mathbf{s})^T + N\,\hat{\Sigma}\,]. \tag{14.85}$$

Similarly, scalar α is estimated by maximization of the joint-likelihood ratio, which leads to the same result as the SMF:

$$\hat{\alpha} = \frac{\mathbf{s}^T\hat{\Sigma}^{-1}\mathbf{x}}{\mathbf{s}^T\hat{\Sigma}^{-1}\mathbf{s}}. \tag{14.86}$$

The resulting GLRT detector can now be derived by substituting these MLEs into joint-likelihood ratio $l(\mathbf{x})$. To make it more analytically tractable, however, it is useful to apply a detector of the form

$$r(\mathbf{x}) = 1 - \frac{1}{[\,l(\mathbf{x})\,]^{1/N+1}}, \tag{14.87}$$

which is equivalent to $l(\mathbf{x})$ in terms of detector performance because the transformation from $l(\mathbf{x})$ to $r(\mathbf{x})$ is monotonic, assuming that the detector threshold is transformed in an equivalent manner. The resulting detector, known as the Kelly detector, has the form

$$r_K(\mathbf{x}) = \frac{(\mathbf{s}^T \hat{\boldsymbol{\Sigma}}^{-1} \mathbf{x})^2}{(\mathbf{s}^T \hat{\boldsymbol{\Sigma}}^{-1} \mathbf{s})\,(N + \mathbf{x}^T \hat{\boldsymbol{\Sigma}}^{-1} \mathbf{x})}. \tag{14.88}$$

By considering application of this detector to hyperspectral imagery, the training data would correspond to the image itself. Assuming that the targets are rare enough that they do not contaminate the estimate for $\boldsymbol{\Sigma}$, and that the number of samples N is large, we see that the Kelly algorithm converges to the two-sided SMF, leaving little if any practical advantage of the theoretically correct GLRT formulation. On the other hand, this formulation can be extended to a situation where an additional scalar parameter is incorporated to account for unknown scaling of background/noise statistics, giving rise to the adaptive coherence/cosine estimator (discussed further in Section 14.3.4). This detector provides practical advantages, as the model can account for subpixel targets where a portion of the background is linearly mixed with the target spectrum.

14.3.3 Constrained energy minimization

As the SMF is a linear filter, it is reasonable to consider the general linear function

$$r(\mathbf{x}) = \mathbf{h}^T \mathbf{x} \tag{14.89}$$

and ask whether vector $\mathbf{h}$ can be chosen to optimize target detection performance by some other detection criterion. One such optimization criterion is to minimize the projected background energy, subject to the constraint that the detection statistic for $\mathbf{x} = \mathbf{s}$ is unity. Energy is defined as

$$E = \frac{1}{N} \sum_{i=1}^{N} r^2(\mathbf{x}_i)$$

$$= \frac{1}{N} \sum_{i=1}^{N} \mathbf{h}^T \mathbf{x}_i \mathbf{x}_i^T \mathbf{h}$$

$$= \mathbf{h}^T \left(\frac{1}{N} \sum_{i=1}^{N} \mathbf{x}_i \mathbf{x}_i^T \right) \mathbf{h}$$

$$= \mathbf{h}^T \mathbf{R} \mathbf{h}, \tag{14.90}$$

where $\mathbf{R}$ is the sample correlation matrix. The optimization problem based on this criterion becomes

$$\min_{\mathbf{h}} \mathbf{h}^T \mathbf{R} \mathbf{h} \quad \text{subject to } \mathbf{h}^T \mathbf{s} = 1, \tag{14.91}$$

which has the solution

$$\mathbf{h} = \frac{\mathbf{R}^{-1} \mathbf{s}}{\mathbf{s}^T \mathbf{R}^{-1} \mathbf{s}}. \tag{14.92}$$

The resulting detection statistic,

$$r_{CEM}(\mathbf{x}) = \frac{\mathbf{s}^T \mathbf{R}^{-1} \mathbf{x}}{\mathbf{s}^T \mathbf{R}^{-1} \mathbf{s}}, \tag{14.93}$$

is known as the constrained energy minimization (CEM) detector (Harsanyi, 1993). If the sample mean vector is subtracted from the image data prior to application of the SMF in Eq. (14.78), then the one-sided SMF can be written as

$$r_{SMF}(\mathbf{x}) = \frac{\mathbf{s}^T \mathbf{R}^{-1} \mathbf{x}}{\sqrt{\mathbf{s}^T \mathbf{R}^{-1} \mathbf{s}}}, \tag{14.94}$$

since the mean-subtracted sample correlation matrix equals the sample covariance matrix. These detection statistics are the same to within the normalization constant in the denominator, meaning that detection performance will be identical. When the mean vector is not removed from the data, however, there can be a difference between CEM and SMF results.

14.3.4 Adaptive coherence/cosine estimator

The additive model forming the basis of the SMF is problematic when applied to hyperspectral data because of the presence of background clutter under the alternative hypothesis. If targets are spatially resolved by an imaging spectrometer, a replacement or substitution model is more appropriate. When adopting such a model, it is necessary to incorporate an appropriate model for the uncertainty or randomness present under the alternative hypothesis; this generally requires information concerning target signature variance that is often unavailable. On the other hand, the possibility of capturing a fraction of background clutter in target pixels would introduce a random component as well. This would occur if the targets are subpixel, meaning that their physical size is less than the GRD of the imaging spectrometer, or when a hyperspectral pixel falls on the edge of a target and captures some of the surrounding background clutter.

One possible model that can be used to account for subpixel targets has the form

$$H_0 : \mathbf{x} = \mathbf{b}$$
$$H_1 : \mathbf{x} = \alpha \mathbf{s} + \beta \mathbf{b}, \tag{14.95}$$

where $\mathbf{b} \sim N(\boldsymbol{\mu}, \boldsymbol{\Sigma})$, $\mathbf{s}$ is a known reference spectrum, and α and β are unknown parameters. To simplify the mathematics, it is common to assume that the sample mean vector $\boldsymbol{\mu}$ is removed from the imagery prior to processing, such that $\mathbf{b} \sim N(0, \boldsymbol{\Sigma})$. If $\boldsymbol{\mu}$ as removed, it is necessary to remove the mean vector from reference spectrum $\mathbf{s}$ as well. Therefore, we can derive the detector under the zero-mean assumption if the following substitutions are made:

$$\mathbf{x} \rightarrow \mathbf{x} - \hat{\boldsymbol{\mu}}, \tag{14.96}$$

and

$$\mathbf{s} \rightarrow \mathbf{s} - \hat{\boldsymbol{\mu}}. \tag{14.97}$$

Suppose that data are further whitened by the operation $\mathbf{z} = \boldsymbol{\Sigma}^{-1/2}\mathbf{x}$. In this case, the null and alternative hypotheses remain the same as in Eq. (14.95), but with background clutter $\mathbf{b}$ now being a zero-mean, white random vector process. Except for unknown parameter β under the alternative hypothesis, this resembles the SAM detector model in Eq. (14.63).

Another detector, developed again for radar signal processing applications but that has proven useful for hyperspectral target detection applications, is the adaptive coherence/cosine estimator (ACE) described by Manolakis (2005) and Manolakis and Shaw (2002). It has been shown by Kraut and Scharf (1999) that the ACE detector can be derived by applying the GLRT formulation from Kelly outlined in Section 14.3.2. In this extension, the signal model includes an unknown scaling β of the noise/background term $\mathbf{b}$ under both the null and alternative hypotheses.

The signal model takes the form

$$H_0 : \mathbf{x} = \beta \mathbf{b}$$
$$H_1 : \mathbf{x} = \alpha \mathbf{s} + \beta \mathbf{b}, \tag{14.98}$$

which is similar to the subpixel model in Eq. (14.95), but with an unknown scalar in both the null and alternative hypotheses. It is assumed that $\mathbf{b} \sim N(0, \Sigma)$.

ACE detector derivation proceeds in the same manner as derivation of the Kelly detector, including the assumption of a training set of noise/background data, with the exception that now unknown parameter β must also be estimated under both hypotheses. All unknown parameters except α are again estimated based on maximizing the joint probability density function of the training and test spectra. In this case, the parameters become

$$\hat{\Sigma}_0 = \frac{1}{N+1} \left[\frac{1}{\beta_0^2} \mathbf{x}\mathbf{x}^T + N \hat{\Sigma} \right], \tag{14.99}$$

$$\hat{\Sigma}_1 = \frac{1}{N+1} \left[\frac{1}{\beta_1^2} (\mathbf{x} - \alpha \mathbf{s})(\mathbf{x} - \alpha \mathbf{s})^T + N \hat{\Sigma} \right], \tag{14.100}$$

$$\hat{\beta}_0^2 = \frac{N-K+1}{NK} \mathbf{x}^T \hat{\Sigma}^{-1} \mathbf{x}, \tag{14.101}$$

and

$$\hat{\beta}_1^2 = \frac{N-K+1}{NK} (\mathbf{x} - \alpha \mathbf{s})^T \hat{\Sigma}^{-1} (\mathbf{x} - \alpha \mathbf{s}), \tag{14.102}$$

where $\hat{\Sigma}$ is again the sample covariance matrix of N training spectra. Unknown scalar α is given again by Eq. (14.86). The GLRT detector based on these estimates becomes the ACE detector:

$$r_{ACE}(\mathbf{x}) = \frac{(\mathbf{s}^T \hat{\Sigma}^{-1} \mathbf{x})^2}{(\mathbf{s}^T \hat{\Sigma}^{-1} \mathbf{s})(\mathbf{x}^T \hat{\Sigma}^{-1} \mathbf{x})}, \tag{14.103}$$

which is a limiting form of the Kelly detector, where N is zero.

When background clutter conforms to an assumed normal model, the cotangent form of the ACE detection statistic exhibits a central F distribution with degrees of freedom $n = 1$ and $p = K - 2$ under the null hypothesis,

$$\frac{r_{ACE}(\mathbf{x})}{1 - r_{ACE}(\mathbf{x})} \bigg| H_0 \sim F_{n,p}(r), \tag{14.104}$$

where the probability density function for a central F distribution is given by

$$p(r; n, m) = \frac{\Gamma\left(\frac{n+m}{2}\right) n^{n/2} m^{m/2}}{\Gamma\left(\frac{n}{2}\right) \Gamma\left(\frac{m}{2}\right)} \frac{r^{n/2-1}}{(m + n\,r)^{(n+m)/2}}. \qquad (14.105)$$

Under the alternative hypothesis,

$$\frac{r_{ACE}(\mathbf{x})}{1 - r_{ACE}(\mathbf{x})} \Big| H_1 \sim F_{n,p}\left(r; \lambda^2\right), \qquad (14.106)$$

where the probability density function for a noncentral F distribution is given by

$$
\begin{aligned}
p(r; n, m, \lambda^2) &= \sum_{p=0}^{\infty} e^{-\lambda^2/2} \frac{(\lambda^2/2)^p}{p!} \frac{\Gamma\left(\frac{n+m}{2} + p\right)}{\Gamma\left(\frac{n}{2} + p\right) \Gamma\left(\frac{m}{2}\right)} \frac{n^{n/2+p} m^{m/2} r^{n/2-1+p}}{(m + n\,r)^{[(n+m)/2]+p}} \\
&= \frac{\Gamma\left(\frac{n}{2}\right) \Gamma\left(1 + \frac{m}{2}\right)}{B\left(\frac{n}{2}, \frac{m}{2}\right) \Gamma\left(\frac{n+m}{2}\right)} \frac{n^{n/2} m^{m/2} r^{n/2-1}}{(m + n\,r)^{(n+m)/2}} \\
&\quad \times e^{-(\lambda^2/2)+(n\lambda^2 r)/[2(m+nr)]} L_{m/2}^{n/2-1}\left(-\frac{n\lambda^2 r}{2(m + nr)}\right), \qquad (14.107)
\end{aligned}
$$

where $B(\alpha, \beta)$ is the beta function, $L_n^m(z)$ is a generalized Laguerre polynomial, and noncentrality parameter λ^2 is the squared SCR:

$$\lambda^2 = \text{SCR}^2 = \alpha^2 \mathbf{s}^T \Sigma^{-1} \mathbf{s}. \qquad (14.108)$$

Equation (14.108) is identical to Eq. (14.83) if the reference spectrum is adjusted to account for mean subtraction as in Eq. (14.97). This is often called the signal-to-clutter plus noise ratio, because sensor noise is incorporated into the background clutter model.

An ACE detection statistic image for the HYDICE Forest Radiance example is illustrated in Fig. 14.32, with corresponding ROC curves in Fig. 14.33. When using the in-scene reference signature, detection performance is exceptionally good for this image. This is consistent with reports in the literature and is likely due to the appropriateness of the substitution model that appears to make the ACE detector more spectrally selective than the SMF detector. However, these results are inferior to those obtained with the SMF detector when using normalized laboratory signatures. A penalty with enhanced spectral selectivity appears in this example to be greater sensitivity to errors in the reference signature. This could be due to the zero-mean assumption in the ACE detector derivation.

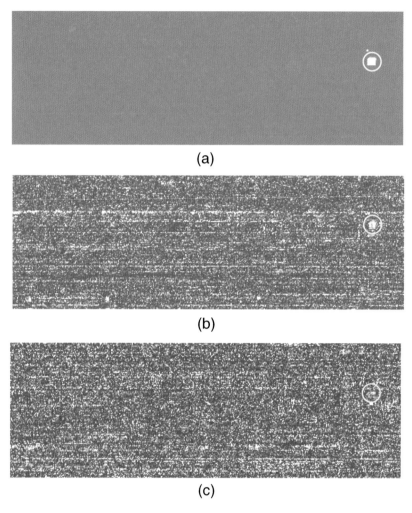

(a)

(b)

(c)

Figure 14.32 Detection statistic images of the HYDICE Forest Radiance image using an ACE detector with three different reference signatures: (a) in-scene mean spectrum, (b) empirical line method, and (c) vegetation normalization. Location of the target is circled in each image.

Note that the direction of the mean-removed signature vector in Eq. (14.97) differs from that of the original signature; thus, the signature variation with scaling parameter α in the alternative hypothesis is not properly modeled.

14.3.5 Subpixel spectral matched filter

When considering applications for hyperspectral imagery, the underlying signal model for the ACE detector in Eq. (14.98) is arguably not well matched to the expected characteristics of the data in three respects. First, it ignores the nonzero mean vector of the background clutter that would need

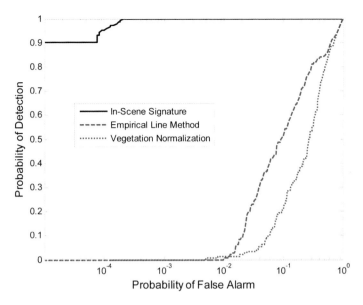

Figure 14.33 ROC curves from applying an ACE detector to the HYDICE Forest Radiance image using three different reference signatures.

to be estimated in GLRT formulation under both hypotheses, as is done for the covariance matrix. Simply subtracting the mean from the entire image and redefining the image and reference spectra by Eqs. (14.96) and (14.97) does not correctly account for the mean vector. This can be seen by recognizing that the direction of the mean-subtracted reference spectrum is different from that of the actual reference spectrum, meaning that scaled versions observed under the alternative hypothesis would differ in the mean-subtracted model relative to what linear mixing would produce. Second, the subpixel model in Eq. (14.95) is arguably a more appropriate model than that in Eq. (14.98). The ACE formulation in Eq. (14.98) includes an unknown scalar β in the null hypothesis that does not seem appropriate. Finally, no background training data are typically available, and statistics must be estimated from the data, which can include the presence of targets among background samples.

By retaining the mean vector in the normal model for **b**, a detector based on the subpixel hypothesis tests expressed in Eq. (14.95) can be approached in the same way as an SMF derivation, but with the inclusion of additional unknown scalar β in the alternative hypothesis (Meola, 2011a). Mean vector μ and covariance matrix Σ are estimated from image sample statistics and are assumed to represent random vector **b** under both hypotheses. MLEs for α and β under the alternative hypothesis can be determined by maximizing the log-likelihood function

$$\frac{\partial}{\partial \alpha} p(\mathbf{x} \mid H_1, \alpha, \beta) = \frac{\partial}{\partial \beta} p(\mathbf{x} \mid H_1, \alpha, \beta) = 0, \tag{14.109}$$

where

$$p(\mathbf{x} \mid H_1, \alpha, \beta) = \frac{1}{\beta^K (2\pi)^{K/2}} \frac{1}{|\hat{\Sigma}|^{1/2}}$$

$$\times \exp\left\{-\frac{1}{2\beta^2}(\mathbf{x} - \alpha \mathbf{s} - \beta \hat{\boldsymbol{\mu}})^T \hat{\Sigma}^{-1} (\mathbf{x} - \alpha \mathbf{s} - \beta \hat{\boldsymbol{\mu}})\right\}. \tag{14.110}$$

By substituting Eq. (14.110) into Eq. (14.109) and solving, MLEs become

$$\hat{\alpha} = \frac{\mathbf{s}^T \hat{\Sigma}^{-1} (\mathbf{x} - \hat{\beta} \boldsymbol{\mu})}{\mathbf{s}^T \hat{\Sigma}^{-1} \mathbf{s}} \tag{14.111}$$

and

$$\hat{\beta} = \frac{-a_1 \pm \sqrt{a_1^2 - 4a_2 a_0}}{2a_2}, \tag{14.112}$$

where

$$a_0 = (\mathbf{x}^T \hat{\Sigma}^{-1} \mathbf{x}) (\mathbf{s}^T \hat{\Sigma}^{-1} \mathbf{s}) - (\mathbf{s}^T \hat{\Sigma}^{-1} \mathbf{x})^2, \tag{14.113}$$

$$a_1 = (\mathbf{s}^T \hat{\Sigma}^{-1} \mathbf{x}) (\mathbf{s}^T \hat{\Sigma}^{-1} \hat{\boldsymbol{\mu}}) - (\mathbf{s}^T \hat{\Sigma}^{-1} \mathbf{s}) (\hat{\boldsymbol{\mu}}^T \hat{\Sigma}^{-1} \mathbf{x}), \tag{14.114}$$

and

$$a_2 = -K (\mathbf{s}^T \hat{\Sigma}^{-1} \mathbf{s}). \tag{14.115}$$

The subpixel SMF based on these MLEs then becomes

$$r_{SP-SMF}(\mathbf{x}) = (\mathbf{x} - \hat{\boldsymbol{\mu}})^T \hat{\Sigma}^{-1} (\mathbf{x} - \hat{\boldsymbol{\mu}})$$
$$- \frac{(\mathbf{x} - \hat{\alpha} \mathbf{s} - \hat{\beta} \hat{\boldsymbol{\mu}})^T \hat{\Sigma}^{-1} (\mathbf{x} - \hat{\alpha} \mathbf{s} - \hat{\beta} \hat{\boldsymbol{\mu}})}{\hat{\beta}^2} - 2K \ln \hat{\beta}. \tag{14.116}$$

The detection statistic image for the HYDICE Forest Radiance example is illustrated for the subpixel SMF detector in Fig. 14.34, with the corresponding ROC curves in Fig. 14.35. When using the in-scene reference signature, detection performance is slightly inferior to both the ACE and SMF detectors. The performance holds up much better

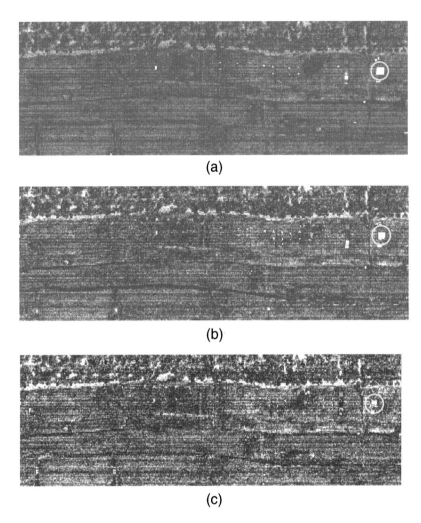

(a)

(b)

(c)

Figure 14.34 Detection statistic images of the HYDICE Forest Radiance image using a subpixel SMF detector with three different reference signatures: (a) in-scene mean spectrum, (b) empirical line method, and (c) vegetation normalization. The location of the target is circled in each image.

than ACE with normalized reflectance signatures and provides detection performance commensurate with (but still inferior to) SMF. This is not necessarily surprising in this case, because the target is well resolved spatially; the penalty for possibly better representation of edge pixels by the subpixel model is a moderately increased FAR. It is interesting that false alarms in the subpixel SMF case tend to be less random, and they appear more correlated to other manmade objects in the scene that could perhaps be filtered out using false-alarm mitigation approaches discussed in Section 14.4. In fact, formulation of the subpixel SMF is similar to that

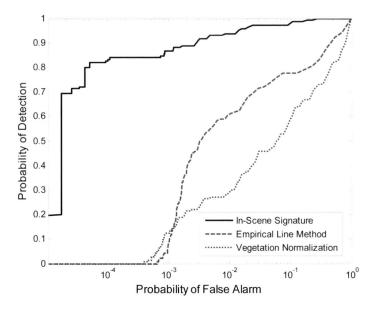

Figure 14.35 ROC curves from applying a subpixel GLRT detector to the HYDICE Forest Radiance image using three different reference signatures.

of the finite target matched filter described in Section 14.4.4. In that case, the mixture coefficients are further constrained, such that $\beta = 1 - \alpha$, and the target reference is assumed to exhibit some randomness described by a scaled version of the background covariance matrix.

14.3.6 Spectral matched filter with normal mixture model

The SMF can be extended to a case where a normal mixture model is used to describe the background as opposed to using unimodal normal distribution. In this case, the detection problem has the same form as Eq. (14.63) but with **b** modeled by a normal mixture distribution. Based on SMF assumptions, including renormalizing the reference spectrum to remove the clutter mean estimate, the conditional probability density functions are

$$p(\mathbf{x} \mid H_0, q) = \frac{1}{(2\pi)^{K/2}} \frac{1}{|\Sigma_q|^{1/2}} \exp\left\{ -\frac{1}{2}(\mathbf{x} - \boldsymbol{\mu}_q)^T \Sigma_q^{-1} (\mathbf{x} - \boldsymbol{\mu}_q) \right\}$$

$$(14.117)$$

and

$$p(\mathbf{x} \mid H_1, q) = \frac{1}{(2\pi)^{K/2}} \frac{1}{|\Sigma_q|^{1/2}} \exp\left\{ -\frac{1}{2}(\mathbf{x} - \mathbf{s})^T \Sigma_q^{-1} (\mathbf{x} - \mathbf{s}) \right\}.$$

$$(14.118)$$

Using class parameter estimates from whatever image classification method is employed, the class-conditional (CC) SMF becomes

$$r_{CCSMF}(\mathbf{x}) = (\mathbf{s} - \hat{\boldsymbol{\mu}}_q)^T \hat{\boldsymbol{\Sigma}}_q^{-1} (\mathbf{x} - \hat{\boldsymbol{\mu}}_q). \qquad (14.119)$$

The CCSMF assumes background classification as *a priori* information for the detector. An alternate approach is to treat class index q as an unknown parameter of the background model. In that case, the GLRT can be used with the probability density functions given in Eqs. (14.108) and (14.109). Again, assuming that the covariance matrix for the target hypothesis is always the same as that for the background hypothesis, this results in the pair-wise adaptive linear matched (PALM) filter detection statistic,

$$r_{PALM}(\mathbf{x}) = \min_q (\mathbf{s} - \hat{\boldsymbol{\mu}}_q)^T \hat{\boldsymbol{\Sigma}}_q^{-1} (\mathbf{x} - \hat{\boldsymbol{\mu}}_q); \qquad (14.120)$$

that is, the SMF for the most similar background class is used. One modification of the GLRT approach is to limit the range of classes for which Eq. (14.120) is minimized to those in the local spatial neighborhood for the pixel under test.

Ideally, the SMF based on a normal mixture model would outperform a standard SMF based on the unimodal normal model because of the more sophisticated background model. In practice, however, such an improvement is not always seen, and performance can even degrade. One reason for this behavior is that parameters of the normal mixture model must be estimated from data and can be contaminated by target spectra in the scene. This can make the background model more target-like, and thus diminish target-to-background separability. Potential for contamination is less severe for a simple multivariate normal model because of the rarity of target pixels relative to background pixels. As the image is divided into classes, however, targets can become more significant contributors to individual class statistics, especially for less populated classes.

A comparison of the CCSMF and PALM algorithms applied to the HYDICE Forest Radiance image is given by detection statistic images in Fig. 14.36, along with ROC curves in Fig. 14.37. These results are based on the four-class, k-means classification map depicted in Fig. 14.21 and the in-scene reference signature. The CCSMF results are very poor due to a large number of false alarms caused by spectra that are outliers from their assigned classes. This is caused by the mismatch between the linear classifier and detection statistic based on second-order class statistics. The PALM ROC curve in this case is closer but still inferior to SMF results.

Corresponding CCSMF and PALM results, based on the four-class SEM classification map depicted in Fig. 14.16, are provided in Figs. 14.38 and

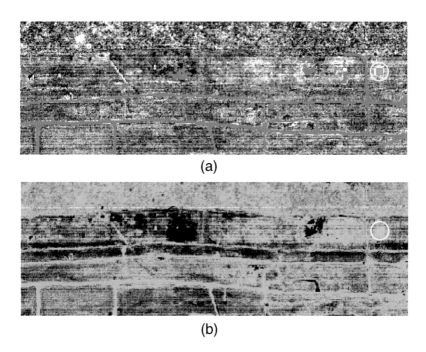

(a)

(b)

Figure 14.36 Detection statistic images of the HYDICE Forest Radiance image using (a) CCSMF and (b) PALM detectors with the in-scene reference spectrum and four-class k-means classification. The location of the target is circled in each image.

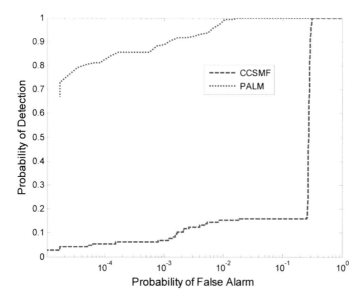

Figure 14.37 ROC curves from applying CCSMF and PALM detectors to the HYDICE Forest Radiance image using the in-scene reference spectrum and four-class k-means classification.

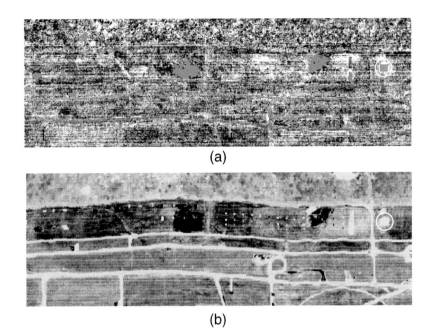

(a)

(b)

Figure 14.38 Detection statistic images of the HYDICE Forest Radiance image using (a) CCSMF and (b) PALM detectors with the in-scene reference spectrum and four-class SEM classification. The location of the target is circled in each image.

14.39. CCSMF results are again very poor, due in this case to overlapping SEM classes. That is, the classification algorithm still assigns spectra to classes when the spectra are somewhat distant and anomalous to the classes. Using SEM classification improves the PALM results slightly, to the point where detection performance is commensurate with that of an SMF detector. The benefit of the added complexity of SEM classification, however, is questionable in this example.

14.3.7 Orthogonal subspace projection

The orthogonal subspace projection algorithm is based on a subspace model for background data as opposed to a statistical model. Consider the detection problem

$$
\begin{aligned}
H_0 &: \mathbf{x} = \mathbf{B}\,\boldsymbol{\beta} + \mathbf{n} \\
H_1 &: \mathbf{x} = \alpha\,\mathbf{s} + \mathbf{B}\,\boldsymbol{\beta} + \mathbf{n},
\end{aligned}
\tag{14.121}
$$

where $\mathbf{B}$ represents a background subspace, $\boldsymbol{\beta}$ is the background basis coefficient vector, α is an unknown scalar, and $\mathbf{n}$ is zero-mean, spectrally independent, normally distributed noise with a standard deviation σ_n.

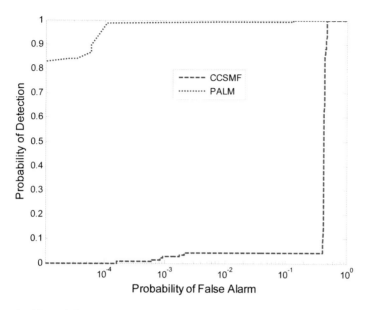

Figure 14.39 ROC curves from applying CCSMF and PALM detectors to the HYDICE Forest Radiance image using the in-scene reference spectrum and four-class SEM classification.

Assume that background subspace $\mathbf{B}$ is estimated from the data and $\mathbf{s}$ is a known reference spectrum. Define subspace projection operator $\mathbf{P}_B$ as

$$\mathbf{P_B} = \mathbf{B}(\mathbf{B}^T\mathbf{B})^{-1}\mathbf{B}^T, \tag{14.122}$$

and its orthogonal complement as

$$\mathbf{P_B^\perp} = \mathbf{I} - \mathbf{P_B}. \tag{14.123}$$

It can be shown that the MLE for unknown constant α is

$$\hat{\alpha} = \frac{\mathbf{s}^T\mathbf{P_B^\perp}\mathbf{x}}{\mathbf{s}^T\mathbf{P_B^\perp}\mathbf{s}}. \tag{14.124}$$

Since Eq. (14.107) characterizes a relative amount of reference spectrum that a data vector is estimated to contain, it is a reasonable detection statistic. The orthogonal subspace projection (OSP) algorithm (Harsanyi and Chang, 1994) merely uses the numerator of the estimate as a discriminant function, or

$$r_{OSP}(\mathbf{x}) = \mathbf{s}^T\mathbf{P_B^\perp}\mathbf{x}. \tag{14.125}$$

The OSP algorithm can be alternatively derived by finding optimal linear filter $\mathbf{h}$ in the orthogonal background subspace that maximizes the SNR. Since

$$\mathbf{h}^T \mathbf{P}_{\mathbf{B}}^\perp \mathbf{x} = \mathbf{h}^T \mathbf{P}_{\mathbf{B}}^\perp \alpha \mathbf{s} + \mathbf{h}^T \mathbf{P}_{\mathbf{B}}^\perp \mathbf{n} \qquad (14.126)$$

under the alternative hypothesis, the SNR can be expressed as

$$\begin{aligned}
\text{SNR} &= \frac{\mathbf{h}^T \mathbf{P}_{\mathbf{B}}^\perp \mathbf{s} \alpha^2 \mathbf{s}^T (\mathbf{P}_{\mathbf{B}}^\perp)^T \mathbf{h}}{\mathbf{h}^T \mathbf{P}_{\mathbf{B}}^\perp E(\mathbf{n}\mathbf{n}^T)(\mathbf{P}_{\mathbf{B}}^\perp)^T \mathbf{h}} \\
&= \frac{\alpha^2}{\sigma_n^2} \frac{\mathbf{h}^T \mathbf{P}_{\mathbf{B}}^\perp \mathbf{s}\mathbf{s}^T (\mathbf{P}_{\mathbf{B}}^\perp)^T \mathbf{h}}{\mathbf{h}^T \mathbf{P}_{\mathbf{B}}^\perp (\mathbf{P}_{\mathbf{B}}^\perp)^T \mathbf{h}}.
\end{aligned} \qquad (14.127)$$

Maximization of Eq. (14.127) results in the simple solution $\mathbf{h} = \mathbf{s}$.

A GLRT approach can also be taken to derive detectors based on the underlying model expressed in Eq. (14.121) in cases where noise variance is either known or unknown (Scharf and Friedlander, 1994). First, consider a known-noise case where the parameter estimate for α is given by Eq. (14.116) and the parameter estimate for $\boldsymbol{\beta}$ is given by

$$\hat{\boldsymbol{\beta}}_0 = (\mathbf{B}^T \mathbf{B})^{-1} \mathbf{B}^T \mathbf{x} \qquad (14.128)$$

under the null hypothesis, and

$$\hat{\boldsymbol{\beta}}_1 = (\mathbf{B}^T \mathbf{P}_{\mathbf{s}}^\perp \mathbf{B})^{-1} \mathbf{B}^T \mathbf{P}_{\mathbf{s}}^\perp \mathbf{x} \qquad (14.129)$$

under the alternative hypothesis, where

$$\mathbf{P}_{\mathbf{s}} = \mathbf{s}\mathbf{s}^T \qquad (14.130)$$

and

$$\mathbf{P}_{\mathbf{s}}^\perp = \mathbf{I} - \mathbf{s}\mathbf{s}^T. \qquad (14.131)$$

To constrain α to positive values under the alternative hypothesis, parameter estimates are modified to

$$\hat{\alpha}_1 = \max(0, \hat{\alpha}) \qquad (14.132)$$

and

$$\hat{\boldsymbol{\beta}}_1 = \begin{cases} \hat{\boldsymbol{\beta}}_0 & \hat{\alpha} \leq 0 \\ (\mathbf{B}^T \mathbf{P}_{\mathbf{s}}^\perp \mathbf{B})^{-1} \mathbf{B}^T \mathbf{P}_{\mathbf{s}}^\perp \mathbf{x} & \hat{\alpha} > 0 \end{cases}. \qquad (14.133)$$

When noise variance is known, a detection statistic of the form

$$r_1(\mathbf{x}) = 2 \ln \hat{l}(\mathbf{x}) = \frac{1}{\sigma_n^2}(\| \hat{\mathbf{n}}_0 \|^2 - \| \hat{\mathbf{n}}_1 \|^2) \qquad (14.134)$$

can be used. Noise estimates under the two hypotheses composing Eq. (14.134) are

$$\hat{\mathbf{n}}_1 = \begin{cases} \mathbf{P}_{\mathbf{B}}^{\perp}\mathbf{x} & \hat{\alpha} \leq 0 \\ \mathbf{P}_{\mathbf{sB}}^{\perp}\mathbf{x} & \hat{\alpha} > 0 \end{cases} \qquad (14.135)$$

and

$$\hat{\mathbf{n}}_0 = \mathbf{P}_{\mathbf{B}}^{\perp}\mathbf{x}, \qquad (14.136)$$

where $\mathbf{sB}$ corresponds to the matrix formed by adding an additional column $\mathbf{s}$ to background subspace matrix $\mathbf{B}$. Inserting Eqs. (14.135) and (14.136) into Eq. (14.134) results in the GLRT detection statistic

$$r_1(\mathbf{x}) = \begin{cases} 0 & \hat{\alpha} \leq 0 \\ \dfrac{1}{\sigma_n^2}\mathbf{x}^T(\mathbf{P}_{\mathbf{B}}^{\perp} - \mathbf{P}_{\mathbf{sB}}^{\perp})\,\mathbf{x} & \hat{\alpha} > 0 \end{cases}. \qquad (14.137)$$

When the noise variance is not known, and, therefore cannot be assumed to be spatially stationary, an adaptive form of the GLRT,

$$r_2(\mathbf{x}) = [\,\hat{l}(\mathbf{x})\,]^{2/K} - 1 = \frac{\| \hat{\mathbf{n}}_0 \|^2}{\| \hat{\mathbf{n}}_1 \|^2} - 1, \qquad (14.138)$$

can be used such that the unknown noise variance cancels out. In this case, the GLRT detection statistic becomes

$$r_2(\mathbf{x}) = \begin{cases} 0 & \hat{\alpha} \leq 0 \\ \dfrac{\mathbf{x}^T(\mathbf{P}_{\mathbf{B}}^{\perp} - \mathbf{P}_{\mathbf{sB}}^{\perp})\,\mathbf{x}}{\mathbf{x}^T\mathbf{P}_{\mathbf{sB}}^{\perp}\,\mathbf{x}} & \hat{\alpha} > 0 \end{cases}. \qquad (14.139)$$

Known-noise version $r_1(\mathbf{x})$ measures the difference in projected energy between the signal-plus-clutter and clutter-only subspaces and compares it to the noise level. Noise-adaptive version $r_2(\mathbf{x})$ estimates noise level by an orthogonal signal-plus-clutter projection.

Both forms of the GLRT OSP detectors, $r_1(\mathbf{x})$ and $r_2(\mathbf{x})$, are compared to the OSP detector in Eq. (14.125) for the HYDICE Forest Radiance example, with background subspace determined by the ten leading principal-component eigenvectors. Detection statistic images are compared in Fig. 14.40, and an ROC comparison is provided in Fig. 14.41. The in-scene reference signature is used for all three cases. For this example, performance is slightly inferior to both the SMF and ACE detectors. The OSP and GLRT detectors yield almost identical results, while the adaptive GLRT result is somewhat comparable. The inferior detection performance is due to background clutter leakage outside the ten-dimensional subspace used to model it.

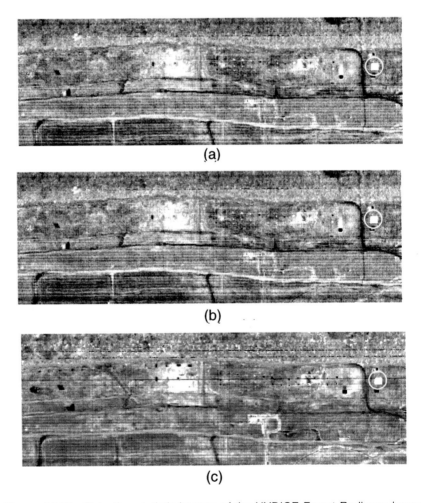

(a)

(b)

(c)

Figure 14.40 Detection statistic images of the HYDICE Forest Radiance image based on the in-scene reference signature using three different orthogonal subspace projection detectors: (a) OSP, (b) GLRT, and (c) adaptive GLRT. The location of the target is circled in each image.

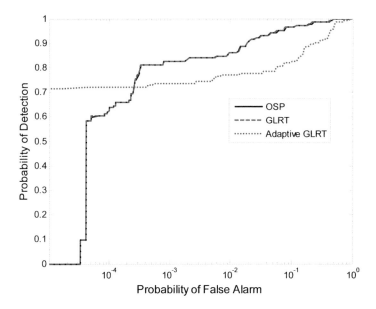

Figure 14.41   ROC comparison of three orthogonal subspace detectors for the HYDICE Forest Radiance image based on the in-scene reference spectrum.

14.4 False-Alarm Mitigation

A limitation of signature-matched detectors such as SMF is that linear detection statistics can be prone to high FARs due to anomalous background spectra that are poorly described by background models underlying the detector, even though these anomalous spectra also are not a good match to the reference spectrum. Since there is no model that accounts for uncertainty or randomness in a target, there is no basis for deciding whether such spectra are indeed anomalous background spectra or perhaps anomalous target spectra. Both can have strong projection in the direction of the reference spectrum but can also be a poor fit to the background model. Thus, the detection statistic is large in both cases, and the spectra are declared targets, even when they might exhibit little spectral resemblance to the reference. This is illustrated for the case of an SMF in Fig. 14.42, which depicts three spectra x_1, x_2, and x_3 with an equal SMF detection statistic, even as they successively deviate further from reference s. Intuitively, it seems that one could employ some process, perhaps after applying the detection threshold, to filter out detections having a large spectral deviation from the reference. This is the concept termed false-alarm mitigation (FAM) that is explored in this section.

14.4.1 Quadratic matched filter

A model that is physically more appropriate for hyperspectral target detection compared to the additive model underlying the SMF (especially

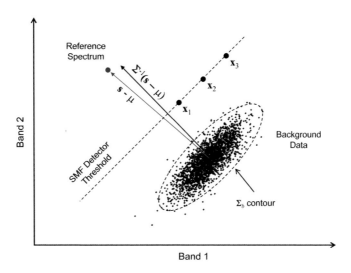

Figure 14.42 Anomalous false-alarm sources using a linear matched filter detector.

for detecting spatially resolved targets) is the substitution model, described by the hypotheses

$$H_0 : \mathbf{x} = \mathbf{b} + \mathbf{n}$$
$$H_1 : \mathbf{x} = \mathbf{s} + \mathbf{n},$$

(14.140)

where $\mathbf{b} \sim N(\boldsymbol{\mu}_b, \boldsymbol{\Sigma}_b), \mathbf{s} \sim N(\boldsymbol{\mu}_s, \boldsymbol{\Sigma}_s)$, and $\mathbf{n} \sim N(0, \boldsymbol{\Sigma}_n)$. If statistics of the background, target, and noise processes are either known or can be estimated from the data, then the GLRT detection statistic would be the quadratic spectral filter (QSF), given by

$$\begin{aligned}
r_{QSF}(\mathbf{x}) = {} & (\mathbf{x} - \boldsymbol{\mu}_b)^T (\boldsymbol{\Sigma}_b + \boldsymbol{\Sigma}_n)^{-1} (\mathbf{x} - \boldsymbol{\mu}_b) \\
& - (\mathbf{x} - \boldsymbol{\mu}_s)^T (\boldsymbol{\Sigma}_s + \boldsymbol{\Sigma}_n)^{-1} (\mathbf{x} - \boldsymbol{\mu}_s) \\
& + \log \frac{|\boldsymbol{\Sigma}_b + \boldsymbol{\Sigma}_n|}{|\boldsymbol{\Sigma}_s + \boldsymbol{\Sigma}_n|}.
\end{aligned}$$

(14.141)

This is identical to the decision surface for a quadratic classifier, where the ratio of prior probability is folded into the detection threshold. It is an intriguing detection statistic because it becomes large only when the spectrum under test is both statistically distant from background clutter through the first term, and similar to the reference through the second term. The incorporation of target covariance matrix $\boldsymbol{\Sigma}_s$, in particular, provides the necessary additional ingredient to alleviate the false-alarm problem discussed earlier. Mathematically, the result is a quadratic decision contour, as illustrated in Fig. 14.43.

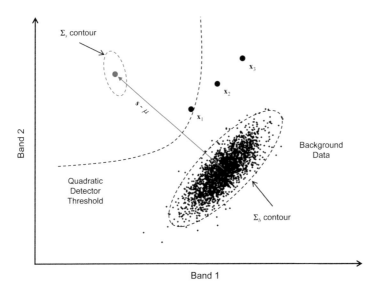

Figure 14.43   False-alarm mitigation using a quadratic spectral filter.

A practical problem with the QSF is that a target covariance matrix is not usually known *a priori* and cannot be estimated from the data because the targets have yet to be detected. One could simply set $\Sigma_s = 0$, such that the noise covariance matrix would represent randomness under the alternative hypothesis; however, this strategy approaches the ED detector due to the low noise level typical of hyperspectral imagery. That is, the noise covariance matrix can significantly underestimate the inherent target signature variability and serve as a poor estimate of the target covariance matrix. A common solution is to assume that $\Sigma_s = \Sigma_b$ and estimate the background covariance matrix from the data, but this simply leads to the SMF detection statistic.

14.4.2 Subpixel replacement model

An FAM strategy devised by DiPietro *et al.* (2010) uses an SMF in conjunction with a second detection statistic that measures the statistical distance from an adaptive target model to filter out anomalous spectra. The FAM test statistic is based on a subpixel replacement model of the form

$$
\begin{aligned}
H_0 &: \mathbf{x} = \mathbf{b} \\
H_1 &: \mathbf{x} = \alpha\,\mathbf{s} + (1 - \alpha)\,\mathbf{b},
\end{aligned}
\tag{14.142}
$$

where $\mathbf{b} \sim N(\boldsymbol{\mu}_b, \Sigma_b)$ and $\mathbf{s} \sim N(\boldsymbol{\mu}_s, \Sigma_s)$. Under $H_0, \mathbf{x}$ is distributed as $\mathbf{b}$, where it is assumed that statistical parameters $\boldsymbol{\mu}_b$ and Σ_b can be estimated from the data. Under $H_1, \mathbf{x}$ is normally distributed with mixture statistics dependent on α, given by

$$\boldsymbol{\mu}(\alpha) = \alpha\,\boldsymbol{\mu}_s + (1 - \alpha)\,\boldsymbol{\mu}_b \qquad (14.143)$$

and

$$\boldsymbol{\Sigma}(\alpha) = \alpha\,\boldsymbol{\Sigma}_s + (1 - \alpha)\,\boldsymbol{\Sigma}_b. \qquad (14.144)$$

Suppose, as shown again in Fig. 14.44, that spectra $\mathbf{x}_1, \mathbf{x}_2$, and $\mathbf{x}_3$ are all detected by an SMF. For each, we can estimate α by the normalized projection given by Eq. (14.64), and then use this estimate to derive the mixture statistics in Eqs. (14.143) and (14.144). In the case of the three illustrated spectra, α estimates will be identical. However, it is clear that $\mathbf{x}_3$ is farther from normal distribution based on mixture statistics depicted in the figure. Thus, a reasonable FAM strategy could consist of a statistical distance measure relative to this mixture distribution. Another solution is to set $\boldsymbol{\Sigma}_s = \alpha\boldsymbol{\Sigma}_b$, where α is a scalar between zero and one.

The obvious candidate for such a test is found by computing the MD from the expected H_1 distribution based on the α parameter estimate for the test spectrum,

$$r_{FAM}(\mathbf{x}) = \left[\mathbf{x} - \boldsymbol{\mu}\left(\frac{\mathbf{s}^T\mathbf{x}}{\mathbf{s}^T\mathbf{s}}\right)\right]^T \left[\boldsymbol{\Sigma}\left(\frac{\mathbf{s}^T\mathbf{x}}{\mathbf{s}^T\mathbf{s}}\right)\right]^{-1} \left[\mathbf{x} - \boldsymbol{\mu}\left(\frac{\mathbf{s}^T\mathbf{x}}{\mathbf{s}^T\mathbf{s}}\right)\right]. \quad (14.145)$$

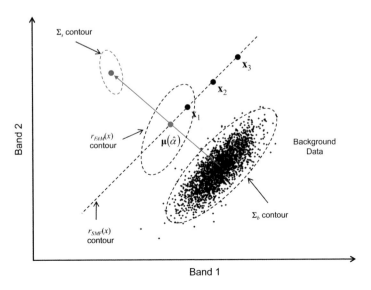

Figure 14.44 FAM based on statistical distance to the mixture density.

Since this statistic measures the distance from the expected distribution under the alternative hypothesis, a high value indicates a false alarm. This approach seems to have no more practical utility than the QSF because the mixture covariance in Eq. (14.144) again depends on the target covariance matrix that is usually unknown. However, if we assume that $\Sigma_s = \Sigma_b$ and $\mu_s = s$, the approach can still exhibit enhanced target selectivity, even though it is unlikely to be as good as the QSF would achieve if Σ_s were known. The FAM test statistic in this case is given as

$$r_{FAM}(\mathbf{x}) = \left[(\mathbf{x} - \mu_b) - \left(\frac{\mathbf{s}^T \mathbf{x}}{\mathbf{s}^T \mathbf{s}}\right)(\mathbf{s} - \mu_b)\right]^T$$

$$\times \Sigma_b^{-1}\left[(\mathbf{x} - \mu_b) - \left(\frac{\mathbf{s}^T \mathbf{x}}{\mathbf{s}^T \mathbf{s}}\right)(\mathbf{s} - \mu_b)\right]. \qquad (14.146)$$

To illustrate this FAM strategy, the results of its application to the HYDICE Forest Radiance test example are shown in Figs. 14.45 and 14.46. The SMF detection statistic in Fig. 14.45(a) indicates potential targets, while the FAM test statistic in Fig. 14.45(b) indicates potential false alarms. When combined in a scatter plot as in Fig. 14.46, we see that there are numerous false alarms that can be removed by the FAM threshold. These rejected false alarms that exceed both SMF and FAM thresholds are due to some of the road pixels and manmade panels in the image. Estimation of ROC performance requires a method of optimally fusing these two test statistics, a subject of continued research.

14.4.3 Mixture-tuned matched filter

The FAM strategy described in the previous section is similar in form to a mixture-tuned matched filter (MTMF) concept developed by Boardman (1998). Like the FAM strategy, the MTMF succeeds the SMF with a test statistic, in this case called an infeasibility metric, which excludes false alarms based on mixture statistics. Unlike the FAM test statistic, the infeasibility metric requires an estimate of the noise covariance matrix to compute it. The target covariance matrix is assumed to be equal to the noise covariance matrix.

The MTMF procedure is as follows. First, the MNF transform is applied to data based on noise covariance matrix Σ_n and the image sample statistics so that the background covariance matrix for the transformed image becomes diagonal $\mathbf{D}_b$ and the noise covariance matrix becomes the identity matrix. Next, data are transformed again to remove the mean vector and whiten the background covariance matrix. Reference spectrum $\mathbf{s}$ is adjusted for mean removal and whitening as described by Eq. (14.81). This operation has the effect of making the noise covariance matrix the

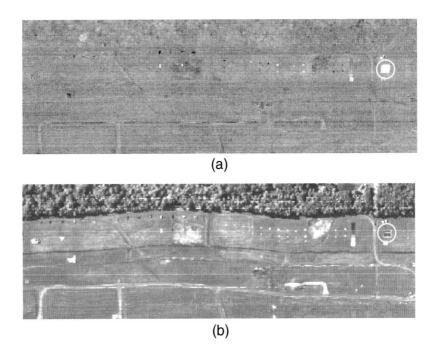

(a)

(b)

Figure 14.45 Test statistics for the HYDICE Forest Radiance image using (a) SMF and (b) FAM. The location of the target is circled in each image.

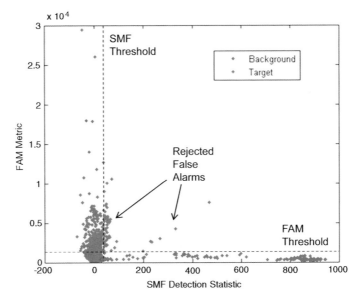

Figure 14.46 Scatter plot of SMF and FAM test statistics for the HYDICE Forest Radiance image.

inverse of diagonalized background covariance matrix $\mathbf{D}_b$. Initial target detection is then performed by applying the simplified SMF in Eq. (14.82) to each clutter-whitened spectrum $\mathbf{z}$.

The infeasibility metric is defined as a statistical distance from the mixture covariance matrix in an orthogonal subspace from the target. Mathematically, it is given by

$$r_{INF}(\mathbf{x}) = \mathbf{z}^T [\Sigma(\hat{\alpha})]^{-1} \mathbf{z}, \qquad (14.147)$$

where

$$\Sigma(\hat{\alpha}) = \hat{\alpha}^2 \mathbf{P}_\mathbf{s}^\perp \mathbf{D}_b^{-1} \mathbf{P}_\mathbf{s}^\perp + (1 - \hat{\alpha}^2)\mathbf{P}_\mathbf{s}^\perp + \varepsilon \mathbf{I}, \qquad (14.148)$$

$$\hat{\alpha} = r_{SMF}(\mathbf{z}) = \mathbf{s}_w^T \mathbf{z}, \qquad (14.149)$$

and ε is a regularization parameter to support the singular mixture covariance matrix inversion in Eq. (14.147). The infeasibility metric is represented graphically by the roughly conical decision surface depicted in Fig. 14.47. A similar metric could be achieved using the FAM test statistic in Eq. (14.145), with the estimated-noise covariance matrix in place of the target covariance matrix in Eq. (14.144).

14.4.4 Finite target matched filter

The finite target matched filter (FTMF) algorithm (Schaum and Stocker, 1997) is a GLRT detection statistic derived from the subpixel replacement

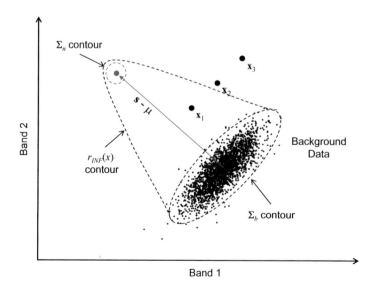

Figure 14.47   Graphical depiction of the infeasibility metric.

model in Eq. (14.142). By using the log-likelihood form of the GLRT, folding the noise statistics into background and target covariance matrices, and eliminating some constant terms, the FTMF detection statistic becomes

$$r_{FTMF}(\mathbf{x}) = (\mathbf{x} - \boldsymbol{\mu}_b)^T \boldsymbol{\Sigma}_b^{-1} (\mathbf{x} - \boldsymbol{\mu}_b)$$
$$- [\mathbf{x} - \boldsymbol{\mu}(\hat{\alpha})]^T [\boldsymbol{\Sigma}(\hat{\alpha})]^{-1} [\mathbf{x} - \boldsymbol{\mu}(\hat{\alpha})] - \log |\boldsymbol{\Sigma}(\hat{\alpha})|, \qquad (14.150)$$

where

$$\hat{\alpha} = \arg \min_{\alpha} \{ [\mathbf{x} - \boldsymbol{\mu}(\alpha)]^T [\boldsymbol{\Sigma}(\alpha)]^{-1} [\mathbf{x} - \boldsymbol{\mu}(\alpha)] + \log |\boldsymbol{\Sigma}(\alpha)| \},$$
$$(14.151)$$

and mixture statistics $\boldsymbol{\mu}(\alpha)$ and $\boldsymbol{\Sigma}(\alpha)$ are given in Eqs. (14.143) and (14.144). Given some estimate for $\boldsymbol{\Sigma}_s$, Eq. (14.151) can be numerically optimized over the interval $\alpha \in [0, 1]$ to find the maximum-likelihood parameter estimate for α to compute the detection statistic. This provides a single test statistic firmly grounded in the GLRT, capable of FAM. The approach again assumes that $\boldsymbol{\Sigma}_s$ is known, which is a major practical limitation. It also requires a numerically challenging minimization to be performed for each test spectrum.

It has been shown by DiPietro *et al.* (2010) that numerical estimation can be solved in closed form if the target covariance matrix is assumed to be a scaled replica of the background covariance matrix; that is,

$$\boldsymbol{\Sigma}_t = \gamma^2 \boldsymbol{\Sigma}_b. \qquad (14.152)$$

In this case, the mixture covariance matrix becomes

$$\boldsymbol{\Sigma}(\alpha) = [\alpha^2 \gamma^2 + (1 - \alpha)^2] \boldsymbol{\Sigma}_b, \qquad (14.153)$$

and parameter α can be found as the solution to the cubic equation

$$A\alpha^3 + B\alpha^2 + C\alpha + D = 0, \qquad (14.154)$$

where

$$A = K (\gamma^2 + 1)^2, \qquad (14.155)$$
$$B = [r_{SMF}(\mathbf{x}) - 3K] (\gamma^2 + 1)^2 - \mathrm{SCR}^2, \qquad (14.156)$$
$$C = -r_{MD}(\mathbf{x}) (\gamma^2 + 1) + K\gamma^2 + 3K + \mathrm{SCR}^2, \qquad (14.157)$$

and

$$D = -K - r_{SMF}(\mathbf{x}) + r_{MD}(\mathbf{x}). \qquad (14.158)$$

The SCR, MD, and SMF statistics in Eqs. (14.155) through (14.158) are specifically defined in this case as

$$SCR^2 = (\mathbf{s} - \boldsymbol{\mu}_b)^T \boldsymbol{\Sigma}_b^{-1} (\mathbf{s} - \boldsymbol{\mu}_b), \qquad (14.159)$$

$$r_{MD}(\mathbf{x}) = (\mathbf{x} - \boldsymbol{\mu}_b)^T \boldsymbol{\Sigma}_b^{-1} (\mathbf{x} - \boldsymbol{\mu}_b), \qquad (14.160)$$

and

$$r_{SMF}(\mathbf{x}) = (\mathbf{s} - \boldsymbol{\mu}_b)^T \boldsymbol{\Sigma}_b^{-1} (\mathbf{x} - \boldsymbol{\mu}_b). \qquad (14.161)$$

There are three possible solutions for α, but a physically realizable solution must fall in the range of zero to one.

14.4.5 Least-angle regression

Another method for achieving enhanced selectivity in a target detection algorithm is to perform a numerical fit (or regression analysis) of the candidate target spectrum for the reference signature and eliminate false alarms by thresholding the goodness of a fit metric. Such a processing stage, sometimes called material identification, is driven by the reference target signature with little regard to the background clutter model. A straightforward approach is to simply use the squared error,

$$\varepsilon^2 = \sum_{k=1}^{K} (x_k - s_k)^2 = \| \mathbf{x} - \mathbf{s} \|^2, \qquad (14.162)$$

as a fit metric, recognized as the ED metric. There are two primary weaknesses to this simple approach. First, it is essential that excellent normalization of the collected radiance spectra to the library reflectance spectra is performed to achieve an adequately good fit when a true target is detected. This implies the need for excellent atmospheric compensation and, in the LWIR spectral region, temperature–emissivity separation. This issue is pervasive for all regression-based material identification methods. The second problem unique to this simple least-squares regression is that target spectra in the image are often partially mixed with background clutter, and such mixing cannot be ignored in many applications. Otherwise, only completely pure target spectra can be detected.

The least-squares regression method proposed by Villeneuve *et al.* (2010) addresses this second problem by using a constrained LMM for target reference to accommodate subpixel or other mixed target spectra. Application of the additivity constraints of the LMM avoids the overfitting that can result by using an unconstrained subspace model. Specifically, we assume that the background is described under the null hypothesis H_0 by

$$\mathbf{x} \mid H_0 = \mathbf{B}\,\boldsymbol{\beta}, \tag{14.163}$$

where $\mathbf{B}$ is the $M \times K$ endmember matrix and $\boldsymbol{\beta}$ is the M-element random coefficient constrained by

$$\sum_{m=1}^{M} |\beta_m| \le 1, \tag{14.164}$$

where M is the number of endmembers. Note that this constraint is slightly different from the typical additivity constraint for the LMM described in Chapter 12. This difference is due to the numerical solver used for this method, which is discussed later.

Under the alternative hypothesis, target signature $\mathbf{s}$ is added along with M background endmembers to produce an $(M + 1) \times K$ target-plus-background endmember matrix $\mathbf{S}$, such that

$$\mathbf{x} \mid H_1 = \mathbf{S}\,\boldsymbol{\alpha}, \tag{14.165}$$

where $\boldsymbol{\alpha}$ is the $(M + 1)$'th element random coefficient, constrained by

$$\sum_{m=1}^{M+1} |\alpha_m| \le 1. \tag{14.166}$$

For each spectrum under test that passes the threshold of the initial detector, (SMF, ACE, etc.), it is necessary to estimate coefficient vectors $\boldsymbol{\alpha}$ and $\boldsymbol{\beta}$ under the competing hypotheses. This can be performed using the least-angle regression (LARS) algorithm developed by Efron *et al.* (2003) and Tibshirani (1996) that efficiently solves the optimization problem

$$\hat{\beta} = \arg \min_{\beta} \| \mathbf{x} - \mathbf{B}\,\boldsymbol{\beta} \|^2 : \sum_{m=1}^{M} |\beta_m| \le \gamma, \tag{14.167}$$

which matches the form of the assumed subpixel mixing model expressed in Eqs. (14.163) through (14.166). By performing this optimization under both hypotheses, constrained estimates of $\boldsymbol{\alpha}$ and $\boldsymbol{\beta}$ are provided.

Target identification is performed by comparing the resultant fit error under the competing hypotheses, and if desired, among multiple target hypotheses. In doing so, we first quantify the fit errors as

$$\chi^2_{H_0} = \| \mathbf{x} - \mathbf{B} \hat{\boldsymbol{\beta}} \|^2 \tag{14.168}$$

and

$$\chi^2_{H_1} = \| \mathbf{x} - \mathbf{S} \hat{\boldsymbol{\alpha}} \|^2. \tag{14.169}$$

Normalizing each fit error by the model degrees of freedom, a significance ratio,

$$\phi = \frac{M + 1}{M} \frac{\chi^2_{H_0}}{\chi^2_{H_1}}, \tag{14.170}$$

measures the modeling improvement provided by the incorporation of the target signature to the background mixture model. This ratio is related to the goodness of fit F-statistic (Triola, 2008). A high significance ratio indicates that incorporation of the target signature is essential for modeling the measured spectrum and provides a basis for detection, or in the case of multiple target hypotheses, material identification. A ratio on the order of unity indicates that the target signature adds little value to modeling the test spectrum, and it is therefore deemed a false alarm. If the spectrum is not well described under either the null or alternative hypotheses, the significance ratio will again be near unity, and the test spectrum rejected as a false alarm. In this way, the method is able to reject anomalous spectra in the tails of background clutter distribution.

14.5 Matched Subspace Detection

As described in the previous section, situations often arise where it is inadequate to use a single reference spectrum to describe the target because this spectrum can exhibit some inherent variability due, for example, to differences in material characteristics across a target class, internal target spectral variation, and changes in illumination or observation geometry. Also, environmental unknowns can manifest as target class variation through errors in atmospheric and environmental normalization or sensor calibration. In these situations, the detector should account for such expected variances in the target. One methodology that supports this is to describe the target in terms of a subspace, where basis vectors of the subspace correspond to the expected directions of target variance. In this section, target detection approaches are extended to incorporate a target subspace model. The underlying assumptions for the detectors to be discussed are outlined in Table 14.3.

Table 14.3 Matched subspace detection algorithm taxonomy.

Algorithm	Models			Parameters	
	Signal	Background	Noise	Known	Unknown
Simple signal subspace detector	Subspace	None	Normal	S, Σ_n	α
Subspace adaptive coherence/cosine estimator	Subspace	Normal	Part of background	S	$\alpha, \beta, \mu, \Sigma$
Joint subspace detector	Subspace	Subspace	Normal	S, Σ_n	α, β, B
Adaptive subspace detector	Subspace	Subspace	White	S	$\alpha, \beta, B, \sigma_n$

14.5.1 Target subspace models

A major challenge associated with detection algorithms based on subspace target representation is determining the target subspace matrix that appropriately represents the uncertainty in a target reference spectrum. If the subspace is too small, the algorithm is unable to detect targets under circumstances where they fall outside of the representative subspace. On the other hand, if the subspace is too large, it results in false alarms due to the presence of background clutter spectra that fall within the target subspace. To appropriately define the subspace, it is necessary to understand the sources of observed target spectral variation. Certainly, variations in the reflectance signature (either across a target or within a target class) are one such component and can be captured by proper reference signature measurements. Another major contributor, however, is uncertainty in the radiative transfer processes altering the pupil-plane spectral radiance spectrum for a given target apparent reflectance. The focus of this section is to discuss methodologies for deriving a target subspace matrix that handles this particular signature variation.

We begin the discussion by outlining a procedure initially developed by Healey and Slater (1999) for target subspace modeling in the VNIR/SWIR spectral region. The procedure is based on forward modeling using the MODTRAN radiative transfer code, followed by dimensionality reduction. The main idea of the method is to predict all possible observed pupil-plane radiance spectra associated with a particular target reflectance spectrum across a range of possible atmospheric and observation conditions. Standard dimensionality-reduction techniques such as SVD are then used to determine basis vectors of a low-order subspace that captures most of the variability over a large set of model realizations. This basis set defines a target subspace matrix that is used in the alternative hypothesis in a subspace detector. MLEs of the basis coefficients are made for each test spectrum in the image to provide a best fit of the spectrum to the target subspace model. Error in the fit is the target detection metric, with a small error indicating a target, and a large error indicating a nontarget.

The target subspace modeling approach advocated by Healey, sometimes called an invariant filter, is based on a diffuse facet model for the target. Written in vector form, an observed spectral radiance signature **s** is related to the underlying spectral reflectance vector $\boldsymbol{\rho}$, according to the linear model

$$\mathbf{s} = \frac{1}{\pi} \boldsymbol{\rho} \otimes \boldsymbol{\tau}_p \otimes (\alpha_s \mathbf{E}_s \cos \theta_s + \mathbf{E}_d) + \mathbf{L}_p + \mathbf{n}, \qquad (14.171)$$

where $\boldsymbol{\tau}_p$ is the scene-to-sensor atmospheric transmission, $\mathbf{E}_s$ is the direct solar irradiance from a zenith angle θ_s, $\mathbf{E}_d$ is the integrated downwelling irradiance, $\mathbf{L}_p$ is the scene-to-sensor path radiance, and **n** is the random sensor noise. The $K \times 1$ spectral vectors $\boldsymbol{\rho}, \boldsymbol{\tau}_p, \mathbf{E}_s, \mathbf{E}_d, \mathbf{L}_p$, and **n** have the same format as image spectrum **x**, and the symbol $\otimes$ denotes element-by-element vector multiplication. The scalar α_s accounts for scaling of the direct solar irradiance component due to shadowing variations. Except for the additional noise term, this is essentially the diffuse facet model described in Chapter 5 converted from continuous to discrete vector form.

The target subspace is estimated from Eq. (14.171) by modeling a large number of possible observed pupil-plane radiance spectra realizations {**s**} for a given reflectance reference $\boldsymbol{\rho}$ by varying the parameters in a radiative transfer model (such as MODTRAN) affecting $\theta_s, \boldsymbol{\tau}_p, \mathbf{E}_s, \mathbf{E}_d$, and $\mathbf{L}_p$. Table 14.4 provides an example set of model parameters used by Healey and Slater (1999), the permutations of which generated an ensemble of 17,920 realizable spectra. From these, the sample covariance matrix can be computed, and SVD performed to derive the basis vectors to form the columns of subspace matrix **S**. In Healey's initial work, all spectra could be fit to the leading nine eigenvectors with less than 0.02% RMS error. From this estimated subspace matrix, each target spectrum **s** in the image can be represented by

$$\mathbf{s} = \mathbf{S}\boldsymbol{\alpha} + \mathbf{n} \qquad (14.172)$$

to within the stated error, where $\boldsymbol{\alpha}$ is a basis coefficient vector unique to each sample.

For each image spectrum **x**, the likelihood function associated with an estimate of unknown basis coefficient vector $\boldsymbol{\alpha}$ for zero-mean, normally distributed noise with a covariance matrix Σ_n is given by

$$L(\mathbf{x} \mid \boldsymbol{\alpha}) = \frac{1}{(2\pi)^{K/2}} \frac{1}{|\Sigma_n|^{1/2}} \exp\left\{-\frac{1}{2}(\mathbf{x} - \mathbf{S}\boldsymbol{\alpha})^T \Sigma_n^{-1} (\mathbf{x} - \mathbf{S}\boldsymbol{\alpha})\right\}.$$

$$(14.173)$$

Table 14.4 Example model parameters for MODTRAN target subspace generation.

Parameter	Values
H_2O (cm)	0.88, 1.44, 3.11, 4.33
O_3 (atm cm)	0.07, 0.11, 0.14
O_2 (atm cm)	8407.9, 8604.0, 9453.2, 10536.8
CH_4 (atm cm)	0.85, 0.86, 0.87
N_2O (atm cm)	0.199, 0.202, 0.209, 0.214
CO (atm cm)	0.064, 0.065, 0.067, 0.070
CO_2 (atm cm)	15.23, 16.17, 17.33, 17.76
Solar zenith (deg)	5, 15, 25, 35, 45, 55, 65, 75
Sensor altitude (km)	1 through 7
Aerosol model	Rural, urban, maritime, desert

The MLE for $\boldsymbol{\alpha}$ is

$$\hat{\boldsymbol{\alpha}} = \arg \max_{\boldsymbol{\alpha}} L(\mathbf{x} \mid \boldsymbol{\alpha}) = (\mathbf{S}^T \boldsymbol{\Sigma}_n^{-1} \mathbf{S})^{-1} \mathbf{S}^T \boldsymbol{\Sigma}_n^{-1} \mathbf{x}. \qquad (14.174)$$

The argument of Eq. (14.173),

$$r_{SS}(\mathbf{x}) = -\frac{1}{2} (\mathbf{x} - \mathbf{S} \hat{\boldsymbol{\alpha}})^T \boldsymbol{\Sigma}_n^{-1} (\mathbf{x} - \mathbf{S} \hat{\boldsymbol{\alpha}}), \qquad (14.175)$$

can serve as a detection statistic, with the MLE in Eq. (14.174) inserted for $\boldsymbol{\alpha}$. In fact, this is the GLRT detection statistic under the scenario where there is no background clutter knowledge, the noise covariance matrix is known, and the target is represented by the subspace matrix. We refer to this as the simple subspace (SS) detector. If the noise is actually white, then Eq. (14.174) reduces to

$$\hat{\boldsymbol{\alpha}} = (\mathbf{S}^T \mathbf{S})^{-1} \mathbf{S}^T \mathbf{x}, \qquad (14.176)$$

and the detection statistic becomes

$$r_{SS}(\mathbf{x}) = -\min_{\boldsymbol{\alpha}} \| \mathbf{x} - \mathbf{S} \boldsymbol{\alpha} \|^2; \qquad (14.177)$$

that is, the residual RMS error in the subspace fit to test spectrum $\mathbf{x}$ is the GLRT detection statistic. A good fit (small error) indicates a target, while a poor fit (large error) indicates background clutter.

To achieve good detection performance, it is important to optimally tailor subspace $\mathbf{S}$ to the uncertainty of the reference spectrum for a given image observation. The example model parameters given in Table 14.4 are overly general in that regard, since one would likely have better knowledge of atmospheric and observation conditions under which the data were collected, especially concerning parameters such as sensor altitude and solar zenith angle. A more refined parameter space for atmospheric conditions could be derived from global weather forecasting data, an approach implemented by Kolodner (2008). An effective subspace model should not only incorporate these expected variances in the radiative transfer, but also factor in inherent variation due to target signature (e.g., spatial variability, weathering, and contamination), apparent reflectance changes due to aspect variation, and uncertainties in sensor calibration. The first two factors can be incorporated into the signature reflectance term in Eq. (14.171), while the last factor would require an imaging spectrometer model.

The MODTRAN model-based approach for synthesizing a target subspace matrix requires absolute sensor calibration to match the ensemble of predicted pupil-plane radiance spectra to the hyperspectral image data. The efficacy of the approach degrades if miscalibration exists and is not somehow incorporated. An alternate approach is to use a scene-based atmospheric compensation method, such as vegetation normalization, and extend it to incorporate variances in the target and vegetation signatures on which it is based. In this case, the underlying signal model can be simply written as

$$s = a \otimes \rho + b + n, \tag{14.178}$$

where a and b incorporate both sensor and radiative transfer components of the linear observation model. For VNIR/SWIR imagery, offset b can be estimated from the darkest spectra in the image, as outlined in Chapter 11. Typically, gain a can be estimated by comparing the sample mean from high-NDVI image spectra with a single vegetation reference spectrum. As depicted in Fig. 14.48, vegetation actually exhibits a fair amount of spectral variation, which can be a source of errors in atmospheric compensation when using vegetation normalization, as was discovered in the example results from Chapter 11.

A scene-based target subspace modeling approach developed by Smetek (2007) incorporates a database of both target reflectance signatures $\{\rho_t\}$ and vegetation reflectance signatures $\{\rho_v\}$ in generation of the ensemble of target signatures $\{s\}$ on which the target subspace is based. Vegetation spectra impact the model through estimated sensor gain

$$\hat{a} = \rho_v^{-1} \otimes (\hat{\mu}_v - \hat{b}), \tag{14.179}$$

where μ_v is the mean of vegetation spectra in the image, and the vector inverse is performed by an element-by-element reciprocal. Given a combination of particular target reference reflectance ρ_t and a vegetation

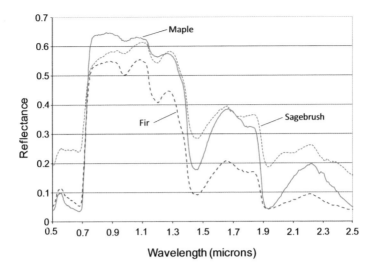

Figure 14.48   Variability of vegetation reflectance spectra.

library reflectance ρ_v, a single observed target spectrum realization is generated by

$$\mathbf{s} = \rho_v^{-1} \otimes (\hat{\mathbf{\mu}}_v - \hat{\mathbf{b}}) \otimes \rho_t + \hat{\mathbf{b}}. \qquad (14.180)$$

Figure 14.49 illustrates an example target reflectance ensemble for the F2 target material in the panel array section of the HYDICE Forest Radiance image, along with vegetation reflectance spectra extracted from USGS and Johns Hopkins University spectral libraries. Figure 14.50 compares the ensemble of modeled target spectra using Eq. (14.180) based on all combinations of target and vegetation reference spectra to actual image spectra. Target subspace matrix **S** can be determined from the generated spectra by identifying the leading eigenvectors in the same manner used for MODTRAN-generated spectra. Figure 14.51 shows a subspace target detection statistic based on estimated target subspace matrix **S** applied to the panel array section of the HYDICE Forest Radiance image. The F2 target panels are the brightest in the image, forming a column of three panels, with the largest at the top and smallest at the bottom.

14.5.2 Subspace adaptive coherence/cosine estimator

The signal subspace detection approach can be extended to incorporate a normal background clutter model (Manolakis and Shaw, 2002). In this case, targets can be subpixel in extent, and the detector formulation accommodates this potential mixture of target and background spectra within pixels. For an unstructured background, the detection problem can

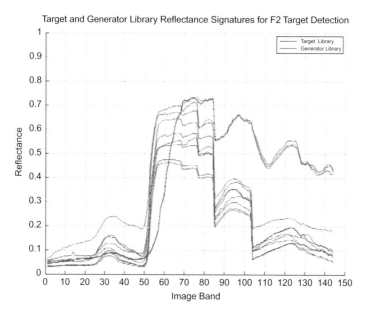

Figure 14.49 Target and vegetation reflectance spectra corresponding to HYDICE Forest Radiance image (Smetek, 2007).

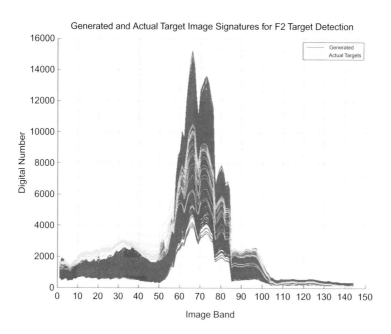

Figure 14.50 Ensemble of modeled and actual target spectra for the HYDICE Forest Radiance image. Blue spectra are generated by the model, and green spectra are extracted from the image (Smetek, 2007).

Figure 14.51 Example subspace target detection statistic image for the panel region of the Forest Radiance image based on the in-scene subspace modeling method for the F2 target panels, indicated by the rectangle (Smetek, 2007).

be cast as

$$H_0 : \mathbf{x} = \beta\,\mathbf{b}$$
$$H_1 : \mathbf{x} = \mathbf{S}\,\boldsymbol{\alpha} + \beta\,\mathbf{b}, \tag{14.181}$$

where **b** is a normally distributed random vector with covariance matrix $\boldsymbol{\Sigma}$ that represents the combination of background and sensor noise, **S** and $\boldsymbol{\alpha}$ are the target subspace matrix and basis coefficient vector, and β is an unknown parameter that represents the potential mixture of background in target pixels. The formulation assumes a zero-mean background **b**, such that the resulting detector should be applied to mean-removed data, with **S** modified for the mean removal, as given in Eq. (14.97); that is, the background mean vector is subtracted from each column of the target subspace matrix, based on reference signature data.

Under the given assumptions, the conditional probability density functions corresponding to the two hypotheses are

$$p(\mathbf{x} \mid H_0) = \frac{1}{(2\pi)^{K/2}} \frac{1}{\beta\,|\boldsymbol{\Sigma}|^{1/2}} \exp\left\{-\frac{1}{2}\beta^2\,\mathbf{x}^T\boldsymbol{\Sigma}^{-1}\,\mathbf{x}\right\}, \tag{14.182}$$

and

$$p(\mathbf{x} \mid H_1) = \frac{1}{(2\pi)^{K/2}} \frac{1}{\beta\,|\boldsymbol{\Sigma}|^{1/2}} \exp\left\{-\frac{1}{2}\beta^2\,(\mathbf{x} - \mathbf{S}\,\boldsymbol{\alpha})^T\boldsymbol{\Sigma}^{-1}\,(\mathbf{x} - \mathbf{S}\,\boldsymbol{\alpha})\right\}. \tag{14.183}$$

Applying the GLRT to this formulation in the same manner as for the ACE detector leads to the detection statistic

$$r(\mathbf{x}) = \frac{\mathbf{x}^T \hat{\boldsymbol{\Sigma}}^{-1} \mathbf{S} \, \hat{\boldsymbol{\alpha}}}{\mathbf{x}^T \hat{\boldsymbol{\Sigma}}^{-1} \mathbf{x}}, \tag{14.184}$$

where the MLE for target abundance gain is given by

$$\hat{\boldsymbol{\alpha}} = (\mathbf{S}^T \hat{\boldsymbol{\Sigma}}^{-1} \mathbf{S})^{-1} \mathbf{S}^T \hat{\boldsymbol{\Sigma}}^{-1} \mathbf{x}. \tag{14.185}$$

By inserting Eq. (14.185) into Eq. (14.184),

$$r_{SS-ACE}(\mathbf{x}) = \frac{\mathbf{x}^T \hat{\boldsymbol{\Sigma}}^{-1} \mathbf{S}(\mathbf{S}^T \hat{\boldsymbol{\Sigma}}^{-1} \mathbf{S})^{-1} \mathbf{S}^T \hat{\boldsymbol{\Sigma}}^{-1} \mathbf{x}}{\mathbf{x}^T \hat{\boldsymbol{\Sigma}}^{-1} \mathbf{x}}. \tag{14.186}$$

This is known as the subspace adaptive coherence/cosine estimator (SS-ACE) detector.

14.5.3 Joint subspace detector

If a subspace model for both target and background is assumed, the detection problem can be given as

$$\begin{aligned} H_0 &: \mathbf{x} = \mathbf{B}\,\boldsymbol{\beta} + \mathbf{n} \\ H_1 &: \mathbf{x} = \mathbf{S}\,\boldsymbol{\alpha} + \mathbf{n}, \end{aligned} \tag{14.187}$$

where $\mathbf{S}$ and $\mathbf{B}$ are matrices containing columns that span target and background subspaces, respectively, and $\mathbf{n}$ is assumed to be normally distributed with zero mean and known covariance matrix $\boldsymbol{\Sigma}_n$. As shown in Section 14.2.4, this leads to the detection statistic

$$r_{JSD}(\mathbf{x}) = \mathbf{x}^T \boldsymbol{\Sigma}_{\mathbf{n}}^{-1} \mathbf{S} \, (\mathbf{S}^T \boldsymbol{\Sigma}_{\mathbf{n}}^{-1} \mathbf{S})^{-1} \mathbf{S}^T \boldsymbol{\Sigma}_{\mathbf{n}}^{-1} \mathbf{x} - \mathbf{x}^T \boldsymbol{\Sigma}_{\mathbf{n}}^{-1} \mathbf{B} \, (\mathbf{B}^T \boldsymbol{\Sigma}_{\mathbf{n}}^{-1} \mathbf{B})^{-1} \mathbf{B}^T \boldsymbol{\Sigma}_{\mathbf{n}}^{-1} \mathbf{x}. \tag{14.188}$$

Under the condition that noise is white with standard deviation σ_n in each band, this reduces to

$$r_{JSD}(\mathbf{x}) = \frac{1}{\sigma_n^2} [\, \mathbf{x}^T \mathbf{S} \, (\mathbf{S}^T \mathbf{S})^{-1} \mathbf{S}^T \mathbf{x} - \mathbf{x}^T \mathbf{B} \, (\mathbf{B}^T \mathbf{B})^{-1} \mathbf{B}^T \mathbf{x}], \tag{14.189}$$

which can be written as

$$r_{JSD}(\mathbf{x}) = \frac{1}{\sigma_n^2} \mathbf{x}^T (\mathbf{P}_S - \mathbf{P}_B) \, \mathbf{x}. \tag{14.190}$$

If the target is potentially subpixel, then the model in Eq. (14.187) can be extended to

$$H_0 : \mathbf{x} = \mathbf{B}\,\boldsymbol{\beta} + \mathbf{n}$$
$$H_1 : \mathbf{x} = \mathbf{S}\,\boldsymbol{\alpha} + \mathbf{B}\,\boldsymbol{\beta} + \mathbf{n} \qquad (14.191)$$

to account for the portion of background clutter that can exist in the observed spectrum under the alternative hypothesis. In this case, noise estimates under the two hypotheses are

$$\hat{\mathbf{n}}_0 = \mathbf{P}_B^{\perp}\,\mathbf{x} \qquad (14.192)$$

and

$$\hat{\mathbf{n}}_1 = \mathbf{P}_{SB}^{\perp}\,\mathbf{x}. \qquad (14.193)$$

If we assume white noise with known standard deviation σ_n, or we prewhiten the data based on the known-noise covariance matrix, then we can apply the detection statistic in Eq. (14.134) to derive a subpixel version of the JSD, which we call the subpixel subspace detector (SSD),

$$r_{SSD}(\mathbf{x}) = \frac{1}{\sigma_n^2}\mathbf{x}^T(\mathbf{P}_B^{\perp} - \mathbf{P}_{SB}^{\perp})\,\mathbf{x} = \frac{1}{\sigma_n^2}\mathbf{x}^T(\mathbf{P}_{SB} - \mathbf{P}_B)\,\mathbf{x}, \qquad (14.194)$$

where matrix $\mathbf{SB}$ is formed by concatenating columns $\mathbf{S}$ and $\mathbf{B}$ together into a single joint subspace matrix.

If the noise statistics and any clutter outside the background subspace are truly normally distributed and white, then the detection statistic should be chi-squared distributed,

$$r_{SSD}(\mathbf{x} \mid H_0) \sim \chi_M^2(0) \qquad (14.195)$$

and

$$r_{SSD}(\mathbf{x} \mid H_1) \sim \chi_M^2(\mathrm{SCR}^2), \qquad (14.196)$$

where M is the target subspace dimensionality, and the centrality parameter is the squared SCR,

$$\mathrm{SCR}^2 = \frac{1}{\sigma_n^2}\mathbf{x}^T\mathbf{P}_B^{\perp}\mathbf{x}. \qquad (14.197)$$

ROC performance can be estimated by

$$P_D = 1 - P[\chi_M^2(\mathrm{SCR}^2) \le \eta] \qquad (14.198)$$

and

$$P_{FA} = 1 - P[\chi_M^2(0) \le \eta]. \qquad (14.199)$$

As described previously, hyperspectral image data do not typically exhibit normal statistics. Even after projecting out the structured background component from the leading principal-component subspace, the remaining data distribution can be non-normal and is almost never white due to residual background clutter that is not captured within the leading subspace. Therefore, it is common to consider this residual background clutter as part of the noise, and instead of performing noise whitening as shown in Eq. (14.188), one can replace the noise covariance matrix with a sample background-plus-noise covariance matrix estimated from image data after orthogonal background subspace projection.

14.5.4 Adaptive subspace detector

The adaptive subspace detector (ASD) is based on the subpixel subspace model expressed in Eq. (14.191), but noise variance is assumed unknown and potentially nonstationary. In this case, it is preferable to use the adaptive form of the GLRT in Eq. (14.138). Inserting the noise estimates in Eqs. (14.192) and (14.193) into this GLRT expression results in the detection statistic

$$r_{ASD}(\mathbf{x}) = \frac{\mathbf{x}^T (\mathbf{P}_B^\perp - \mathbf{P}_{SB}^\perp) \mathbf{x}}{\mathbf{x}^T \mathbf{P}_{SB}^\perp \mathbf{x}}. \qquad (14.200)$$

Under the same assumptions as discussed for SSD, the ASD detection statistic will be F-distributed according to

$$r_{ASD}(\mathbf{x} \mid H_0) \sim \frac{M}{K - L - M} F_{M,K-L-M}(0) \qquad (14.201)$$

and

$$r_{ASD}(\mathbf{x} \mid H_1) \sim \frac{M}{K - L - M} F_{M,K-L-M}(\mathrm{SCR}^2), \qquad (14.202)$$

where L is the background subspace dimensionality, M is the target subspace dimensionality, and the noncentrality parameter is again the SCR given in Eq. (14.197). Theoretical ROC performance is given by

$$P_D = 1 - P[F_{M,K-L-M}(\text{SCR}^2) \leq \eta] \qquad (14.203)$$

and

$$P_{FA} = 1 - P[F_{M,K-L-M}(0) \leq \eta]. \qquad (14.204)$$

Because there is likely to be background clutter leakage outside the subspace defined by **B** that deviates from normal distribution, the theoretical ROC expressions given in Eqs. (14.198) and (14.199) for SSD and in Eqs. (14.203) and (14.204) for ASD are optimistic projections, perhaps by a large margin. Actual ROC performance must be estimated from large areas of real background clutter data to be a realistic measure.

14.6 Change Detection

Change detection is a common remote sensing technique used to detect features in a scene that are either too difficult to find in a single image or relate specifically to a time-varying change in scene context. A common example is the detection of subtle changes in land cover over time using remotely sensed panchromatic or multispectral images. Methods such as band ratios, vegetation indices, spatial features, and contextual information have all been used to accentuate such land-cover changes. An important feature of these methods is the ability to be invariant to illumination and environmental changes between images, yet sensitive to these local changes of interest. An alternate approach is to use some sort of predictor from prior images to the current observation that compensates for illumination and environmental changes, and then use a subtractive process to accentuate local changes. This is the primary methodology discussed in this section.

The focus of the methods described is to detect small objects that have been inserted, deleted, or moved between hyperspectral observations but are too subtle relative to background clutter to be detected in a single observation. As seen throughout this entire chapter, background clutter can be highly structured and difficult to model using statistical distributions or subspace matrices. In a change-detection scenario, however, the reference image provides a version of the complex clutter distribution, albeit possibly under different illumination and environmental conditions. The basic idea of change detection is to develop some sort of predictor to transform the reference (prior) image or images into the same illumination and environmental conditions of the test (current) image, and then to simply subtract the transformed reference image from the test image to cancel background clutter. This can be followed by some form of detector to locate change targets in the residual noise.

14.6.1 Affine change model

The phenomenological basis of hyperspectral change detection is the invariance of surface reflectance properties over time. With reference again to the diffuse facet model, pupil-plane radiance in the VNIR/SWIR spectral range can be modeled as

$$L_p(\lambda) = a(\lambda)\rho(\lambda) + b(\lambda), \qquad (14.205)$$

where

$$a(\lambda) = \frac{g(\lambda)\,\tau_a(\lambda)}{\pi}[\alpha\,E_s(\lambda)\cos\theta_s + \beta\,E_d(\lambda)] \qquad (14.206)$$

and

$$b(\lambda) = g(\lambda)\,L_a(\lambda) + d(\lambda). \qquad (14.207)$$

The model consists of atmospheric parameters $\tau_a(\lambda), L_a(\lambda), E_s(\lambda), E_d(\lambda)$, and θ_s previously defined, along with the imaging spectrometer gain $g(\lambda)$ and offset $d(\lambda)$. This particular model also includes shadow coefficients α and β to account for potentially local illumination changes due to shadowing from neighboring objects. As shown previously, this can be written in vector form as

$$\mathbf{x} = \mathbf{a} \otimes \boldsymbol{\rho} + \mathbf{b} + \mathbf{n}, \qquad (14.208)$$

where $\mathbf{x}, \boldsymbol{\rho}, \mathbf{a}$, and $\mathbf{b}$ are $K \times 1$ element column vectors representing the measured spectrum, the scene reflectance spectrum, and the model gain and offset spectra [given by Eqs. (14.206) and (14.207) in vector form], respectively, and the final vector $\mathbf{n}$ represents sensor noise. While $\mathbf{a}$ and $\mathbf{b}$ can change due to illumination and environmental influences, any change in surface reflectance $\boldsymbol{\rho}$ is deemed a change of interest to be detected. Although there is little phenomenological basis for mixing across spectral bands, Eq. (14.208) can be generalized to an affine transformation,

$$\mathbf{x} = \mathbf{A}\boldsymbol{\rho} + \mathbf{b} + \mathbf{n}, \qquad (14.209)$$

where $\mathbf{A}$ is theoretically a diagonal matrix with the elements of $\mathbf{a}$ along the diagonal. While the MWIR and LWIR spectral region are not explicitly addressed in this section, a similar model for this spectral region can be developed with the additional complexity of scene temperature variation that impacts the emitted component of scene radiation.

Now consider a case where there are two observations of the same scene, a reference image given by data matrix $\mathbf{X}^{(1)}$ and test image $\mathbf{X}^{(2)}$, between which there are changes of interest to detect. The superscript indicates the observation time (1 = reference, 2 = test). Consider first the null hypothesis for which no such changes occur. Based on the affine model, each spectral measurement vector $\mathbf{x}_i^{(1)}$ and $\mathbf{x}_i^{(2)}$ composing data matrices $\mathbf{X}^{(1)}$ and $\mathbf{X}^{(2)}$, respectively, are related to invariant reflectance spectrum $\boldsymbol{\rho}_i$ according to

$$\mathbf{x}_i^{(1)} = \mathbf{A}^{(1)} \boldsymbol{\rho}_i + \mathbf{b}^{(1)} + \mathbf{n}^{(1)} \tag{14.210}$$

and

$$\mathbf{x}_i^{(2)} = \mathbf{A}^{(2)} \boldsymbol{\rho}_i + \mathbf{b}^{(2)} + \mathbf{n}^{(2)}, \tag{14.211}$$

where parameters of the affine model are superscripted in correspondence with the observation. In the absence of noise $[\mathbf{n}^{(1)} = \mathbf{n}^{(2)} = 0]$, the spectra should correspond according to the affine transformation

$$\mathbf{x}_i^{(2)} = \mathbf{A}^{(1,2)} \mathbf{x}_i^{(1)} + \mathbf{b}^{(1,2)}, \tag{14.212}$$

where

$$\mathbf{A}^{(1,2)} = \mathbf{A}^{(2)}[\mathbf{A}^{(1)}]^{-1} \tag{14.213}$$

and

$$\mathbf{b}^{(1,2)} = \mathbf{b}^{(2)} - \mathbf{A}^{(1,2)}\mathbf{b}^{(1)}. \tag{14.214}$$

In the presence of noise, we consider the residual difference between the optimum affine transformation of reference image spectrum $\mathbf{x}_i^{(1)}$ according to the right-hand side of Eq. (14.212) and test image spectrum $\mathbf{x}_i^{(2)}$,

$$\boldsymbol{\varepsilon}_i = \mathbf{x}_i^{(2)} - (\mathbf{A}^{(1,2)} \mathbf{x}_i^{(1)} + \mathbf{b}^{(1,2)}), \tag{14.215}$$

as an indicator of change. Specifically, if we assume a potential reflectance change $\Delta\boldsymbol{\rho}$ only under the alternative hypothesis, the change-detection problem becomes

$$\begin{aligned} H_0 &: \boldsymbol{\varepsilon} = \mathbf{n}^{(2)} - \mathbf{A}^{(1,2)}\mathbf{n}^{(1)} \\ H_1 &: \boldsymbol{\varepsilon} = \mathbf{A}^{(2)}\Delta\boldsymbol{\rho} + \mathbf{n}^{(2)} - \mathbf{A}^{(1,2)}\mathbf{n}^{(1)}; \end{aligned} \tag{14.216}$$

that is, the change-detection problem becomes one of detecting the reflectance difference in the residual sensor noise with the background clutter effectively cancelled out.

To perform change detection, it should be apparent that it is first necessary to establish a one-to-one correspondence between spectral measurements that make up the two observations $\mathbf{X}^{(1)}$ and $\mathbf{X}^{(2)}$. Establishing this correspondence is known as image registration and is typically accomplished by remapping datasets from their original sensor coordinates onto a common set of ground-referenced coordinates, where they can be compared at the pixel level. A two-step registration process can be used that consists first of orthorectification on geocoordinates, followed by scene-based registration (Eismann *et al.*, 2009). The orthorectification process utilizes time-referenced navigation data, sensor pointing data, a sensor model, and a digital elevation model of the terrain, and is usually capable of achieving registration down to the level of a few pixels. Scene-based registration using local image phase correlation and scene warping and resampling achieves higher-precision image coalignment to the fractional pixel level necessary for spectral change analysis.

14.6.2 Change detection using global prediction

The standard methods of change detection attempt to estimate affine transformation parameters $\mathbf{A}^{(1,2)}$ and $\mathbf{b}^{(1,2)}$ from the joint global statistics of two hyperspectral images, transform the reference image to a test image using Eq. (14.212), perform the subtraction indicated in Eq. (14.215), and detect the change in the residual image based on the model given in Eq. (14.216). This global-prediction approach, depicted in Fig. 14.52, assumes that affine transformation parameters are space invariant, such that global sample statistics of $\mathbf{X}^{(1)}$ and $\mathbf{X}^{(2)}$ are sufficient to estimate transformation parameters. Either anomaly or signature-matched detection algorithms can be used to detect changes in the residual image, dependent on whether *a priori* spectral signature information is available for the expected changes.

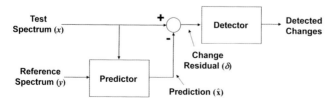

Figure 14.52 Flow diagram of hyperspectral change detection using global prediction.

A minimum-MSE predictor can be formed based on the cross-covariance matrix estimate from the sample statistics,

$$\hat{\Sigma}^{(1,2)} = \frac{1}{N} \sum_{i=1}^{N} (\mathbf{x}_i^{(2)} - \hat{\boldsymbol{\mu}}^{(2)})(\mathbf{x}_i^{(1)} - \hat{\boldsymbol{\mu}}^{(1)})^T, \qquad (14.217)$$

where $\boldsymbol{\mu}^{(1)}$ and $\boldsymbol{\mu}^{(2)}$ are mean vectors for $\mathbf{x}^{(1)}$ and $\mathbf{x}^{(2)}$. Determining optimum transformation from a minimum-MSE perspective follows the standard Wiener filter derivation, requiring error $\boldsymbol{\varepsilon}$ to be orthogonal to reference spectra $\mathbf{x}^{(1)}$, or

$$E\{ [\mathbf{x}^{(2)} - (\mathbf{A}^{(1,2)}\mathbf{x}^{(1)} + \mathbf{b}^{(1,2)})] \, (\mathbf{x}^{(1)})^T \} = 0. \qquad (14.218)$$

We also require the residual error to be zero mean,

$$E[\mathbf{x}^{(2)} - (\mathbf{A}^{(1,2)}\mathbf{x}^{(1)} + \mathbf{b}^{(1,2)})] = 0. \qquad (14.219)$$

From Eq. (14.219), the estimate for $\mathbf{b}^{(1,2)}$ is found in terms of $\mathbf{A}^{(1,2)}$ as

$$\hat{\mathbf{b}}^{(1,2)} = \boldsymbol{\mu}^{(2)} - \mathbf{A}^{(1,2)}\boldsymbol{\mu}^{(1)}. \qquad (14.220)$$

From Eq. (14.218), we find that

$$E[\mathbf{x}^{(2)}(\mathbf{x}^{(1)})^T] - \mathbf{A}^{(1,2)}E[\mathbf{x}^{(1)}(\mathbf{x}^{(1)})^T] - \mathbf{b}^{(1,2)}E[(\mathbf{x}^{(1)})^T] = 0. \qquad (14.221)$$

Substituting the estimate from Eq. (14.220) into Eq. (14.221) gives

$$E[\mathbf{x}^{(2)}(\mathbf{x}^{(1)})^T] - \boldsymbol{\mu}^{(2)}(\boldsymbol{\mu}^{(1)})^T = \mathbf{A}^{(1,2)}\{E[\mathbf{x}^{(1)}(\mathbf{x}^{(1)})^T] - \boldsymbol{\mu}^{(1)}(\boldsymbol{\mu}^{(1)})^T\}, \qquad (14.222)$$

from which the estimate for $\mathbf{A}^{(1,2)}$ is found:

$$\hat{\mathbf{A}}^{(1,2)} = \hat{\Sigma}^{(1,2)}(\hat{\Sigma}^{(1)})^{-1}. \qquad (14.223)$$

Therefore,

$$\hat{\mathbf{b}}^{(1,2)} = \boldsymbol{\mu}^{(2)} - \hat{\Sigma}^{(1,2)}(\hat{\Sigma}^{(1)})^{-1}\boldsymbol{\mu}^{(1)}. \qquad (14.224)$$

Use of this global predictor is known as the chronochrome change-detection method (Schaum and Stocker, 2003).

An alternate global predictor that avoids estimation of the cross-covariance matrix is the covariance-equalization method that normalizes second-order statistics of the reference image to the test image (Schaum and Stocker, 2004). Normalization of the covariance matrix can be accomplished through the transformation

$$\hat{\mathbf{A}}^{(1,2)} = (\hat{\boldsymbol{\Sigma}}^{(2)})^{1/2}(\hat{\boldsymbol{\Sigma}}^{(1)})^{-1/2}, \qquad (14.225)$$

which is symbolic notation for the whitening/dewhitening transformation

$$\hat{\mathbf{A}}^{(1,2)} = \mathbf{V}^{(2)}(\mathbf{D}^{(2)})^{1/2}(\mathbf{V}^{(2)})^{T}\mathbf{V}^{(1)}(\mathbf{D}^{(1)})^{-1/2}(\mathbf{V}^{(1)})^{T}, \qquad (14.226)$$

where $\mathbf{D}^{(1)}$ and $\mathbf{D}^{(2)}$ are diagonalized covariance matrices for test and reference images, and $\mathbf{V}^{(1)}$ and $\mathbf{V}^{(2)}$ are corresponding eigenvector matrices. To equalize the mean vectors, the estimate for $\mathbf{b}^{(1,2)}$ is given again by Eq. (14.224).

Theoretically, chronochrome is a superior linear predictor, because it minimizes the MSE, even when image statistics are non-normal, by incorporating the cross-covariance matrix. Both methods are illustrated by applying them to the image pair from the Hyperspec VNIR sensor, pictured in Fig. 14.53. The reference image was collected on 2 November 2005, while the test image was collected on 14 October 2005. These images are characterized by some significant illumination changes (evidenced by the strong shadow differences) and contain two small tarp bundles indicated in the test image that are not present in the reference. Finding these two tarp bundles is the objective of change detection.

The MD detector is used for single-image target detection as well as change detection by applying it to the chronochrome and covariance-equalization residual images. Detection statistic images are compared visually in Fig. 14.54, and the resulting ROC curves are depicted for each

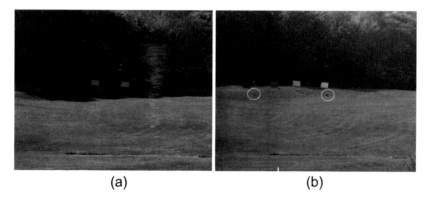

(a)   (b)

Figure 14.53 Illustration of tower-based scenes used for change detection performance analysis: (a) reference image and (b) test image.

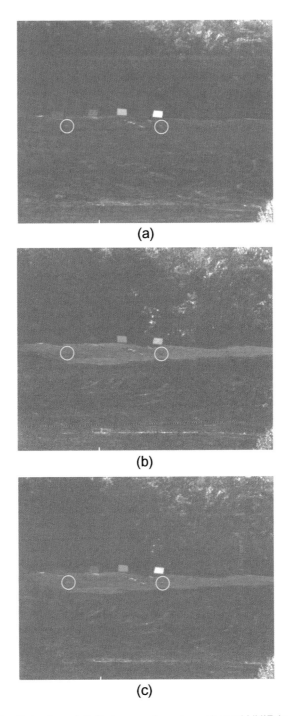

(a)

(b)

(c)

Figure 14.54 Detection statistic images for a tower-based VNIR image pair with (a) single-frame detection using MD, (b) change detection using global covariance equalization and MD, and (c) change detection using chronochrome and MD.

case in Fig. 14.55. The SCR for the tarp bundles is on the order of 2.6, causing the poor detection performance using a single-image MD detector. Chronochrome change-detection performance is only slightly better, and covariance-equalization change-detection performance is actually worse. This is due to the strong space-varying illumination changes in this image. In more-typical airborne imagery with less-substantial shadowing differences, these change-detection approaches have been reported to provide more than an order of magnitude reduction in FARs compared to single-image detectors.

Because chronochrome is based on the cross-covariance matrix, it is expected to be more sensitive than covariance equalization to misregistration between test and reference images that can be expected in real airborne or space-based remote sensing situations. This is certainly true in the prediction step, although misregistration should equally affect both methods in the process of subtracting the prediction from the test image to produce the residual image from which changes are detected. The impact of spatial registration was studied (based on data discussed earlier) by synthetically applying different levels of spatial misregistration (Eismann *et al.*, 2008b). Change-detection performance results are shown in Fig. 14.56 in terms of the area under the ROC curve (AUC) as a function of misregistration. The simulated GSD in this case was 1 m. The results indicate similar detection performance up to about one-quarter-pixel misregistration, beyond which the chronochrome method degrades

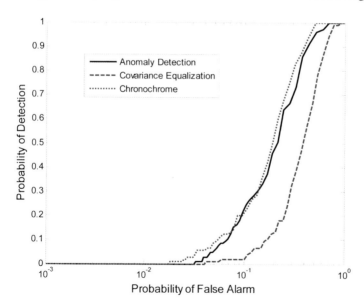

Figure 14.55 ROC curve comparison for a tower-based VNIR image pair with single frame and change detection, using global prediction based on the MD detector.

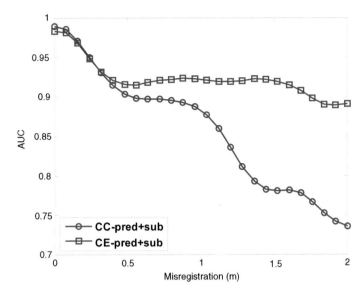

Figure 14.56 Impact of image misregistration on change detection using global prediction (Eismann *et al.*, 2008b). (CC − pred + sub = chronochrome prediction and subtraction; CE − pred + sub = covariance equalization prediction and subtraction).

more significantly than covariance equalization. Thus, the covariance-equalization method can be more robust to residual image misregistration greater than one-quarter of a pixel.

To illustrate change detection in a realistic airborne setting, consider the change pair shown in Fig. 14.57 captured by an airborne VNIR imaging spectrometer, with performance similar to the ARCHER system discussed in Chapter 1. The changes present in this pair consist of an insertion of panels in areas denoted by red boxes. The chronochrome and covariance-equalization detection statistic images shown in Fig. 14.58 provide good response to the changes but also exhibit significant background clutter leakage consistent with that seen in ground-based results, likely due to spatially varying illumination effects. There appears to be little performance difference between the chronochrome and covariance-equalization results, indicating that image registration on the order of one-quarter pixel or better is nominally achieved.

14.6.3 Change detection using spectrally segmented prediction

A limitation of global prediction methods is that they assume atmospheric and environmental parameters of the observation model to be space invariant, such that a single affine transformation can be used for all corresponding pairs of image spectra. This assumption is violated in regions where radiative transfer parameters exhibit spatial variation, such

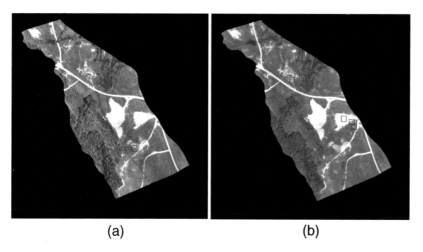

(a)   (b)

Figure 14.57 Hyperspectral change pair collected by airborne VNIR imaging spectrometer: (a) reference image and (b) test image (Eismann *et al.*, 2008b).

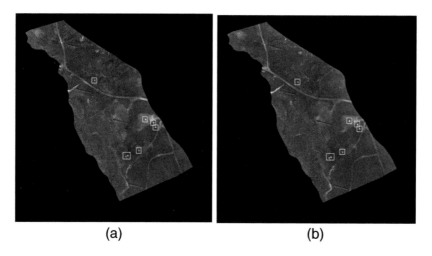

(a)   (b)

Figure 14.58 Detection statistic images of airborne VNIR hyperspectral image data using (a) chronochrome and (b) covariance-equalization global prediction methods (Eismann *et al.*, 2008b).

as areas of partial and total shadow. With reference to the observation model in Eqs. (14.205), (14.206), and (14.207), shadow parameters α and β in particular are likely to vary spatially, causing the global affine relationship between $\mathbf{X}^{(1)}$ and $\mathbf{X}^{(2)}$ to break down. It is also possible that some of the other atmospheric and environmental terms exhibit some spatial dependence, although one would expect such dependence to be more slowly varying across an image, relative to potential shadow changes.

One way to handle such space-varying transformational model changes is to employ segmentation to break the image into subsets of spectra that transform similarly between observations. Suppose that such a classification map existed that segmented both reference and test images into Q classes, for which the corresponding spectra were assigned to the same class. For each image subset ω_q, where q is the class index, we can define class-conditional transformation parameters $\mathbf{A}^{(1,2)} \mid q$ and $\mathbf{b}^{(1,2)} \mid q$ that can be estimated from sample statistics (mean vectors, covariance matrices, and cross-covariance matrix) of the subset of corresponding spectra assigned to the class indexed by q. The class-conditional predictor,

$$\hat{\mathbf{x}}_i^{(2)} \mid q = (\hat{\mathbf{A}}^{(1,2)} \mid q) \, \mathbf{x}_i^{(1)} + (\mathbf{b}^{(1,2)} \mid q), \qquad (14.227)$$

then provides additional degrees of freedom to compensate for space-varying illumination and other model variations across the scene, as long as these variations can be properly segmented into classes.

A challenge to using this approach is devising a method that can segment an image pair into subsets of spectra for which transformation parameters are common. Several variations of segmentation methods have been investigated and compared (Eismann *et al.*, 2008a). In general, the reported results indicate that performing SEM clustering on a joint image, formed by concatenating reference and test imagery on a pixel-by-pixel basis, according to

$$\mathbf{z}_i = \left[(\mathbf{x}_i^{(2)})^T \ (\mathbf{x}_i^{(1)})^T \right]^T, \qquad (14.228)$$

provides better results than segmenting either of the images individually or using mean-based clustering. It is important to recognize, however, that spectral classification of the image is not necessarily a good basis for estimating how space-varying model parameters should be segmented. Apparently, classification of the joint image is influenced more by space-varying illumination differences (such as changes in shadows) than is single image classification and seems to work well enough to support a performance advantage relative to global prediction. The overall spectrally segmented change-detection approach is depicted in Fig. 14.59.

Detection statistic images for the Hyperspec change-detection example with spectrally segmented prediction are shown in Fig. 14.60. The images are based on six joint image classes derived from the concatenation of the ten leading principal components of test and reference images, with 20 SEM iterations, segmented transformation of the entire reference image, and application of the MD detector on the prediction residual. Background clutter is effectively suppressed by the segmented predictor, allowing small targets to be detected in the residual clutter and noise. ROC performance is

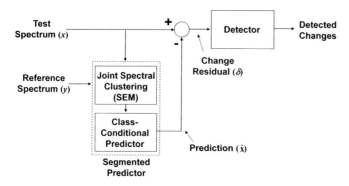

Figure 14.59 Flow diagram of hyperspectral change detection using spectrally segmented prediction.

Figure 14.60 Detection statistic image for a tower-based VNIR image pair with spectrally segmented change detection, based on the chronochrome predictor and the MD detector.

compared to global prediction methods in Fig. 14.61. The FAR is reduced by roughly two orders of magnitude, relative to global change detectors and the single-frame MD detector.

The impact of class-conditional prediction and filtering on airborne VNIR data is illustrated in Fig. 14.62, based on five classes derived from the six leading principal components of the 40-band reference image, using ten SEM iterations. Segmentation results in suppression of the background clutter leakage seen in Fig. 14.58 when using corresponding global predictors. Targets are still detected, in this case with an appreciable reduction in FAR.

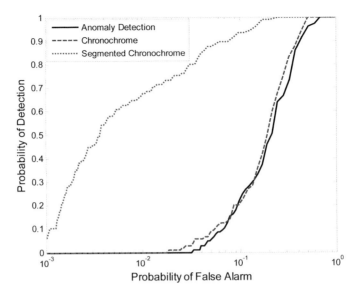

Figure 14.61 ROC curve comparison for a tower-based VNIR image pair with spectrally segmented change detection, based on the MD detector.

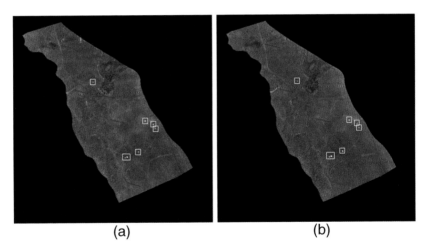

Figure 14.62 Detection statistic images of airborne VNIR hyperspectral image data using spectrally segmented (a) chronochrome and (b) covariance equalization (Eismann *et al.*, 2008b).

14.6.4 Model-based change detection

Another change-detection approach pursued by Meola *et al.* (2011b) is to employ a radiometric model to differentiate changes that are attributable to illumination and environmental changes from those that are most likely due to a change in underlying reflectance. The method is based on the diffuse facet model from Eq. (14.162). In this case, however, the model is extended to account for the pupil-plane spectral radiance measurements at two observation times,

$$\mathbf{x}_i^{(1)} = \frac{1}{\pi}\boldsymbol{\rho}_i^{(1)} \otimes \boldsymbol{\tau}_p^{(1)} \otimes (\alpha_i^{(1)}\,\mathbf{E}_s^{(1)}\cos\theta_s^{(1)} + \beta_i^{(1)}\,\mathbf{E}_d^{(1)}) + \mathbf{L}_p^{(1)} + \mathbf{n}_i^{(1)}$$

(14.229)

and

$$\mathbf{x}_i^{(2)} = \frac{1}{\pi}\boldsymbol{\rho}_i^{(2)} \otimes \boldsymbol{\tau}_p^{(2)} \otimes (\alpha_i^{(2)}\,\mathbf{E}_s^{(2)}\cos\theta_s^{(2)} + \beta_i^{(2)}\,\mathbf{E}_d^{(2)}) + \mathbf{L}_p^{(2)} + \mathbf{n}_i^{(2)},$$

(14.230)

where the parenthetical superscript indicates observation time (1 = reference, 2 = test), and subscript i indicates a spatial position in the image. Relative to Eq. (14.171), this model incorporates scalar shadow coefficients β_i for integrated downwelling irradiance as well as for direct solar irradiance. It is important to note that $\boldsymbol{\tau}_p$, $\mathbf{E}_s$, $\mathbf{E}_d$, and $\mathbf{L}_p$ are spectrally varying and can change between observations but are assumed to be space invariant. Shadow coefficients α_i and β_i are scalar quantities that vary spatially and can also change in time, while solar zenith angle θ_s is a scalar that can vary both spatially (due to surface orientation variations) and between observations. We can therefore combine the cosine of the solar zenith into shadow coefficients α_i. Only reflectance $\boldsymbol{\rho}$ and noise $\mathbf{n}$ can vary spectrally, spatially, and between observations, but the noise is assumed to be normally distributed, $\mathbf{n} \sim N(0, \boldsymbol{\Sigma}_\mathrm{n})$, and spatially and temporally uncorrelated.

As described in Section 14.5.1, a subspace model can be used to constrain expected variation in the atmospheric and illumination parameters composing Eqs. (14.229) and (14.230). In this case, the coupled subspace modeling approach (Chandra and Healey, 2008) can be used to preserve correlation between the expected realizations of $\boldsymbol{\tau}_p$, $\mathbf{E}_s$, $\mathbf{E}_d$, and $\mathbf{L}_p$. This approach, described further in Chapter 11, generates numerous realizations of these parameters using MODTRAN, places the realizations in a joint matrix that concatenates each combination of $\boldsymbol{\tau}_p$, $\mathbf{E}_s$, $\mathbf{E}_d$, and $\mathbf{L}_p$ into a single vector, and then determines the basis vectors of this joint data matrix in the same manner as previously discussed. The result is a low-order, linear subspace approximation of these parameters according to

$$\frac{1}{\pi}\boldsymbol{\tau}_p \otimes \mathbf{E}_s \approx \sum_{m=1}^{M} \varepsilon_m\,\mathbf{s}_m = \mathbf{S}\,\boldsymbol{\varepsilon},$$

(14.231)

$$\frac{1}{\pi}\boldsymbol{\tau}_p \otimes \mathbf{E}_d \approx \sum_{m=1}^{M} \varepsilon_m \, \mathbf{d}_m = \mathbf{D}\,\boldsymbol{\varepsilon}, \tag{14.232}$$

and

$$\mathbf{L}_p \approx \sum_{m=1}^{M} \varepsilon_m \, \mathbf{p}_m = \mathbf{P}\,\boldsymbol{\varepsilon}. \tag{14.233}$$

Note that basis coefficient vector $\boldsymbol{\varepsilon}$ is the same for all aggregated parameters, which is what maintains their correlation in accordance with the coupled subspace model. By placing these subspace approximations into the observation model, Eqs. (14.229) and (14.230) can be written as

$$\mathbf{x}_i^{(1)} \approx \boldsymbol{\rho}_i^{(1)} \otimes (\alpha_i^{(1)} \, \mathbf{S}\,\boldsymbol{\varepsilon}^{(1)} + \beta_i^{(1)} \, \mathbf{D}\,\boldsymbol{\varepsilon}^{(1)}) + \mathbf{P}\,\boldsymbol{\varepsilon}^{(1)} + \mathbf{n}_i^{(1)} \tag{14.234}$$

and

$$\mathbf{x}_i^{(2)} \approx \boldsymbol{\rho}_i^{(2)} \otimes (\alpha_i^{(2)} \, \mathbf{S}\,\boldsymbol{\varepsilon}^{(2)} + \beta_i^{(2)} \, \mathbf{D}\,\boldsymbol{\varepsilon}^{(2)}) + \mathbf{P}\,\boldsymbol{\varepsilon}^{(2)} + \mathbf{n}_i^{(2)}. \tag{14.235}$$

If hyperspectral images are composed of N spectra each composed of K spectral bands, and the coupled subspace model has a dimensionality of M, then the observation model in Eqs. (14.234) and (14.235) represents $2NK$ measurements and $2NK + 4N + 2M$ unknown parameters. There is insufficient data from which to estimate all of the unknown parameters. Therefore, without further constraining the model, it is not possible to estimate unknown surface reflectance spectra from which to make a change detection decision.

If there is no *a priori* information concerning specific reflectance changes of interest, the change detection problem can be cast as an anomalous change detection hypothesis test:

$$\begin{aligned} H_0 &: \boldsymbol{\rho}_i^{(1)} = \boldsymbol{\rho}_i^{(2)} \\ H_1 &: \boldsymbol{\rho}_i^{(1)} \neq \boldsymbol{\rho}_i^{(2)}. \end{aligned} \tag{14.236}$$

Under uncorrelated noise assumptions, the GLRT is based on the log-likelihood

$$l(\mathbf{x,y}) = \log \frac{\displaystyle\max_{\theta_1} p(\mathbf{x} \mid H_1)\, p(\mathbf{y} \mid H_1)}{\displaystyle\max_{\theta_0} p(\mathbf{x} \mid H_0) p(\mathbf{y} \mid H_0)}, \tag{14.237}$$

where

$$\theta_0 = \{\, \boldsymbol{\varepsilon}^{(1)}, \boldsymbol{\varepsilon}^{(2)}, \boldsymbol{\rho}_i, \alpha_i^{(1)}, \alpha_i^{(2)}, \beta_i^{(1)}, \beta_i^{(2)} \quad \forall i = 1, 2, \ldots, N\} \quad (14.238)$$

and

$$\theta_1 = \{\, \boldsymbol{\varepsilon}^{(1)}, \boldsymbol{\varepsilon}^{(2)}, \boldsymbol{\rho}_i^{(1)}, \boldsymbol{\rho}_i^{(2)}, \alpha_i^{(1)}, \alpha_i^{(2)}, \beta_i^{(1)}, \beta_i^{(2)} \quad \forall i = 1, 2, \ldots, N\}.$$

$$(14.239)$$

Under the null hypothesis, a constant $\boldsymbol{\rho}_i$ for the two observations is included in θ_0, reducing the number of unknowns to $NK + 4N + 2M$, which can theoretically be estimated from the $2NK$ measurements to obtain MLEs for $\mathbf{x}^{(1)}$ and $\mathbf{x}^{(2)}$. Under the alternative hypothesis, the parameter estimation problem is ill-conditioned, and there is no *a priori* information concerning the surface reflectance difference signature. Therefore, MLEs for $\mathbf{x}^{(1)}$ and $\mathbf{x}^{(2)}$ are the measurements, and the numerator of Eq. (14.237) is a constant. Under the aforementioned noise assumptions, the GLRT change-detection statistic for the model-based change detector (MBCD) becomes

$$r_{MBCD}(\mathbf{x}^{(1)}, \mathbf{x}^{(2)}) = [\, \mathbf{x}^{(1)} - (\hat{\mathbf{x}}^{(1)} \mid \hat{\theta}_0)\,]^T \Sigma_n^{-1} [\, \mathbf{x}^{(1)} - (\hat{\mathbf{x}}^{(1)} \mid \hat{\theta}_0)\,]$$
$$+ [\, \mathbf{x}^{(2)} - (\hat{\mathbf{x}}^{(2)} \mid \hat{\theta}_0)\,]^T \Sigma_n^{-1} [\, \mathbf{x}^{(2)} - (\hat{\mathbf{x}}^{(2)} \mid \hat{\theta}_0)\,].$$

$$(14.240)$$

The maximum-likelihood parameter estimate for θ_0 in Eq. (14.240) is obtained by the optimization problem

$$\hat{\theta}_0 = \arg \min_{\theta_0} f(\mathbf{X}^{(1)}, \mathbf{X}^{(2)}; \theta_0), \qquad (14.241)$$

where we define the cost function as

$$f(\mathbf{X}^{(1)}, \mathbf{X}^{(2)}; \theta_0) = \sum_{i=1}^{N} \left\{ \begin{array}{l} [\, \mathbf{x}_i^{(1)} - (\hat{\mathbf{x}}_i^{(1)} \mid \hat{\theta}_0)\,]^T \Sigma_n^{-1} [\, \mathbf{x}_i^{(1)} - (\hat{\mathbf{x}}_i^{(1)} \mid \hat{\theta}_0)\,] \\ + [\, \mathbf{x}_i^{(2)} - (\hat{\mathbf{x}}_i^{(2)} \mid \hat{\theta}_0)\,]^T \Sigma_n^{-1} [\, \mathbf{x}_i^{(2)} - (\hat{\mathbf{x}}_i^{(2)} \mid \hat{\theta}_0)\,] \end{array} \right\}.$$

$$(14.242)$$

The change detection problem has been turned into a matter of fitting the underlying observation model expressed in Eqs. (14.234) and (14.235) under the null hypothesis, with the error in the model fit providing the change-detection statistic. It should be noted that this optimization problem cannot be performed on an individual pixel basis because atmospheric parameters $\boldsymbol{\varepsilon}^{(1)}$ and $\boldsymbol{\varepsilon}^{(2)}$ are common to all spectra across the image. A numerically tractable approach that has been pursued for this very large-scale, constrained nonlinear optimization problem is to break parameter set θ_0 into more manageable parts and use an iterative, alternating optimization procedure, where at each step, the component part is optimized separately, assuming that the others are constant. A physically motivated separation of the parameter set divides it into components containing atmospheric basis coefficients, reflectance spectra, and shadow coefficients. The consequences of this separation are that the model in Eqs. (14.234) and (14.235) becomes linear when two of the three component parameter sets are assumed to be constant, and that optimization can be performed on a pixel-by-pixel basis for the reflectance spectrum and shadow coefficient components. A succession of quadratic programming implementations is employed to perform overall optimization. Atmospheric coefficients and reflectance spectra can be initialized by some form of atmospheric compensation, with a scene-based brightness metric used to initialize shadow coefficients.

Again, using the Hyperspec VNIR change pair example to illustrate the method, a detection statistic image is given in Fig. 14.63, with corresponding ROC curve comparisons in Fig. 14.64. It is seen that the model-based approach performs very well in detecting the green-tarp bundle but has difficulty in detecting the camouflaged-tarp bundle. This camouflaged tarp has a low reflectance, and its insertion appears to be consistent with a model transformation from full illumination to shadow. This represents a limitation of this method, in that target changes consistent with model parameter changes are not detected. When the ROC curve is modified to focus only on detection of the green tarp, the model-based method significantly outperforms global predictive methods, with more than a two-order-of-magnitude reduction in false-alarm probability for the same probability of detection. However, efficacy of this approach across a broad set of hyperspectral imaging situations has yet to be characterized.

The MBCD formulation offers future possibilities for addressing the remaining challenges confronting the practical application of hyperspectral change detection. First, the underlying model can be extended to incorporate spatial correlation in shadow coefficients; this could ameliorate the problem encountered with the dark change targets in the previous example. Furthermore, a local reflectance mixture model could be incorporated to accommodate potential image-to-image misregistration, or a smoothness criterion could be incorporated into the cost function to inhibit reflectance changes from being modeled by atmospheric parameters. Finally, a geometric model could be incorporated to address potential parallax errors between observations. Thus, while the model-based method introduces added complexity to perform necessary large-scale nonlinear optimization relative to statistical approaches, it

Figure 14.63 Detection statistic images of ground-based VNIR hyperspectral image data using the model-based change detection method (Meola *et al.*, 2011b).

offers greater flexibility in incorporating knowledge of complex physical processes that can cause image changes that are not of interest and are difficult to accommodate statistically. This conjecture into future possibilities of MBCD must be supported by future research and development.

14.7 Summary

A wide variety of algorithms have been developed to detect small targets in hyperspectral imagery. Most can be understood based on the theory of the GLRT and differ in the level of *a priori* target and background information for a particular detection application, as well as assumed background and noise models and characteristics. This chapter attempts to provide a theoretical foundation for the variety of detectors reported in the literature (with application to an image example) to gain some insight into the characteristics of detection statistic images and ROC curves. As with the example image classification results from the preceding chapter, the reader should exercise caution in generalizing relative and absolute detection performance results, as results can depend greatly on the nature of the hyperspectral imagery and the target detection problem.

Hyperspectral target detection remains an active area of research, as the reported achievable false-alarm rates often exceed the demands of the intended applications, especially when real-time, fully automated target detection is desired. The seemingly fertile areas of research in this regard focus on the topics of the last three sections: false-alarm mitigation, subspace matched detection, and change detection. The key to the first

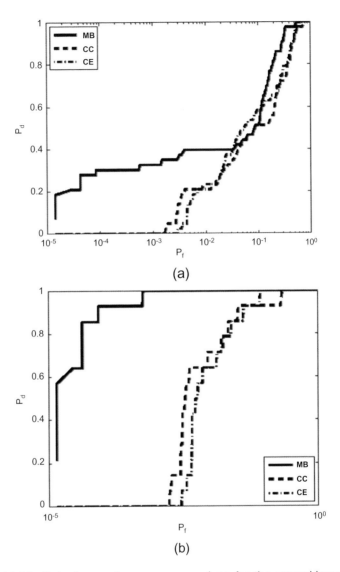

Figure 14.64 Detection performance comparison for the ground-based VNIR hyperspectral image data, using the MBCD method: (a) including camouflaged tarp and (b) not including camouflaged tarp (Meola *et al.*, 2011b).

two areas is developing viable methods to model target behavior in some accurate statistical manner to cope with the variability of the target signature and to achieve improved target selectivity against anomalous background spectra. Change detection can suppress complex background clutter that is correlated between observation times but suffers from background changes due to shadows, misregistration, and parallax. New algorithms are being developed to tackle these problems as this final chapter is being concluded.

14.8 Further Reading

The beginning section of this chapter follows the treatment of target detection theory provided by Scharf (1991), which is recommended as a reference for the reader seeking a more in-depth understanding of detection and estimation. Citations are provided for the specific target detection algorithms described in this chapter, even though the original developers may have approached detector derivation in a manner that differs from what was presented here.

References

Boardman, J. W., "Leveraging the high dimensionality of AVIRIS data for improved sub-pixel target unmixing and rejection of false positives: mixture-tuned matched filtering," *Sum. 7th Ann. JPL Airborne Geosci.,* JPL Publication 97-21, vol. 1, p. 55 (1998).

Caefer, C. E., Silverman, J., Orthal, O, Antonelli, D., Sharoni, Y., and Rotman, S. R., "Improved covariance matrices for point target detection in hyperspectral imagery," *Opt. Eng.* **47**, 076402-1–8 (2008) [doi:10.1117/1.2965814].

Carlotto, M. J., "A cluster-based approach for detecting man-made objects and changes in imagery," *IEEE T. Geosci. Remote Sens.* **43**, 374–387 (2005).

Chandra, K. and Healey, G., "Using coupled subspace models for reflectance/illumination separation," *IEEE T. Geosci. Remote Sens.* **46**(1), 284–290 (2008).

DiPietro, R. S., Manolakis, D., Lockwood, R., Cooley, T., and Jacobsen, J., "Performance evaluation of hyperspectral detection algorithms for sub-pixel objects," *Proc. SPIE* **7695**, 76951W-1–11 (2010) [doi:10.1117/12.850036].

Efron, B., Hastie, T., Johnstone, I., and Tibshirani, R., "Least angle regression," *Ann. Stat.* **32**, 407–451 (2003).

Eismann, M. T., Meola, J., and Hardie, R. C., "Hyperspectral change detection in the presence of diurnal and seasonal variations," *IEEE T. Geosci. Remote Sens.* **46**, 237–249 (2008a).

Eismann, M. T., Meola, J., Stocker, A., Beaven, S., and Schaum, A., "Airborne hyperspectral change detection of small targets," *Appl. Optics* **47**, F27–F45 (2008b).

Eismann, M. T., Stocker, A. D., and Nasrabadi, N. M., Automated hyperspectral cueing for civilian search and rescue," *Proc. IEEE* **97**, 1031–1055 (2009).

Harsanyi, J. C., "Detection and Classification of Subpixel Spectral Signatures in Hyperspectral Image Sequences," Ph.D. Dissertation, Univ. of Maryland (1993).

Harsanyi, J. C. and Chang, C-I., "Hyperspectral image classification and dimensionality reduction: an orthogonal subspace projection approach," *IEEE T. Geosci. Remote Sens.* **32**, 779–785 (1994).

Healey, G. and Slater, D., "Models and methods for automated material identification in hyperspectral imagery acquired under unknown illumination and atmospheric conditions," *IEEE T. Geosci. Remote Sens.* **37**, 2706–2717 (1999).

Hytla, P. C., Hardie, R. C., Eismann, M. T., and Meola, J., "Anomaly detection in hyperspectral imagery: comparison of methods using diurnal and seasonal data," *J. Appl. Remote Sens.* **3**, 033546 (2010) [doi:10.1117/1.3236689].

Kelly, E. J., "An adaptive detection algorithm," *IEEE T. Aero. Elec. Syst.* **AES-22**, 115–127 (1986).

Kraut, S. and Scharf, L. L., "The CFAR adaptive subspace detector is a scale-invariant GLRT," *IEEE T. Signal Proces.* **47**, 2538–2541 (1999).

Lebedev, N. N., *Special Functions and Their Applications*, Dover, New York (1972).

Kolodner, M. A., "Automated target detection system for hyperspectral imaging sensors," *Appl. Optics* **47**, F61–F70 (2008).

Manolakis, D., "Taxonomy of detection algorithms for hyperspectral imaging applications," *Opt. Eng.* **44**, 066403 (2005) [doi:10.1117/1.1930927].

Manolakis, D. and Shaw, G., "Detection algorithms for hyperspectral imaging applications," *IEEE Sign. Proces. Mag.* **19**(1), 29–43 (2002).

Meola, J., "Generalized likelihood ratio test for hyperspectral subpixel target detection," Personal communication, Jan. 2011a.

Meola, J., Eismann, M. T., Moses, R. L., and Ash, J. N., "Detecting changes in hyperspectral imagery using a model-based approach," *IEEE T. Geosci. Remote Sens.* **49**, 2647–2661 (2011b).

Reed, I. S. and Yu, X. "Adaptive multiband CFAR detection of an optical pattern with unknown spectral distribution," *IEEE T. Acoust. Speech Signal Proces.* **38**(10), 1760–1770 (1990).

Scharf, L. L. and Friedlander, B., "Matched subspace detectors," *IEEE T. Signal Proces.* **42**(8), 2146–2157 (1994).

Scharf, L. L., *Statistical Signal Processing: Detection, Estimation, and Time Series Analysis*, Addison Wesley, Reading, MA (1991).

Schaum, A., "Spectral subspace matched filtering," *Proc. SPIE* **4381**, 1–17 (2001) [doi:10.1117/12.436996].

Schaum, A. and Stocker, A., "Hyperspectral change detection and supervised matched filtering based on covariance equalization," *Proc. SPIE* **5425**, 77–90 (2004) [doi:10.1117/12.544026].

Schaum, A. and Stocker, A., "Spectrally-selective target detection," *Proc. Int. Symp. Spectral Sens. Res.* (1997).

Schaum, A. and Stocker, A. D., "Linear chromodynamics models for hyperspectral target detection," p. 23, *Proc. 2003 IEEE Aerospace Conf.*, 77–90 (2003).

Smetek, T. E., "Hyperspectral Imagery Detection Using Improved Anomaly Detection and Signature Matching Methods," Ph.D. Dissertation AFIT/DS/ENS/07-07, Air Force Institute of Technology, June 2007.

Stein, D. W., Beaven, S. G., Hoff, L. E., Winter, E. M., Schaum, A. P., and Stocker, A. D., "Anomaly detection from hyperspectral imagery, *IEEE Sign. Proces. Mag.* **19**(1), 58–69 (2002).

Tibshirani, R., "Regression shrinkage and selection via the LASSO," *J. Royal Stat. Soc. B* **58**, 267–288 (1996).

Triola, M. F., *Elementary Statistics*, 11th Edition, Addison Wesley, Reading, MA (2008).

Villeneuve, P. V., Boisvert, A. R., and Stocker, A. D., "Hyperspectral sub-pixel target identification using least-angle regression," *Proc. SPIE* **7695**, 76951V-1–11 (2010) [doi:10.1117/12.850563].

Index

A

a posteriori probability, 564
Abbe number, 251
aberrations, 250
absorbance, 134
absorption coefficient, 55, 75
absorption cross-section, 75
abundance, 530
abundance matrix, 531
abundance vector, 555
acousto-optic tunable filter (AOTF), 400
adaptive coherence/cosine estimator (ACE), 659
Adaptive Infrared Imaging Spectroradiometer (AIRIS), 398
adaptive subspace detector (ASD), 694
additive model, 507
additivity, 535, 545
adjacency effect, 201, 232
affine transformation, 507
Airborne Hyperspectral Imager (AHI), 357
Airborne Imaging Spectrometer (AIS), 22
Airborne Real-Time Cueing Hyperspectral Enhanced Reconnaissance (ARCHER), 27
Airborne Visible/Infrared Imaging Spectrometer (AVIRIS), 22, 349

B

background spectral radiance, 288
background-limited performance (BLIP), 285
backscatter cross-section, 78
bacteriorhodopsin, 165
bare-soil index (BI), 466
bathymetry, 31
Bayes theorem, 565
Bayesian classification rule, 565

Airy radius, 249
albedo, 140
albedo, spherical, 232
algal plumes, 29
allowed transitions, 90
alpha emissivity, 480
alternative hypothesis, 617
angular frequency, 39
anomaly detection, 624
aplanatic, 251
apodization function, 372
apparent spectral properties, 9, 133
argon gas discharge lamp, 420
astigmatism, 250
atmospheric compensation, 451
atmospheric path radiance, 206
atmospheric path transmission, 205
Atmospheric Removal Program (ATREM), 485

Beard–Maxwell model, 149
Beer's law, 55
beta-carotene, 165
between-class scatter matrix, 571
bidirectional reflectance
   distribution function (BRDF),
   147
biquadratic model, 418
blackbody, 150
blackbody calibration source, 434
blaze angle, 328
Bloch wave function, 129
Bohr radius, 97
Boltzmann's constant, 119
bond strength, 111, 116
Born–Oppenheimer
   approximation, 90

C

carbon dioxide, 157
carbon monoxide, 119, 158
carbonate, 172
cavity blackbody, 434
cellulose, 169
centering, 514
change detection, 695
characteristic equation, 512
charge velocity, 40
charge-coupled readout device,
   266
chi-squared distribution, 627
chlorophyll, 165
Christiansen peak, 178
chromatic aberration, 250
chronochrome, 699
class index, 565
class-conditional probability, 565
class-conditional spectral
   matched filter (CCSMF), 667
classical mechanics, 45
classification, 546, 564
cluster-based anomaly detector
   (CBAD), 642

clustering, 546
cold shield, 264
Compact Airborne Spectral
   Sensor (COMPASS), 354
complementary metal-oxide
   semiconductor (CMOS), 266
complementary subspace
   detector, 637
complex dielectric constant, 43,
   48
complex index of refraction, 44,
   51
complex propagation vector, 43
conductivity, 41
confusion matrix, 568
constant FAR (CFAR), 632
constitutive relations, 41
constrained energy minimization
   (CEM) detector, 658
correlation diameter, 306
$\cos^4$ law, 290
coupled-subspace model, 492
covariance equalization, 699
covariance matrix, 509
curl, 38
cutoff spatial frequency, 256
cutoff wavelength, 130, 269

D

damping coefficient, 46
damping constant, 49
dark current, 272, 274, 282
data matrix, 504
detection statistic, 619
detector array, 266
detector shunt resistance, 283
deuterium lamp, 431
diagonal matrix, 512
diffraction efficiency, 329
diffuse, 146
diffuse spectra, 128
Digital Image and Remote
   Sensing Image Generation
   (DIRSIG), 229

dimensionality, 510
dipole moment, 89
direct solar irradiance, 201
direct thermal emission, 201
directional hemispherical
  reflectance (DHR), 147
disassociation energy, 116
disaster management, 31
discriminant function, 565
dispersion, angular, 316, 325
dispersion, material, 316
dispersion, prism, 315
distortion, 250
divergence, 38
Doppler broadening, 95
downwelling radiance, 201
Drude theory, 52

E
edge height overshoot, 308
eigenvalue, 512
eigenvector, 512
Einstein coefficients, 92
electric displacement, 40
electric permittivity, 38, 41
electromagnetic radiation, 37
electromagnetic wave, 38
electron charge, 46
electron rest mass, 46
electron volt, 119
electronic charge, 40
emission spectrum, 100
emissive spectral region, 201
emissivity, 153
empirical line method (ELM),
  454
endmember, 530
endmember matrix, 531
Euclidean distance (ED), 546
Euclidean distance (ED) detector,
  652
exitance, 69
exo-atmospheric solar irradiance,
  221

expectation maximization (EM),
  596
extinction coefficient, 76
extrinsic photoconductor, 278

F
F distribution, 661
Fabry–Pérot interferometer, 395
false color, 4
false-alarm mitigation (FAM),
  674
false-alarm rate (FAR), 622
fast line-of-sight atmospheric
  analysis of spectral hypercubes
  (FLAASH), 490
FastICA, 522
feature extraction, 564
field curvature, 250
Field-Portable Imaging
  Radiometric Spectrometer
  Technology (FIRST), 387
finite target matched filter
  (FTMF), 680
Fischer's linear discriminant, 571
fixed pattern noise, 292
f-number, 247
focal plane array (FPA), 243
focal plane spectral irradiance,
  289
forbidden transitions, 90
Fourier transform, 368
Fourier Transform Hyperspectral
  Imager (FTHSI), 391
Fourier transform infrared
  (FTIR), 363
Fourier transform spectrometer,
  20
Franck–Condon factor, 127
free-charge current density, 40
free-charge density, 40
fuzzy cluster-based anomaly
  detector (FCBAD), 643

G

gaseous effluent model, 233
gaseous effluents, 27
Gaussian mixture model (GMM) detector, 639
Gaussian mixture Reed–Xiaoli (GMRX) detector, 639
gelbstoff absorption, 237
generalized distance formula, 504
generalized image-quality equation (GIQE), 307
generalized likelihood ratio test (GLRT), 623
generation rate, 271
geology, 22
geometric support function, 342
Geosynchronous Imaging Fourier Transform Spectrometer (GIFTS), 388
Gerchberg–Saxton algorithm, 412
grating, 324
grating equation, 325
grating spectrometer, 19
ground-plane radiance, 230
ground-resolved distance (GRD), 301
ground-sampled distance (GSD), 296

H

Hagen–Rubens formula, 191
Hamiltonian, 85
harmonic oscillator, 46, 111
Heisenberg uncertainty principle, 94
hemispherical directional reflectance (HDR), 147
high-resolution transmission molecular absorption (HITRAN), 206
HydroLight, 235
hydroxide, 172
Hyperion, 353

Hyperspectral Digital Imagery Collection Experiment (HYDICE), 351
Hyperspectral Imager for the Coastal Ocean (HICO^{TM}), 31
hyperspectral imaging, 6
hypothesis test, 617

I

ideal geometric image, 247
image registration, 698
imaging spectrometry, 14
improved split-and-merge clustering (ISMC), 586
in-scene atmospheric compensation (ISAC), 468
independent component analysis (ICA), 520
index of refraction, 44
index structure parameter distribution, 306
indicator function, 618
indium antimonide detector, 277
indium gallium arsenide detectors, 276
inherent dimensionality, 526
inscribed simplex, 532
integrated photoelectrons, 291
integrating sphere, 433
intensity, 71
interferogram, 366
internuclear distance, 111
intrinsic spectral properties, 9
irradiance, 69
iterative self-organizing data analysis technique (ISODATA), 584

J

Johnson noise, 284
joint subspace detector, 636

K

k nearest neighbor (kNN), 606

Karhunen–Loeve transformation, 511
Kelly detector, 657
kernel function, 606, 613
kernel trick, 612
keystone, 340
Kirchoff's law, 154
k-means algorithm, 580
Kolmogorov–Smirnov (KS) test, 471
krypton gas discharge lamp, 420
Kubelka–Munk, 139

L
labeling, 564
Lambertian, 72, 146
Landsat, 5
law of reflection, 57
least-angle regression (LARS), 683
lignin, 169
likelihood function, 508
likelihood ratio, 618
likelihood ratio test (LRT), 619
Linde–Buzo–Gray (LBG) clustering, 547, 580
line shape, 94
linear classifier, 567
linear mixing model (LMM), 529
linear variable filter (LVF), 404
load-resistance circuit, 274
local normal model, 540
log-likelihood function, 508
log-likelihood ratio, 619
longwave infrared, 2
Lorentz force, 40
Lorentz model, 46, 49
Lorentzian line-shape function, 94

M
magnetic field amplitude, 38
magnetic induction, 40
magnetic permeability, 38, 41

magnification, 247
Mahalanobis distance (MD), 546, 625
margin, 591
maximum likelihood, 508, 596
maximum noise fraction (MNF) transformation, 518
Maxwell–Boltzmann distribution, 118
Maxwell's equations, 37, 40
mean vector, 509
Mercer's condition, 613
mercury cadmium telluride detector, 278
mercury–neon gas discharge lamp, 420
methane, 158
methanol, 123
Michelson interferometer, 364
midwave infrared, 2
Mie scattering, 76
mixture scatter matrix, 571
mixture-tuned matched filter (MTMF), 678
model-based change detector (MBCD), 710
moderate-resolution atmospheric transmission and radiance code (MODTRAN), 206
modulation depth, 365
modulation transfer function (MTF), 256
molecular decomposition, 122
molecular orbital, 123
monochromator, 419
Morse potential, 112
multiplexer, 266
multispectral imaging, 4

N
negentropy, 521
Neyman–Pearson lemma, 619
Night Vision Imaging Spectrometer (NVIS), 31

nitrous oxide, 158

noise estimation, 519

noise gain factor, 308

noise-adjusted
  principal-component (NAPC)
  transformation, 518

noise-equivalent irradiance (NEI),
  285

noise-equivalent power (NEP),
  285

noise-equivalent radiance (NEL),
  294

noise-equivalent reflectance
  difference NE$\Delta\rho$, 294

noise-equivalent spectral radiance
  (NESR), 294

noise-equivalent temperature
  difference ($rmNE\Delta T$), 295

noncentrality parameter, 628, 661

nonlinear dispersion, 340

nonmargin support vector, 592

nonuniformity, 292

normal compositional model
  (NCM), 553

normal mixture model, 544

normal modes, 120

Normalized Differential
  Vegetation Index (NDVI), 168

Normalized Image Interpretability
  Rating Scale (NIIRS), 307

null hypothesis, 617

numerical aperture (NA), 290

Nyquist frequency, 297

O

oblique projection, 497

optical constants, 44

optical depth, 486

optical path difference (OPD),
  250, 364

optical system, 247

optical throughput, 290

optical transmission, 264

orthogonal subspace projection
  (OSP), 670

orthorectification, 698

oscillator strength, 49, 93

ozone, 157

P

P branch, 118

pair-wise adaptive linear matched
  (PALM) filter, 667

panchromatic imaging, 3

Parzen density, 606

path radiance, 201

pectin, 169

photoconductive gain, 272

photoconductor, 270

photocurrent, 272, 274

photodiode, 272

phycobiliprotein, 165

phytoplankton absorption, 237

Pixel Purity Index^{TM}(PPI^{TM}), 531

plane wave, 39, 42

Plank's constant, 85

plasma frequency, 48

point source, 71

point spread function (PSF), 248

polarization, 47, 79

positivity, 535, 545

potential function, 86

Poynting vector, 54

principal-component analysis
  (PCA), 510

prior probability, 565

prism, 314

probability density function, 507

probability of detection, 618

probability of false alarm, 618

projection operator, 525

projection vector, 496

pupil-plane radiance, 230

pushbroom scanning, 244

Q

Q branch, 118

quadratic classifier, 566
quadratic spectral filter, 675
quantization noise, 294
quantum detector, 268
quantum efficiency, 269
quantum mechanics, 83
quartz–tungsten (QTH) lamp, 431
quasi-local covariance matrix, 632
quick atmospheric compensation (QUAC), 467

R
R branch, 118
radial wave function, 97
radiance, 70
radiative transfer, 9, 199
radiometric calibration, 423
Rayleigh criterion, 300
Rayleigh scattering, 78
readout integrated circuit (ROIC), 266
readout noise, 292
receiver operating characteristic (ROC), 618
red-edge inflection, 168
red-edge reflectance, 168
reduced mass, 46
Reed–Xiaoli (RX) algorithm, 630
reflectance, 134
reflective spectral region, 201
reflectivity, amplitude, 58
reflectivity, power, 58, 135
relative edge response, 308
relative gain nonuniformity, 427
relative nonlinearity, 429
remote sensing, 1
residual spatial nonuniformity, 428
resolving power, 318
resolving power, Fabry–Perot interferometer, 397
resolving power, Fourier transform spectrometer, 370

resolving power, grating, 326
resolving power, prism, 318
resolving power, spatial Fourier transform spectrometer, 378
resonance frequency, 49
resonant frequency, 47
responsivity, 270
reststrahlen, 175
reststrahlen band, 67

S
Sagnac interferometer, 379
scattering coefficient, 76
scattering cross-section, 73
scattering efficiency, 74
scattering phase function, 74
scene-based calibration, 446
Schrödinger wave equation, 85
score function, 593
search and rescue, 27
separability, 570
separating matrix, 521
shortwave infrared, 2
shot noise, 281
shrinkwrapping, 533
signal spectral radiance, 288
signal-to-clutter ratio (SCR), 622
signal-to-noise ratio (SNR), 284
significance ratio, 684
silicate, 172
silicon detector, 275
Silverstein distribution, 527
simplex maximization, 533
simplex volume, 532
single-facet model, 229
smile, 340
Snell's law, 57
solid angle, 72
spatial noise, 292
spatial position vectors, 39
Spatially Enhanced Broadband Array Spectrograph System (SEBASS), 357

Spatially Modulated Imaging
   Fourier Transform Spectrometer
   (SMIFTS), 390
spatial–spectral distortion, 339
specific detectivity, 285
spectral angle, 505
spectral angle mapper (SAM),
   648
spectral apparent temperature,
   468
spectral calibration, 340, 417
spectral irradiance, 69
spectral matched filter (SMF),
   652
spectral measurement vector, 504
spectral mixture, 505
spectral radiance, 71
spectral range, Fabry–Perot
   interferometer, 397
spectral range, Fourier transform
   spectrometer, 371
spectral range, grating
   spectrometer, 326
spectral resolution, 345
spectral resolution, Fourier
   transform spectrometer, 370
spectral resolution, geometric,
   345
spectral response function (SRF),
   338
Spectrally Enhanced Broadband
   Array Spectrograph System
   (SEBASS), 22
Spectralon®, 431
spectrometer,
   chromotomographic imaging,
   407
spectrometer, Czerny–Turner, 335
spectrometer, double aplanatic
   prism, 323
spectrometer, Dyson, 337
spectrometer, Fourier transform,
   363

spectrometer, grating, 331
spectrometer, Littrow, 322, 334
spectrometer, Offner, 336
spectrometer, Paschen–Runge,
   335
spectrometer, prism, 319
spectrometer, Schmidt imaging,
   323
spectrometer, wedge imaging,
   404
spectroscopy, 8, 95
spectroscopy, rotational, 107
spectroscopy, vibrational, 111
spectrum, 6
specular, 146
speed of light, 39
spherical aberration, 250
spontaneous emission, 91
spot diameter, 254
stationary state, 85
Stefan–Boltzmann law, 153
step-stare pointing, 246
stimulated absorption, 91
stimulated emission, 91
stochastic expectation
   maximization (SEM), 547, 598
stochastic mixing model (SMM),
   551
subpixel replacement model, 676
subspace, 523
subspace model, 524
subsurface reflectance, 236
sulfate, 172
superscribed simplex, 533
supervised classifier, 564
support vector, 591
support vector machine (SVM),
   590

T
target subspace, 685
t-distribution, 550
temperature–emissivity
   separation, 453

transfer calibration, 433
transfer matrix, 143
transimpedance-amplifier circuit, 274
transmission, 55
transmissivity, power, 60
transmittance, 134
true color, 4
two-point radiometric calibration, 425
two-stream method, 136

U
unsupervised classifier, 564
upwelling radiance, 468

V
vegetation, 163
vegetation mapping, 25
vegetation normalization, 458
vegetation science, 25
vibronic, 127

vicarious calibration, 446
virtual dimensionality, 527
visible, 2
volume reflectance, 139

W
wave equation, 38, 42
wave function, 83
wave vector, 39
wavelength, 41
wavenumber, 91
whiskbroom scanning, 244
whitening, 514
Wien displacement law, 153
within-class scatter matrix, 570

X
xenon gas discharge lamp, 420

Z
zero path difference (ZPD), 364